Sustainability Solutions

An Interdisciplinary and Transdisciplinary Nexus

Sustainability Solutions addresses a wide range of global challenges with applied solutions in the interdisciplinary and transdisciplinary sustainability nexus. Titles in the series present a strong sustainable development focus, and topics which connect environment, economic and social applications are encouraged. Topics could include environmental impacts, vulnerability, risk assessment and reduction, human-environmental systems (e.g. food, water, health) and innovative solutions to climate change. Volumes in the series present novel solutions to societal challenges using innovative application, for example demonstrating a solutions-based approach using machine learning, artificial intelligence, decision support systems, advance geospatial technology. Case studies and shorter report volumes which present knowledge transfer for policymaking and governance are also encouraged.

Keywords:

Interdisciplinary environmental solutions, transdisciplinary knowledge transfer, geospatial technologies, natural resource management, decision support system, artificial intelligence, deep learning, smart technology, environmental science, economic, social and spatial, human-environment interactions, sustainable development

Hukum Singh
Editor

Forestry for a Carbon-Neutral and Sustainable Future

A Scientific Perspective

Springer

Editor
Hukum Singh
ICFRE-Forest Research Institute
Dehradun, Uttarakhand, India

ISSN 3059-278X ISSN 3059-2798 (electronic)
Sustainability Solutions
ISBN 978-3-032-06426-4 ISBN 978-3-032-06427-1 (eBook)
https://doi.org/10.1007/978-3-032-06427-1

This Springer imprint is published by the registered company Springer Nature Switzerland AG
The registered company address is: Gewerbestrasse 11, 6330 Cham, Switzerland

Competing Interests The author has no competing interests to declare that are relevant to the content of this manuscript.

About This Book

This book presents a comprehensive examination of forestry's pivotal role in addressing climate change and advancing sustainability. It highlights the scientific foundations of forests' contributions to carbon neutrality, emphasizing their ability to mitigate greenhouse gas emissions and enhance climate resilience. Drawing on the latest research and innovative practices, the book offers in-depth insights into effective forest management strategies that support a carbon-neutral future. Key concepts and principles of carbon neutrality and sustainability are thoroughly discussed, alongside sustainable forestry practices that overcome existing barriers and chart future pathways. The book also explores the critical intersection of forestry with the Sustainable Development Goals (SDGs), demonstrating its influence on poverty reduction, food security, clean energy, biodiversity conservation, and climate action. This volume underscores forestry's potential to address multiple global challenges and promote resilient, sustainable communities by integrating scientific evidence with practical approaches. Intended for policymakers, forest managers, and researchers, this book serves as an essential resource for harnessing forestry in global climate strategies and shaping a carbon-neutral, sustainable future.

About This Book

Contents

About the Editor

Dr. Hukum Singh is a forestry and climate change scientist associated with the ICFRE-Forest Research Institute, India. His research encompasses various dimensions of forest and climate change, including eco-physiology, climate-smart forestry, forest fires, climate change adaptation and mitigation, urban forestry, and urban pollution. Dr. Singh is the recipient of various national and international awards and recognitions for his contribution in the field of his expertise. Dr. Singh has authored over 150 publications, including research articles, review articles, books, book chapters, and conference proceedings, with journals and publishers of international repute.

Chapter 1
Carbon Neutrality and Environmental Sustainability: Concept and Significance

Hukum Singh, Sachin Shah, Krishna Giri, Rama Pal, and Trisha Roy

Abstract Climate change is widely recognized as a global physical phenomenon and remains a subject of extensive debate among scientists, policymakers, and the public. The continuous rise in atmospheric CO_2 concentrations, primarily driven by human-induced activities, is a major contributor to this crisis. In response, global efforts are underway to balance carbon emissions through diverse strategies aimed at achieving a carbon-neutral future. Carbon neutrality has thus emerged as a critical pathway for mitigating climate change and advancing environmental sustainability. This chapter examines the conceptual framework, interconnections, and practical implications of achieving carbon neutrality and promoting environmental sustainability. It offers a critical analysis of key drivers, sector-specific strategies, and governance mechanisms essential for attaining net-zero emissions. These include the deployment of renewable energy, the implementation of nature-based solutions, the promotion of sustainable land use, and the decarbonization of the industrial and transportation sectors. Additionally, it addresses the socio-economic and institutional challenges that hinder progress toward carbon neutrality, particularly in terms of equity, technological access, and policy coherence. It further emphasizes the pivotal role of scientific advancements, technological innovations, and integrated policy frameworks in driving systemic transformation to support effective climate action and sustainable development. Ultimately, it underscores the synergies between carbon neutrality and the Sustainable Development Goals (SDGs), advocating for comprehensive and inclusive approaches that align with the objectives of environmental sustainability and broader development priorities.

H. Singh · S. Shah
ICFRE-Forest Research Institute, Dehradun, India

K. Giri
Centre of Excellence on Sustainable Land Management, Indian Council of Forestry Research and Education, PO New Forest, Dehradun, Uttarakhand, India

R. Pal (✉) · T. Roy
ICAR-Indian Institute of Soil and Water Conservation, Dehradun, Uttarakhand, India
e-mail: rama.env@gmail.com

H. Singh (ed.), *Forestry for a Carbon-Neutral and Sustainable Future*, Sustainability Solutions, https://doi.org/10.1007/978-3-032-06427-1_1

Keywords Circular economy · Climate change mitigation · Decarbonization · Environmental governance · Natural climate solutions · Sustainable development goals (SDGs)

1.1 Introduction

Climate change is one of the most alarming global challenges of the twenty-first century, primarily driven by the accumulation of greenhouse gases (GHGs), including carbon dioxide (CO_2), methane (CH_4), and nitrous oxide (N_2O), in the atmosphere (IPCC, 2023). Anthropogenic activities, such as the combustion of fossil fuels, industrialization, deforestation, and unsustainable land-use practices, have significantly altered the Earth's carbon balance. The Intergovernmental Panel on Climate Change has underscored the need to limit global warming to 1.5 °C above pre-industrial levels to prevent catastrophic climate risks (IPCC, 2023). Against this backdrop, the pursuit of carbon neutrality has emerged as a crucial mechanism to mitigate global climate change.

Carbon neutrality is intended to achieve net-zero carbon emissions by balancing anthropogenic emissions with equivalent removals through biological or technological interventions (UNFCCC, 2021). It involves reducing emissions at the source, increasing energy efficiency, adopting renewable energy systems, and enhancing carbon sinks such as forests, soils, and oceans. Carbon offsets, including afforestation and carbon capture technologies, play a key role in neutralizing GHG emissions (Griscom et al., 2017). This concept is widely adopted as a measurable and goal-oriented approach to achieving carbon neutrality.

Besides, environmental sustainability encompasses the responsible use and conservation of natural resources to maintain ecosystem functions, biodiversity, and human well-being for present and future generations (WCED, 1987). It involves maintaining the integrity of air, water, soil, and biological systems whilst minimizing pollution and ecological degradation. Environmental sustainability is a foundational concept in global sustainability science and policy, promoting long-term ecological resilience (Rockström et al., 2009).

There is a strong interconnection between carbon neutrality and environmental sustainability. Carbon neutrality focuses specifically on managing the carbon cycle to mitigate climate change; however, environmental sustainability provides the broader ecological context in which carbon strategies are implemented (Singh et al., 2023).

The activities related to afforestation contribute both to carbon sequestration and biodiversity conservation, offering co-benefits for environmental sustainability (Seddon et al., 2020). However, there may be trade-offs associated with large-scale bioenergy plantations, which could threaten water resources and biodiversity if not properly managed.

International agreements, particularly the Paris Agreement (2015), the UN 2030 Agenda for Sustainable Development, and the Kunming-Montreal Global Biodiversity Framework, provide an integrated policy framework for achieving carbon neutrality and environmental sustainability. The Paris Agreement encourages undertaking efforts to limit the temperature rise and achieve net-zero emissions by the second half of the twenty-first century (UNFCCC, 2015). However, the Sustainable Development Goals (SDGs), particularly SDG 13 (Climate Action), SDG 7 (Affordable and Clean Energy), and SDG 15 (Life on Land), emphasize the need for coordinated environmental stewardship.

A movement toward carbon neutrality requires transformative changes in various sectors, including energy, transportation, industry, agriculture, and land use. Scientific innovation and technological advancements in renewable energy, carbon capture and storage (CCS), nature-based solutions, and circular economy models are key enablers that significantly contribute to environmental sustainability (IPCC, 2022). Furthermore, policy coherence and governance mechanisms, such as carbon pricing, green finance, and international cooperation, are equally vital to ensure that mitigation efforts align with broader sustainability goals (Sachs et al., 2019).

Additionally, there is a need to account for regional disparities in emissions, vulnerabilities, and capacities within the framework of carbon neutrality and sustainability transition (Singh, 2024b). Developing countries, while contributing less to historical emissions, often face higher climate risks and resource constraints. Therefore, equity, justice, and capacity-building are vital to designing inclusive climate strategies (Roberts & Parks, 2007). Moreover, ensuring a transition that supports livelihoods, Indigenous rights, and community participation is fundamental to achieving a sustainable and resilient future.

This chapter examines the conceptual underpinnings, interlinkages, and practical significance of carbon neutrality and environmental sustainability, exploring definitions, principles, and metrics concerning climate change mitigation and the Sustainable Development Goals (SDGs). This chapter also discusses key drivers, sectoral strategies, and governance mechanisms essential for achieving net-zero emissions, as well as the socio-economic and institutional challenges that hinder progress. It offers a future perspective to integrate carbon neutrality into broader sustainability frameworks, ensuring long-term resilience, ecological integrity, and intergenerational equity.

1.2 Conceptual Framework

The concepts of carbon neutrality and environmental sustainability are central to contemporary discourse on climate action and ecological governance. While distinct in their focus, both represent essential strategies to address the multiple, interconnected crises facing the planet, including global warming, biodiversity loss, and resource depletion.

1.2.1 Carbon Neutrality: Meaning and Metrics

Carbon neutrality, also referred to as "net-zero carbon emissions," is defined as a state in which the amount of anthropogenic carbon dioxide (CO_2) released into the atmosphere is balanced by an equivalent amount of CO_2 removal or offset, either through natural processes or technological interventions (UNFCCC, 2021). The ultimate goal of carbon neutrality is to halt the rise in atmospheric CO_2 concentrations, thereby mitigating global warming and stabilizing the climate system.

Achieving carbon neutrality involves a two-pronged approach, i.e., reducing emissions at their source (transitioning to renewable energy, improving energy efficiency, and modifying land-use practices) and enhancing carbon sinks (afforestation, reforestation, soil carbon sequestration, and direct air capture technologies). Carbon accounting mechanisms, mainly the Greenhouse Gas Protocol and ISO 14064, offer methodologies for measuring, reporting, and verifying emissions across three scopes:

- **Scope 1**: Direct emissions from owned or controlled sources.
- **Scope 2**: Indirect emissions from the generation of purchased energy.
- **Scope 3**: All other indirect emissions across the value chain.

Key metrics used to assess carbon neutrality include total greenhouse gas (GHG) emissions (in CO_2-equivalent), carbon intensity (emissions per unit of GDP or product), and the carbon footprint of individuals, institutions, or nations. Moreover, Nationally Determined Contributions (NDCs), carbon budgets, and net-zero pledges are policy instruments linked to the pursuit of carbon neutrality under the Paris Agreement framework (IPCC, 2022).

1.2.2 Environmental Sustainability: Principles and Pillars

Environmental sustainability refers to the ability of natural systems to support life and human activity over the long term without undergoing irreversible degradation. It emphasizes the responsible management of natural resources to maintain ecological integrity, resilience, and the provisioning of ecosystem services across generations (Rockström et al., 2009; WCED, 1987).

There are several foundational pillars of environmental sustainability, including ecological integrity, intergenerational equity, resource efficiency, pollution prevention, and resilience and adaptive capacity (Fig. 1.1). Frameworks such as the planetary boundaries (Rockström et al., 2009) and the Doughnut Economics Model (Raworth, 2017) provide conceptual lenses for maintaining human development within safe environmental limits. These frameworks underscore the importance of biodiversity conservation, sustainable land and water management, and climate stability as essential components of sustainability.

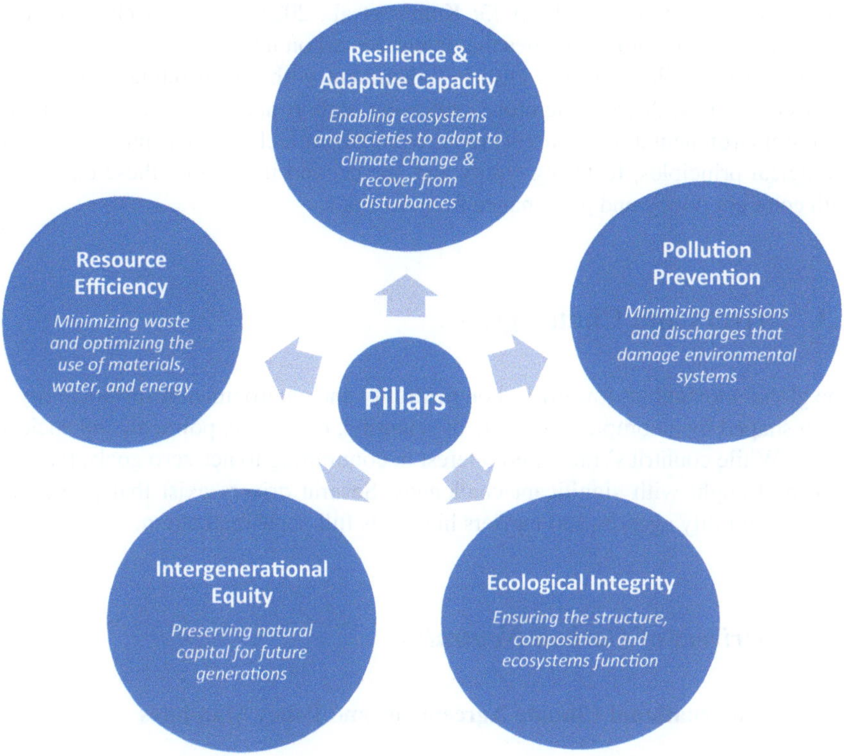

Fig. 1.1 Foundational pillars of environmental sustainability

1.2.3 Interlinkages Between Carbon Neutrality and Sustainability

Although carbon neutrality and environmental sustainability are often addressed through separate policy lenses, their goals and implementation strategies are deeply intertwined. Carbon neutrality represents a measurable climate target, while environmental sustainability serves as a guiding philosophy for harmonizing human development with ecological limits. Their convergence is particularly evident in integrated approaches such as climate-smart agriculture and forestry, green infrastructure, and nature-based solutions (Barwa & Singh, 2024).

For instance, afforestation and forest conservation contribute to carbon sequestration, biodiversity enhancement, water regulation, and soil stability, which are central goals of environmental sustainability (Seddon et al., 2020). Similarly, transitioning to renewable energy reduces GHG emissions and lowers air pollution, enhances public health, and reduces ecological risk from fossil fuel extraction (Ruhil et al., 2024).

However, synergies must be carefully managed to avoid trade-offs. A poorly planned bioenergy project may offset carbon emissions and harm biodiversity and water resources (Singh et al., 2023; Kumar et al., 2021). It is essential to adopt systemic, cross-sectoral strategies that integrate carbon neutrality targets within the broader sustainability agenda, ensuring alignment with the Sustainable Development Goals (UN, 2015). Therefore, achieving carbon neutrality without compromising environmental sustainability requires balancing technological innovation with ecological principles, fostering equity and justice, and integrating these objectives with coherent policy and governance frameworks.

1.3 Drivers and Challenges

The global momentum toward carbon neutrality and environmental sustainability is being shaped by a complex interplay of scientific, economic, political, and societal forces. While countries' increased interest in committing to net-zero goals, the path remains fraught with significant challenges. Several drivers exist that propel the carbon neutrality agenda, and barriers hinder its full-scale realization.

1.3.1 Drivers of Carbon Neutrality

1.3.1.1 International Climate Agreements and Policy Mandates

Global climate governance frameworks, particularly the Paris Agreement (UNFCCC, 2015), serve as a key driver, urging nations to commit to net-zero emissions by the second half of the twenty-first century. These agreements have catalyzed the formulation of Nationally Determined Contributions (NDCs), carbon pricing mechanisms, and net-zero roadmaps guiding both public and private sector efforts.

1.3.1.2 Scientific Consensus and IPCC Assessments

The Intergovernmental Panel on Climate Change (IPCC) has emphasized the need for profound and rapid emissions cuts through various measures. The Sixth Assessment Report (AR-6) emphasizes that global carbon neutrality must be achieved by mid-century to limit warming to 1.5 °C, underscoring the irreversible consequences of inaction (IPCC, 2023).

1.3.1.3 Technological Innovation and Clean Energy Transition

Rapid advancements in renewable energy technologies, energy storage, carbon capture and storage (CCS), and low-emission transportation systems have made carbon-neutral solutions increasingly viable and cost-effective (IEA, 2022). The low costs of solar and wind energy have been particularly transformative in achieving carbon neutrality and environmental sustainability.

1.3.1.4 Economic Incentives and Market Mechanisms

Carbon markets, emissions trading systems (ETS), green finance, and corporate Environmental, Social, and Governance (ESG) frameworks are pushing industries to adopt low-carbon pathways. Financial institutions and investors are increasingly aligning portfolios with climate targets, incentivizing sustainable business models (OECD, 2021).

1.3.1.5 Public Awareness and Societal Pressure

Increased attention over climate change, amplified by civil society, youth movements, and the media, has created a social license for climate action. This societal momentum is pressuring governments and corporations to set and meet carbon neutrality goals (UNEP, 2022a, 2022b, 2022c).

1.3.1.6 Co-benefits and Ecosystem Services

Carbon neutrality strategies often deliver multiple co-benefits, including improved air quality, enhanced public health, biodiversity conservation, and job creation in green sectors. These synergies further strengthen the case for a transition to net-zero emissions (Griscom et al., 2017).

1.3.2 Barriers and Challenges

1.3.2.1 Technological and Infrastructure Limitations

Despite progress in achieving carbon neutrality and environmental sustainability, many carbon-neutral technologies, such as carbon capture and storage (CCS), hydrogen fuels, and negative emissions technologies, remain expensive, underdeveloped, or not yet scalable. Existing infrastructure in fossil fuel-dependent economies poses additional hurdles (IEA, 2021).

1.3.2.2 Policy Incoherence and Weak Governance

Barriers, including fragmented policies, a lack of long-term planning, and weak enforcement mechanisms, undermine the effective implementation of activities aimed at progressing carbon neutrality. Besides, inconsistencies between national and subnational climate policies, as well as competing sectoral priorities, often delay coordinated action (Sachs et al., 2019).

1.3.2.3 Financial Constraints and Economic Risks

In many developing countries, the high upfront costs of transitioning to low-carbon technologies and the perceived risks for investors limit access to climate finance. Additionally, existing subsidies for fossil fuels distort market signals and delay decarbonization (UNDP, 2022).

1.3.2.4 Equity and Just Transition Issues

Carbon neutrality strategies may inadvertently marginalize vulnerable communities if they fail to integrate social equity and justice considerations into their approaches. Workers in carbon-intensive sectors, indigenous peoples, and low-income populations may bear disproportionate burdens if inclusive planning is not implemented (Roberts & Parks, 2007).

1.3.2.5 Political Resistance and Short-Termism

Political cycles, vested interests, and lobbying by carbon-intensive industries often result in resistance to bold climate policies. The focus on short-term economic gains may overshadow the long-term benefits of carbon neutrality (Newell & Paterson, 2010).

1.3.2.6 Data Gaps and Monitoring Limitations

Accurate measurement, reporting, and verification (MRV) of emissions and mitigation outcomes is a significant challenge, particularly in regions with limited technical capacity. This undermines transparency and accountability in assessing progress toward net-zero targets (UNFCCC, 2021).

1.4 Sectoral Strategies for Achieving Carbon Neutrality

Achieving carbon neutrality requires systemic transformations across the major sectors of the economy. Sector-specific strategies play a pivotal role in reducing GHG emissions while enabling sustainable development. Key mitigation approaches in the energy, forestry, land use, agriculture, industry, and transport sectors, as well as urban systems, that contribute significantly to national and global emissions are discussed below.

1.4.1 Energy Sector: Decarbonization and Renewable Transition

Globally, the energy sector is a dominant source of anthropogenic GHG emissions, accounting for approximately 73% of total emissions, primarily from the combustion of fossil fuels for electricity generation, industrial processes, heating, and transportation (IEA, 2022). Therefore, decarbonizing the energy sector is central to any credible strategy for achieving carbon neutrality and meeting international climate goals, particularly the Paris Agreement, which aims to limit the global temperature rise to well below 2 °C, and preferably to 1.5 °C (UNFCCC, 2015). The shift from a fossil fuel-dominated energy system to a low-carbon, renewable-based system is essential for climate stabilization, reducing air pollution, improving energy security, and achieving broader sustainable development outcomes.

The large-scale deployment of renewable energy technologies is a crucial component of decarbonizing the energy sector. Solar photovoltaic (PV) and wind power have emerged as the fastest-growing sources of electricity, driven by dramatic cost reductions, supportive policies, and technological innovation. Solar PV and onshore wind are the cheapest sources of power generation in most regions, making them economically competitive with fossil-based alternatives. Further, hydropower, bioenergy, geothermal, and offshore wind diversify the renewable energy portfolio, offering flexible and dispatchable low-carbon options. Integrating these alternative energy sources into energy systems requires smart grid infrastructure, advanced forecasting, and robust storage capacity to ensure grid reliability and efficiency (IRENA, 2023).

The electrification of end-use sectors, particularly transport, buildings, and industry, is a parallel pathway to reducing emissions across the broader economy. Electric vehicles (EVs), powered by renewable electricity, are a crucial mitigation strategy in the transportation sector, offering zero tailpipe emissions and lower life-cycle emissions than internal combustion engine vehicles. Similarly, electric heat pumps, induction cookers, and high-efficiency lighting and appliances are transforming the residential and commercial building sectors. In industrial applications, clean electricity is an alternative approach for low- and medium-temperature processes. At the same time, harder-to-abate sectors may benefit from green hydrogen and direct electrification in the long term (IPCC, 2022).

Energy efficiency is a crucial pillar of energy decarbonization, often referred to as the "first fuel" in clean energy transitions. Efficiency improvements reduce energy demand and emissions while delivering economic and social benefits, such as lower energy bills and reduced infrastructure costs. Measures such as building insulation, high-efficiency appliances, waste heat recovery in industry, and energy-efficient lighting (LEDs) have a significant role in reducing energy intensity. Additionally, digital technologies such as smart meters and automated demand response systems enhance energy management and encourage consumer participation in energy savings.

To accommodate increasing shares of renewable electricity, energy systems must be modernized through smart grids and energy storage solutions. Smart grids integrate digital technologies to monitor and manage the flow of electricity, enabling real-time balancing of supply and demand. Battery storage, particularly lithium-ion, has become essential for smoothing fluctuations in solar and wind output. Meanwhile, large-scale options such as pumped hydro storage and emerging technologies like flow batteries and hydrogen storage are expanding the toolbox for enhancing grid stability and integrating renewable energy (IRENA, 2023).

In parallel, phasing out fossil fuel infrastructure is indispensable for aligning energy systems with net-zero targets. This involves retiring or repurposing coal-fired power plants, ending subsidies for fossil fuels, and halting new investments in carbon-intensive projects. The International Energy Agency (IEA, 2021) has made clear that no new coal, oil, or gas infrastructure can be developed if the world is to achieve net-zero emissions by 2050.

Carbon pricing instruments, such as carbon taxes and emissions trading systems (ETS), internalize the environmental costs of emissions, incentivizing cleaner technologies and practices. Government policies, including renewable energy targets, feed-in tariffs, tax credits, and green public procurement, have catalyzed private investment in low-carbon energy systems. International cooperation and climate finance, primarily through mechanisms such as the Green Climate Fund (GCF), are crucial for supporting clean energy transitions in developing countries, where energy access and poverty alleviation remain pressing challenges (UNDP, 2022).

Despite significant progress, several barriers continue to hinder the rapid decarbonization of the energy sector. These include technical limitations such as energy storage capacity, institutional inertia, regulatory uncertainty, and the entrenched political economy of fossil fuels. In developing countries, challenges related to investment risk, limited grid infrastructure, and lack of technical capacity further complicate the transition. Addressing these constraints requires systemic change, including fostering innovation ecosystems, capacity building, public–private partnerships, and integrated planning across various scales and sectors.

1.4.2 Forestry and Land Use: Natural Climate Solutions

Natural Climate Solutions (NCS), such as forests and land-use systems, are integral to global carbon dynamics, playing dual roles as both sources and sinks of greenhouse gases (GHGs). Forest degradation releases a significant amount of carbon dioxide (CO_2) into the atmosphere (Singh, 2024a). Conversely, healthy and well-managed ecosystems sequester more carbon, making forests essential tools for achieving carbon neutrality (Singh et al., 2018). NCS encompasses conservation, restoration, and land management strategies, providing cost-effective and scalable approaches to mitigate climate change, enhance biodiversity, and support rural livelihoods (Griscom et al., 2017).

Globally, land use, land-use change, and forestry (LULUCF) activities account for 13–18% of anthropogenic greenhouse gas (GHG) emissions, primarily through deforestation, forest degradation, peatland drainage, and unsustainable agricultural expansion (IPCC, 2022). However, these systems absorb up to 30% of CO_2 emissions annually if managed sustainably. Practices such as avoided deforestation, reforestation, afforestation, agroforestry, and soil carbon enhancement play a transformative role in mitigating global emissions while delivering co-benefits for water security, food systems, and biodiversity conservation (Jangra et al., 2024).

Programs such as REDD+ (Reducing Emissions from Deforestation and Forest Degradation) under the UNFCCC framework incentivize developing countries to conserve forests by offering financial rewards for verified reductions in emissions (UNFCCC, 2018). Beyond carbon, these efforts contribute to preserving ecological integrity, regulating hydrological cycles, and supporting indigenous and forest-dependent communities (Joshi et al., 2021).

Afforestation and reforestation are vital components of NCS. These strategies increase terrestrial carbon stocks and help restore ecological functions such as erosion control, microclimate regulation, and pollination (Chazdon & Brancalion, 2019). Agroforestry systems offer multifunctional benefits, sequestering carbon while enhancing soil fertility, improving farm productivity, and providing additional income sources for farmers (Singh, 2024a). Agroforestry has demonstrated significant potential for climate mitigation, particularly when combined with sustainable land management and traditional knowledge (Joshi et al., 2020). Similarly, silvopastoral systems that integrate trees with pasture have been shown to improve animal welfare, increase biomass productivity, and reduce methane emissions from livestock (Singh & Kumar, 2022).

Soil carbon sequestration is another crucial strategy. Practices such as cover cropping, no-till farming, organic amendments (including biochar and compost), and rotational grazing may help restore degraded soils and enhance below-ground carbon storage. Peatlands, which cover only 3% of the Earth's land surface, store twice as much carbon as the entire world's forests combined, and are especially important. Preventing their drainage and restoring hydrology are among the most cost-effective climate actions (Leifeld & Menichetti, 2018).

Despite their promise, the potential of forestry and land-use strategies is constrained by challenges such as land tenure insecurity, conflicting land-use pressures, insufficient funding, and weak governance mechanisms (Singh et al., 2023). Effective implementation requires clear policy frameworks, participatory land-use planning, and transparent monitoring systems. The integration of remote sensing technologies, geospatial mapping, and MRV protocols enhances accountability and supports results-based financing.

Importantly, NCS must be aligned with broader development and equity goals. Ensuring the rights of Indigenous Peoples and local communities is crucial for the ethical and practical implementation of conservation efforts. Moreover, empowering women and marginalized groups in forest governance strengthens the resilience and inclusivity of these strategies (FAO, 2020).

1.4.3 Agriculture: Low-Emission Practices

Agriculture is a significant contributor to and a potential solution for mitigating climate change. Globally, the agricultural sector contributes 20–25% of total anthropogenic GHG emissions, including methane (CH_4) and nitrous oxide (N_2O), from enteric fermentation in ruminant livestock, rice paddy cultivation, manure management, and synthetic fertilizer application (FAO, 2021; IPCC, 2022). However, agriculture is more challenging to decarbonize than other sectors due to its biological and dispersed nature. It offers a broad array of low-emission practices and nature-based solutions that significantly reduce its carbon footprint while ensuring food security, ecosystem services, and the well-being of rural communities (Singh et al., 2024).

Precision agriculture employs technologies such as satellite imagery, GPS-guided machinery, remote sensing, and data analytics to optimize the application of water, fertilizers, and pesticides. Matching input use to crop needs may minimize nitrogen losses to the atmosphere in the form of nitrous oxide (IPCC, 2022). Precision nutrient management, particularly site-specific nitrogen application and the use of slow-release or enhanced-efficiency fertilizers, has proven effective in reducing emissions while maintaining crop yields (Kumar et al., 2020).

Methane emissions from rice paddies, a significant source of greenhouse gases (GHGs), can be substantially reduced by modifying water management practices. One of the well-documented strategies is Alternate Wetting and Drying (AWD), where fields are intermittently drained instead of continuously flooded. AWD not only lowers methane production by limiting anaerobic conditions but also conserves water and reduces energy inputs for irrigation (Linquist et al., 2015). Similarly, improved residue management, reduced tillage, and timely transplanting may support emissions reductions in rice-based systems.

Livestock production is a significant source of emissions, particularly through enteric fermentation in ruminants and inadequate management of manure. Strategies to reduce livestock emissions include:

- Dietary improvements, such as adding feed additives (e.g., fats, nitrates, seaweed) or high-digestibility forages to reduce methane generation in the rumen.
- Genetic selection for animals with higher feed conversion efficiency and lower methane yield per unit of output.
- Manure management systems, such as anaerobic digestion and composting, not only reduce methane and nitrous oxide emissions but also generate renewable energy (biogas) and organic fertilizers.

Agroecological and conservation agriculture practices provide synergistic benefits for mitigation, adaptation, and productivity. These include reduced or zero tillage, diverse crop rotations, cover cropping, and intercropping, which enhance soil organic carbon, improve nutrient cycling, and increase resilience to climate shocks. Agroforestry systems, which integrate trees with crops or livestock, serve as effective carbon sinks while enhancing biodiversity and improving microclimates (Joshi et al., 2024).

Regenerative agricultural practices such as the use of organic amendments (e.g., compost, biochar), reduced soil disturbance, and maintaining soil cover may lead to increased soil carbon sequestration. Sustainable soil management may sequester 1–3 metric tons of CO_2-equivalent per hectare per year, depending on soil type and climatic conditions (Lal, 2020). However, the permanence of this carbon sink is vulnerable to changes in land use and management, highlighting the need for consistent policy and monitoring frameworks.

The promotion of climate-smart agriculture (CSA) encompasses mitigation, adaptation, and food security objectives. CSA emphasizes tailored interventions based on agroecological conditions and integrates technological innovations with traditional knowledge. Integrated crop-livestock systems, water harvesting, and agrometeorological advisories enhance adaptive capacity while reducing emissions.

Despite these promising strategies, several barriers limit the widespread adoption of low-emission practices in agriculture. These include limited access to knowledge, finance, inputs, and markets; weak extension services; and socio-cultural resistance to change. In developing countries, smallholder farmers often lack the capital or security to invest in climate-smart technologies. Addressing these challenges requires enabling policies, investment in research and development, capacity building, and the inclusion of farmers in participatory decision-making.

1.4.4 Industry and Transport: Circular Economy and Electrification

Achieving carbon neutrality requires transformative changes in both the industrial and transport sectors, which together account for nearly 40% of global greenhouse gas (GHG) emissions (IPCC, 2022). These sectors are historically reliant on fossil fuels and resource-intensive practices. Therefore, decarbonization strategies in these sectors must involve not only a shift in energy sources but also structural changes in production, consumption, and waste management. Two major pathways toward sectoral transformation are electrification, powered by renewable energy, and the adoption of circular economy principles, which prioritize resource efficiency, reuse, and waste minimization.

The industrial sector is responsible for approximately 24% of global CO_2 emissions, primarily from energy use and industrial processes such as cement, steel, aluminum, and chemical production (IEA, 2021). These are considered "hard-to-abate" sectors due to their high-temperature heat requirements and process-related emissions. Industries must pursue a combination of technological innovation, electrification, fuel switching, and material circularity to achieve net-zero goals.

The electrification of industrial processes using renewable electricity is a central pathway to decarbonization. Technologies such as electric arc furnaces in steel-making, electrical resistance heating, and inductive heating are increasingly replacing fossil fuel-based systems. For high-temperature applications where electrification is challenging, green hydrogen, produced through electrolysis using renewable energy, is emerging as a viable alternative fuel, particularly in sectors such as steel, ammonia, and cement production (IRENA, 2022).

Carbon capture, utilization, and storage (CCUS) is being explored as a means to mitigate emissions that are difficult to eliminate, particularly in the cement and chemical industries. While still expensive and technologically complex, CCUS may capture CO_2 at the source and either store it underground or convert it into valuable products, such as building materials or synthetic fuels (IEAGHG, 2021).

In addition to fuel and process shifts, circular economy strategies offer systemic approaches to industrial decarbonization. This involves rethinking product design, extending product life cycles, promoting recycling and remanufacturing, and substituting carbon-intensive materials with more sustainable alternatives. For example, using recycled aluminum requires only 5% of the energy used in primary production, drastically reducing emissions. Industrial symbiosis, where waste or by-products from one industry serve as raw materials for another, is another promising model demonstrated in eco-industrial parks worldwide (Ellen MacArthur Foundation, 2019).

The transport sector accounts for approximately 14% of global greenhouse gas (GHG) emissions, with the majority originating from road vehicles, aviation, and shipping (IPCC, 2022). Emissions continue to rise due to increasing demand for mobility, urbanization, and globalization of trade. Therefore, decarbonizing transport is a critical priority for national climate policies and net-zero roadmaps.

Electrification is at the forefront of low-carbon mobility solutions. Battery electric vehicles (BEVs) and plug-in hybrid electric vehicles (PHEVs), when powered by renewable energy sources, may significantly reduce emissions compared to internal combustion engine vehicles. In parallel, the deployment of charging infrastructure, battery recycling systems, and supportive policies (e.g., tax incentives, fuel economy standards, and vehicle mandates) is accelerating the transition in many countries. Notably, global EV sales surpassed 10 million units in 2022, and this number is expected to grow rapidly in the coming decade (IEA, 2023).

For long-distance freight, aviation, and maritime sectors, where batteries are less practical, alternative fuels are gaining traction. These include biofuels, green ammonia, and sustainable aviation fuels (SAFs), as well as hydrogen-based fuels, which offer lower lifecycle emissions. Efforts are also underway to enhance vehicle efficiency, including the development of aerodynamic designs, the use of lightweight materials, and the implementation of smart logistics systems that reduce fuel consumption and emissions.

Urban transport decarbonization further requires a shift toward low-emission options, such as public transportation, cycling, walking, and shared mobility systems. Transit-oriented development (TOD) and innovative mobility solutions integrate land use and transport planning to minimize commuting distances and reduce reliance on private vehicles. These strategies may reduce emissions, improve air quality, decrease congestion, and enhance urban livability.

A circular economy perspective is also applicable to transportation, focusing on vehicle design for disassembly and reuse, remanufacturing of parts, and closed-loop supply chains for critical raw materials, such as lithium and rare earth elements, used in batteries. As the transport sector becomes increasingly electrified, addressing the environmental footprint of battery production and disposal will be essential for ensuring overall sustainability.

The transition of industry and transport to carbon-neutral systems requires comprehensive policy frameworks, public–private partnerships, and innovation ecosystems. Carbon pricing, green procurement, emissions regulations, and subsidies for clean technologies are powerful levers to guide investments and behavior. Equally important are research and development (R&D) investments in next-generation technologies and capacity building to foster adoption, especially in developing economies.

Despite the growing momentum, several challenges persist. These include high upfront costs, technology lock-in, supply chain constraints (particularly for green hydrogen and critical minerals), and institutional inertia. In many regions, the lack of grid infrastructure, low public awareness, and inadequate policy coherence further delay progress. Addressing these challenges requires integrated planning, financial innovation, and a just transition approach to ensure that workers and communities dependent on high-carbon industries are not left behind.

1.4.5 Urban Systems: Smart Cities and Sustainable Infrastructure

Urban systems are at the forefront of climate change mitigation and adaptation. With more than 56% of the global population residing in urban areas, cities are responsible for over 70% of global energy consumption and approximately 75% of global CO_2 emissions, primarily from buildings, transportation, energy systems, and waste (IPCC, 2022; UN-Habitat, 2022). As urbanization accelerates, particularly in Asia and Africa, cities face a dual challenge, i.e., reducing their carbon footprints while meeting the infrastructure and service demands of growing populations. Therefore, reimagining urban development through smart cities and sustainable infrastructure has become a cornerstone strategy for achieving carbon neutrality (Singh, 2024c).

Smart cities are characterized by the integration of digital technologies and data analytics to optimize the efficiency of urban services, reduce resource consumption, and lower emissions. Intelligent transport systems, automated energy management, and real-time pollution monitoring are core components (Singh, 2025). Smart grids, for example, enable dynamic balancing of energy demand and supply, facilitating the integration of renewable energy and energy storage into urban energy systems (Garg et al., 2016). Moreover, building automation systems in commercial and residential properties significantly reduce energy use by optimizing lighting, heating, cooling, and appliance operation based on occupancy and time-of-day patterns (IEA, 2021). The synergy between digitalization and decarbonization is thus a key enabler of urban carbon neutrality.

The transformation of urban infrastructure is equally essential. Buildings are among the most significant sources of urban emissions, accounting for nearly 30% of global energy-related CO_2 emissions (IEA, 2022). Transitioning to net-zero energy buildings involves utilizing energy-efficient construction materials, high-performance insulation, passive solar design, and integrating rooftop solar photovoltaic (PV) systems. Retrofitting existing building stock is particularly important in developed cities. In contrast, newly developing cities in the Global South have the opportunity to leapfrog directly into low-carbon infrastructure by utilizing green building codes, performance standards, and sustainability certifications such as LEED, EDGE, and IGBC.

Sustainable urban mobility is a critical pillar of carbon-neutral cities. Transportation is a significant source of urban air pollution and GHG emissions. Strategies such as electrifying public transport fleets, promoting non-motorized transport (cycling, walking), and developing transit-oriented development (TOD) may significantly reduce the transport sector's carbon intensity. Cities like Amsterdam, Bogotá, and Copenhagen have successfully implemented multimodal, low-emission transport systems that prioritize public and active transport over private car use. Additionally, digital platforms that enable ride-sharing and integrated transport services, such as mobility-as-a-service (MaaS), contribute to reducing vehicle kilometers traveled and emissions.

Cities also play a crucial role in waste management and circular economy practices. Organic waste, if improperly managed, releases methane, a potent greenhouse gas. Integrated waste management systems, which encompass waste segregation, recycling, composting, and waste-to-energy technologies, enable cities to reduce emissions and decrease their reliance on landfills. Circular urban metabolism frameworks aim to minimize resource input and waste output by closing material loops and extending product lifecycles, thus contributing to emissions reduction and resource efficiency.

Urban green infrastructure and nature-based solutions (NbS) are gaining prominence as strategies that simultaneously address mitigation, adaptation, and biodiversity goals. Urban forests, green roofs, vertical gardens, wetlands, and permeable pavements contribute to carbon sequestration, reduce the urban heat island effect, and enhance climate resilience. Pataki et al. (2011) demonstrated that urban vegetation not only absorbs CO_2 but also improves thermal comfort and reduces building energy demand. These interventions are especially valuable in rapidly growing, climate-vulnerable cities.

Water-energy nexus management and resilient urban planning are central to sustainability. Efficient water supply, treatment, and recycling reduce energy use and emissions. In contrast, climate-resilient infrastructure, including flood defenses, heat-resistant roads, and distributed energy systems, enables cities to withstand climate extremes. Holistic urban planning that integrates land use, transportation, housing, and ecosystem services is crucial for mitigating spatial inefficiencies and reducing carbon leakage.

The success of carbon-neutral urban systems relies on effective governance, sound financial management, and active community participation. Local governments are often best positioned to implement climate action plans, yet they require enabling national policies, financial autonomy, and technical capacity. Instruments such as green municipal bonds, climate finance mechanisms, and public–private partnerships (PPPs) are essential for mobilizing resources. Furthermore, urban climate solutions must be inclusive and equitable, ensuring that vulnerable populations are not excluded from the benefits of clean infrastructure and green jobs (UNEP, 2022a, 2022b, 2022c).

Cities are both significant sources of emissions and key places for innovations and implementation to fight against climate change. Smart city technologies, low-carbon infrastructure, and participatory urban governance together offer a transformative pathway to carbon neutrality. Therefore, embracing integrated, inclusive, and future-ready strategies may help urban systems to lead the global transition to a sustainable and climate-resilient future.

1.5 Policy Instruments and Governance Mechanisms

Achieving carbon neutrality and transitioning toward environmental sustainability require robust and coherent policy instruments and governance mechanisms that can drive profound structural changes across sectors and scales. The complexity and urgency of the climate crisis demand ambitious emission reduction targets, as well as an integrated policy ecosystem that aligns environmental goals with economic development and social equity.

At the core of climate policy architecture are carbon pricing mechanisms, which internalize the external cost of greenhouse gas emissions and provide economic signals that influence investment and behavior. Two principal instruments dominate this approach, i.e., carbon taxes and emissions trading systems (ETS). A carbon tax directly sets a price on emissions, offering certainty in cost but uncertainty in emissions reduction. Conversely, ETS, such as the European Union Emissions Trading System (EU ETS), cap the total allowable emissions and allow market trading of emission permits, thereby ensuring environmental certainty while letting the market determine the price (World Bank, 2023). As of 2022, over 70 carbon pricing initiatives have been implemented or scheduled across jurisdictions, covering more than 20% of global emissions (World Bank, 2022).

Besides, regulatory standards and mandates remain essential tools in sectors where market signals alone may be insufficient to drive change. These include fuel economy standards for vehicles, renewable portfolio standards (RPS), building energy codes, and performance benchmarks for appliances and industrial equipment. For example, India's Perform, Achieve, and Trade (PAT) scheme sets energy-saving targets for energy-intensive industries and allows the trading of efficiency credits, blending regulation with market flexibility (BEE, 2022).

Subsidy reforms are also crucial, particularly in economies heavily reliant on fossil fuels. Globally, fossil fuel subsidies amounted to over $1 trillion in 2022, distorting markets, encouraging overconsumption, and discouraging investment in clean energy alternatives (IEA, 2023). Redirecting these subsidies toward renewable energy, energy access programs, and just transition initiatives may support climate action and social development simultaneously.

Beyond economic and regulatory instruments, planning frameworks and strategic roadmaps play a crucial role in guiding long-term decarbonization efforts. Nationally Determined Contributions (NDCs), submitted under the Paris Agreement, articulate each country's climate ambition and outline sectoral mitigation targets. Some countries have taken further steps by developing long-term low-emission development strategies (LT-LEDS) or net-zero roadmaps, which integrate emissions reductions with broader development priorities, such as energy security, poverty reduction, and public health (UNFCCC, 2021).

Multilevel governance is crucial for aligning national goals with sub-national implementation. Cities and regions are often at the forefront of innovation in climate policy, particularly in sectors such as transportation, buildings, and waste management. Frameworks such as ICLEI's Global Covenant of Mayors for Climate and Energy and the C40 Cities initiative demonstrate the growing influence of urban governance in climate action. Effective coordination between federal, state, and local levels, along with horizontal coordination across ministries and sectors, is essential to avoid policy fragmentation and regulatory overlap.

Institutional capacity and legal frameworks underpin the effective design, implementation, and enforcement of climate policies. Environmental protection laws, climate acts, and institutional reforms that empower dedicated climate agencies strengthen accountability and ensure compliance. The UK's Climate Change Act (2008) sets legally binding emission reduction targets and establishes an independent Committee on Climate Change to provide oversight and recommendations.

Market-based instruments, such as green bonds, sustainability-linked loans, and climate risk disclosure regulations, are mobilizing private finance toward low-carbon investments. The rise of environmental, social, and governance (ESG) criteria in corporate governance is also reshaping business behavior and investor expectations, especially as climate-related financial risks become more visible and material (TCFD, 2021).

A critical and often underemphasized dimension of climate governance is equity and justice. Climate policies must consider the differential responsibilities and capacities of countries, communities, and individuals, taking into account their varying abilities to mitigate and adapt to climate change. Mechanisms such as Just Transition frameworks are necessary to ensure that workers and vulnerable groups are not left behind during the transition to a low-carbon economy. Social protection, retraining programs, and participatory decision-making processes are crucial to establishing trust and a social license for transformative policies (ILO, 2015).

International cooperation and climate diplomacy are essential for fostering trust, mobilizing financial resources, and exchanging technology. Mechanisms under the United Nations Framework Convention on Climate Change (UNFCCC), such as the Green Climate Fund (GCF) and the Technology Mechanism, facilitate technology transfer and capacity building in developing countries. Regional alliances (EU Green Deal and African Union's climate strategies) and bilateral partnerships (Just Energy Transition Partnerships) further enhance policy coherence and cross-border collaboration.

It is now clear that a combination of economic instruments, regulatory frameworks, long-term planning, inclusive governance, and international cooperation forms the backbone of effective climate policy. To achieve carbon neutrality in a timely and equitable manner, these instruments must be coordinated, adaptive, and grounded in scientific evidence while also being responsive to local realities and global commitments.

1.6 Innovations and Emerging Solutions

Achieving carbon neutrality requires widespread adoption of existing technologies and the development and deployment of innovative and emerging solutions that accelerate deep decarbonization across sectors. As the global community confronts the limitations of conventional approaches, technological breakthroughs, digital transformation, and nature-integrated innovations are playing an increasingly vital role in shaping a low-carbon, climate-resilient future. These innovations span energy systems, industrial processes, carbon management technologies, and systemic approaches to sustainability, often guided by enabling policy and market instruments.

The most promising technological frontier is the development of green hydrogen, produced through the electrolysis of water using renewable electricity. Green hydrogen serves as a clean and flexible energy carrier, helping to decarbonize hard-to-abate sectors such as steel manufacturing, shipping, aviation, and high-temperature industrial processes (IRENA, 2022). Unlike grey or blue hydrogen, green hydrogen generates zero carbon emissions during production. Countries such as Germany, Japan, and India have launched national hydrogen missions, and several pilot projects are demonstrating their potential at scale. However, challenges, including cost, infrastructure, and storage, must be addressed through coordinated investment, research, and development, as well as market incentives.

The advancement of digital technologies, such as artificial intelligence (AI), the Internet of Things (IoT), blockchain, and big data analytics, is also reshaping carbon neutrality pathways. Smart grids, AI-powered energy management systems, and predictive maintenance tools enable real-time optimization of resource use and emissions reduction. For instance, IoT sensors may monitor energy flows and GHG emissions across industrial processes, enabling targeted interventions and automated controls. Blockchain is explored globally for transparent carbon accounting, the trading of renewable energy certificates, and the verification of carbon credits in voluntary carbon markets (UNEP, 2022a, 2022b, 2022c). These digital tools enhance the accuracy, traceability, and accountability of decarbonization efforts.

Nature-based and biotechnological innovations are also central to the carbon neutrality agenda. Bioenergy with carbon capture and storage (BECCS) combines renewable energy generation from biomass with CO_2 capture and storage, effectively delivering negative emissions. Similarly, innovations in enhanced weathering, soil carbon sequestration, and microalgae-based CO_2 fixation are being explored for their potential to draw down atmospheric carbon while offering co-benefits for agriculture and ecosystem health (Minx et al., 2018). In urban environments, the use of carbon-sequestering construction materials, such as bio-concrete and carbon-negative bricks, represents a shift toward net-zero infrastructure.

Additionally, circular economy models are driving innovations in design, consumption, and waste management. Innovations in materials science, such as biodegradable plastics, recyclable composites, and low-carbon cement, are helping to reduce the life-cycle emissions of products. New business models, such as product-as-a-service, sharing platforms, and extended producer responsibility (EPR) systems, help encourage people to use resources more sustainably and generate less waste. These models are supported by digital platforms and regulatory reforms that align market incentives with environmental goals.

Policy-driven innovation ecosystems are playing a pivotal role in shaping the future. Governments and international bodies are increasingly using mission-oriented innovation strategies, climate innovation hubs, and green public procurement to catalyze the development of low-carbon solutions. The Mission Innovation initiative, launched alongside the Paris Agreement, brings together over 20 countries to invest in clean energy innovation, with a focus on breakthroughs in storage, fuels, and grid technologies (Mission Innovation, 2021). Multilateral funding instruments, including the Green Climate Fund (GCF) and Clean Technology Fund (CTF), also support demonstration projects and technology transfer in developing countries.

Social and institutional innovations such as participatory governance, climate citizen assemblies, and behavioral nudges are increasingly recognized as critical complements to technological solutions. These approaches help shift societal norms, improve the acceptability of policies, and foster collaborative climate action at local and regional levels. Integrating equity, inclusion, and just transition principles into innovation ecosystems ensures that no one is left behind in the pursuit of carbon neutrality.

1.7 Implications for SDGs and Planetary Health

Carbon neutrality and environmental sustainability have far-reaching implications for the Sustainable Development Goals (SDGs) and the broader concept of planetary health. The SDGs, adopted in 2015 as part of the United Nations' 2030

Agenda for Sustainable Development, provide a comprehensive framework that integrates economic growth, social inclusion, and environmental protection. Carbon neutrality serves as a foundational enabler for multiple SDGs, including climate action (SDG 13), affordable and clean energy (SDG 7), sustainable cities (SDG 11), and responsible consumption and production (SDG 12).

Transitioning to low-carbon energy systems directly advances SDG 7 (Affordable and Clean Energy) by expanding access to renewable energy and reducing reliance on fossil fuels. This shift reduces emissions and enhances energy security, lowers energy costs, and creates jobs in green sectors (IRENA, 2023). Promoting decentralized energy systems, such as solar mini-grids in rural areas, supports carbon neutrality strategies that contribute to achieving energy equity and help meet SDG 1 (No Poverty) and SDG 10 (Reduced Inequalities) by enabling inclusive development.

Furthermore, low-emission practices in agriculture and sustainable land use contribute to SDG 2 (Zero Hunger) and SDG 15 (Life on Land) by enhancing food security, improving soil health, and conserving biodiversity. Forest-based natural climate solutions, such as reforestation and agroforestry, offer significant carbon sequestration potential while supporting ecosystem services and rural livelihoods (Griscom et al., 2017). These synergies highlight the importance of integrating climate mitigation and adaptation into sectoral development policies, ensuring that progress in one domain reinforces outcomes in others.

In urban areas, strategies such as smart city planning, energy-efficient buildings, and sustainable transportation systems directly support SDG 11 (Sustainable Cities and Communities). Simultaneously, improving air quality and reducing GHG emissions has positive effects on SDG 3 (Good Health and Well-being) by lowering the incidence of respiratory and cardiovascular diseases. These health co-benefits are especially significant in rapidly urbanizing and industrializing regions, where air pollution remains a leading cause of mortality (UNEP, 2022a, 2022b, 2022c).

Beyond the SDGs, the transition toward carbon neutrality is deeply intertwined with the principles of planetary health, which recognize the interdependence of human health and the state of natural systems. As defined by the Rockefeller Foundation–Lancet Commission, planetary health focuses on safeguarding the Earth's life-supporting systems, including climate, biodiversity, water, and soil, to ensure the well-being of current and future generations (Whitmee et al., 2015). Human-induced climate change poses a significant threat to the stability of planetary boundaries, leading to increased disease burdens, food insecurity, extreme weather events, and ecosystem collapse.

Mitigating emissions through sustainable energy, land use, and industrial practices contributes directly to maintaining these boundaries and preserving planetary health. For example, reducing deforestation helps stabilize local climates, regulate hydrological cycles, and prevent the spread of zoonotic diseases. Similarly, transitioning to plant-based diets and reducing food waste can lower methane and nitrous oxide emissions from agriculture, thereby supporting both climate targets and human health outcomes (Springmann et al., 2018).

However, these benefits will not be realized automatically. There is a risk of trade-offs and unintended consequences if climate policies are implemented in isolation or without considering social equity, local contexts, and systemic interactions. For instance, large-scale bioenergy production or afforestation without safeguards could threaten food security, displace communities, or reduce biodiversity. Therefore, aligning carbon neutrality strategies with SDG and planetary health objectives requires integrated, inclusive, and participatory approaches that balance environmental imperatives with social justice and development priorities.

Achieving carbon neutrality presents a transformative pathway for realizing the Sustainable Development Goals (SDGs) and protecting planetary health. It provides a unique opportunity to restructure human systems in ways that are low-carbon and equitable, resilient, and ecologically balanced. As such, embedding sustainability science, intersectoral collaboration, and long-term systems thinking into climate strategies is essential for achieving a safe future.

1.8 Conclusion

Achieving carbon neutrality is a climate mitigation target that represents a profound systemic shift toward a more sustainable, equitable, and resilient future. Transitioning to a carbon-neutral planet requires transformative changes across various sectors, including energy, land use, agriculture, industry, and urban systems. Decarbonizing the energy sector through renewable technologies, energy efficiency, and electrification forms the backbone of this transition. Likewise, sustainable land and forest management, low-emission agricultural practices, circular industrial models, and smart urban infrastructure are essential for reducing emissions while enhancing ecological integrity and socio-economic resilience. Importantly, the pursuit of carbon neutrality is intrinsically linked to achieving the Sustainable Development Goals (SDGs) and safeguarding planetary health. Strategies that reduce greenhouse gas emissions, expand energy access, improve air quality, and protect natural ecosystems yield broad co-benefits for public health, equity, biodiversity, and long-term human well-being. These interlinkages underscore the need to integrate climate action within broader development and environmental frameworks. However, the pathway to carbon neutrality is fraught with complex challenges, including technological, financial, institutional, and socio-political constraints. Overcoming these requires coherent policy frameworks, inclusive governance, international cooperation, and sustained investment in science, innovation, and capacity-building. A just transition is also imperative, one that protects vulnerable populations and ensures social equity throughout the transformation process. Integrating carbon neutrality within the broader context of environmental sustainability is crucial for staying within planetary boundaries and ensuring intergenerational justice. This highlights the need to reduce emissions and rethink how humans interact with the environment. Aligning climate strategies with the Sustainable Development Goals (SDGs) and the principles of planetary health can help lead the world toward a carbon-neutral and sustainable future for the next generations.

References

Barwa, S., & Singh, H. (2024). Climate-induced changes in forest ecosystems of Indian Himalayan region: A meta-analysis. In *Urban forests, climate change and environmental pollution: Physio-biochemical and molecular perspectives to enhance urban resilience* (pp. 839–866). Springer.

BEE. (2022). *Perform, achieve and trade (PAT) scheme.* Bureau of Energy Efficiency. https://beeindia.gov.in

Chazdon, R. L., & Brancalion, P. H. (2019). Restoring forests as a means to many ends. *Science, 365*(6448), 24–25. https://doi.org/10.1126/science.aax9539

Ellen MacArthur Foundation. (2019). *Completing the picture: How the circular economy tackles climate change.* https://ellenmacarthurfoundation.org

FAO. (2020). *The state of the world's forests 2020: Forests, biodiversity and people.* Food and Agriculture Organization of the United Nations. https://www.fao.org

FAO. (2021). *Global greenhouse gas emissions from agriculture, forestry and other land use.* Food and Agriculture Organization of the United Nations.

Garg, A., Pal, D., Singh, H., & Pandey, D. C. (2016). A comparative study of NDBI, NDISI and NDII for extraction of urban impervious surface of Dehradun [Uttarakhand, India] using Landsat 8 imagery. In *2016 International conference on emerging trends in communication technologies (ETCT)* (pp. 1–5).

Griscom, B. W., et al. (2017). Natural climate solutions. *Proceedings of the National Academy of Sciences, 114*(44), 11645–11650. https://doi.org/10.1073/pnas.1710465114

IEA. (2021). *Net zero by 2050: A roadmap for the global energy sector*. https://www.iea.org

IEA (2022). *World Energy Outlook 2022*, IEA, Paris. https://www.iea.org/reports/world-energy-outlook-2022

IEA. (2023). *Fossil fuel subsidies 2023.* International Energy Agency. https://www.iea.org

IEAGHG. (2021). *Global status of CCS 2021.* https://www.globalccsinstitute.com

ILO. (2015). *Guidelines for a just transition towards environmentally sustainable economies and societies for all.* International Labour Organization.

IPCC. (2022). *Climate change 2022: Mitigation of climate change.* Contribution of Working Group III to the Sixth Assessment Report of the Intergovernmental Panel on Climate Change.

IPCC. (2023). *Climate change 2023: Synthesis report.*

IRENA. (2022). *Green hydrogen cost reduction: Scaling up electrolysers to meet the 1.5 °C climate goal.* International Renewable Energy Agency. https://www.irena.org

IRENA. (2023). *Renewable capacity statistics 2023.* International Renewable Energy Agency. https://www.irena.org

Jangra, A., Kumar, K., Maikhuri, S., Bhandari, M. S., Pandey, S., Singh, H., & Barthwal, S. (2024). Unveiling stress-adapted endophytic bacteria: Characterizing plant growth-promoting traits and assessing cross-inoculation effects on *Populus deltoides* under abiotic stress. *Plant Physiology and Biochemistry, 210*, 108610.

Joshi, R., Sharma, B., Singh, H., Dhakal, N., Ayer, S., & Maraseni, T. (2024). Poplar plantation as an agroforestry approach: Economic benefits and its role in carbon sequestration in North India. *Journal of Resources and Ecology, 15*(4), 880–888.

Joshi, R., Singh, H., Chhetri, R., & Yadav, K. (2020). Assessment of carbon sequestration potential in degraded and non-degraded community forests in Terai region of Nepal. *Journal of Forest and Environmental Science, 36*(2), 113–121.

Joshi, R., Singh, H., Chhetri, R., Poudel, S. R., & Rijal, S. (2021). Carbon sequestration potential of community forests: A comparative analysis of soil organic carbon stock in community managed forests of Far-Western Nepal. *Eurasian Journal of Soil Science, 10*(2), 96–104.

Kumar, A., Dwivedi, G. K., Tewari, S., Sah, V. K., Singh, H., Kumar, P., Kumar, N., & Kaushal, R. (2020). Soil organic carbon pools under T*erminalia chebula* Retz. based agroforestry system in Himalayan foothills, India. *Current Science, 118*(7), 1098–1103.

Kumar, N., Singh, H., & Kumar, A. (Eds.). (2021). *Renewable energy and green technology: Principles and practices.* CRC Press.

Lal, R. (2020). Soil organic carbon sequestration and agronomic productivity. *Soil and Tillage Research, 204*, 104719.

Leifeld, J., & Menichetti, L. (2018). The underappreciated potential of peatlands in global climate change mitigation strategies. *Nature Communications, 9*, 1071. https://doi.org/10.1038/s41467-018-03406-6

Linquist, B. A., van Groenigen, K. J., Adviento-Borbe, M. A., Pittelkow, C. M., & van Kessel, C. (2015). An agronomic assessment of greenhouse gas emissions from major cereal crops. *Global Change Biology, 21*(1), 107–117.

Minx, J. C., et al. (2018). Negative emissions—Part 1: Research landscape and synthesis. *Environmental Research Letters, 13*(6), 063001. https://doi.org/10.1088/1748-9326/aabf9b

Mission Innovation. (2021). *Accelerating the clean energy revolution*. https://www.mission-innovation.net

Newell, P., & Paterson, M. (2010). *Climate capitalism: Global warming and the transformation of the global economy*. Cambridge University Press.

OECD. (2021). *Green recovery and beyond: How to get climate goals back on track.*

Pataki, D. E., et al. (2011). Urban ecosystems and the North American carbon cycle. *Global Change Biology, 17*(2), 755–768.

Raworth, K. (2017). *Doughnut economics: Seven ways to think like a 21st-century economist*. Chelsea Green Publishing.

Roberts, J. T., & Parks, B. C. (2007). *A climate of injustice: Global inequality, north-south politics, and climate policy*. MIT Press.

Rockström, J., et al. (2009). A safe operating space for humanity. *Nature, 461*(7263), 472–475. https://doi.org/10.1038/461472a

Ruhil, T., Singh, H., & Barthwal, S. (2024). Long-term physiological monitoring for detecting climate change impacts on forest ecosystems. In *Forests and climate change: Biological perspectives on impact, adaptation, and mitigation strategies* (pp. 817–834). Springer.

Sachs, J. D., et al. (2019). Six Transformations to achieve the sustainable development goals. *Nature Sustainability, 2*(9), 805–814. https://doi.org/10.1038/s41893-019-0352-9

Seddon, N., et al. (2020). Understanding the value and limits of nature-based solutions to climate change and other global challenges. *Philosophical Transactions of the Royal Society B, 375*(1794), 20190120. https://doi.org/10.1098/rstb.2019.0120

Singh, H. (2024a). Exploring adaptive modulation in plant functional traits and its impact on the productivity of *Acacia auriculiformis* under CO_2-enriched environment. *Industrial Crops and Products, 210*, 118186.

Singh, H. (2024b). *Forests and climate change: Biological perspectives on impact, adaptation, and mitigation strategies*. Springer Nature.

Singh, H. (2024c). *Urban forests, climate change and environmental pollution: Physio-biochemical and molecular perspectives to enhance urban resilience*. Springer Nature.

Singh, H. (2025). The role of plant functional traits in mitigating urban climate and environmental pollution. In *Plant functional traits* (pp. 297–308). Elsevier.

Singh, H., & Kumar, M. (2022). Climate change and its impact on Indian Himalayan forests: Current status and research needs. In *Climate change: Impacts, responses and sustainability in the Indian Himalaya* (pp. 223–242). Springer.

Singh, H., Kumar, N., & Kumar, A. (2024). Enhancing resource use efficiency in crops through plant functional traits. In *Plant functional traits for improving productivity* (pp. 97–117). Springer.

Singh, H., Kumar, N., Singh, R., & Kumar, M. (2023). Assessing the climate change impact on the habitat suitability of the range-restricted bird species (*Catreus wallichii*) in the Indian Himalayan ecosystem. *Environmental Science and Pollution Research, 30*(57), 121224–121235.

Singh, H., Verma, A., & Kumar, M. (2018). Can elevated CO_2 alter the growth dynamics, morphology, physiology and biochemistry of the forestry tree species? In *Climate change and agriculture* (pp. 288–305). Discovery Publishing House.

Springmann, M., et al. (2018). Options for keeping the food system within environmental limits. *Nature, 562*(7728), 519–525. https://doi.org/10.1038/s41586-018-0594-0

TCFD. (2021). *Task force on climate-related financial disclosures: 2021 status report*. https://www.fsb-tcfd.org
UN. (2015). *Transforming our world: The 2030 agenda for sustainable development*. https://sdgs.un.org/2030agenda
UNEP. (2022a). *Air pollution and human health: The silent killer*. United Nations Environment Programme. https://www.unep.org
UNEP. (2022b). *Emissions gap report 2022*. https://www.unep.org/resources/emissions-gap-report-2022
UNEP. (2022c). *Emissions gap report 2022: The closing window—Climate crisis calls for rapid transformation of societies*. United Nations Environment Programme. https://www.unep.org
UNFCCC. (2015). *The Paris agreement*. United Nations Framework Convention on Climate Change.
UNFCCC. (2018). *REDD+ frequently asked questions*. United Nations Framework Convention on Climate Change. https://unfccc.int/topics/land-use/workstreams/redd/faq-on-redd
UNFCCC. (2021). *Long-term low emission development strategies*. https://unfccc.int/lt-leds
UN-Habitat. (2022). *World cities report 2022: Envisaging the future of cities*. https://unhabitat.org
United Nations Development Programme. (2022). *Human Development Report 2021/2022: Uncertain Times*. Unsettled Lives: Shaping our Future in a Transforming World.
WCED. (1987). *Our common future*. Oxford University Press.
Whitmee, S., et al. (2015). Safeguarding human health in the Anthropocene epoch: Report of the Rockefeller Foundation-Lancet Commission on planetary health. *The Lancet, 386*(10007), 1973–2028. https://doi.org/10.1016/S0140-6736(15)60901-1
World Bank. (2022). *State and trends of carbon pricing 2022*. https://carbonpricingdashboard.worldbank.org
World Bank. (2023). *Carbon pricing and the climate crisis: A policy primer*. https://www.worldbank.org

Chapter 2
The Climate Crisis and Forestry Contribution Towards Sustainable Development Goals (SDGs)

Alok Kumar Singh, Rohini Thakur, Simran Thakur, Himank Thakur, Savita Dewangan, Rupali Singh, Hukum Singh, and Vivek Chauhan

Abstract The planet is undergoing a rapid and unprecedented transformation, with climate change emerging as one of the most critical global challenges. The Intergovernmental Panel on Climate Change (IPCC) has warned of a significant rise in global temperatures, leading to more frequent and intense extreme weather events such as heatwaves, droughts, floods, and storms. To mitigate the impacts of climate change and ensure a sustainable future, it is essential to harness the potential of forests. Often referred to as the "lungs of the planet," forests play a vital role in regulating the Earth's climate by absorbing carbon dioxide from the atmosphere, thereby reducing greenhouse gas concentrations and contributing to climate change mitigation. In addition to their role in carbon sequestration, forests provide a wide range of ecosystem services, including biodiversity conservation, regulation of the water cycle, prevention of soil erosion, and the supply of timber, firewood, and non-timber forest products. Further, it explores the critical role of forests in advancing the Sustainable Development Goals (SDGs), particularly SDG 13 (Climate Action), SDG 15 (Life on Land), SDG 6 (Clean Water and Sanitation), SDG 7 (Affordable and Clean Energy), SDG 2 (Zero Hunger), and SDG 1 (No Poverty). Forest-based carbon sequestration not only aids in climate mitigation but also underpins human well-being and sustainable development by providing essential ecosystem goods and services. The chapter also examines the current state of global climate change and the progress made toward achieving the Sustainable Development Goals (SDGs). Despite concerted global efforts, numerous challenges persist, including deforestation, population pressure, prolonged droughts, increasing forest fires, inadequate funding, and policy gaps. Furthermore, it highlights the crucial role of forests in mitigating climate change and promoting sustainable development. It seeks to inspire action and promote policies that prioritize forest conservation and sustainable forest management.

A. K. Singh (✉) · R. Thakur · S. Thakur · H. Thakur · S. Dewangan
Rajiv Gandhi South Campus, Banaras Hindu University, Mirzapur, Uttar Pradesh, India
e-mail: alokkj63@gmail.com; alokkj63@bhu.ac.in

R. Singh · H. Singh · V. Chauhan
ICFRE-Forest Research Institute, Dehradun, Uttarakhand, India

H. Singh (ed.), *Forestry for a Carbon-Neutral and Sustainable Future*, Sustainability Solutions, https://doi.org/10.1007/978-3-032-06427-1_2

Keywords Biodiversity conservation · Food security · Carbon sink · Deforestation · Agroforestry

2.1 Introduction

The planet is experiencing unprecedented and accelerating climate change. The burning of fossil fuels releases carbon dioxide into the atmosphere, contributing significantly to an increase in greenhouse gas emissions. This human-induced warming trend is undeniable, with the Intergovernmental Panel on Climate Change (IPCC, 2022) projecting significant global temperature increases. These rising temperatures are already causing a cascade of detrimental effects, including more frequent and intense extreme weather events such as tsunamis, floods, heatwaves, cyclones, and storms. The United Nations (UN, 2019) warns of a potential temperature rise exceeding three degrees Celsius, which would cause irreversible damage to the Earth's ecosystems. This warming trend has profound implications worldwide. Rising temperatures are contributing to environmental degradation, an increase in natural disasters, extreme weather patterns, food and water insecurity, economic disruption, and even social unrest and conflict (Singh, 2024b). The capacity of vital carbon sinks, such as Canada's boreal forests and the temperate mixed forests of Germany and Russia, to absorb carbon dioxide is diminishing. The boreal forest region, in particular, witnessed some of the most significant global temperature increases in the late twentieth century (Martínez & Laudon, 2024). This warming is disproportionately impacting winter temperatures, with daily lows rising more rapidly than daily highs during the summer. This shift has led to a decrease in the number of freezing days in most boreal regions, enhancing the survival of insects that damage trees (Barwa & Singh, 2024).

The impacts of climate change extend beyond terrestrial ecosystems. Coral reefs, which are crucial for marine biodiversity, are highly susceptible to environmental changes. As climate change warms and acidifies the oceans, coral reefs face a severe threat. These delicate ecosystems rely on calcium carbonate to build their skeletons, a process that becomes increasingly difficult in acidic waters (Garg et al., 2016). Consequently, coral reefs are experiencing mass bleaching events and declining health, with devastating consequences for marine life. The Amazon rainforest, once a significant carbon sink, is now becoming an increasingly considerable carbon source. In recent years, this vital ecosystem has been emitting more carbon dioxide than it can absorb, a worrying trend accelerated by climate change. The rainforest is nearing a tipping point, where it may no longer be able to generate sufficient rainfall to sustain itself (Nobre et al., 2016). This could lead to the transformation of the rainforest into a fire-prone savannah, further eroding its capacity to sequester carbon and exacerbating climate change.

The Polar Regions are experiencing some of the most dramatic and visible impacts of climate change. The melting of glaciers and ice sheets is accelerating, contributing significantly to rising sea levels. Even with significant reductions in greenhouse gas

Fig. 2.1 17 UN sustainable development goals (UNEP, 2020)

emissions, the loss of a substantial portion of these ice masses by 2100 is inevitable. The Arctic Ocean has experienced a dramatic decline in its oldest and thickest ice, with over 95% of it already lost (Ivanov, 2023). Rising sea levels increase the risk of coastal erosion and intensify the impact of coastal storms, such as hurricanes and typhoons, as warmer air and ocean temperatures fuel more frequent and intense events (Fig. 2.1).

The United Nations Conference on the Human Environment, held in Stockholm in 1972, drew the world's attention to the importance of sustainable development. The World Commission on Environment and Development published the report entitled "Our Common Future," also known as the Brundtland Report, in 1987. This report introduces the concept of sustainable development and describes the procedures for achieving SDGs (Brundtland, 1987).

Further progress was made at the Earth Summit 1992 Rio, where 178 countries adopted Agenda 21, a non-binding action plan for sustainable human development and environmental protection. In 2000, the Millennium Summit at the UN Headquarters in New York led to the adoption of eight Millennium Development Goals (MDGs), focusing on reducing extreme poverty by 2015. In 2002, during the World Summit on Sustainable Development in South Africa, the Johannesburg Declaration on Sustainable Development was adopted. This deceleration emphasized the eradication of poverty, environmental protection, and multilateral partnerships, advancing Agenda 21 and the Millennium Development Goals (MDGs).

However, the MDGs faced various criticisms, including a top-down approach, lack of inclusivity, oversimplification of complex issues, a narrow focus on gender equality, and promotion of the economic agendas of developed nations (Briant, 2017).

The MDGs failed to address systemic inequalities and the structural roots of poverty, limiting their long-term impact. This gap led to the formation of a new set of Global Development Goals. In 2015, during the United Nations' Sustainable Development Summit, all member states of the United Nations adopted the 2030 Agenda for Sustainable Development, comprising 17 Sustainable Development Goals and 169 targets (UNEP, 2014). The "2030" in the title refers to the target year by which the 17 Sustainable Development Goals (SDGs) are intended to be achieved (Table 2.1).

These goals act as a blueprint and provide guidelines on how to achieve a sustainable future for the governments of both developed and developing countries. The goals encompass a wide range of global issues, including poverty, hunger, gender equality, access to clean water, conservation of life on land and below water, and climate action. These goals were framed under five 'areas of critical importance,' which are also known as the 5 'P's: i.e., People, Planet, Prosperity, Peace, and Partnerships. The goals were launched with a motto of 'Ensuring that no one is left behind' (Morton et al., 2017). The United Nations Department of Economic and Social Affairs (UNDESA), within the Division for Sustainable Development Goals (DSDG), serves as the Secretariat for the Sustainable Development Goals (SDGs), providing substantive support and capacity-building for the goals.

2.2 Climate Change Mitigation and Promoting Sustainable Development

Forests cover nearly a third of the Earth's land surface, equivalent to the combined areas of Brazil, Canada, China, and the United States. They regulate ecosystems, protect biodiversity, play a crucial part in the carbon cycle, support livelihoods, and provide goods and services that can drive sustainable growth (Kumar et al., 2023; Singh et al., 2025a; Singh et al., 2024a). Healthy forests are powerful carbon sinks, meaning they absorb and store carbon dioxide.

Forests are also one of the most important solutions to addressing the effects of climate change. Approximately 2.6 billion tonnes of carbon dioxide, one-third of the CO_2 released from burning fossil fuels, is absorbed by forests every year (Joshi et al., 2021). Estimates indicate that, globally, between 2001 and 2019, forests absorbed twice as much carbon as they emitted, resulting in 7.6 billion metric tonnes of CO_2 being sequestered per year (Bhatt, 2024). Therefore, forests provide a crucial capacity to remove greenhouse gases (GHG) from the atmosphere and help us avoid the worst impacts of the climate crisis (Galos et al., 2012). Thus, forests mitigate climate change by enhancing carbon sequestration in growing trees and soils, as well as by increasing the sustainable supply of renewable raw materials (Patel et al., 2022; Singh et al., 2024b). These materials can store carbon for extended periods and serve as alternatives to fossil-based and other carbon-intensive sources. Forests play a vital role in environmental conservation and economic sustainability and also provide livelihoods to indigenous people and communities (Minz et al., 2021). The forest

Table 2.1 Sustainable development goals

SDG	Goal
Goal 1: no poverty	End of poverty in all forms everywhere
Goal 2: zero hunger	End hunger, achieve food security and improved nutrition, and promote sustainable agriculture
Goal 3: good health and well-being	Ensure healthy lives and promote well-being for all at all ages
Goal 4: quality education	Ensure inclusive and equitable quality education and promote lifelong learning opportunities for all
Goal 5: gender equality	Achieve gender equality and empower all women and girls
Goal 6: clean water and sanitation	Ensure the availability and sustainable management of water and sanitation for all
Goal 7: affordable and clean energy	Ensure access to affordable, reliable, sustainable, and modern energy for all
Goal 8: decent work and economic growth	Promote sustained, inclusive, and sustainable economic growth, full and productive employment, and decent work for all
Goal 9: industry innovation and infrastructure	Build resilient infrastructure, promote inclusive and sustainable industrialization, and foster innovation
Goal 10: reduce inequality	Reduce inequality within and among countries
Goal 11: sustainable cities and communities	Make cities and human settlements inclusive, safe, resilient, and sustainable
Goal 12: responsible consumption and production	Ensure sustainable consumption and production patterns
Goal 13: climate action	Take urgent action to combat climate change and its impacts by regulating emissions and promoting developments in renewable energy
Goal 14: life below water	Conserve and sustainably use the oceans, seas, and marine resources for sustainable development
Goal 15: life on land	Protect, restore, and promote sustainable use of terrestrial ecosystems, sustainably manage forests, combat desertification, and halt and reverse land degradation and biodiversity loss
Goal 16: peace, justice, and strong institutions	Promote peaceful and inclusive societies for sustainable development, provide access to justice for all, and build effective, accountable, and inclusive institutions at all levels
Goal 17: partnership for goals	Strengthen the means of implementation and revitalize the global partnership for sustainable development

also makes a major contribution to achieving various sustainable development goals. They directly contribute to SDG 13 (Climate Action) by absorbing carbon dioxide, which helps mitigate climate change. They play a role in biodiversity conservation (SDG 15), provision of clean water (SDG 6), access to affordable and clean energy (SDG 7), food security (SDG 2), and poverty alleviation (SDG 1) (Joshi et al., 2020).

2.3 Key Causes of Climate Change

2.3.1 Rising Atmospheric Greenhouse Gases

Carbon dioxide is the primary greenhouse gas contributing to recent climate change. Carbon dioxide enters the atmosphere through the burning of fossil fuels, solid waste, trees, and other biological materials, as well as through certain chemical reactions, such as those involved in cement manufacturing (Anderson et al., 2016). Carbon dioxide is naturally absorbed and emitted as part of the carbon cycle through plant and animal respiration, volcanic eruptions, and the exchange of gases between the ocean and the atmosphere (Schmitz et al., 2014). Another important greenhouse gas is methane. Many natural and human activities produce methane. For example, natural wetlands, agricultural activities, and the extraction and transportation of fossil fuels. Nitrous Oxide is produced mainly through agricultural fertilizers, fossil fuel burning, and other natural biological processes (Delmas et al., 1997). Many fluorides like Chlorofluorocarbons, hydrochlorofluorocarbons, hydrofluorocarbons, perfluorocarbons, and sulphur hexafluoride, together called F-gases, which are often used in coolants, foaming agents, fire extinguishers, solvents, pesticides, and aerosol propellants (Sheldon & Crimmin, 2022). Ground-level ozone is emitted through a chemical reaction between nitrogen oxides and volatile organic compounds in the presence of sunlight (Yildizhan et al., 2024). This ozone trap heats and acts as a greenhouse gas. Water vapour also acts as a greenhouse gas. Water vapour molecules absorb heat (infrared radiation) emitted by Earth's surface. Aerosols in the atmosphere can impact the climate.

Natural Causes of Climate Change

Some amount of climate change can be attributed to natural phenomena. Globally, carbon emissions range between 54.33 and 75.50 Gt CO_2-eq, with natural emissions contributing between 18.13 and 39.30 Gt CO_2-eq, and approximately equal to 29.07 Gt CO_2-eq (Xi-Liu & Qing-Xian, 2018). Various phenomena, such as volcanic eruptions, fluctuations in solar radiation (Bhargawa & Singh, 2019), tectonic shifts, and even small changes in Earth's orbit, have an impact on its warming and cooling patterns. Changes in the Earth's orbit and its axis of rotation have had a significant effect on the climate in the past. These cycles influence the distribution and intensity of solar energy received by Earth, affecting long-term climate patterns such as ice ages and interglacial periods. The variation in solar activity influences the intensity of the sunlight that reaches the earth's surface. These changes can affect the Earth's climate and have a minor impact on climate change. Additionally, volcanic eruptions have played a significant role in climate change, releasing large quantities of carbon dioxide in the distant past (Stenchikov, 2021). Some explosive volcano eruptions can also release particles into the upper atmosphere, where they act as cooling aerosols and reflect enough sunlight to space to cool the Earth's surface for several years.

2.3.2 Human-Driven Causes of Climate Change

Carbon is retained at the surface or below ground by both vegetation and soil, which serve as reservoirs. However, anthropogenic activities are eliminating or altering this vegetative biomass through logging and other development operations, which release the carbon stored in the biomass and release it into the atmosphere (Kumar et al., 2020). The vital yet delicate relationship between soil and climate has been severely upset by the quick development of agricultural methods. This industry is particularly notable for emitting significant amounts of methane and nitrous oxide, two potent greenhouse gases that effectively trap heat (Wadanambi et al., 2020). Due to the extensive use of chemical fertilizers and certain crop management practices that prioritize yield over soil health, agriculture is now a significant contributor to the release of greenhouse gases (GHGs) (Verma et al., 2023; Chahal et al., 2022a).

Furthermore, methane emissions from large-scale industrial livestock farming remain a significant source of atmospheric methane, primarily originating from the digestive systems of cattle and other ruminants. Additionally, a substantial source of greenhouse gas emissions is the industrial sector and manufacturing facilities that create consumer goods. A considerable amount of carbon was stored and released into the atmosphere (Singh & Kumar, 2022).

The vital yet delicate relationship between soil and climate has been severely upset by the quick development of agricultural methods. The manufacturing of a small number of carbon-intensive goods, including basic chemicals, iron and steel, cement, aluminum, glass, and paper, is a significant contributor to this sector's industrial emissions. Emissions from the oil and gas sectors are produced at every stage of their lifecycle, including during the original drilling procedures and combustion (Table 2.2).

2.4 Climate Change Impact on Biodiversity

Climate change not only influences the growth and survival of trees but also affects the prevalence and distribution of pests and diseases, further complicating forest management and conservation efforts. It is a significant threat to forest biodiversity, affecting species population dynamics, altering species distribution, and impacting ecosystem structure and function. Approximately 15% of the world's land is designated as terrestrial protected areas, which are still insufficient to address global biodiversity

Table 2.2 Key causes of climate change

Cause	Details
Green house	The natural process is intensified by human activities, trapping more heat and causing global warming
Key greenhouse gases	
Carbon dioxide	Released by burning fossil fuels (coal, oil, natural gas), deforestation, and industrial processes (e.g., cement production)
Methane	Emitted from agriculture (livestock digestion), natural wetlands, fossil fuel extraction, and transport
Nitrous oxide (N_2O)	Released through fertilizer use, soil management, and industrial processes
F gases	Synthetic gases, such as CFCs and HFCs, are used in refrigeration systems, aerosol products, and fire extinguishers
Aerosols	Microscopic particles from fossil fuel burning, volcanic eruptions, and biomass burning can either cool (reflect sunlight) or warm (absorb sunlight) the atmosphere
Natural factors	
Volcanic activity	Eruptions release particles (e.g., SO_2) that temporarily cool the climate by reflecting sunlight
Earth's orbit	Changes in orbit and axis rotation have caused historical ice age cycles
Solar activity	Variations in solar energy output influence climate but have minimal effect on recent warming trends
Human activities	
Fossil fuel	Burning coal, oil, and gas for energy and transport is the primary source of CO_2 and methane
Deforestation	Cutting forests for agriculture and urbanization releases stored carbon, reducing the capacity of forests to act as carbon sinks
Agriculture	Livestock emits methane; fertilizer use releases nitrous oxide; unsustainable practices degrade soil
Industry	Produces carbon-intensive materials (e.g., steel, cement, chemicals), emitting significant greenhouse gases
Oil and gas	Methane leaks from drilling, fracking, refining, and transportation; highly potent in trapping heat

loss. Economic growth has been influenced by increased production and consumption, thereby indirectly affecting biodiversity within ecosystems and exacerbating the impact of climate change (Habibullah et al., 2022).

Climate change has a significant impact on forest productivity and biodiversity through alterations in environmental conditions, such as temperature, CO_2 levels, and nutrient availability. Forest ecosystems are among the most diverse habitats on earth, home to most of the world's terrestrial species (Jangra et al., 2024). Forest biodiversity not only encompasses species diversity but also genetic diversity and ecosystem diversity. Forests support complex food webs, where primary producers (trees and plants) capture energy from sunlight and transfer it to herbivores and top consumers, maintaining ecological balance (Anandita & Jeganathan, 2024). The deterioration of forests often leads to the disruption of local food chains, habitat loss, species extinction, and a reduction in ecosystem services, underscoring the importance of forest conservation in maintaining biodiversity. The warming of climates is causing gradual shifts in the distribution of forest ecosystems as species are moving to higher latitudes and altitudes (Parmesan & Yohe, 2013). This migration indicates the species' adaptability to climate change, which affects its survival and distribution, particularly at the margins of its range and in the core areas of its habitat.

Forests play a crucial role in the ecosystem, balancing biodiversity, mitigating climate change, acting as a carbon sink, and providing livelihoods for humans. They also influence the abiotic components that have a profound effect on their growth and development. Approximately 31% of the world's land area serves as a habitat for three-quarters of terrestrial biodiversity and meets the needs of millions of people.

Forest ecosystems are vulnerable to pests and diseases, as changing climatic conditions favour the growth of these threats, with projections showing a potential increase in pest outbreaks. In the Western Himalayas of India, forests in Himachal Pradesh have exhibited significant shifts in susceptibility to changing climate conditions, with vulnerability scores increasing by 15–20% under future climate scenarios compared to the current climate (Upgupta et al., 2015). Similarly, Canada's boreal forests, as well as those in Germany and Russia's temperate regions, are witnessing changes in carbon dynamics due to depleting and degraded forest productivity and management practices.

2.5 Forests as Nature's Solution to Climate Change

2.5.1 Forests as Carbon Sinks

Trees use sugar to build wood, branches, and roots. Wood is an incredible carbon sink because it is mainly composed of carbon (about 50%). It lasts for years as a standing tree and takes years to decompose after the tree dies. While trees primarily store carbon, they also release some carbon, such as when their leaves decompose, or their roots burn sugar to capture nutrients and water. Forests capture and store different amounts of carbon at different speeds depending on the average age of the trees in the stand and the number of trees in the stand (Singh, 2024a). High densities of trees characterize young forests and are highly effective at capturing carbon. Young trees proliferate and can rapidly absorb carbon. Not every small sapling becomes a large tree due to competition for light, resources, and growing space, but when they die and decompose, little carbon is released. The trees that remain continue to grow and sequester more carbon as the forest matures (Singh et al., 2018).

Tropical rainforests collectively sequester more carbon from the atmosphere than temperate or boreal forests, but they're also increasingly destroyed for agricultural expansion. While temperate and tropical regrowth forests have experienced increases in carbon sinks, boreal and intact tropical forests have experienced declines (Pan et al., 2024).

Forests Offer Ecosystem Services

Forests are vital ecosystems that provide a wide range of services critical to sustaining life and enhancing human well-being. Ecosystem services are defined as "the benefits provided by the ecosystem". In 2005, the Millennium Ecosystem Assessment was published, categorizing ecosystem services into four categories: provisioning, regulating, cultural, and supporting (Fig. 2.2).

Climate Regulation

Forests have a significant impact on regional and global climates by regulating water cycles, contributing to cloud formation and precipitation patterns, and thereby influencing both temperature and humidity levels (Raihan et al., 2022). Forests create their microclimate and help reduce the overall temperature in their surrounding area. Carbon sequestration mitigates climate change by reducing the concentration of CO_2, a significant greenhouse gas, in the atmosphere (Singh, 2024b). Forests act as carbon sinks, absorbing carbon dioxide from the atmosphere and storing it until it is incorporated into their life cycle.

Fig. 2.2 Forest ecosystem services (Martínez Pastur et al., 2018)

2.5.1.1 Water Cycle Regulation

The dense canopy of forests intercepts rainfall, some of which evaporates back into the atmosphere while the remainder nourishes the forest and replenishes rivers and groundwater (François et al., 2024). The roots of trees bind soil particles, improving soil structure and reducing erosion (Vishwakarma et al., 2024). This enhances water holding capacity and improves water infiltration. Thus, forests play an important role in regulating water, groundwater recharge, and precipitation processes.

2.5.1.2 Soil Conservation

Trees reduce soil erosion caused by wind and water by binding soil particles together, thereby stabilizing the soil. The addition of leaf litter to soil increases the organic matter content, improves soil fertility, and enhances nutrient cycling. Forests enhance the microbial and fungal network, thereby detoxifying and maintaining the health and structure of the soil (Bengtsson et al., 2019).

2.5.1.3 Biodiversity

Biodiversity refers to the variety of life at each ecological scale, ranging from the genome to the biome (Gaggiotti et al., 2018). Biodiversity plays a vital role in providing direct services and goods, as well as in the proper functioning of ecosystem services. Biodiversity can enhance the resilience of forest ecosystems, enabling them to adapt to environmental changes and disturbances and thereby secure ecosystem services for future generations (Joshi et al., 2024).

Moreover, community evenness mediates the relationship between biodiversity and productivity, with higher evenness leading to higher productivity (Hordijk et al., 2023).

2.5.1.4 Socio-economic Values

Changes in forest health and productivity have a significant impact. Besides direct employment, forests serve as a source of livelihood for local communities, providing fuelwood and fodder (FAO, 2022). The economic contribution of forests is substantially undervalued, with the formal forest sector adding more than USD 1.5 trillion to global economies. This underestimation overlooks the vast array of ecosystem services provided by forests, including carbon sequestration, water regulation, and biodiversity conservation (FAO, 2022).

2.6 Forests and the Sustainable Development Goals

Forests play a critical rolc in achieving the Sustainable Development Goals. These goals aim to address global challenges, including poverty, inequality, climate change, and environmental degradation while ensuring the prosperity and well-being of all (Ruhil et al., 2024). The forest can play a key role in achieving the following goals (Fig. 2.3).

2.6.1 SDG 1: No Poverty: To End Poverty in All Its Forms Everywhere by 2030

The alleviation of global poverty is a significant objective of the 2030 UN Sustainable. Forests are the most productive land-based ecosystems in the world and have many direct and indirect roles in poverty elevation (Wang et al., 2023). Globally, the forest-dwelling population ranges from 1 million to over 1 billion individuals. The World Bank approximates that 240 million people reside in predominantly forested ecosystems. It is estimated that approximately one-quarter of the world's poor and 90% of the poorest populations rely significantly on forests for their livelihoods (Chao, 2012). Globally, forests have two key roles in poverty alleviation. First, they act as an essential safety net, helping rural communities avoid poverty or providing

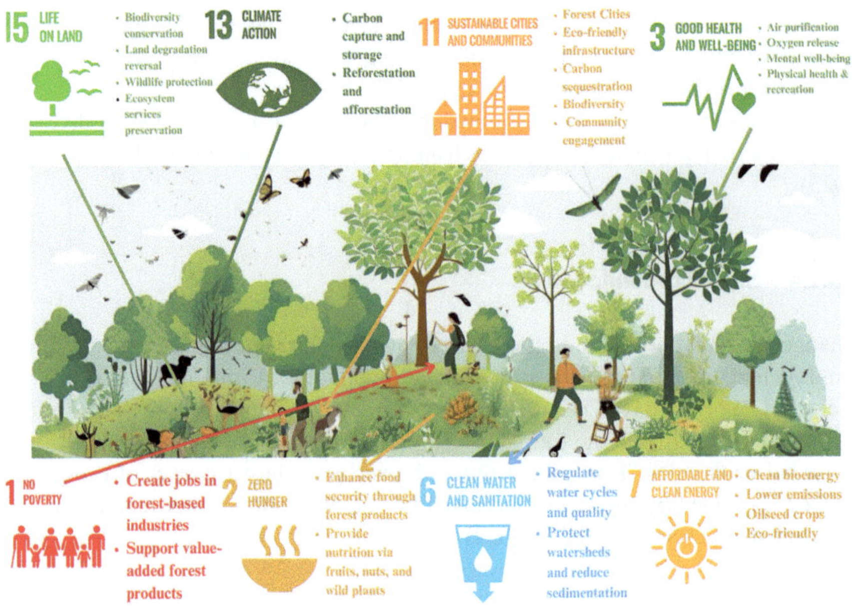

Fig. 2.3 Role of forests in achieving SDG (UNEP, 2020)

support to those who are already impoverished. Second, forests have untapped potential to actively lift some rural populations out of poverty (Sunderlin et al., 2003). The role of forests in poverty alleviation can be understood as:

The Forest Service's services derived from the forest in the context of poverty alleviation can be classified into provisioning services, which refer to the tangible products that forests provide, such as timber, i.e., industrial wood products like wood, teak wood, pulpwood, and fuelwood, as well as non-timber forest products such as medicinal plants, fruits, nuts, bamboo, canes, gums, resin, tendu leaves, fiber, lac, and silk (Kumar & Singh, 2020). The NTFP plays a crucial role in rural livelihoods. Approximately 100 million people living in and around the forest depend heavily on NTP collection (Saxena, 2003).

Forests play an important role in employment generation. Approximately 33 million people in the forest sector globally were engaged from 2017 to 2019, which accounts for approximately 1% of the total global employment (Lippe et al., 2022). The forestry sector is responsible for providing approximately 10 million real jobs in developing countries, in addition to delivering between 30 and 50 million informal jobs in the wood industry. Approx. 1.6 billion people worldwide rely heavily on forest resources for their livelihoods (Dubois, 2003).

SDG 2: Zero Hunger: To End Hunger, Achieve Food Security, and Improved

Extreme hunger and malnutrition remain significant barriers to sustainable development. This goal aims to create a hunger-free world by 2030. In 2022, approximately 735 million people, or 9.2% of the world's population, were experiencing chronic hunger. Around 2.4 billion people faced moderate to severe food insecurity in 2022 (SDG, 2024). An estimated 45 million children under the age of 5 suffered from wasting, 148 million had stunted growth, and 37 million were overweight (Sachs et al., 2023). The solution to addressing extreme hunger and malnutrition lies in global food security. Forests contribute to food security by providing supplementary and complementary resources to agricultural products, enhancing nutrition and resilience (FAO, 1997). Forests, trees, and agroforestry play a key role in supporting food security and nutrition through its four dimensions: availability, access, utilization, and stability. Providing essential resources that contribute to a reliable and nutritious food supply enables communities to meet their needs in a sustainable manner (Gitz et al., 2021).

The forests directly contribute to food security. Forests, trees, and agroforestry provide non-wood forest products, which serve as a source of food and fuel for cooking purposes (Kumar & Singh, 2020). The Non-Wood Forest Products consist of goods of biological origin other than wood, derived from forests and other wooded land, which includes mushrooms, fruits, nuts, herbs, aromatic plants, game, fibers (used in construction, clothing, or handicrafts), resins, gums, saps, and products used for medicinal, cosmetic, or cultural purposes (Aluma, 2002). Globally, around 50% of all fruit consumed by humans originates from trees (Retamales, 2011). Arable crops, vegetables, forest trees and shrubs, and pasture grasses are the major components, providing diversity and productivity to the system (Chahal et al., 2022b). The forest also serves as a significant source of biofuels for cooking and boiling water.

According to an estimate, approximately 2.4 billion people cook with wood fuel, which accounts for about 40% of the population in less developed countries. In addition, 764 million of these people may also use this fuel wood for boiling water (Rametsteiner & Whiteman, 2014). Trees also provide animal fodder for livestock, fulfilling their nutritional needs for milk and meat.

The indirect role of Forests in food security is that forests and trees provide ecosystem services that help promote sustainable agriculture and food production. Trees with deep roots bring nutrients from the deeper layers and make them available for crops, which helps enhance agricultural productivity (Nair, 1993). Forest ecosystems also harbor larger biodiversity, which provides various pollinating agents (Garibaldi et al., 2011).

2.6.2 SDG 3: Good Health and Well-Being: Ensure Healthy Lives and Promote Well-Being for All at All Ages

Only around one-third to half of the global population was covered by essential health services in 2017. Recent trends indicate that only 39–63% of the population will be covered under essential health services by 2024 (SDG). In 1948, the WHO defined human health as a state of complete physical, mental, and social well-being, and not merely the absence of disease or infirmity. Forests and trees contribute to a variety of ecosystem services, helping to form a healthy living environment and restore degraded lands. They provide a range of tangible benefits, including flood and drought mitigation, noise pollution control, erosion prevention, and water purification (Karjalainen et al., 2010).

The role of forests in promoting physiological health in urban environments can harm the physical and mental well-being of people. Living in a building-dominated climate causes a rise in stress levels (Rajoo et al., 2020). Forest ecosystems play a significant role in the mental and physical well-being of individuals through various mechanisms. They play a crucial role in alleviating stress and facilitating recovery from attentional fatigue, which improves both psychological and physical rehabilitation outcomes. Research indicates that forests play a pivotal role in helping people alleviate mental health issues and enhance their overall mental well-being (Bratman et al., 2012). Brief walks in forested areas help reduce tiredness and contribute to alleviating negative emotions, such as anxiety, anger, and confusion. Forest therapy plays a role in enhancing the effectiveness of the human immune system, particularly by boosting the activity of natural killer cells (Tsao et al., 2018).

Medicinal herbs play a crucial role in providing cost-effective healthcare and ensuring livelihood stability and are among the most valuable NTFPs (Astutik et al., 2019). Approximately 80% of the population in developing countries relies mainly on plant-based medicines (Karunamoorthi et al., 2013). People have been using forest medicines for over 5,000 years; more than 1 billion people use forest-based medicinal remedies. In India, forest plants are commonly used to treat snake bites, asthma, jaundice, dropsy, gynecological problems, piles, elephantiasis, bronchitis, rheumatism, leprosy, diabetes, cancer, pneumonia, paralysis, pharyngitis, ulcers, dysentery, cough, skin diseases, fever, and lactation insufficiency (FAO, 2020).

Forests have a variety of preventive roles. Their role in mitigating various natural disasters is widely acknowledged as cost-effective and dependable (Perzl et al., 2022). Trees and forests help remove air pollutants like ozone, Sulfur dioxide, and nitrous oxide, primarily by taking them up through stomata, which helps reduce air pollution (Nowak et al., 2014). Forests also play a crucial role in reducing disaster risks by regulating water flow to prevent flooding, stabilizing soils to avoid landslides, and mitigating the effects of climate change, such as droughts. Coastal forests, such as mangroves, protect communities from storm surges and coastal erosion. Forest ecosystems help mitigate and adapt to extreme weather events by providing buffering functions, such as cooling, water storage, and wind protection, which reduce the impact of floods, droughts, and temperature increases.

SDG 6: Clean Water and Sanitation: To Ensure Access to Safe Water Sources and Sanitation for All

Water is a common and essential resource for life, not only for health but also for poverty reduction, food security, peace, human rights, ecosystem preservation, and education. Despite this 2.2 billion people around the world still lack safely managed drinking water, including 785 million without basic drinking water (SDG, 2024). Forests are a significant source of the world's drinking water, with millions of people relying on the high-quality freshwater they provide. For example, the Uluguru Mountains are a source of drinking water for the 2.5 million residents of the Tanzanian capital. Forests play a vital role in ensuring clean water and sanitation by regulating water cycles, maintaining watershed integrity, and filtering pollutants.

Forests play a crucial role in the hydrological cycle by influencing water flow, quality, and climate regulation, as well as rainfall interception, evaporation, and transpiration, which affect stream flow and groundwater levels. Forests play a crucial role in water storage, preventing soil erosion, and regulating the climate. Changes in forest cover can have a significant impact on local water cycles and precipitation (Roberts, 2009). Forested watersheds typically provide superior water quality compared to other land uses such as agriculture, industrial activities, and urban development. This is primarily because these land uses tend to have elevated levels of pollutants (Calder et al., 2007). Forests act as natural filters, trapping sediments, nutrients, and impurities from runoff, while their root systems stabilize the soil, reducing erosion and protecting water quality.

2.6.3 SDG 7 Affordable and Clean Energy: To Ensure Access to Affordable, Reliable, Sustainable, and Modern Energy for All

Approximately 548 million people worldwide lack access to electricity (SDG, 2024). Lack of clean cooking leads to exposure of women and children to harmful gases and smoke. Countries can accelerate the transformation to an affordable, reliable, and sustainable energy system by investing in renewable energy resources, prioritizing energy-efficient practices, and adopting clean energy technologies. Forests can act as a potential source of bioenergy and biofuel. Plant-based biofuels can help lower environmental threats as they have a low net release of carbon dioxide and sulfur into the atmosphere (Singh & Singh, 2023). Therefore, plant-based biofuels are the only form of renewable energy that can completely replace fossil fuels (Kumar et al., 2021).

Oilseed from various forest trees can be used for the production of liquid biofuels, such as biodiesel. For example, the seeds and fruit pulp of Simarouba (*Simarouba glauca*) can be used for ethanol production (Sharma et al., 2016). Ethanol serves as a sustainable biofuel; engines operating on ethanol exhibit a potential reduction in emissions (Mendiburu et al., 2022). Additionally, lignocellulosic biomass can be utilized for the production of second-generation ethanol. This will help in clean energy production.

Plant biomass can be utilized in a gasification system to provide fuel for machinery operation (Mandal & Sharma, 2023). Gasification is a process of converting biomass into energy that produces high-value fuels while emitting fewer greenhouse gases. This not only saves fossil fuels but also helps reduce greenhouse gas (GHG) emissions in the atmosphere.

SDG 11: Sustainable Cities and Communities: To Make Cities Inclusive, Safe, Resilient and Sustainable

Urban areas constitute merely 3% of the Earth's terrestrial surface; however, they are responsible for 60–80% of global energy consumption and 75% of carbon emissions. Furthermore, numerous cities exhibit heightened susceptibility to the impacts of climate change and natural disasters (Singh, 2024c). Around half of the world's population resides in cities (UN, 2019). Urban forests in such areas play a crucial role in creating sustainable cities and communities. Urban forests can be defined as networks or systems comprising all woodlands, groups of trees, and individual trees located in urban and peri-urban areas; they include, therefore, forests, street trees, trees in parks and gardens, and trees in derelict corners (Konijnendijk et al., 2024). They provide essential ecological services, including improving air quality, mitigating climate change, supporting biodiversity, and enhancing human health.

Urban forests in and around cities play a vital role in addressing various environmental challenges. Field studies have demonstrated vegetation significantly reduces noise pollution, providing an effective natural sound barrier (Dwyer et al., 1992). Urban forests help in improving the air quality as trees and vegetation absorb many harmful pollutants like particulate matter (PM2.5 and PM10), nitrogen oxides (NOx), sulphur dioxide (SO_2), carbon monoxide (CO), and ozone (O_3) and also in carbon sequestration and reduction of GHG. Another critical benefit of urban forests is the enrichment of biodiversity, as cities and urban areas are often characterized by a biodiversity deficit (Ventura et al., 2024). These green spaces around cities host a diverse range of flora and fauna, from towering trees to tiny insects. The diverse vegetation in urban forests provides food, shelter, and breeding grounds for countless species, contributing to the overall biodiversity of the area (Singh et al., 2024a). They also aid in temperature regulation (Ornelas et al., 2023). Incorporating various green spaces, such as rooftop gardens, urban forests, parks, and roadside vegetation, can help cities significantly mitigate temperature fluctuations, thereby promoting a more comfortable and sustainable urban environment (Kale et al., 2024).

These roles of urban forests align with SDG 11, which aims to create inclusive, safe, resilient, and sustainable cities and human settlements, emphasizing the importance of green spaces in promoting environmental and biodiversity health (Singh, 2025).

2.6.4 SDG 13: Climate Action: Taking Urgent Action to Tackle Climate Change and Its Impacts

If immediate actions are not taken, climate change will cause an adverse increase in average global temperatures beyond 3 °C and will adversely affect every ecosystem (SDG, 2024). As per the Paris Agreement 2015, the countries are legally bound to prevent "the increase in the global average temperature to well below 2 °C above pre-industrial levels" and pursue efforts "to limit the temperature increase to 1.5 °C above pre-industrial levels" (UNFCCC, 2016).

Forests are a major component of the global carbon cycle, removing at least 3 billion tons of carbon released due to human activities and acting as a significant carbon sink in terrestrial ecosystems. The 4 billion tons of the forest ecosystem act as a reservoir of carbon and are responsible for storing almost twice the amount of carbon than the atmosphere (Canadell & Michael, 2008). Forests play a vital role in mitigating climate change by capturing and storing carbon emitted from human activities (Kasim & Ghassan, 2023). Through photosynthesis, plants capture carbon dioxide and convert it into living biomass, such as tree trunks, roots, shoots, branches, leaves, and fruit. The carbon stock for 2023 in Indian forest has been estimated 7,285.5 Mt (ISFR, 2023). In addition to carbon storage, forests impact climate systems in various ways, such as regulating atmospheric moisture and rainfall patterns through the process of evapotranspiration. At least 40% of rainfall is attributed to evapotranspiration (Brack, 2019). These services help reduce the vulnerability of sectors and people beyond the forestry sector, including provisioning services such as food and fuel, cultural services like recreational, spiritual, or religious activities, and regulatory services like maintaining the water supply.

2.6.5 SDG 15 Life on Land: To Sustainably Manage Forests, Combat Desertification, Halt and Reverse Land Degradation, and Halt Biodiversity Loss

Forests are vital to maintaining life on land, as they provide essential habitat, regulate the climate, sustain biodiversity, and offer numerous ecosystem services. Forest covers 31% of the total land area (Singh et al., 2024a, 2024b, 2024c). Forests are the most diverse and productive ecosystems on land, as they harbor the vast majority of the world's terrestrial species. Forest biological diversity refers to all life forms found within forested areas and the ecological roles they perform (FAO, 2020).

Forests are Earth's most important and diverse hotspots. They play a crucial role in maintaining life on land by providing tangible benefits, such as food and firewood, as well as intangible benefits, including oxygen (Kaltenborn et al., 2020).

Forests are the most diverse ecosystems on Earth, and out of 70,000 known vertebrate species, forests are home to 5,000 amphibians, 7,500 birds, and 3,700 mammals. Additionally, approximately 1.3 million invertebrate species have been described, with an estimated total of 5–10 million, primarily insects that inhabit forests (Singh et al., 2023). There are approximately 391,000 known vascular plant species, of which 60% are found in tropical forests (FAO, 2020).

2.7 Challenges

Forests cover nearly one-third of the land area on our planet and are home to most of the world's life on land. They are also essential to human health, as they purify our water and air. Maintaining forests and their biodiversity is both complex and sensitive, and the natural and human impacts on forest ecosystems are making this increasingly difficult. Forests provide humans with numerous tangible and intangible benefits. Minor forest products continue to play a significant role in the lives of people living in impoverished states to this day. Human activities generally pose the threat to forests. Agriculture has been the primary reason for the conversion of forest land into non-forestry purposes.

2.7.1 Deforestation and Forest Degradation

Deforestation refers to the conversion of forest land to other uses. The leading cause of deforestation is agriculture. Deforestation can have severe negative impacts on forest biodiversity. Since 1990, an estimated 420 million hectares of forest have been lost due to deforestation. Africa had the largest net forest loss of 3.9 million hectares annually (2010–2020), with a rising loss rate since 1990, while South America's loss decreased to 2.6 million hectares annually in 2010–2020, half the rate of 2000–2010; Asia had the highest net forest gain, and Oceania saw losses in both the 1990–2000 and 2000–2010 decades (FAO, 2020). Deforestation is a particular concern in tropical rainforests because these forests are home to a significant portion of the world's biodiversity (Morris, 2010). In the Amazon forest, approximately 17% of the forest has been lost over the last 50 years, primarily due to forest conversion for cattle ranching (Barreto et al., 2005). Deforestation in this region is particularly rampant near populated areas, roads, and rivers; however, remote areas have also been encroached upon with the discovery of valuable resources, such as gold and oil. Agriculture is generally acknowledged as the primary driver of deforestation, as forests are cleared to make space for crops or cattle farms. In comparison, only a small proportion of global deforestation is caused directly by the forest sector through the establishment of plantations and overexploitation of timber.

2.7.2 Climate Change

Climate change is one of the most significant challenges currently facing humanity. The increased severity and frequency of natural disasters, changing weather patterns, polar ice melt, sea level rise, and drought are just some of the consequences already experienced by populations around the world. While forests can help to combat climate change, they are also highly vulnerable to changing climatic conditions (Singh et al., 2025b).

The climate at a given location determines the type of forest that can become established. When climate conditions change, forests must adapt to these changes. However, the adaptation process usually requires more time than the changing climate conditions allow. This often results in the loss of forests, their biodiversity, and their ability to mitigate the impacts of climate change. Climate change will impact several natural disturbances that threaten forest health. These include insect outbreaks, invasive species, wildfires, and storms. Some disturbances, such as wildfires, occur quickly. Others, such as changes in animal or plant populations, occur over decades to centuries. Some of their effects may be temporary, allowing a forest to recover. In other cases, a forest may suffer lasting impacts. Climate change is expected to impact the ability of forests to provide key ecosystem services, including carbon storage, clean air and water supply, recreation, and wildlife habitat.

2.7.3 Policy and Governance Challenges

Inadequate legislation and enforcement are a significant challenge for sustainable forest management. Various developed and developing countries around the world lack robust and effective legal structures for safeguarding forests and promoting their sustainable use. In instances where these policies exist, various shortcomings in enforcement mechanisms result in activities such as illegal logging, land appropriation, and unsustainable exploitation of forest resources. Forests are also open resources that serve multiple purposes, including timber production, biodiversity conservation, and local livelihoods. Balancing these competing interests often can lead to conflicts and unsustainable practices. The increasing effects of climate change, such as forest fires, droughts, and pests, complicate forest management. Policymakers often face difficulty in adapting forest management strategies to cope with these dynamic challenges. Many regions suffer from weak governance structures, a lack of technical expertise, and fragmented institutions that are unable to coordinate forest management efforts effectively across different sectors and levels of government.

2.8 Conclusion

Forests are vital ecosystems and indispensable allies in addressing the climate crisis and achieving the Sustainable Development Goals (SDGs). The escalating climate crisis—driven by greenhouse gas emissions, deforestation, and human activities—threatens ecosystems, biodiversity, and human well-being. Forests offer a natural solution by acting as carbon sinks, supporting biodiversity, and providing essential ecosystem services. They contribute directly to multiple SDGs, including poverty reduction (SDG 1), food security (SDG 2), good health (SDG 3), clean water (SDG 6), affordable clean energy (SDG 7), climate action (SDG 13), and life on land (SDG 15). Their roles in carbon sequestration, water regulation, and renewable energy make them central to sustainable development and climate resilience. However, forests face growing pressures from deforestation, degradation, and climate impacts. To safeguard their benefits, it is imperative to adopt sustainable forest management, strengthen policy frameworks, foster global cooperation, and invest in innovation. Recognizing forests as foundational to both ecological integrity and human prosperity is essential. Increased investment in their conservation is urgently needed to ensure a sustainable, resilient, and equitable future for all.

References

Aluma, J. (2002). Non-wood forest products no. 13. Resource assessment of non-wood forest products; experience and biometric principles. *African Journal of Ecology*, *40*(4), 413–413. https://doi.org/10.1046/j.1365-2028.2002.00393.x

Anandita, S. K. A., & Jeganathan, C. (2024). Understanding and mitigating climate change impacts on ecosystem health and functionality. *Rendiconti Lincei, Scienze Fisiche e Naturali, 35*, 793–818. https://doi.org/10.1007/s12210-024-01259-4

Anderson, T. R., Hawkins, E., & Jones, P. D. (2016). CO_2, the greenhouse effect and global warming: From the pioneering work of Arrhenius and Callendar to today's earth system models. *Endeavour, 40*(3), 178–187. https://doi.org/10.1016/j.scitotenv.2024.173359

Astutik, S., Pretzsch, J., Ndzifon, & Kimengsi, J. (2019). Asian medicinal plants' production and utilization potentials: A review. *Sustainability, 11*(19), 5483. https://doi.org/10.3390/su11195483

Barreto, P., Souza, J. C., Anderson, A., Salomao, R., Wiles, J., & Nogueron, R. (2005). *Human pressure in the Brazilian Amazon* (Vol. 3, pp. 1–6). Instituto do Homem e Meio ambiente da Amazônia.

Barwa, S., & Singh, H. (2024). Climate-induced changes in forest ecosystems of Indian Himalayan region: A meta-analysis. In *Urban forests, climate change and environmental pollution: Physio-biochemical and molecular perspectives to enhance urban resilience* (pp. 839–866). Springer.

Bengtsson, J., Bullock, J. M., Egoh, B., Everson, C., Everson, T., O'Connor, T., O'Farrell, P. J., Smith, H. G., & Lindborg, R. (2019). Grasslands more important for ecosystem services than you might think. *Ecosphere, 10*(2), e02582. https://doi.org/10.1002/ecs2.2582

Bhargawa, A., & Singh, A. K. (2019). Solar irradiance, climatic indicators and climate change—An empirical analysis. *Advance in Space Research, 64*(1), 271–277. https://doi.org/10.1016/j.asr.2019.03.018

Bhatt, R. P. (2024). Restoration of global forests and biodiversity through green energy. *International Journal of Biological Innovations (IJBI), 6*(1), 25–38. https://doi.org/10.46505/IJBI.2024.6104

Brack, D. (2019). Forests and climate change; background study prepared for the fourteenth session of the United Nations Forum on Forests. *Global Forest Goals, 2*, 5–12.

Bratman, G. N., Hamilton, J. P., & Daily, G. C. (2012). The impacts of nature experience on human cognitive function and mental health. *Annals of the New York Academy of Sciences, 1249*(1), 118–36. https://doi.org/10.1111/j.1749-6632.2011.06400.x

Briant, C. J. (2017). Unheard voices: A critical discourse analysis of the millennium development goals' evolution into the sustainable development goals. *Third World Quarterly, 38*(1), 16–41. https://doi.org/10.1080/01436597.2016.1166944

Brundtland, G. H. (1987). *Report of the World Commission on Environment and Development: Note by the secretary-general, United Nations* (Vol. 634, pp. 1385–1397). https://doi.org/10.1016/j.scitotenv.2018.04.084

Calder, I., Hofer, T., Vermont, S., & Warren, P. (2007). Towards a new understanding of forests and water. *Unasylva, 58*(229), 3–10.

Canadell, J. G., & Michael, R. (2008). Managing forests for climate change mitigation. *Science, 320*(5882), 1456–1457. https://doi.org/10.1126/science.1155458

Chahal, R., Verma, S. K., & Singh, A. K. (2022a). Effect of biofertilizers nutrient uptake by green gram [*Vigna radiata* (L.) Wilczek] under agri-horti system in Vindhyan region of Eastern Uttar Pradesh. *International Journal of Plant & Soil Science, 34*(23), 686–691. https://doi.org/10.9734/IJPSS/2022/v34i232476

Chahal, R., Verma, S. K., & Singh, A. K. (2022b). Effect of fertilizers on yield attributes of green gram [*Vigna radiata* (L.) Wilczek] under Guava (*Psidium guajava* L.) based Agri-horti system in Vindhyan region of Eastern Uttar Pradesh. *Ecology, Environment and Conservation, 28*(4), 1773–1777.

Chao, S. (2012). *Forest peoples: Numbers across the world*. Forest Peoples Programme (Vol. 10, pp. 2017).

Delmas, R., Serça, D., & Jambert, C. (1997). Global inventory of NOx sources. *Nutrient Cycling in Agroecosystems, 48*, 51–60. https://doi.org/10.1023/A:1009793806086

Dubois, O. (2003). Forest-based poverty reduction: A brief review of facts, figures, challenges and possible ways forward. In T. Oksanen, B. Pajari, & T. Tuomasjukka (Eds.), *Forests in poverty reduction strategies: Capturing the potential* (pp. 65–81).

Dwyer, J. F., McPherson, E. G., Schroeder, H. W., & Rowntree, R. A. (1992). Assessing the benefits and costs of the urban forest. *Arboriculture & Urban Forestry, 18*(5), 227–234. https://doi.org/10.48044/jauf.1992.045

FAO. (1997). *Agriculture, food and nutrition for Africa: A resource book for teachers of agriculture.* Agriculture Organization of the United Nations. Food and Nutrition Division.

FAO. (2020). *Forests for human health and well-being—Strengthening the forest–health–nutrition nexus.* https://openknowledge.fao.org/handle/20.500.14283/cb1491en

FAO. (2022). *The state of the world's forests.* Food and Agriculture Organization of the United Nations. http://www.fao.org/documents/card/en/c/cb6533en

François, M., de Aguiar Jr, T. R., Mielke, M. S., Rousseau, A. N., Faria, D., & Mariano-Neto, E. (2024). Interactions between forest cover and watershed hydrology: A conceptual meta-analysis. *Water, 16*(23), 3350. https://doi.org/10.3390/w16233350

Gaggiotti, O. E., Chao, A., Peres-Neto, P., Chiu, C. H., Edwards, C., Fortin, M. J., & Selkoe, K. A. (2018). Diversity from genes to ecosystems: A unifying framework to study variation across biological metrics and scales. *Evolutionary Applications, 11*(7), 1176–1193. https://doi.org/10.1111/eva.12593

Galos, B., Haensler, A., Rechid, D., Sieck, K., & Jacob, D. (2012). The role of forests in mitigating climate change—A case study for Europe. *Acta Silvatica et Lignaria Hungarica, 8*(1), 87–102. https://doi.org/10.2478/v10303-012-0007-2

Garg, A., Pal, D., Singh, H., & Pandey, D. C. (2016). A comparative study of NDBI, NDISI and NDII for extraction of urban impervious surface of Dehradun [Uttarakhand, India] using Landsat 8 imagery. In *2016 International conference on emerging trends in communication technologies (ETCT)* (pp. 1–5).

Garibaldi, L. A., Steffan-Dewenter, I., Kremen, C., Morales, J. M., Bommarco, R., Cunningham, S. A., Carvalheiro, L. G., Chacoff, N. P., Dudenhöffer, J. H., Greenleaf, S. S., & Holzschuh, A. (2011). Stability of pollination services decreases with isolation from natural areas despite honey bee visits. *Ecology Letters, 14*(10), 1062–1072. https://doi.org/10.1111/j.1461-0248.2011.01669.x

Gitz, V., Pingault, N., Meybeck, A., Ickowitz, A., McMullin, S., Sunderland, T. C., Vinceti, B., Powell, B., Termote, C., Jamnadass, R. H., & Dawson, I. K. (2021). *Contribution of forests and trees to food security and nutrition* (FTA brief no. 5). The CGIAR Research Program on Forests, Trees and Agroforestry (FTA). https://doi.org/10.17528/cifor/008006

Habibullah, M. S., Din, B. H., Tan, S. H., & Zahid, H. (2022). The impact of economic growth on biodiversity: An empirical assessment. *Ecological Indicators, 121*, 107060. https://doi.org/10.1016/j.ecolind.2020.107060

Hordijk, W., Kauffman, S., & Steel, M. (2023). Community evenness mediates the relationship between biodiversity and productivity. *Nature Communications, 14*, 1234. https://doi.org/10.1038/s41467-023-26745-6

IPCC. (2022). *Climate change: A threat to human wellbeing and health of the planet. Taking action now can secure our future.* The International Panel on Climate Change. https://www.ipcc.ch/2022/02/28/pr-wgii-ar6/

ISFR. (2023). *The India state of forest report, Forest Survey of India (FSI).* Ministry of Environment Forest and Climate Change.

Ivanov, V. (2023). Arctic Sea ice loss enhances the oceanic contribution to climate change. *Atmosphere, 14*(2), 409. https://doi.org/10.3390/atmos14020409

Jangra, A., Kumar, K., Maikhuri, S., Bhandari, M. S., Pandey, S., Singh, H., & Barthwal, S. (2024). Unveiling stress-adapted endophytic bacteria: Characterizing plant growth-promoting traits and assessing cross-inoculation effects on *Populus deltoides* under abiotic stress. *Plant Physiology and Biochemistry, 210*, 108610.

Joshi, R., Sharma, B., Singh, H., Dhakal, N., Ayer, S., & Maraseni, T. (2024). Poplar plantation as an agroforestry approach: Economic benefits and its role in carbon sequestration in North India. *Journal of Resources and Ecology, 15*(4), 880–888.

Joshi, R., Singh, H., Chhetri, R., Poudel, S. R., & Rijal, S. (2021). Carbon sequestration potential of community forests: A comparative analysis of soil organic carbon stock in community managed forests of Far-Western Nepal. *Eurasian Journal of Soil Science, 10*(2), 96–104.

Joshi, R., Singh, H., Chhetri, R., & Yadav, K. (2020). Assessment of carbon sequestration potential in degraded and non-degraded community forests in Terai region of Nepal. *Journal of Forest and Environmental Science, 36*(2), 113–121.

Kale, I., Kumar, N. M., Singh, H., Ali, S., Singh, A. K., & Hareesh, D. (2024). Urban ecosystem services. In *Futuristic trends in agriculture engineering & food sciences* (Vol. 3, Book 2, IIP Series, pp. 117–128). e-ISBN: 978-93-5747-340-8. https://doi.org/10.58532/V3BCAG2P1CH13

Kaltenborn, B, P., Linnell, J. D., & Gómez-Baggethun, E. (2020). Can cultural ecosystem services contribute to satisfying basic human needs. A case study from the Lofoten archipelago, northern Norway. *Applied Geography, 120*, 102229. https://doi.org/10.1016/j.apgeog.2020.102229

Karjalainen, E., Sarjala, T., & Raitio, H. (2010). Promoting human health through forests: Overview and major challenges. *Environmental Health and Preventive Medicine, 15*(1), 1–8. https://doi.org/10.1007/s12199-008-0069-2

Karunamoorthi, K., Jegajeevanram, K., Vijayalakshmi, J., & Mengistie, E. (2013). Traditional medicinal plants: A source of phytotherapeutic modality in resource-constrained health care settings. *Journal of Evidence-Based Complementary and Alternative Medicine, 18*(1), 67–74. https://doi.org/10.1177/2156587212460241

Kasim, S., & Ghassan, M. (2023). Carbon sequestration through forestry: A differential equation approach. *Medical Journal of Babylon, 15*, 88–92. https://doi.org/10.1111/brv.12990

Konijnendijk, C. C., Sadio, S., Randrup, T. B., & Schipperijn, J. (2024). Urban and peri-urban forestry in a development context-strategy and implementation. *Journal of Arboriculture, 30*(5), 269–276.

Kumar, A., Dwivedi, G. K., Tewari, S., Sah, V. K., Singh, H., Kumar, P., Kumar, N., & Kaushal, R. (2020). Soil organic carbon pools under *Terminalia chebula* Retz. based agroforestry system in Himalayan foothills, India. *Current Science, 118*(7), 1098–1103.

Kumar, M., & Singh, H. (2020). Agroforestry as a nature-based solution for reducing community dependence on forests to safeguard forests in rainfed areas of India. In *Nature-based solutions for resilient ecosystems and societies* (pp. 289–306). Springer.

Kumar, N., Singh, H., & Kumar, A. (Eds.). (2021). *Renewable energy and green technology: Principles and practices.* CRC Press.

Kumar, R., Kumar, R., Karmakar, S., Kumar, A., Singh, A. K., Kumar, A., & Singh, J. (2023). Impact of amide fertilizer on carbon sequestration under the agroforestry system in the eastern plateau region of India. *Sustainability, 15*(12), 9775. https://doi.org/10.3390/su1512977510.3390/su15129775

Lippe, R. S., Schweinle, J., Cui, S., Gurbuzer, Y., Katajamäki, W., Villarreal-Fuentes, M., & Walter, S. (2022). *Contribution of the forest sector to total employment in national economies: Estimating the number of people employed in the forest sector*. Food and Agricultural Organization of the United Nations.

Mandal, S., & Sharma, R. K. (2023). Application of gasification technology in agriculture for power generation. In *Handbook of energy management in agriculture* (pp. 261–282). Springer Nature Singapore.

Martínez Pastur, G., Perera, A. H., Peterson, U., & Iverson, L. R. (2018). Ecosystem services from forest landscapes: An overview. In *Ecosystem services from forest landscapes: Broadscale considerations* (Vol. 2018, p. 1).

Martínez, G. E., Nilsson, M. B., & Laudon. (2024). Drought response of the boreal forest carbon sink is driven by understorey-tree composition. *Nature Geoscience, 17*, 197–204. https://doi.org/10.1038/s41561-024-01374-9

Mendiburu, A. Z., Lauermann, C. H., Hayashi, T. C., Mariños, D. J., da Costa, R. B., Coronado, C. J., Roberts, J. J., & de Carvalho Jr, J. A. (2022). Ethanol as a renewable biofuel: Combustion characteristics and application in engines. *Energy, 257*, 124688. https://doi.org/10.1016/j.energy.2022.124688

Minz, S. D., Singh, A. K., Kumar, N. M., & Singh, B. K. (2021). Effect of crop geometry and nitrogen management on growth attributes of pearl millet (*Pennisetum glaucum* L.) under guava based Agri-Horti system. *Journal of Pharmaceutical Innovation, 10*(9), 2191–2195.
Morris, R. J. (2010). Anthropogenic impacts on tropical forest biodiversity: A network structure and ecosystem functioning perspective. *Philosophical Transactions of the Royal Society B: Biological Sciences, 365*(1558), 3709–3718. https://doi.org/10.1098/rstb.2010.0273
Morton, S., Pencheon, D., & Squires, N. (2017). Sustainable development goals (SDGs), and their implementation: A national global framework for health, development and equity needs a systems approach at every level. *British Medical Bulletin, 24*(1), 81–90.
Nair, P. K. R. (1993). Agroforestr. In L. Panser (Ed.), *Tropical forestry handbook* (Ch. 16, pp. 983–1053). Springer Verlag.
Nobre, C. A., Sampaio, G., Borma, L. S., Castilla-Rubio, J. C., Silva, J. S., & Cardoso, M. (2016). Land-use and climate change risks in the Amazon and the need of a novel sustainable development paradigm. *Proceedings of the National Academy of Sciences, 113*(39), 10759–10768. https://doi.org/10.1073/pnas.1605516113
Nowak, D. J., Hirabayashi, S., Bodine, A., & Greenfield, E. (2014). Tree and forest effects on air quality and human health in the United States. *Environmental Pollution, 193*, 119–129. https://doi.org/10.1016/j.envpol.2014.05.028
Ornelas, A., Cordeiro, A., & Lameiras, J. M. (2023). Thermal comfort assessment in urban green spaces: Contribution of thermography to the study of thermal variation between tree canopies and air temperature. *Land, 12*(8), 1568. https://doi.org/10.3390/land12081568
Pan, Y., Birdsey, R. A., & Phillips, O. L. (2024). The enduring world forest carbon sink. *Nature, 631*, 563–569. https://doi.org/10.1038/s41586-024-07602-x
Parmesan, C., & Yohe, G. (2013). A globally coherent fingerprint of climate change impacts across natural systems. *Nature, 421*(6918), 37–42. https://doi.org/10.1038/nature01286
Perzl, F., Bono, A., Garbarino, M., & Motta, R. (2022). Protective effects of forests against gravitational natural hazards. In *Protective forests as ecosystem-based solution for disaster risk reduction (ECO-DRR)* (pp. 1–23). https://doi.org/10.5772/intechopen.99506
Patel, A. K., Bohra, J. S., Dewangan, S., & Singh, A. K. (2022). Effect of fertility levels on sesame (Sesamum indicum) varieties under agri-horti system. *The Indian Journal of Agricultural Sciences, 92*(8), 1015–1017. https://doi.org/10.56093/ijas.v92i8.98465
Raihan, F., Li, G., & Liu, S. (2022). The role of forests in regulating local and global climates. *Journal of Climate, 35*(6), 1825–1842. https://doi.org/10.1175/JCLI-D-21-0289.1
Rajoo, K. S., Karam, D. S., & Abdullah, M. Z. (2020). The physiological and psychosocial effects of forest therapy: A systematic review. *Urban Forestry & Urban Greening, 54*, 126744. https://doi.org/10.1016/j.ufug.2020.126744
Rametsteiner, E. & Whiteman, A. (2014). *State of the world's forests. Enhancing the socio-economic benefits from forests*. FAO.
Retamales, J. B. (2011). World temperate fruit production: Characteristics and challenges. *Revista Brasileira de Fruticultura, 33*, 121–130. https://doi.org/10.1590/S0100294520110005000015
Roberts, J. M. (2009). The role of forests in the hydrological cycle. *Forests & Forest Plants, 3*, 42–76.
Ruhil, T., Singh, H., & Barthwal, S. (2024). Long-term physiological monitoring for detecting climate change impacts on forest ecosystems. In *Forests and climate change: Biological perspectives on impact, adaptation, and mitigation strategies* (pp. 817–834). Springer.
Sachs, J. D., Lafortune, G., Fuller, G., Drumm, E. (2023). Sustainable development report. In *Implementing the SDG stimulus*.
Saxena, N. C. (2003). Livelihood diversification and non-timber forest products in Orissa: Wider lessons on the scope for policy change (pp. 48–59). Overseas Development Institute.
Schmitz, O. J., Raymond, P. A., Estes, J. A., Kurz, W. A., Holtgrieve, G. W., Ritchie, M. E., Schindler, D. E., Spivak, A. C., Wilson, R. W., Bradford, M. A., & Christensen, V. (2014) Animating the carbon cycle. *Ecosystem, 17*, 344–359. https://doi.org/10.1007/s10021-013-9715-7

SDG. (2024). *Progress towards the sustainable development goals report of the secretary general assembly seventy-ninth session A/79/79 E/2022/55.*

Sharma, N., Bohra, B., Pragya, N., Ciannella, R., Dobie, P., & Lehmann, S. (2016). Bioenergy from agroforestry can lead to improved food security, climate change, soil quality, and rural development. *Food and Energy Security, 5*(3), 165–183. https://doi.org/10.1002/fes3.87

Sheldon, D. J., & Crimmin, M. R. (2022). Repurposing of F-gases: Challenges and opportunities in fluorine chemistry. *Chemical Society Reviews, 51*(12), 4977–4995. https://doi.org/10.1039/d1cs01072g

Singh, H., Verma, A., & Kumar, M. (2018). Can elevated CO_2 alter the growth dynamics, morphology, physiology and biochemistry of the forestry tree species? In *Climate change and agriculture* (pp. 288–305). Discovery Publishing House.

Singh, H., & Kumar, M. (2022). Climate change and its impact on Indian Himalayan forests: Current status and research needs. In *Climate change: Impacts, responses and sustainability in the Indian Himalaya* (pp. 223–242). Springer.

Singh, S., & Singh, S. (2023). Plant-based biofuels: An overview. In *Green approach to alternative fuel for a sustainable future* (Vol. 1, pp. 433–442). https://doi.org/10.1016/B978-0-12-824318-3.00024-2

Singh, H., Kumar, N., Singh, R., & Kumar, M. (2023). Assessing the climate change impact on the habitat suitability of the range-restricted bird species (*Catreus wallichii*) in the Indian Himalayan ecosystem. *Environmental Science and Pollution Research, 30*(57), 121224–121235.

Singh, H. (2024a). Exploring adaptive modulation in plant functional traits and its impact on the productivity of *Acacia auriculiformis* under CO_2-enriched environment. *Industrial Crops and Products, 210*, 118186.

Singh, H. (2024b). *Forests and climate change: Biological perspectives on impact, adaptation, and mitigation strategies*. Springer Nature.

Singh, H. (2024c). *Urban forests, climate change and environmental pollution: Physio-biochemical and molecular perspectives to enhance urban resilience*. Springer Nature.

Singh, A. K., Shukla, S., Singh, M. K., Singh, A., Singh, R., & Singh, B. K. (2024a). Urban forests: Importance, challenges and opportunities. In *Urban forests, climate change and environmental pollution* (pp. 3–45). https://doi.org/10.1007/978-3-031-67837-0_2

Singh, H., Kumar, N., & Kumar, A. (2024b). Enhancing resource use efficiency in crops through plant functional traits. In *Plant functional traits for improving productivity* (pp. 97–117). Springer.

Singh, R., Chavan, S. B., Kakade, V. D., Morade, A. S., Singh, A. K., Rawale, G. B., Uthappa, A. R., Keerthika, A., Chichaghare, A. R., Gurav, S., & Reddy, K. S. (2024c). Ecosystem services provided by urban and peri-urban forests. In *Urban forests, climate change and environmental pollution* (pp. 417–445). Springer. https://doi.org/10.1007/978-3-031-67837-0_20

Singh, H. (2025). The role of plant functional traits in mitigating urban climate and environmental pollution. In *Plant functional traits* (pp. 297–308). Elsevier.

Singh, A. K., Shukla, S., Dey, A., Singh, A., & Tantuway, G. (2025a). Agroforestry: climate-smart agriculture technique for semi-arid climate. In *Sustainable agriculture management in semi-arid climates: volume 1* (pp. 51–68). Cham: Springer Nature Switzerland. https://doi.org/10.1007/978-3-031-94062-0_4

Singh, R., Prakash, P., Singh, A. K., Shukla, S., Sharma, P., Chauhan, V., & Singh, B. K. (2025b). Altitudinal variation in carbon stocks of *Quercus leucotrichophora* forests in the Western Himalayas: Implications for climate change mitigation. *Trees, Forests and People,* 101059. https://doi.org/10.1016/j.tfp.2025.101059

Stenchikov, G. (2021). The role of volcanic activity in climate and global changes. In *Climate change* (Ch. 29, pp. 607–643).

Sunderlin, W. D., Angelsen, A. U., & Wunder, S. (2003). Forests and poverty alleviation. In *State of the world's forests* (Vol. 2003, pp. 61–73).

Tsao, T. M., Tsai, M. J., Hwang, J. S., Cheng, W. F., Wu, C. F., Chou, C. C., & Su, T. C. (2018). Health effects of a forest environment on natural killer cells in humans: An observational pilot study. *Oncotarget, 9*(23), 16501. https://doi.org/10.18632/oncotarget.24741

UN. (2019). *The climate crisis—A race we can win.* The United Nations. https://www.un.org/en/un75/climate-crisis-race-we-can-win

UNEP. (2014). *How halting deforestation can help counter the climate crisis.* United Nation Environment Programme. https://www.unep.org/news-and-stories/story/how-halting-deforestation-can-help-counter-climate-crisis

UNEP. (2020). *UN report: As the world's forests continue to shrink, urgent action is needed to safeguard their biodiversity.* United Nation Environment Programme. https://www.unep.org/news-and-stories/press-release/un-report-worlds-forests-continue-shrink-urgent-action-needed

UNFCCC. (2016). *Booklet of the Paris agreement: 2016.* United Nations Framework Convention on Climate Change (UNFCCC).

Upgupta, S., Ravindranath, N. H., Sharma, J., & Chaturvedi, R. K. (2015). Climate change impact and vulnerability assessment of forests in the Indian Western Himalayan region: A case study of Himachal Pradesh, India. *Climate Risk Management, 10*, 63–76. https://doi.org/10.1016/j.crm.2015.08.001

Ventura, L., Strubbe, D., & Shwartz, A. (2024). Beyond the concrete jungle: The value of urban biodiversity for regional conservation efforts. *Scieence of the Total Environment, 955*, 177222. https://doi.org/10.1016/j.scitotenv.2024.177222

Verma, K., Sharma, P., Bhardwaj, D. R., Kumar, R., Kumar, N. M., & Singh, A. K. (2023). Land and environmental management through agriculture, forestry and other land use (AFOLU) system. In *Land and Environmental Management Through Forestry* (pp. 247–271). https://doi.org/10.56093/ijas.v92i8.98465

Vishwakarma, S. P., Singh, J. K., Sharma, P., Singh, A. K., & Kumar, N. M. (2024). Sulphur and boron application for enhanced yield of blackgram (*Vigna mungo* L.) under custard apple (Annona squamosa L.) based agroforestry system in Vindhyan region. *Ecology, Environment & Conservation* (0971765X) *2*, 7. https://doi.org/10.53550/EEC.2024.v30i04s.068

Wadanambi, R. T., Wandana, L. S., Chathumini, K. K., Dassanayake, N. P., Preethika, D. D., & Arachchige, U. S. (2020). The effects of industrialization on climate change. *Journal of Research Technology & Engineering, 1*(4), 86–94.

Wang, L., Wang, E., Mao, X., Benjamin, W., & Liu, Y. (2023). Sustainable poverty alleviation through forests: Pathways and strategies. *Science of the Total Environment, 904*, 167336. https://doi.org/10.1016/j.scitotenv.2023.167336

XiLiu, Y., & QingXian, G. (2018). Contributions of natural systems and human activity to greenhouse gas emissions. *Advances in Climate Change Research, 9*(4): 243–252. http://doi.org/10.1016/j.accre.2018.12.003

Yildizhan, H., Udriştioiu, M. T., Pekdogan, T., & Ameen, A. (2024). Observational study of ground-level ozone and climatic factors in Craiova, Romania, based on one-year high-resolution data. *Scientific Reports, 14*(1), 26733. https://doi.org/10.1038/s41598-024-77989-0

Chapter 3
Sustainable Forest Management: Enhancing Carbon Storage and Reducing Greenhouse Gas Emissions for a Carbon-Neutral Future

Chandramohan Singh, Hitesh Gupta, Aman Mahajan, Shubham Thapliyal, Gaurav Kumar, and Shishpal Choudhary

Abstract Forests play a vital role in maintaining ecological balance, providing essential services that support life on Earth. They provide habitat for a diverse range of species, strengthen global biodiversity, and contribute significantly to ecosystem functions such as air and water purification, soil erosion control, and climate regulation. Sustainable forest management (SFM) plays a crucial role in mitigating climate change by enhancing carbon storage and reducing greenhouse gas emissions (GHG). This chapter has explored various strategies and practices that contribute to achieving a carbon-neutral future through responsible forest management. Key findings emphasize the importance of maintaining and expanding forest cover, as forests play a crucial role as significant carbon sinks. Implementing sustainable harvesting techniques, such as selective logging and reduced-impact logging, can minimize disturbance to forest ecosystems while allowing for economic benefits. Additionally, reforestation and afforestation efforts have shown promising results in increasing the potential for carbon sequestration. The chapter has also emphasized the need for integrated approaches that consider multiple ecosystem services provided by forests, including biodiversity conservation and water regulation. The core principle of SFM is to safeguard the vital functions of forests by ensuring that resource use does not exceed the environment's capacity to regenerate and replenish itself. This approach helps to maintain forest health and productivity while mitigating the impacts of human activity. In this context, Initiatives such as REDD+ (Reducing Emissions from Deforestation and Forest Degradation) demonstrate the potential for collaborative efforts in addressing climate change through forest conservation. In this chapter, we explore distinct categories of promising forestry practices, all aimed

C. Singh (✉) · H. Gupta · S. Thapliyal · G. Kumar · S. Choudhary
College of Forestry, Veer Chandra Singh Garhwali Uttarakhand University of Horticulture and Forestry, Ranichauri, Uttarakhand, India
e-mail: chandramohans118@gmail.com

A. Mahajan
Genetics and Tree Improvement Division, ICFRE-Forest Research Institute, Dehradun, Uttarakhand, India

H. Singh (ed.), *Forestry for a Carbon-Neutral and Sustainable Future*, Sustainability Solutions, https://doi.org/10.1007/978-3-032-06427-1_3

at achieving the dual objectives of promoting sustainable forest management while also preserving and capturing carbon.

Keywords Sustainable forest management · Afforestation · Carbon storage · Forest ecosystems

3.1 Introduction

Forest ecosystems provide essential and diverse services to societies globally, serving as critical habitats for numerous species, offering wood resources, supporting biodiversity preservation, and playing a crucial role in the well-being of Indigenous communities and overall human health (Nelson et al., 2009). Second only to the oceans, global forests significantly influence the carbon cycle by facilitating a substantial exchange of carbon between the atmosphere and the terrestrial biosphere, contributing considerably to efforts to mitigate climate change (Boyd et al., 2019). These ecosystems store over 80% of terrestrial aboveground carbon and more than 70% of soil organic carbon (SOC), with carbon absorption playing a crucial role in mitigating ecosystem degradation caused by global warming. The decrease in carbon dioxide levels in the atmosphere relies on either minimizing emissions or enhancing absorption. As of 2011, the global forest carbon stock was estimated at 861 petagrams (Pg), with soil (44%), biomass (42%), deadwood (8%), and litter (5%) constituting the main components. However, between 1990 and 2015, this stock decreased by 13.6 Pg, primarily in living biomass and soils, especially in tropical forests (Kohl et al., 2015). According to recent studies, global forest ecosystems absorb over 30% of human-caused CO_2 emissions annually and remove approximately 2 Pg of carbon from the Earth's atmosphere through net growth (Bellassen & Loyssaert, 2014).

The World Commission on Environment and Development report defined "sustainability" as 'development that meets the needs of the present without compromising the ability of future generations to meet their own needs.' This transformative concept has altered our understanding of economic development, influenced consumer behavior, policy-making, and production practices while embraced the full spectrum of human values (Ascher, 2007). The concept of sustainability integrates three fundamental dimensions: economic prosperity, environmental conservation, and social equity (Kastenhofer & Rammel, 2005). This framework seeks to promote community well-being through sound economic practices while safeguarding natural resources for future generations (Koehler & Hecht, 2006). This concept was further developed into the "triple bottom line" where the sustainability of the 3Ps (profit, people, and planet) was used to measure the financial, social, and environmental performance of corporations over a period of time.

In the long run, employing a sustainable approach to forest management that maintains or increases carbon stocks while ensuring a consistent annual yield of timber, fiber, or energy can yield significant and lasting benefits for mitigation. Many mitigation efforts involve initial investments, with returns and additional benefits accumulating over several years to decades (Singh et al., 2024a, 2024b).

The combined impact of measures such as reduced deforestation, forest management, agro-forestry, afforestation, and bio-energy can grow from the present and extend beyond 2030. Beyond their services, forests play a crucial role in carbon sequestration, as trees absorb carbon dioxide during their growth (Dumitraşcu et al., 2020). Global forests sequester 296 gigatons (Gt) of carbon, encompassing both aboveground and belowground biomass (FAO, 2015). However, forests play a dual role in climate change; they sequester atmospheric carbon through photosynthesis while simultaneously releasing carbon back into the atmosphere when deforested (Matthew et al., 2018). Deforestation and forest degradation account for 13% of the total global CO_2 emissions (IPCC, 2019). According to the latest IPCC special report, a substantial expansion of forests by 950 million hectares is necessary to restrict global temperature rise to 1.5 °C by 2050. The imperative for sustainable development and the mitigation of climate change has prompted a global emphasis on reducing carbon emissions and enhancing environmental quality (Islam et al., 2022). However, it's vital to note that mature forests can become carbon sources due to the decomposition of harvest residues, root emissions, fires, infestations, droughts, and storms. Consequently, the role of forests as carbon sinks and the optimal methods for cultivating them to store carbon and address climate change remain subjects of ongoing debate (Singh & Kumar, 2022).

In a broad sense, carbon neutrality refers to achieving net-zero emissions of all greenhouse gases (GHG) (Chen et al., 2022). However, the IPCC (2019) characterizes carbon neutrality as the achievement of net-zero CO_2 emissions, which entails a balance between CO_2 emissions and their removal over time. It is essential to reduce carbon emissions from food production, industrial processes, and fossil fuel consumption while simultaneously promoting carbon sequestration in both aquatic and terrestrial ecosystems. This approach is vital for achieving carbon neutrality and ensuring the sustainable continuation of human activities (Hong et al., 2023). There are multiple approaches to attaining carbon neutrality across different countries; however, reaching net-zero carbon emissions presents a significant challenge due to the magnitude of the emissions fluxes involved (Hanssen et al., 2020). To achieve the goal of zero CO_2 emissions, it is essential to invest in the development and widespread adoption of carbon-free renewable energy sources, including sunlight, tides, wind, wave power, water power, rain, and geothermal power, as well as biomass and organic materials (Carlotta-Jones et al., 2020). The world will not achieve carbon neutrality and sustainable development by 2050 if the transition from conventional to renewable energy is not accelerated (Tiwari, 2021; Kumar et al., 2021). CO_2 removal from the environment can leverage natural resources and processes, given that fossil fuel-based CO_2 emissions are unlikely to be significantly reduced in the near future (Sun et al., 2022). To mitigate global warming, researchers are exploring methods to enhance industrial carbon capture and sequestration, as well as in land and water ecosystems (Li et al., 2021).

3.2 Concept and Principles of Sustainable Forest Management

3.2.1 Concept

The concept of sustainability gained prominence during the late 1980s and the early 1990s with the publication of the Brundtland Report (1987) and the Conference on Environment and Development held in Rio de Janeiro, Brazil, in 1992 (also known as the Earth Summit). Nevertheless, the importance of conserving natural resources for the benefit of future generations has long been recognized. The negative impact of historical forest resource exploitation, as well as the need for continued use of these resources for future generations, was already noted as early as the seventeenth century (Glacken, 1976, as cited in Wiersum, 1995). However, it was not until the eighteenth century that the concept of sustainability was explicitly referred to, as follows: "every wise forest manager must assess the forest stands promptly, maximizing their utilization while ensuring that future generations derive at least as much benefit as the current generation" (Schmutzenhofer, 1992, as cited in Wiersum, 1995). This first definition was based on the principle of sustainable forest yield, primarily aimed at ensuring sustainable timber production. It was posited that maintaining stands conducive to timber production would also sustain the sustainability of non-wood forest products. This assumption primarily emphasizes the sustainability of the productive functions of forest resources, often neglecting other important aspects such as ecological and socio-economic functions. There is no universally accepted definition of SFM remains elusive, but certain fundamental concepts are broadly accepted: "*the process of managing permanent forest land to achieve one or more clearly specified objectives of management with regard to the production of a continuous flow of desired forest products and services without undue reduction of its inherent values and future productivity and without undue undesirable effects on the physical and social environment*" (proposed by International Tropical Timber Organization: ITTO, 1992), and "*the stewardship and use of forests and forest lands in a way, and at a rate, that maintains their biodiversity, productivity, regeneration capacity, vitality and their potential to fulfil, now and in the future, relevant ecological, economic and social functions, at local, national, and global levels, and that does not cause damage to other ecosystems*" (proposed by the second ministerial conference for the protection of the forest: MCPFE, 1993). This concept integrates ecological and socioeconomic concerns across different scales of management and for different periods. Nevertheless, both concepts refine the definition of sustainable development, as given by the Brundtland Commission (1987), "development that meets the needs of the present without compromising the ability of future generations to meet their own needs," to apply it to forests.

3.2.2 Principles of Sustainable Forest Management

SFM integrates economic, social, and ecological considerations to achieve comprehensive forestry goals (Fig. 3.1). Each aspect must be balanced; neglecting any one of them can compromise sustainability.

3.2.2.1 Nature Protection Principle

Ecologically sustainable forestry tackles the environmental impacts of deforestation by enhancing air quality through oxygen production and pollutant absorption, preserving biodiversity by supporting a wide range of flora and fauna, and mitigating climate change by capturing carbon in soil and trees (which are 50% carbon by dry mass). It also plays a vital role in soil protection by preventing erosion through stable forest floors and robust root systems, and it helps in flood prevention by slowing water flow with tree canopies and roots. A successful forest management plan must consider protection aspects, such as improving air quality and reducing pollution, as well as protecting against soil erosion, increasing soil fertility, safeguarding wildfire and their habitats, preserving biodiversity, and mitigating climate change by reducing CO_2 emissions (Singh et al., 2018).

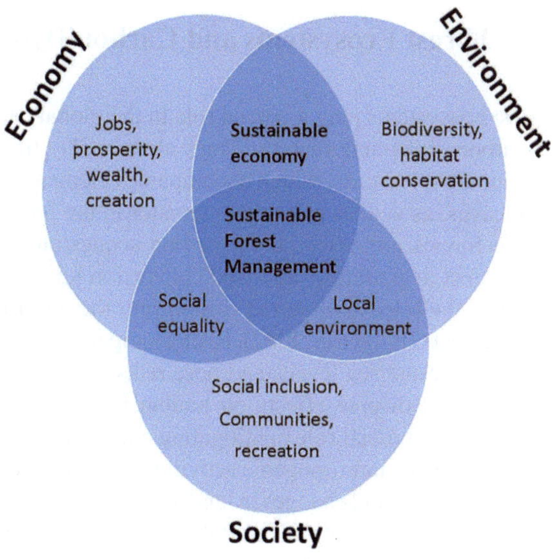

Fig. 3.1 Three main pillars of the sustainable forest management (SFM)

3.2.2.2 Economic Development Principle

Economically sustainable forestry harmonizes tree harvesting with conservation efforts and commercial interests, thereby fostering job creation, boosting local and regional incomes, enhancing international trade relationships, and attracting investment to support overall economic growth. This approach ensures that forest management practices contribute to both environmental preservation and economic development (Singh, 2024c). A successful forest management plan must consider protection aspects such as reducing timber harvesting, revenues from non-timber forest products, investment costs, opportunity costs, and long-term economic benefits.

3.2.2.3 Social Development Principle

Socially sustainable forestry focuses on enhancing the quality of life for communities that rely on forests, reducing unemployment by creating forestry-related jobs, ensuring work safety by adhering to safety standards in forest operations, and promoting equity by upholding gender and racial fairness, labour rights, and social security. This approach aims to enhance the well-being of individuals and communities while supporting fair and safe working conditions.

3.3 Forest Ecosystems and Carbon Dynamics

Forest ecosystems play a crucial role in the global carbon cycle by functioning both as carbon sinks and sources. Trees absorb CO_2 from the atmosphere and store it within forest biomass, which encompasses elements such as trunks, branches, roots, and leaves, as well as soil organic matter in the form of dead wood (Singh, 2024c). While forests are essential for carbon sequestration, practices such as deforestation, forest degradation, and land conversion hinder this function. Such disruptions result in the release of stored carbon into the atmosphere, exacerbating the greenhouse effect (Houghton, 2012). Rising temperatures and climate change alter weather patterns, elevate sea levels, and pose risks to ecosystems. These changes also exert intricate and diverse effects on human health and overall well-being. Moreover, forests play a crucial role in mitigating climate change; however, their emissions can fluctuate due to anthropogenic activities, including deforestation, wildfires, management strategies, and changes in land use. Therefore, accurately evaluating the sources and global warming potential (GWP) of greenhouse gases (GHGs) can effectively prioritize mitigation strategies and reduce the overall impact of these emissions.

Additionally, the global warming potential (GWP) of greenhouse gas (GHG) emissions varies based on several factors, including the source of the emissions, the type of GHG, the timescale, and spatial extent. Therefore, it is essential to pay special attention to the dynamics of CO_2 emissions from organic soils and other types of land. Organic soil CO_2 or carbon dioxide emissions associated with soils rich in organic matter, typically found in peatlands and wetlands, are enormous carbon reservoirs when untouched. However, activities such as peat extraction or drainage can alter these into significant CO_2 sources (Chen et al., 2024). In contrast, other land CO_2 emissions refer to carbon emissions from various land types, such as emissions or removals from grasslands, shrublands, non-forested wetlands, and areas experiencing urbanization or other land-use changes (Souza et al., 2020). The strong positive relationship between organic soil CO_2 and forestland CO_2, along with its negative correlation to fire and deforestation indicators, underscores its crucial role in carbon sequestration. The sensitivity of organic soils to temperature variations makes them particularly relevant to climate change discussions, as warming can accelerate the release of CO_2 (Singh, 2025).

3.4 Strategies for Enhancing Carbon Storage

Terrestrial ecosystems act as a sponge for atmospheric CO_2, storing it in both living tree biomass and decomposed organic matter, which significantly impacts the global carbon cycle (Gupta et al., 2024). The Intergovernmental Panel on Climate Change (IPCC) categorizes carbon storage methods into three distinct types: geological storage, oceanic storage, and mineralization. CO_2 sequestration is a process aimed at reducing atmospheric CO_2 levels by capturing and storing it in various types of carbon sinks (Singh, 2024b). CO_2 is a primary greenhouse gas (GHG) that plays a significant role in the phenomenon of global warming. According to the International Energy Agency (IEA, 2021), carbon dioxide emissions associated with fossil fuels reached an unprecedented level of 36.3 gigatons in 2021, accounting for approximately 79% of total anthropogenic CO_2 emissions. Approximately 50% of total anthropogenic CO_2 emissions recorded from 1750 to 2021 occurred within the last three decades, primarily due to energy and industrial activities such as power generation, transportation, and cement manufacturing (Dods et al., 2021). CO_2 can be captured from significant sources, such as power plants, natural gas processing facilities, and certain industrial processes, using various technologies that effectively separate CO_2 from flue gas or process streams (Bains et al., 2017). It is also possible to capture CO_2 from the atmosphere using techniques that extract it from the ambient air. The captured CO_2 is subsequently compressed and transported to an appropriate deep subterranean site, where it can be securely stored or buried within geological formations, such as depleted oil and gas reservoirs, deep saline aquifers, or coal seams (Joshi et al., 2021).

The storage of CO_2 within these geological formations relies on several trapping mechanisms that prevent or delay the migration of CO_2 to the surface, including residual, solubility, structural, and mineral trapping (Mahajan et al., 2021).

3.4.1 Conservation Agriculture

The Food and Agriculture Organization (FAO) characterizes conservation agriculture (CA) as an agricultural system that emphasizes minimal soil disturbance (no-tillage or minimum tillage), encourages crop diversification, and ensures the maintenance of a continuous soil cover (Pittelkow et al., 2015). The three principles of CA contribute to the enhancement of biodiversity and soil quality, the improvement of crop yields, and the increase of SOC (Kumar et al., 2020). CA is frequently equated with reduced-tillage or no-tillage practices, leading to questions about its effectiveness in improving soil carbon and crop yields due to data variability. A meta-analysis comparing no-tillage to conventional tillage found that no-tillage resulted in an average yield reduction of 5.7%. However, the effects of no-tillage were variable, with some conditions showing equivalent or even greater yields compared to conventional tillage (Pittelkow et al., 2015). The individual and combined effects of CA components in tropical agroecosystems recorded the rate of SOC increase, ranging from 0.16 to 1.01 tonnes (t) hectare $(ha)^{-1}$ $year^{-1}$, contributing to carbon mitigation only by a small degree (Powlson et al., 2016). Since the three components of CA are often studied or applied separately in research or practice, their actual effect on SOC remains unclear. Additionally, the amount of carbon input significantly influences the increase of SOC through CA. The type of crop, along with the intensity and duration of cultivation, plays a critical role in determining the carbon input (Joshi et al., 2020). The incorporation of deep-rooted plant species characterized by substantial aboveground and belowground biomass, along with an increase in the number of crops per year, may enhance the ability of CA systems to store carbon (Table 3.1).

3.4.2 Agroforestry

Agroforestry serves as a strategy for both mitigation and adaptation by incorporating trees into agricultural systems (Singh et al., 2024a, 2024b). The conversion of land use from pastures to plantations, from native forests to crops, and from pastures to crops results in a substantial reduction of soil organic carbon (SOC) stocks (Jangra et al., 2024). Agroforestry has the potential to recover up to 35% of the carbon stock lost from forests. The incorporation of woody biomass in agroforestry practices plays a substantial role in enhancing the agricultural biomass carbon reservoir. The aboveground biomass of row crops is typically less than 25 tons of carbon per hectare ($t\ C\ ha^{-1}$) (Singh et al., 2023).

In contrast, agroforestry systems increase this biomass to a range of 50–75 $t\ C\ ha^{-1}$ (Zomer et al., 2016). He revised the global estimate of agricultural biomass carbon from 11.1 gigatons of carbon (Gt C) to 45.3 Gt C, with trees contributing more than 75%. Trees, specifically, have a greater input in SOC by promoting deeper deposition and decreasing the rate of decomposition. The alteration of soil carbon levels in

Table 3.1 Soil-carbon boosting approaches. *Source* Nayak et al. (2022)

Sequestration strategy	Potential/estimated SOC increase	Advantages	Disadvantages	References
Conservation agriculture	1.2% to 1.01 t C ha^{-1} $year^{-1}$	• Promote biodiversity • Improve soil nutrient • Increase water retention	• Varying data concerning effects on SOC • Varying data concerning effects on crop yield	Pittelkow et al. (2015) Powlson et al. (2016)
Agroforestry	5.3 Gt C $year^{-1}$ globally	• Increased above-ground biomass • Enhanced soil carbon inputs	• SOC sequestration rates vary based on climate, soil type, management, tree age, and organic inputs	Zomer et al. (2016)
Biochar	8–18% at 60–90 t ha^{-1} application rates, respectively 6.6 Mt C $year^{-1}$ at \$50 t^{-1} CO_2	• Retains 50% of initial carbon • Serves as both a carbon source and sink	• Variable impacts on soil quality, priming effects, and SOC levels depending on biochar type and age	Kammann et al. (2015) El-Naggar et al. (2018)
Afforestation	0.75 Gt C $year^{-1}$ to 5.8 Gt C $year^{-1}$, depending on land price, region and time	• Increase in above-ground biomass • Increased soil carbon input	• Variable effects on SOC stock depending on soil type, climate, tree species	• Sathaye et al. (2006) • Humpenöder et al. (2014) • Doelman et al. (2020) • Hong et al. (2023)
Restoration of wetlands	0.35–1.10 t C ha^{-1} $year^{-1}$ depending on landscape position and depth	• Higher SOC content compared to cultivated wetlands	• Natural and anthropogenic factors influence mitigation potential	• Xu et al. (2019) • Taillardat et al. (2020)

agroforestry is influenced by factors such as the type of tree species, their density, and the climate; however, it is generally found to be between that of agriculture and forest.

3.4.3 *Biochar*

Plant residues, a crucial source of soil carbon, undergo rapid decomposition, which quickly releases the added carbon back into the atmosphere. The thorough burning of organic matter creates biochar via pyrolysis (or other methods), which converts the carbon into a stable and resistant substance that can last from several hundred to several thousand years (Lehmann, 2007). Biochar is capable of retaining around 50% of initial carbon, in contrast to combustion and decomposition processes, which retain merely about 3% and 10–20% of carbon, respectively (El-Naggar et al., 2018). Thus, biochar functions as both a carbon source and a carbon sink within the soil ecosystem. It is estimated that the application of biochar can offset approximately 12% of the annual net anthropogenic CO_2 emissions (Woolf et al., 2010). There is an increasing focus on enhancing the efficacy of biochar through modifications to the production process. These modifications encompass physicochemical alterations that control the pyrolytic process, as well as composting biochar with organic or composted materials. Some research indicates that co-composted biochar enhances soil fertility and promotes the sequestration of soil carbon, resulting in up to a 305% increase in crop yield (Kammann et al., 2015). However, the assertion that biochar's ability to increase SOC is often exaggerated in small-scale greenhouse and laboratory conditions. The potential for global mitigation through biochar production utilizing current pyrolysis facilities is projected to reach 6.6 Mt C year^{-1}, with a net revenue of \$177 million at a carbon price of \$50 t^{-1} CO_2 (Han et al., 2021). They reported an average increase in SOC content following biochar application up to 45.8% but with large regional variations (Joshi et al., 2024).

3.4.4 *Afforestation or Reforestation*

The concept of expanding forested areas has been a pivotal focus within the discourse on climate mitigation, as it presents a cost-effective alternative to other, more expensive policy measures (electrification of industry, large-scale deployment of renewable resources). Since the 1990s, afforestation efforts have been extensively undertaken worldwide, resulting in an increase of approximately 1.05 × 108 ha in the area of planted forests (FAO, 2016). However, achieving mitigation objectives via afforestation depends on the area and the carbon sequestration potential of each afforestation (IPCC, 2014). Differing model characteristics and scenario assumptions range the estimated sequestration potential of afforestation/reforestation from 176 Gt CO_2 using 231 Mha of land at \$220 t^{-1} CO_2 (Sathaye et al., 2006) to 700 Gt CO_2 using

2,800 Mha of land at \$1,165 t^{-1} CO_2 (Humpenöder et al., 2014). This broad range leads to uncertainty over the feasibility of afforestation. The mitigation potential of afforestation is estimated to be moderate (0.75 Gt CO_2 $yearr^{-1}$ in 2030, 1.5 Gt $year^{-1}$ in 2050, 1.1 Gt $year^{-1}$ in 2100) at low CO_2 prices (\$50 t^{-1} CO_2), which limits mitigation to tropical regions with low land prices and high forest growth rates. At higher CO_2 prices (\$200 t^{-1} CO_2), afforestation mitigation potential increases significantly (2.7 Gt CO_2 $year^{-1}$ in 2030, 4.9 Gt CO_2 $year^{-1}$ in 2050, 5.8 Gt CO_2 $year^{-1}$ in 2100) with the inclusion of temperate regions having high land prices and low forest growth rates (Doelman et al., 2020).

Afforestation of barren land with low initial soil organic carbon stock has a greater potential to serve as a carbon sink compared to cropland or grassland. Afforestation on land with a higher initial soil organic carbon (SOC) stock can lead to an initial loss of SOC. This occurs because the process disrupts the soil's equilibrium, making the organic carbon more vulnerable to decomposition and erosion. However, the addition of fresh SOC from tree litter can recharge the SOC pool, leading to priming effects. The fresh SOC input (tree litter) can replenish the soil's organic carbon reservoir and trigger priming effects. The excessive availability of carbon in carbon-rich soils amplifies priming effects, resulting in a greater loss of soil organic carbon (Guo et al., 2021). Carbon sequestration is higher in moist, cold climates than in moist, warm climates, and it occurs more in broad-leaved deciduous forests compared to coniferous forests. Furthermore, forests that are 20 years old exhibit an increased SOC content down to a depth of 100 cm (Guo et al., 2021). Afforestation increases biomass carbon, with a recent study suggesting that it could potentially sequester 205 gigatons of carbon in ecosystems, offsetting 68% of global carbon emissions (Bastin et al., 2019).

3.4.5 Restoration of Wetlands

For millennia, wetlands have been drained and utilized for agriculture, with notable examples including crop fields on river floodplain soils and rice fields. Approximately half of the world's wetlands have been lost to agriculture, industrial activities, or urbanization (Verhoeven & Setter, 2010). The loss of a diverse, species-rich habitat, such as a wetland, not only significantly impacts biodiversity but also changes it from a CO_2 reservoir to a source of it. Wetland restoration, achieved through the cessation of land-use activities and the subsequent recovery of soil organic carbon (SOC) stocks, is a potential means of mitigating climate change (Ruhil et al., 2024). Natural environmental factors, including climate, soil properties, hydrological regimes, and plant and microbial communities, regulate the dynamics of carbon flux within wetland ecosystems. These factors are influenced by anthropogenic factors (e.g., nutrient input, leveling, drainage, altered hydrologic conditions). A meta-analysis of 41 studies examined the soil carbon dynamics associated with wetland restoration, revealing that restored wetlands have approximately 14% greater soil carbon content than cultivated wetlands (Xu et al., 2019). The sequestration of soil organic carbon (SOC) in restored wetlands is influenced by various factors, including water supplementation, seasonal hydrology before cultivation, climate, and the duration of restoration and cultivation (Xu et al., 2019). An analysis of the costs associated with wetland restoration and the carbon sequestered found that restoring coastal wetlands, such as mangroves, is more cost-effective than restoring inland wetlands. For the latter scenario, conservation is a more suitable approach than restoration (Taillardat et al., 2020).

3.4.6 Increasing Off-Site Carbon Stocks in Wood Products and Enhancing Product and Fuel Substitution

Wood products sourced from forests managed sustainably contribute to mitigating the saturation of carbon stocks in forests. By aligning the annual harvest with or keeping it below the forest's yearly increment, these practices enable the maintenance or growth of carbon stocks in the forest while concurrently supplying a continuous carbon stream to fulfill societal demands for fiber, timber, and energy (Fig. 3.2). The timeframe of carbon retention in wood products spans from short durations, such as days for biofuels, too much longer periods, reaching centuries in the case of structures like houses and furniture (Singh, 2024a).

Utilizing wood-based materials can effectively displace more carbon-intensive construction alternatives, such as concrete, steel, aluminum, and plastics, potentially yielding significant reductions in greenhouse gas (GHG) emissions (Barwa & Singh,

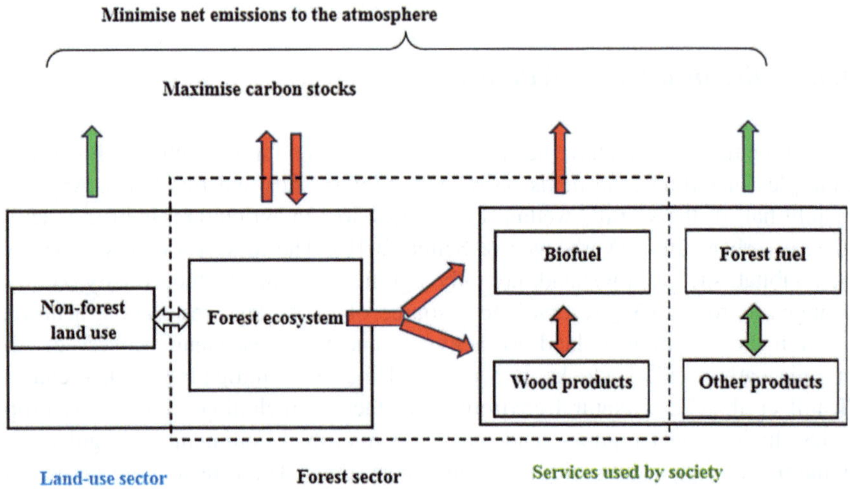

Fig. 3.2 Forest sector mitigation strategies necessitate an evaluation of their impacts on carbon storage in forest ecosystems, particularly in relation to sustainable harvest rates and net greenhouse gas (GHG) emissions across all sectors

2024). Logging residues, treatments to mitigate fire risks in forests, efforts to enhance forest health, fuelwood harvesting, waste from the forest products industry, urban wood residues, and energy plantations could collectively yield hundreds of millions of tons of biomass annually (US Department of Energy, 2016). Biomass can be utilized by burning it directly, blending it with coal, or incorporating it into oil- and gas-based electricity generation processes to mitigate greenhouse gas emissions (Xu et al., 2019).

3.4.7 Evenaged Versus Uneven-Aged System

Considering industrial purposes and operational efficiency, forests are managed as even-aged systems under integrated forest management (IFM) (Ruiz-Peinado et al., 2017). However, silvicultural systems influence forest carbon sequestration and storage (CSS) differently. Even-aged forests sequester and store more carbon in biomass than uneven-aged systems, primarily used in Ecological Forest Management (EFM), but these store more carbon in soils (Parajuli & Chang, 2012). According to Nilsen and Strand (2013), long-term timber production in uneven-aged stands was estimated to be 95% of that in even-aged stands. Additionally, over 81 years, the net carbon sequestration difference favoured the even-aged stands by 37 Mg ha^{-1}. During this time, mineral soil carbon increased by 21 Mg ha^{-1} in uneven-aged stands compared to even-aged stands, which was primarily attributed to differences in the O horizon. Notably, the carbon content in the humus layer of the uneven-aged stand was 20% higher than that in the even-aged stand (Nilsen & Strand, 2013).

Additionally, model simulations indicate that uneven-aged systems may be a preferable strategy. This should be considered when aiming to enhance CCS rates over the long term. Seidl et al. (2008) concluded that transitioning to uneven-aged forestry practices has significant potential to enhance carbon storage within forest ecosystems while simultaneously achieving diverse management objectives. A simulation by Taylor et al. (2008) compared the management effects of continuous cover and clear-cutting with reserves in the Canadian boreal forest and observed that total ecosystem carbon increased in uneven-aged stands.

3.4.8 Old Growth Forest Conservation

Old-growth and intact forests play a crucial role in stabilizing carbon storage on land, preserving biodiversity, and providing numerous ecosystem functions. These forests cover 22% of the world's forested land area (12 million km^2), where three countries—Russia, Brazil, and Canada—account for nearly two-thirds of this total. Old-growth forests have been found to sequester approximately 1.36 petagrams of carbon per year (Luyssaert et al., 2008). McGarvey et al. (2015) reported that old-growth forests possess a 30% greater total carbon density (154 Mg ha^{-1}) and an 1800% higher deadwood carbon density (120 Mg ha^{-1}) compared to the surrounding younger stands (5 Mg ha^{-1}). Consequently, old-growth forests significantly contribute to increased soil carbon through the accumulation of deadwood, more so than either intact forest management (IFM) or even early forest management (EFM) (McGarvey et al., 2015). On the one hand, stand age exhibits a positive correlation with biomass and carbon accumulation up to a certain point in the tree's lifespan. However, this relationship changes when trees reach an advanced age, at which point the net carbon uptake is believed to reach equilibrium with respiratory processes and increased mortality rates (Xu et al., 2012). While older tree populations demonstrate diminished carbon storage efficiency in their living woody structures, they retain the capacity for ongoing carbon accumulation (Luyssaert et al., 2008). This highlights the necessity of a preservation strategy for old-growth forests, focusing on minimizing deforestation and forest degradation. These issues require urgent attention, particularly in tropical regions, where prompt action is essential.

3.4.9 Selective Logging and Reduced Impact Logging (RIL)

Selective logging practices are crucial for Sustainable Forest Management (SFM), particularly in natural tropical forests, where they are prevalent worldwide. This method, which involves extracting selected commercial trees, affects over 400 million hectares—approximately 10% of the global forest area. Despite its importance in managing forests, selective logging has led to forest degradation, reducing timber supply, carbon stock potential, and habitat quality. This practice is accountable for at least half of the emissions resulting from the degradation of tropical forests (Poudyal et al., 2018). To address these issues, RIL practices have been promoted since the early 1980s, aiming to minimize forest damage while maintaining timber production. Reduced-impact logging for climate (RIL-C) is a subset of RIL practices explicitly designed to reduce carbon emissions and address climate change concerns. It also aligns with initiatives such as REDD+, voluntary carbon markets, and the UN Paris Climate Agreement (Ellis et al., 2019). RIL practices involve careful planning, mapping, felling, extraction, and monitoring, which significantly reduce forest damage and benefit biodiversity and post-logging recovery compared to conventional methods. However, RIL alone may not be sufficient to sustain financially viable timber production cycles.

3.5 Interventions for Reducing GHG Emissions

3.5.1 Energy Systems

Substantial and deep reductions in GHG demand significant transformations in the energy system. Adaptation strategies can help mitigate climate-related risks to the energy system (Paul et al., 2024). Net-zero carbon dioxide energy systems require a significant decrease in overall fossil fuel consumption, limited use of unabated fossil fuels, and the implementation of carbon capture and storage (CCS) technologies within the remaining fossil fuel-based systems. The IEA estimates that electrification could contribute nearly 19% of the cumulative emissions reductions needed to achieve a net-zero energy system by 2050. Climate change and related extreme events will affect future energy systems, including hydropower generation, bioenergy productivity, thermal power plant efficiencies, and demands for heating and cooling (IPCC, 2021). For instance, studies indicate that integrating renewables into the energy grid can displace fossil fuel generation, resulting in substantial reductions in CO_2 emissions—up to 45% in regions where renewables supply a significant portion of electricity.

The most feasible energy system adaptation options support infrastructure resilience, reliable power systems, and efficient water use for both existing and emerging energy generation systems (Garg et al., 2016). Adaptations for hydropower and thermoelectric power generation are effective in most regions up to 1.5–2 °C, but their effectiveness decreases at higher levels of warming (IPCC, 2021). Diversifying energy generation through wind, solar, and small-scale hydroelectric sources, alongside managing demand with energy storage and efficiency improvements, can enhance energy reliability and mitigate vulnerabilities to climate change, particularly for rural populations (Sojitra et al., 2024). Additionally, advancements in energy efficiency and technology have made renewable energy more accessible and cost-effective. For example, the costs of solar photovoltaic units have decreased by approximately 85% over the past decade, facilitating broader adoption of these units. This shift not only reduces reliance on carbon-intensive fuels but also promotes sustainable development by creating jobs and improving energy security.

3.5.2 Infrastructure, Industry, and Transport

Urban systems play a crucial role in achieving significant emissions reductions and promoting climate-resilient development, especially when integrated planning considers physical, natural, and social infrastructures (Mahajan et al., 2024). Effective mitigation can occur during the design, construction, retrofit, use, and disposal stages of buildings. Mitigation interventions for buildings include low-emission construction materials, a highly efficient building envelope, and the integration of renewable energy solutions at the construction phase; highly efficient appliances/

equipment, the optimization of building use and their supply with low-emission energy sources at the use phase; and recycling and reusing construction materials at the disposal phase. Reducing greenhouse gas emissions in the industry requires coordinated actions across value chains to promote all mitigation options, including demand management, improving energy and materials efficiency, creating circular material flows, utilizing abatement technologies, and implementing transformational changes in production processes. Reducing industrial emissions requires coordinated efforts across value chains to promote various mitigation options, including circular material flows, deploying abatement technologies, demand management, energy, and materials efficiency, and implementing transformational changes in production processes. To reduce emissions from the production and use of chemicals, a life cycle approach is necessary. Additionally, depending on availability, strategies such as Carbon Capture and Utilization (CCU), direct air capture of CO_2, and Carbon Capture and Storage (CCS) should be considered.

Transport-related greenhouse gas (GHG) emissions can be reduced through demand-side strategies and low-GHG technologies. Changing urban layouts helps reallocate street space for cycling and walking, leveraging digitalization (e.g., teleworking), and implementing programs that encourage changes in consumer behavior (e.g., adjusting transport pricing). Furthermore, sustainable biofuels, low-emissions hydrogen, and synthetic fuels can help mitigate CO_2 emissions from shipping, aviation, and heavy-duty land transport. Nonetheless, improvements in production processes and cost reductions are required for these solutions to be effective.

3.5.3 Land, Ocean, Food, and Water

Numerous agricultures, forestry, and other land use (AFOLU) options offer both adaptation and mitigation benefits that can be scaled up in the near term across most regions. Conservation, better management, and restoration of forests and ecosystems offer the most significant potential for economic mitigation, particularly through reduced deforestation in tropical areas, which have the highest overall mitigation potential. Demand-side measures, such as shifting to sustainable healthy diets and reducing food loss and waste, along with sustainable agricultural intensification, minimize ecosystem conversion and emissions of CH_4 and N_2O. These measures can also free up land for reforestation and the restoration of ecosystems. Sustainably sourced agricultural and forestry products, including durable wood products, can be used as alternatives to more greenhouse gas-intensive options in various sectors. Effective adaptation strategies encompass cultivar enhancements, agroforestry practices, community-focused adaptation, diversification of farms and landscapes, and urban farming. These AFOLU response options require the integration of biophysical, socioeconomic, and other enabling factors. The effectiveness of ecosystem-based adaptation and most water-related adaptation options declines with increasing warming. Gupta et al. (2024) emphasize that agri-horticulture-based land use systems (LUSs) can effectively reduce carbon emissions through CO_2 sequestration in Mirzapur (UP). These systems also enhance livelihoods by promoting soil health, reducing soil erosion, and helping to mitigate anthropogenic warming.

3.5.4 Society, Livelihoods, and Economies

Society, livelihoods, and economies play interconnected roles in reducing greenhouse gas emissions by driving awareness, sustainable practices, and systemic change. Societal shifts promote eco-friendly norms, public advocacy, and grassroots initiatives. Sustainable livelihoods in agriculture, renewable energy, and climate-resilient practices lower emissions while supporting green jobs. Economies contribute through green investments, carbon pricing, circular economy models, and innovation in low-carbon technologies. Together, these elements create a synergistic approach where informed societies demand change, economies enable it, and livelihoods adapt to foster sustainable development and emission reductions.

3.6 Successful Sustainable Forest Management Projects

3.6.1 The Amazon Rainforest—The Amazon Region Protected Areas Project (ARPA)—A Regional-Scale Initiative

Amazon Region Protected Areas Program (ARPA) is a collaborative initiative funded by both governmental and non-governmental agencies, focusing on expanding conservation efforts in the Brazilian Amazon rainforest. It is a partnership involving the Brazilian Government, the Brazilian Biodiversity Fund (FUNBIO), the German Development Bank (KFW), the Global Environmental Facility, the World Bank, Greenspan, WWF, Conservation International, and various other organizations. ARPA was launched in 2002 to protect the Amazon's rich natural resources while supporting sustainable development to address urgent social needs. ARPA has established a system of well-managed parks and protected areas that covers approximately 193,000 square miles, which exceeds the total geographic extent of the United States National Park System. The project was granted approximately $370 million in funding.

ARPA aims to achieve clear and ambitious goals:

- Establish 70 million acres of new protected areas for strict conservation use.
- Transform 31 million acres of pre-existing but neglected parks by bringing them up to effective management standards.

- Establish 22 million acres of sustainable use reserves in which local communities will have a stake and will benefit from effective stewardship.
- Set up a long-term Protected Areas Trust Fund to ensure the financial viability and integrity of the park system in perpetuity.

3.6.2 The Plant for the Planet "Billion Tree Campaign"—A Global-Scale Initiative with Local-Scale Influence

The campaign was inspired by Professor Wangari Maathai, the 2004 Nobel Peace Prize laureate and founder of Kenya's Green Belt Movement, which has successfully planted over 30 million trees across 12 African countries since its inception in 1977. The Billion Tree Campaign was launched on November 8, 2006, during the 12th meeting of the Conference of the Parties (COP 12) to the United Nations Framework Convention on Climate Change (UNFCCC), held in Nairobi, Kenya. The objective is to plant at least one billion trees globally each year. The United Nations Environment Programme (UNEP) has announced that more than 12 billion trees have been planted in 193 countries as part of the Billion Tree Campaign. This milestone was achieved following a tree-planting initiative in Kenya organized by the Green Africa Foundation, which has successfully planted over 24 million trees to date. Leading the global effort are China and India, with China planting 2.8 billion trees and India following closely with 2.1 billion trees.

The Billion Tree Campaign aims to improve the quality of life in communities worldwide by harnessing the numerous benefits that trees offer. This includes combating climate change through carbon sequestration, boosting local economies with products such as timber, and providing essential ecosystem services, including soil regulation, erosion control, and cultural benefits. Since then, UNEP has officially transferred the campaign to the Plant for the Planet Foundation, a youth-led environmental organization based in Germany. As the children's aim is now to plant a trillion trees, they transformed the Billion Tree Campaign into the Trillion Tree Campaign.

3.6.3 Forest Stewardship Council (FSC)

The FSC was established in 1994 as an initiative aimed at addressing international concerns about widespread deforestation. The FSC operates as a global, membership-based organization with a governance structure that is based on stakeholder participation, democratic decision-making, equity, and transparency (FSC, 2012). It serves as a platform through which forest owners, environmental organizations, social groups, and timber industries collaborate to develop solutions for enhancing sustainable forest management practices. The FSC is a multi-stakeholder organization that, through its membership, develops forest management and chain of custody standards, delivers trademark assurance, and provides accreditation services to a global network of committed businesses, organizations, and communities (FSC, 2012).

FSC is an international forest certification system established to promote sustainable management of forests and forest products. According to the council, using the FSC logo denotes that a product originates from environmentally, socially, and economically responsible sources. In addition to its global certification standard, FSC also formulates nation-specific standards within certain countries. The FSC's certification framework is based on 10 key Principles and associated Criteria (FSC P&C) that underpin all its standards and certification procedures. The FSC aims to promote the sustainable and equitable management of the world's forests, focusing on environmental responsibility, social benefit, and economic feasibility (FSC, 2012). It claims that forests are managed according to the organization's standards and are purported to offer benefits to local and broader communities, including enhanced air and water quality, as well as a contribution to climate change mitigation. FSC addresses issues such as deforestation, climate change, and illegal logging, and certain reports suggest it has produced beneficial impacts on economic growth, environmental preservation, poverty reduction, and social and political empowerment. The body has published a global strategy with five goals:

- Advancing globally responsible forest management.
- Ensure equitable access to the benefits of Forest Stewardship Council (FSC) systems.
- Ensure integrity, credibility, and transparency of the FSC system.
- Create business value for products from FSC-certified forests.
- Strengthen the global network to deliver on goals 1 through 4.

3.6.4 Miyawaki Forests—Japan and Global

The Miyawaki method, pioneered by Japanese botanist Dr. Akira Miyawaki, is an innovative approach to reforestation and afforestation that focuses on creating dense, biodiverse forests in urban areas. This afforestation technique involves cultivating fast-growing, dense stands of indigenous plant species to establish a natural forest ecosystem. It is also known as the pot plantation method, as it consists of planting trees and shrubs closely together to accelerate their collective growth. This approach has gained global recognition for its effectiveness in natural environments and its role in mitigating the impacts of climate change. By establishing a dense and diverse planting of native species, he aimed to reconstitute the natural forest ecosystems that predated human influence.

3.6.5 REDD+ Program (Reducing Emissions from Deforestation and Forest Degradation)

The REDD+ program is a global initiative formed at the 2007 COP 13 conference in Bali, aimed at mitigating climate change by addressing issues of deforestation and forest degradation while promoting sustainable forest management practices (UNFCCC, 2007). The Cancun Agreements in 2010 established REDD+ as an official UNFCCC mechanism aimed at mitigating climate change through forest-based approaches, including reducing emissions from deforestation and forest degradation, increasing carbon capture in forests, and conserving biodiversity (UNFCCC, 2007). The program promotes sustainable practices, such as agroforestry and responsible logging, while also involving indigenous and local communities in decision-making processes to ensure they receive financial benefits. The REDD+ initiative has evolved into a comprehensive framework that integrates financial incentives, monitoring protocols, and protections for the rights of local communities. Its objectives are twofold: to reduce global emissions from forest loss and to enhance the role of forests in carbon storage, thereby contributing to global climate goals. By promoting forest conservation, REDD+ seeks to balance ecological, economic, and social needs while addressing climate change. REDD+ (also recognized in Article 5 of the Paris Agreement) is a framework designed to guide activities aimed at reducing forest degradation and deforestation (Fig. 3.3). This framework aims to implement REDD+ activities at both national and sub-national levels of government to mitigate anthropogenic pressure on forests. The developing countries will receive payments in exchange for proof that they have reduced deforestation or for demonstrating their forest preservation activities.

3.7 Economic and Social Dimensions

3.7.1 Economic Dimensions

The economic dimension of SFM focuses on ensuring that forest resources are utilized in a manner that supports economic development while preserving the ecological and social functions of forests. Key components of this economic aspect include.

3.7.1.1 Job Creation and Economic Growth

SFM generates employment opportunities across various sectors, including the harvesting, processing, and marketing of forest products. This is particularly crucial in developing countries, where the forestry sector can offer employment opportunities for millions, especially in rural regions with limited alternative livelihood options. It

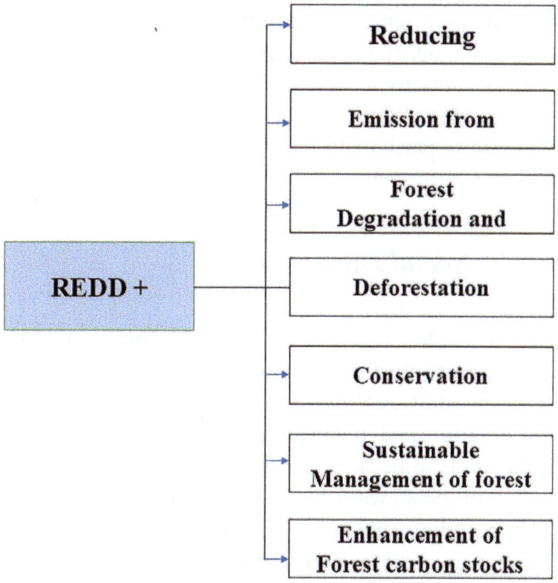

Fig. 3.3 Component of REDD+

is estimated that sustainable forest management could generate $230 billion in business opportunities and create 16 million jobs by 2030, emphasizing its potential to boost economic activity while tackling deforestation challenges. Income generation from sustainable forestry is derived from timber sales and non-timber forest products (NTFPs), including fruits, nuts, and medicinal plants. These resources can be leveraged to create high-value products, such as furniture and paper. By promoting education and training programs tailored to the needs of sustainable forest management, communities can build a skilled workforce capable of adapting to evolving environmental challenges. Overall, integrating job creation and income generation into sustainable forest management not only supports ecological goals but also enhances social resilience and economic stability in communities that rely on forest resources. Moreover, the monetization ecosystem services through mechanisms such as carbon credits and Payment for Ecosystem Services (PES) can enhance the integration of economic development with environmental conservation.

3.7.1.2 Trade and Market Dynamics

Trade plays a crucial role in sustainable forest management by facilitating the exchange of forest products while promoting sustainable practices. This includes assessing the economic benefits and costs associated with forestry outputs to inform better management decisions. It is also influencing production practices, governance structures, and the economic viability of forest resources. The European Union's Forest Law Enforcement, Governance, and Trade (FLEGT) initiative exemplifies how trade agreements can promote Sustainable Forest Management (SFM). By establishing Voluntary Partnership Agreements (VPAs) with timber-exporting countries, FLEGT aims to combat illegal logging while enhancing governance and regulatory frameworks in partner nations. These agreements promote transparency and community participation in forest management decisions.

3.7.2 *Social Dimensions*

The social dimension of SFM focuses on the relationships between communities, forests, and forest management practices. This aspect underscores the significance of social equity, cultural values, and community well-being in the context of forest conservation and management. Community participation plays a crucial role in sustainable forest management, serving as a foundation for effective governance and resource stewardship. This participatory approach empowers local communities, including indigenous peoples and rural stakeholders, to engage actively in the management of forest resources. Involving communities in decision-making processes can help sustainable forest management initiatives leverage traditional knowledge and practices that have been honed over generations. This collaboration not only enhances ecological conservation but also supports social equity and economic development within these communities. Furthermore, community forest management initiatives have demonstrated effectiveness in conserving biodiversity, restoring degraded lands, and improving livelihoods by aligning environmental goals with the socio-economic needs of local populations.

3.8 Challenges and Barriers

Forests are a global resource, and effectively addressing important issues related to their use and management requires global collaboration. To fully understand and address the challenges in forest science, it is essential to establish an appropriate framework and enhance our capacity to generate knowledge for a sustainable future. A balanced approach is crucial for the successful management and development of forests, ensuring their existence at acceptable levels for the benefit of present and future generations. These are some key highlights of challenges and barriers to sustainable forest management:

- The overarching impact of climate change leads to increased frequency of droughts, wildfires, and pest infestations, which contribute to forest degradation and loss. These changes complicate the management of forests and their ecosystems. Climate change hotspots within Indian forests, as identified by the India State of Forest Report (ISFR) (2021), indicate that approximately 45–64% of the country's forests will experience adverse effects from climate change and rising temperatures by 2030.

- Land-use changes, including the conversion of forests for agricultural purposes, urban expansion, and mining, are among the primary drivers of deforestation and forest degradation. Such transformations not only diminish forest cover but also compromise critical ecosystem services, including carbon sequestration and biodiversity preservation.
- Addressing the drivers of forest degradation and deforestation and enabling sustainable forest management (SFM), requires greater innovation and better coordination in global forestry dialogue, national sector planning, and technical analysis. Demonstrating the potential of forests to reduce poverty, support economic growth, and provide environmental services at local and global levels necessitates close collaboration between donors and governments, linking forest sector activities with national strategies. Promoting forest ownership and access rights, as well as emphasizing stakeholder participation in policy formulation and implementation, are essential for poverty reduction and effective forest governance.
- Economic crises exacerbate pressures on forests as communities turn to unsustainable practices for subsistence. Increased unemployment in urban areas leads to forest encroachment as displaced populations return to rural regions, thereby heightening the risk of illegal logging and the overexploitation of resources. Additionally, subsidies for competing land uses can divert investments away from sustainable forestry practices.

3.9 Future Perspective and Recommendations

SFM is increasingly recognized as essential for balancing ecological, economic, and social objectives. As pressures on forest ecosystems intensify due to climate change, urbanization, and economic demands, several emerging strategies and frameworks are shaping the future of Sustainable Forest Management (SFM). These are some highlights of future perspectives and recommendations for Sustainable Forest management:

- Advancements in forestry practices are crucial for achieving Sustainable Forest Management (SFM). Innovations such as improved recycling methods, greater reliance on plantations, and farm forestry are helping reduce the pressure on natural resources while still meeting industrial demands. Additionally, the focus on ecotourism and community involvement in protected area management is fostering a more sustainable relationship between local populations and forest ecosystems.
- Certain forest management tools require further refinement, such as field inventory techniques, remote sensing, and geographic information systems. Additionally, the use of sample plots to monitor forest stands and activities aimed at enhancing

the quality and value of standing wood is essential. Ultimately, maintaining user-friendly and up-to-date forest resource databases is crucial to support informed decision-making in resource management.

- Sustainable agricultural practices are another effective way to reduce emissions. Techniques such as minimizing soil disturbance, enhancing the precision of fertilizer application, and integrating trees into farmland can help lower CH_4 and N_2O emissions from agricultural activities. Additionally, these practices can improve soil carbon storage and reduce emissions associated with fertilizer use.
- Governments and policymakers need to establish more effective strategies to prevent and manage wildfires. Implementing early warning systems, conducting controlled burns under safe conditions, and involving local communities in prevention efforts can significantly reduce emissions from forest fires. Additionally, pricing carbon through carbon taxes or cap-and-trade systems can encourage companies to decrease pollution. The revenue generated from these initiatives can be reinvested in projects such as tree planting and the development of renewable energy sources.
- Ultimately, promoting international collaboration and providing financial assistance to developing nations for climate change mitigation and adaptation is vital. This support may involve technology transfer, capacity building, and funding for sustainable land management and forest conservation initiatives.

3.10 Conclusion

SFM is a holistic and dynamic approach that seeks to harmonize economic growth, social well-being, and environmental stewardship within the forestry sector. At its core, SFM is committed to ensuring that forests continue to provide essential resources such as timber, biodiversity, and ecosystem services without compromising the ability of future generations to benefit from them. The effective execution of SFM depends on several integrated practices, including biodiversity conservation, selective and reduced-impact logging, active community engagement, and the use of certification systems to promote transparency and accountability in forest management. Additionally, it provides a powerful strategy to enhance carbon storage and reduce greenhouse gas emissions. The full potential of forests to pursue a carbon-neutral future can be harnessed by implementing evidence-based practices, leveraging technological advancements, and fostering international cooperation. However, continued research, monitoring, and adaptive management are necessary to address the emerging challenges and optimize the role of forests in climate change mitigation. Global economic pressures drive the unsustainable exploitation of forest resources, whereas weak governance and inconsistent policies allow deforestation and degradation to persist. Insufficient funding, a lack of awareness, and competing land-use priorities often limit the scope and effectiveness of sustainable forest management (SFM) practices. Despite these challenges, SFM remains indispensable in the fight against climate change, the protection of biodiversity, and the support of rural livelihoods.

References

Ascher, W. (2007). Policy sciences contributions to analysis to promote sustainability. *Sustainability Science, 2*(2), 141–149.

Bains, P., Psarras, P., & Wilcox, J. (2017). CO_2 capture from the industry sector. *Progress in Energy and Combustion Science, 63*, 146–172.

Barwa, S., & Singh, H. (2024). Climate-induced changes in forest ecosystems of Indian Himalayan region: A meta-analysis. In *Urban forests, climate change and environmental pollution: Physio-biochemical and molecular perspectives to enhance urban resilience* (pp. 839–866). Springer.

Bastin, J. F., Finegold, Y., Garcia, C., Mollicone, D., Rezende, M., Routh, D., & Crowther, T. W. (2019). The global tree restoration potential. *Science, 365*(6448), 76–79.

Bellassen, V. & Luyssaert, S. (2014). Carbon sequestration: Managing forests in uncertain times. *Nature, 506*(7487), 153–155.

Boyd, P. W., Claustre, H., Levy, M., Siegel, D. A., & Weber, T. (2019). Multi-faceted particle pumps drive carbon sequestration in the ocean. *Nature, 568*(7752), 327–335.

Brundtland, G. H. (1987). World commission on environment and development. *Environmental Policy and Law, 14*(1), 26–30.

Carlotta-Jones, D. I., Purdy, K., Kirwan, K., Stratford, J., & Coles, S. R. (2020). Improved hydrogen gas production in microbial electrolysis cells using inexpensive recycled carbon fibre fabrics. *Bioresource Technology, 304*, 122983.

Chen, L., Msigwa, G., Yang, M., Osman, A. I., Fawzy, S., Rooney, D. W., & Yap, P. S. (2022). Strategies to achieve a carbon neutral society: A review. *Environmental Chemistry Letters, 20*(4), 2277–2310.

Chen, X., Mallik, A. U., Yu, Z., Wang, Z., Wang, S., Dong, Y., & Bu, Z. J. (2024). Drainage-driven loss of carbon sequestration of a temperate peatland in Northeast China. *Ecosystems, 27*(2), 207–221.

Dods, M. N., Kim, E. J., Long, J. R., & Weston, S. C. (2021). Deep CCS: Moving beyond 90% carbon dioxide capture. *Environmental Science & Technology, 55*(13), 8524–8534.

Doelman, J. C., Stehfest, E., van Vuuren, D. P., Tabeau, A., Hof, A. F., Braakhekke, M. C., & Lucas, P. L. (2020). Afforestation for climate change mitigation: Potentials, risks and trade-offs. *Global Change Biology, 26*(3), 1576–1591.

Dumitraşcu, M., Kucsicsa, G., Dumitrică, C., Popovici, E. A., Vrînceanu, A., Mitrică, B., & Şerban, P. R. (2020). Estimation of future changes in aboveground forest carbon stock in Romania A prediction based on forest-cover pattern scenario. *Forests, 11*(9), 914.

Ellis, P. W., Gopalakrishna, T., Goodman, R. C., Putz, F. E., Roopsind, A., Umunay, P. M., & Griscom, B. W. (2019). Reduced-impact logging for climate change mitigation (RIL-C) can halve selective logging emissions from tropical forests. *Forest Ecology and Management, 438*, 255–266.

El-Naggar, A., Awad, Y. M., Tang, X. Y., Liu, C., Niazi, N. K., Jien, S. H., & Lee, S. S. (2018). Biochar influences soil carbon pools and facilitates interactions with soil: A field investigation. *Land Degradation & Development, 29*(7), 2162–2171.

FAO. (2015). Global forest resources assessment, 2015 country report, Malaysia. Food and Agriculture Organization.

FAO. (2016). *Global forest resources assessment 2015 how are the world's forests changing?* (p. 47). Food and Agriculture Organization of the United Nations.

FSC. (2012, November). *Global FSC certificates: type and distribution*. Forest Stewardship Council.

Garg, A., Pal, D., Singh, H., & Pandey, D. C. (2016). A comparative study of NDBI, NDISI and NDII for extraction of urban impervious surface of Dehradun [Uttarakhand, India] using Landsat 8 imagery. *2016 International conference on emerging trends in communication technologies (ETCT)* (pp. 1–5).

Guo, Y., Abdalla, M., Espenberg, M., Hastings, A., Hallett, P., & Smith, P. (2021). A systematic analysis and review of the impacts of afforestation on soil quality indicators as modified by climate zone, forest type and age. *Science of the Total Environment, 757*, 143824.

Gupta, H., Janju, S., Mahajan, A., Singh, C., Sharma, S., Prajapati, A. (2024). Forests and agroforestry: Nature-based solutions for climate change mitigation. In *Forests and climate change: Biological perspectives on impact, adaptation, and mitigation strategies* (pp. 421–443). Springer Nature Singapore.

Han, M., Zhao, Q., Li, W., Ciais, P., Wang, Y. P., Goll, D. S., & Wu, F. (2021). Global soil organic carbon changes and economic revenues with biochar application. *GCB Bioenergy, 14*(3), 364–377.

Hanssen, S. V., Daioglou, V., Steinmann, Z. J. N., Doelman, J. C., Van Vuuren, D. P., & Huijbregts, M. A. J. (2020). The climate change mitigation potential of bioenergy with carbon capture and storage. *Nature Climate Change, 10*(11), 1023–1029.

Hong, H., Lu, Y., Zhu, X., Wu, Q., Jin, L., Jin, Z., & Yu, H. (2023). Cytotoxicity of nitrogenous disinfection byproducts: A combined experimental and computational study. *Science of the Total Environment, 856*, 159273.

Houghton, R. A. (2012). Carbon emissions and the drivers of deforestation and forest degradation in the tropics. *Current Opinion in Environmental Sustainability, 4*(6), 597–603.

Humpenöder, F., Popp, A., Dietrich, J. P., Klein, D., Lotze-Campen, H., Bonsch, M., & Müller, C. (2014). Investigating afforestation and bioenergy CCS as climate change mitigation strategies. *Environmental Research Letters, 9*(6), 064029.

IEA. (2021). *Net zero by 2050—A roadmap for the global energy sector*. International Energy Agency. https://www.iea.org/reports/net-zero-by-2050

India State of Forest Report (Forest Survey of India). (2021). Ministry of Environment and Forests, Government of India.

International Tropical Timber Organization. (1992). *Criteria for the measurement of sustainable tropical forest management*. The Organization.

IPCC. (2014). *Climate change 2014 synthesis report: A report of the intergovernmental panel on climate change Geneva, Switzerland.*

IPCC. (2019). Summary for policymakers. In *Climate change and land: An IPCC special report on climate change, desertification, land degradation, sustainable land management, food security, and greenhouse gas fluxes in terrestrial ecosystems*. Intergovernmental Panel on Climate Change.

IPCC. (2021). In V. Masson-Delmotte et al. (Eds.) *Summary for policymakers climate change 2021: The physical science basis*. Contribution of Working Group I to the Sixth Assessment Report of the Intergovernmental Panel on Climate Change. Cambridge University Press (accepted).

Islam, M. M., Chowdhury, M. A. M., Begum, R. A., & Amir, A. A. (2022). A bibliometric analysis on the research trends of climate change effects on economic vulnerability. *Environmental Science and Pollution Research, 29*(39), 59300–59315.

Jangra, A., Kumar, K., Maikhuri, S., Bhandari, M. S., Pandey, S., Singh, H., & Barthwal, S. (2024). Unveiling stress-adapted endophytic bacteria: Characterizing plant growth-promoting traits and assessing cross-inoculation effects on *Populus deltoides* under abiotic stress. *Plant Physiology and Biochemistry, 210*, 108610.

Joshi, R., Sharma, B., Singh, H., Dhakal, N., Ayer, S., & Maraseni, T. (2024). Poplar plantation as an agroforestry approach: Economic benefits and its role in carbon sequestration in North India. *Journal of Resources and Ecology, 15*(4), 880–888.

Joshi, R., Singh, H., Chhetri, R., Poudel, S. R., & Rijal, S. (2021). Carbon sequestration potential of community forests: A comparative analysis of soil organic carbon stock in community managed forests of Far-Western Nepal. *Eurasian Journal of Soil Science, 10*(2), 96–104.

Joshi, R., Singh, H., Chhetri, R., & Yadav, K. (2020). Assessment of carbon sequestration potential in degraded and non-degraded community forests in Terai region of Nepal. *Journal of Forest and Environmental Science, 36*(2), 113–121.

Kammann, C. I., Schmidt, H. P., Messerschmidt, N., Linsel, S., Steffens, D., Müller, C., & Joseph, S. (2015). Plant growth improvement mediated by nitrate capture in co-composted biochar. *Scientific Reports, 5*(1), 11080.

Kastenhofer, K., & Rammel, C. (2005). Obstacles to and potentials of the societal implementation of sustainable development: A comparative analysis of two case studies. *Sustainability: Science, Practice and Policy, 1*(2), 5–13.

Koehler, D. A., & Hecht, A. D. (2006). Sustainability, wellbeing, and environmental protection: perspectives and recommendations from an environmental protection agency forum. *Sustainability: Science, Practice and Policy, 2*(2), 22–28.

Kohl, M., Lasco, R., Cifuentes, M., Jonsson, Ö., Korhonen, K. T., Mundhenk, P., & Stinson, G. (2015). Changes in forest production, biomass and carbon: Results from the 2015 UN FAO global forest resource assessment. *Forest Ecology and Management, 352*, 21–34.

Kumar, M., & Singh, H. (2020). Agroforestry as a nature-based solution for reducing community dependence on forests to safeguard forests in rainfed areas of India. In *Nature-based solutions for resilient ecosystems and societies* (pp. 289–306). Springer.

Kumar, A., Dwivedi, G. K., Tewari, S., Sah, V. K., Singh, H., Kumar, P., Kumar, N., & Kaushal, R. (2020). Soil organic carbon pools under *Terminalia chebula* Retz. based agroforestry system in Himalayan foothills, India. *Current Science, 118*(7), 1098–1103.

Kumar, N., Singh, H., & Kumar, A. (Eds.). (2021). *Renewable energy and green technology: Principles and practices*. CRC Press.

Lehmann, J. (2007). A handful of carbon. *Nature, 447*(7141), 143–144.

Li, T., Li, M. Y., & Tian, L. (2021). Dynamics of carbon storage and its drivers in Guangdong province from 1979 to 2012. *Forests, 12*(11), 1482.

Luyssaert, S., Schulze, E. D., Börner, A., Knohl, A., Hessenmöller, D., Law, B. E., & Grace, J. (2008). Old-growth forests as global carbon sinks. *Nature, 455*(7210), 213–215.

Mahajan, A., Sojitra, A., Gupta, H., & Arora, D. (2021). Carbon sequestration in forest and agroforestry—A global perspective. In *Advances in forestry and agro-forestry* (p. 437).

Mahajan, A., Paul, B. N., Mehra, A., Gupta, H., & Sarath, S. (2024). *Urban forests: Benefits, barriers, and sustainable solutions. In Urban forests, climate change and environmental pollution* (pp. 481–505). Springer.

Matthew, N. K., Shuib, A., Muhammad, I. S. M. A. I. L., Eusop, M. E. M., Ramachandran, S., Afandi, S. H. M., & Samdin, Z. (2018). Carbon stock and sequestration valuation in a mixed dipterocarp forest of Malaysia. *Sains Malaysiana, 47*(3), 447–455.

McGarvey, J. C., Thompson, J. R., Epstein, H. E., & Shugart, H. H., Jr. (2015). Carbon storage in old-growth forests of the Mid-Atlantic: Toward better understanding the eastern forest carbon sink. *Ecology, 96*(2), 311–317.

MCPFE. (1993). General declaration and resolutions adopted. In *Proceedings of the second ministerial conference on the protection of forest in Europe, Helsinki*. Report. Liaison Unit, Vienna.

Nayak, N., Mehrotra, R., & Mehrotra, S. (2022). Carbon bio sequestration strategies: A review. *Carbon Capture Science & Technology, 4*, 100065.

Nelson, E., Mendoza, G., Regetz, J., Polasky, S., Tallis, H., Cameron, D., & Shaw, M. (2009). Modeling multiple ecosystem services, biodiversity conservation, commodity production, and tradeoffs at landscape scales. *Frontiers in Ecology and the Environment, 7*(1), 4–11.

Nilsen, P., & Strand, L. T. (2013). Carbon stores and fluxes in even-and uneven-aged Norway spruce stands. *Silva Fennica, 47*(4).

Parajuli, R., & Chang, S. J. (2012). Carbon sequestration and uneven-aged management of loblolly pine stands in the Southern USA: A joint optimization approach. *Forest Policy and Economics, 22*, 65–71.

Paul, B. N., Mahajan, A., & Sarath, S. (2024). Climate change and forest ecosystems: Restoration and rehabilitation approaches. In *Forests and climate change: Biological perspectives on impact, adaptation, and mitigation strategies* (pp. 553–571). Springer Nature Singapore.

Pittelkow, C. M., Liang, X., Linquist, B. A., Van Groenigen, K. J., Lee, J., Lundy, M. E., & Van Kessel, C. (2015). Productivity limits and potentials of the principles of conservation agriculture. *Nature, 517*(7534), 365–368.

Poudyal, B. H., Maraseni, T., & Cockfield, G. (2018). Evolutionary dynamics of selective logging in the tropics: A systematic review of impact studies and their effectiveness in sustainable forest management. *Forest Ecology and Management, 430*, 166–175.

Powlson, D. S., Stirling, C. M., Thierfelder, C., White, R. P., & Jat, M. L. (2016). Does conservation agriculture deliver climate change mitigation through soil carbon sequestration in tropical agro-ecosystems? *Agriculture, Ecosystems & Environment, 220*, 164–174.

Ruhil, T., Singh, H., & Barthwal, S. (2024). Long-term physiological monitoring for detecting climate change impacts on forest ecosystems. In *Forests and climate change: Biological perspectives on impact, adaptation, and mitigation strategies* (pp. 817–834). Springer.

Ruiz-Peinado, R., Oviedo, J. A. B., Senespleda, E. L., Oviedo, F. B., & Gaztelurrutia, M. (2017). Forest management and carbon sequestration in the Mediterranean region: A review. *Forest System, 26*(2), 10.

Sathaye, J., Makundi, W., Dale, L., Chan, P., & Andrasko, K. (2006). GHG mitigation potential, costs and benefits in global forests: A dynamic partial equilibrium approach. *The Energy Journal, 27*, 127–162.

Seidl, R., Rammer, W., Lasch, P., Badeck, F., & Lexer, M. J. (2008). Does conversion of even-aged, secondary coniferous forests affect carbon sequestration? A simulation study under changing environmental conditions. *Silva Fennica, 42*(3), 369.

Singh, H., Verma, A., & Kumar, M. (2018). Can elevated CO_2 alter the growth dynamics, morphology, physiology and biochemistry of the forestry tree species? In *Climate change and agriculture* (pp. 288–305). Discovery Publishing House.

Singh, H., & Kumar, M. (2022). Climate change and its impact on Indian Himalayan forests: Current status and research needs. In *Climate change: Impacts, responses and sustainability in the Indian Himalaya* (pp. 223–242). Springer.

Singh, H., Kumar, N., Singh, R., & Kumar, M. (2023). Assessing the climate change impact on the habitat suitability of the range-restricted bird species (*Catreus wallichii*) in the Indian Himalayan ecosystem. *Environmental Science and Pollution Research, 30*(57), 121224–121235.

Singh, H. (2024a). Exploring adaptive modulation in plant functional traits and its impact on the productivity of *Acacia auriculiformis* under CO_2-enriched environment. *Industrial Crops and Products, 210*, 118186.

Singh, H. (2024b). *Forests and climate change: Biological perspectives on impact, adaptation, and mitigation strategies*. Springer Nature.

Singh, H. (2024c). *Urban forests, climate change and environmental pollution: Physio-biochemical and molecular perspectives to enhance urban resilience*. Springer Nature.

Singh, H., Kumar, N., & Kumar, A. (2024a). Enhancing resource use efficiency in crops through plant functional traits. In *Plant functional traits for improving productivity* (pp. 97–117). Springer.

Singh, N., Riyal, M. K., Singh, B., Khanduri, V. P., Rawat, D., Singh, C., Pinto, M. M. C., & Kumar, M. (2024b). Carbon sequestration potential of agroforestry versus adjoining forests at different altitudes in the Garhwal Himalayas. *Atmosphere, 15*(3), 313.

Singh, H. (2025). The role of plant functional traits in mitigating urban climate and environmental pollution. In *Plant functional traits* (pp. 297–308). Elsevier.

Sojitra, A., Arora, D., Singh, K., Malik, A., & Mahajan, A. (2024). Identifying future research and directions to address forest and climate change challenges. In *Forests and climate change: Biological perspectives on impact, adaptation, and mitigation strategies* (pp. 851–877). Springer Nature Singapore.

Souza, C. M., Shimbo, J. Z., Rosa, M. R., Parente, L. L., Alencar, A. A., Rudorff, B. F. T., et al. (2020). Reconstructing three decades of land use and land cover changes in Brazilian biomes with Landsat archive and earth engine. *Remote Sensing, 12*(17), 2735.

Sun, X., Xie, M., Mai, L., & Zeng, E. Y. (2022). Biobased plastic: A plausible solution toward carbon neutrality in plastic industry? *Journal of Hazardous Materials, 435*, 129037.

Taillardat, P., Thompson, B. S., Garneau, M., Trottier, K., & Friess, D. A. (2020). Climate change mitigation potential of wetlands and the cost-effectiveness of their restoration. *Interface Focus, 10*(5), 20190129.

Taylor, A. R., Wang, J. R., & Kurz, W. A. (2008). Effects of harvesting intensity on carbon stocks in eastern Canadian red spruce (*Picea rubens*) forests: An exploratory analysis using the CBM-CFS3 simulation model. *Forest Ecology and Management, 255*(10), 3632–3641.

Tiwari, A. (2021). Advancing materials towards climate neutrality by 2050. *Advanced Materials Letters, 12*(8), 21081650.

UNFCCC. (2007). *Views on the range of topics and other relevant information relating to reducing emissions from deforestation in developing countries*. Submissions from Parties.

US Department of Energy. (2016). *2016-Billion-ton report: Advancing domestic resources for a thriving bioeconomy*, (Vol. 1: Economic availability of feedstock). US Department of Energy, ORNL/TM2016/160, Oak Ridge National Laboratory.

Verhoeven, J. T., & Setter, T. L. (2010). Agricultural use of wetlands: Opportunities and limitations. *Annals of Botany, 105*(1), 155–163.

Wiersum, K. F. (1995). 200 years of sustainability in forestry: Lessons from history. *Environmental Management, 19*, 321–329.

Woolf, D., Amonette, J. E., Street-Perrott, F. A., Lehmann, J., & Joseph, S. (2010). Sustainable biochar to mitigate global climate change. *Nature Communications, 1*(1), 56.

Xu, C. Y., Turnbull, M. H., Tissue, D. T., Lewis, J. D., Carson, R., Schuster, W. S., & Griffin, K. L. (2012). Age-related decline of stand biomass accumulation is primarily due to mortality and not to reduction in NPP associated with individual tree physiology, tree growth or stand structure in a Quercus-dominated Forest. *Journal of Ecology, 100*(2), 428–440.

Xu, S., Liu, X., Li, X., & Tian, C. (2019). Soil organic carbon changes following wetland restoration: A global meta-analysis. *Geoderma, 353*, 89–96.

Zomer, R. J., Neufeldt, H., Xu, J., Ahrends, A., Bossio, D., Trabucco, A., & Wang, M. (2016). Global tree cover and biomass carbon on agricultural land: The contribution of agroforestry to global and national carbon budgets. *Scientific Reports, 6*(1), 29987.

Chapter 4
Enhancing Forest and Tree Cover Through Quality Planting Materials (QPMs): A Pathway to Achieve Carbon Neutrality and Future Sustainability

Maneesh S. Bhandari and Hukum Singh

Abstract Global enhancement of forest and tree cover is a critical aspect for combating climate change, biodiversity loss, and land degradation and for attaining a carbon–neutral and sustainable future. Achieving carbon neutrality and long-term ecological sustainability hinges critically on enhancing forest and tree cover with high-quality and superior planting material. Quality Planting Materials (QPMs) possesses superior genetic, physiological, and morphological traits that actively enhance survival rates, boost growth performance, and improve resilience under diverse climatic and environmental conditions. This chapter examines the crucial role of QPM in accelerating forest restoration, enhancing biomass productivity, and maximizing carbon sequestration potential. QPM not only ensures higher survival rates and faster growth of plantations but also enhances their resilience to climate-induced stressors, pests, and diseases—factors essential for maintaining ecosystem stability and biodiversity. The chapter delves into the science of QPM production, including clonal propagation, tissue culture, and seed orchard development, while emphasizing the importance of provenance selection, genetic diversity, and site-species matching. Moreover, it highlights policy frameworks, institutional mechanisms, and technological interventions required to scale up QPM adoption across afforestation, reforestation, and agroforestry programs. It also highlights the integration of QPM into national forestry strategies and climate action plans of countries, which may effectively contribute to their Nationally Determined Contributions (NDCs), meet the Sustainable Development Goals (SDGs), and foster a resilient green economy. Eventually, the enhancement of forest and tree cover using QPM emerges as a scientifically sound and scalable pathway to combat climate change, promote ecological restoration, and facilitate sustainable development in the wake of global climate change.

Keywords Quality Planting Material (QPM) · Carbon neutrality · Forest restoration · Climate resilience · Sustainable forestry

M. S. Bhandari (✉) · H. Singh
ICFRE-Forest Research Institute, Dehradun, India
e-mail: maneesh31803@gmail.com; bhandarims@icfre.org

H. Singh (ed.), *Forestry for a Carbon-Neutral and Sustainable Future*, Sustainability Solutions, https://doi.org/10.1007/978-3-032-06427-1_4

4.1 Introduction

Forests are among the most effective natural-based solutions for combating climate change. Forests act as vital carbon sinks, absorbing atmospheric carbon dioxide and storing it in biomass and soil. With increasing global commitments to achieving carbon neutrality, expanding forest and tree cover has become a crucial strategy for enhancing carbon sequestration, restoring degraded lands, and preserving ecosystem services. However, the success of these interventions largely depends on the quality of planting material used in afforestation and reforestation programs.

Quality Planting Material (QPM) is seedlings or propagules that are genetically superior, disease-resistant, and well-adapted to specific ecological conditions. Unlike conventional planting stock, QPM ensures higher survival rates, improved growth performance, and greater biomass accumulation over time. These characteristics make QPM crucial in enhancing the efficiency and outcomes of forest restoration, carbon offsetting projects, and tree-based livelihood systems (Arunachalam et al., 2020). In the context of climate-smart forestry, QPM forms the foundation of productive and resilient forest landscapes.

The QPM-based planning significantly amplifies the carbon sequestration potential of planted forests. Scientifically selected and improved planting material grows faster, accumulates more biomass, and stores more carbon than unimproved planting materials. Furthermore, QPM enhances the uniformity and predictability of plantation performance, allowing better planning and monitoring of climate mitigation projects. Thus, investing in the development of QPM is a strategic step toward achieving nationally determined contributions (NDCs) under the Paris Agreement.

QPM offers social and economic advantages due to the higher survival and productivity of plantations, resulting in increased yields of timber, fuelwood, fruits, and other non-timber forest products, thereby supporting rural livelihoods and reducing pressure on natural forests. Moreover, when integrated into agroforestry and community forestry models, QPM may enhance land productivity, improve food and income security, and contribute to rural development, aligning with the United Nations' Sustainable Development Goals (SDGs).

Despite its importance, the adoption of QPM faces several challenges, including limited awareness, inadequate infrastructure for seed collection and nursery management, and a lack of technical capacity among stakeholders. Addressing these issues requires coordinated efforts to strengthen institutional mechanisms, invest in research and development, and implement robust quality control and certification systems. Promoting equitable access to QPM, particularly among smallholders and local communities, is also crucial to ensure an inclusive and widespread impact (Singh, 2025).

This chapter discusses the role of QPM in expanding forest and tree cover as a climate mitigation strategy. It explores the science behind QPM and its contributions to carbon sequestration and sustainable development. The chapter also outlines policy recommendations and strategic pathways for mainstreaming QPM in national forestry programs, ultimately positioning it as a key enabler of carbon neutrality and long-term ecological sustainability.

4.2 Quality Planting Material (QPM)

Quality Planting Material (QPM) is the seedlings, saplings, or propagules that possess desirable genetic, physiological, and morphological characteristics to ensure better survival, vigorous growth, and higher productivity under field conditions (David & Sperling, 1999). QPM is a cornerstone of successful afforestation, reforestation, and agroforestry programs, particularly in the context of climate change mitigation and long-term forest sustainability. Unlike conventional planting stock, QPM is developed through scientific selection, breeding, and propagation techniques to meet site-specific ecological and climatic requirements.

The key fundamental characteristics of QPM include high genetic purity, uniformity, disease resistance, adaptability to local environmental conditions, and superior growth performance. Such material is derived from high-quality seeds from plus trees, clonal propagation of elite genotypes, or advanced techniques such as tissue culture and micropropagation. These sources ensure that the planting material exhibits enhanced growth traits, such as faster biomass accumulation, improved form, and better wood properties, contributing significantly to its carbon sequestration potential and economic value (Maredia et al., 1999).

The key advantage of QPM is its consistency in performance across similar environmental conditions, which reduces uncertainty in plantation outcomes. High-quality seedlings are better equipped to withstand biotic and abiotic stresses, including pests, diseases, drought, and poor soil conditions. This resilience is significant in the face of climate variability, making QPM an integral part of climate-smart forestry and ecosystem-based adaptation strategies.

The development of QPM is a multidisciplinary effort involving tree improvement programs, genetic resource conservation, and nursery management. It requires collaboration between forestry research institutions, seed certification agencies, and nursery operators to ensure the availability and accessibility of verified planting stock. Maintaining a reliable chain of custody from seed collection to nursery propagation and field deployment is crucial for preserving the genetic quality and traceability of QPM.

In India, Institutes of the Indian Council of Forestry Research and Education (ICFRE) across the agroclimatic zones, State Forest Departments, and national-level initiatives such as the National Agroforestry Policy have recognized the importance of QPM in enhancing forest productivity and ecological restoration. Similarly, global frameworks, such as the United Nations Decade on Ecosystem Restoration (2021–2030), emphasize the use of high-QPM to ensure the long-term success of restoration efforts.

Hence, QPM is not just about planting more trees. It is about growing the right trees in the right place using scientifically vetted, resilient, and productive material.

Ensuring the use of QPM may significantly enhance the effectiveness of reforestation and restoration initiatives, thereby contributing to climate mitigation, biodiversity conservation, and sustainable livelihoods.

4.2.1 Global Status of QPM

QPM plays a crucial role in enhancing productivity, economic growth, and climate resilience. High-quality seeds and vegetative propagules contribute to increased crop yields, ensuring food security and improving farmers' incomes, particularly in developing regions. Improved planting materials reduce dependency on chemical inputs and minimize crop losses due to pests and diseases, leading to cost-effective and sustainable farming practices (Singh, 2024b). Furthermore, in the face of climate change, resilient planting material ensures adaptation to extreme weather conditions, promoting ecological restoration and afforestation efforts. The development and distribution of QPM, such as disease-free root and tuber crops, play a significant role in biodiversity conservation while sustaining livelihoods and strengthening rural economies. However, many farmers continue to rely on seeds or vegetative materials saved from previous harvests or obtained informally from adjacent farms, resulting in lower yields and increased vulnerability to pests and diseases. Improving the genetic and physical quality of planting materials has the potential to enhance crop yields by up to 40%, considerably contributing to food security and overall agricultural production, especially when farmers adopt policies that ensure regular renewal of their planting stock (Rohrbach et al., 2002; Bertin et al., 2012). Some of the studies have been listed in Table 4.1.

4.3 Genetic and Physiological Traits of High-Quality Seedlings

The success of afforestation, reforestation, and agroforestry initiatives greatly depends on the genetic and physiological quality of the seedlings used. High-quality seedlings possess specific traits that enable them to survive, establish quickly, and grow vigorously under a range of environmental conditions (Joshi et al., 2021). These traits are critical for maximizing carbon sequestration, enhancing ecosystem resilience, and ensuring the long-term sustainability of planted forests.

Table 4.1 Global status of quality planting materials (QPMs)

Sl. no	Continent	Country	Species name	Number of quality planting materials produced	Establishment place	Interpretation	References
1	Asia	India	*Dalbergia sissoo*	63	Northern India, Southwestern Nepal	Plus tree selection and progeny testing, identification of the superior genotypes for establishing first-generation seed orchards	Bangarwa (2002)
2	Asia	India	*Eucalyptus* sp.	10 million	Forest nurseries	Agroforestry initiatives drive high demand in India	NCCF
3	Asia	India	*Tectona grandis*	Clonal propagation achieved for 63-year-old mature trees	Nilambur, Kerala	Coppice shoots from mature trees and rootstocks successfully rooted with IBA treatment	Palanisamy and Subramanian (2001)
4	Asia	India	*Casuarina* sp.	8 million	Forest nurseries	Agroforestry initiatives drive high demand in India	NCCF (https://www.nccf.in)
5	Asia	China	*Pinus elliotti*	20	Guangdong Province, China	The selection of 38 superior trees resulted in improved seed orchards with high seed yields	Kellison et al. (1982)
6	Asia	Malaysia	*Acacia mangium*, *Acacia crassicarpa*, *Acacia aulococarpa*, *Acacia auriculiformis*	–	Malaysia	Phenotypic selection and screening of superior genotypes based on vegetative growth performance of selected 12-years-old *Acacia* sp.	Kumar et al. (2016)

(continued)

Table 4.1 (continued)

Sl. no	Continent	Country	Species name	Number of quality planting materials produced	Establishment place	Interpretation	References
7	Asia	China	Hybrid Sweetgum (*Liquidambar styraciflua* × *L. formosana*)	3	Beijing, China	This study reveals that downregulation of growth-promoting genes and signal transduction pathways may explain the slower growth of dwarf tetraploid hybrid sweetgum, with IAA and GA_3 enhancing shoot and root elongation	Chen et al. (2022)
8	Europe	Norway	*Picea abies*	1370	Norway	Genomic selection improves genetic gain and reduces breeding cycle length	Chen et al. (2018)
9	Europe	Finland	*Pinus sylvestris*	160	Tuusula, Padasjoki, Punkaharju, Kinnula, Pyhajarvi, Kuhmo (Finland)	Progeny trials confirm growth performance	Haapanen et al. (2016)
10	Europe	Germany, France, Netherlands, Finland, Italy, Austria	*Fagus sylvatica*	10	Germany, France, Netherlands, Finland, Italy, Austria	Seed orchards improve genetic diversity	Buiteveld et al. (2007)

(continued)

Table 4.1 (continued)

Sl. no	Continent	Country	Species name	Number of quality planting materials produced	Establishment place	Interpretation	References
11	Europe	Lithuania	*Betula pendula*	100	Ezerelis forest district	Progeny tests ensure site adaptability	Baliuckienë and Baliuckas (2006)
12	Europe	Sweden	*Betula pendula*	14	Ezerelis forest district	Progeny tests ensure site adaptability	Baliuckienë and Baliuckas (2006)
13	Europe	Italy, Spain	*Castanea sativa*	319	Italy, Spain	Genetic diversity analysis revealed two distinct groups with gene flow between wild and cultivated trees	Alessandri et al. (2022)
14	Europe	Poland	*Quercus robur*	6600 (monoculture), 3300 (double-row), 2400 (group planting) seedlings per hectare	North-eastern Poland	Group planting reduces browsing damage and improves regeneration success under high herbivory pressure	Bolibok et al. (2021)
15	Africa	South Africa	*Eucalyptus nitens*	–	South Africa	Improved seed orchard bulks demonstrated significant growth performance compared to unimproved controls, indicating successful genetic improvement	Swain et al. (2013)

(continued)

Table 4.1 (continued)

Sl. no	Continent	Country	Species name	Number of quality planting materials produced	Establishment place	Interpretation	References
16	Africa	Niger	*Prosopis africana*	256	Niger	Significant genetic variation in wood density; individual tree heritability is higher for wood density than growth traits; larger trees tended to have denser wood	Montes and Weber (2009)
17	Africa	Nigeria	*Prosopis africana*	–	Benue State	Seeds from the derived savannah zone showed the best performance in terms of seedling height, leaf number, and total biomass, as well as improved germination rates following acid scarification	Uleh and Fagbemi (2017)
18	Africa	Niger	*Prosopis africana*	275	Niger	Provenances from drier locations with smaller seedlings and larger root/shoot ratios in the nursery had better growth and survival in the field	Weber et al. (2015)
19	South America	Brazil	*Acacia mearnsii*	32	Brazil	Early selection based on plant height after ~60 days in the nursery and 2 years in the field resulted in an annual genetic gain of ~3% for height and 2% for diameter at breast height	Bisognin et al. (2022)

(continued)

Table 4.1 (continued)

Sl. no	Continent	Country	Species name	Number of quality planting materials produced	Establishment place	Interpretation	References
20	South America	Peru	*Calycophyllum spruceanum*	–	Peruvian Amazon	Phenotypic selection in natural populations showed non-significant differences in progeny growth between selected and non-selected mother trees, suggesting that phenotypic selection alone is limited in its effectiveness	Cornelius et al. (2011)
21	South America	Brazil	*Corymbia maculata*	38	Pedemeiras, SP	Significant genetic variation was observed among and within provenances and progenies, indicating potential for genetic improvement through selective breeding	Sato et al. (2010)
22	South America	Brazil	*Eucalyptus* spp.	–	Brazil	The use of inbred progenies in breeding programs is proposed to release greater genetic variability and expose harmful alleles, facilitating their selection and improving genetic gains	Ramalho et al. (2022)

(continued)

Table 4.1 (continued)

Sl. no	Continent	Country	Species name	Number of quality planting materials produced	Establishment place	Interpretation	References
23	South America	Uruguay	*Pinus taeda*	124	Uruguay	Progeny selection in a clonal seed orchard resulted in improved genetic diversity and potential genetic gains, thereby contributing to the sustainability and productivity of future plantations	Torres-Dini et al. (2024)
24	South America	Brazil	*Eucalyptus grandis* and *Eucalyptus urophylla*	196	Brazil	The origin of the individual in a progeny test had a small effect on the performance of the clone when evaluated in other environments; estimates of realized heritability and the linear regression coefficient were small in all selection strategies	Furtini et al. (2012)
25	Australia	Australia	*Eucalyptus dunnii*	57	Victoria	Progeny trials were established with 57 open-pollinated families, and significant data were collected on growth traits and wood density	Hardner et al. (2016)

(continued)

Table 4.1 (continued)

Sl. no	Continent	Country	Species name	Number of quality planting materials produced	Establishment place	Interpretation	References
26	Australia	Australia	*Eucalyptus nitens*	691	South Island, New Zealand	A third-generation progeny test involving 72 families with 124 individuals per family was used for genomic prediction analysis, for which genomic information and phenotypic trait records were available	Suontama et al. (2019)
27	Australia	Australia	*Eucalyptus globulus*	514	Latrobe in northern Tasmania	Progeny trials were established to assess genetic parameters and G × E interactions, revealing significant genetic variation in growth and wood properties	Stackpole et al. (2010)
28	Australia	Australia	*Corymbia citriodora*	*Corymbia* hybrid families: Female parent -*C. torelliana* Male parent -*C. citriodora* sussp. *Variegata* 162, *C. citriodora* subsp. *Citriodora* 36, *C. henryi* 30, *C. maculata* 15, *Corymbia* hybrid 29	Queensland, Australia	Progeny trials were conducted to assess genetic parameters for growth traits, and the findings were used to inform selection and breeding strategies	Lee (2007)

(continued)

Table 4.1 (continued)

Sl. no	Continent	Country	Species name	Number of quality planting materials produced	Establishment place	Interpretation	References
29	North America	United States	Eastern cottonwood (*Populus deltoides*), loblolly pine (*Pinus taeda*), sweetgum (*Liquidambar styraciflua*), and sycamore (*Platanus occidentalis*)	4 genotype, 3226 trees across 14 sites	US forest	Hybrid sweetgum exhibits strong potential for biofuel cultivation due to its rapid growth, pest resistance, and adaptability; however, further research is needed on its productivity under changing environmental conditions	Ozment (2023a, 2023b)

4.3.1 Genetic Traits

- **Provenance and Genetic Purity**: The seedlings originated from genetically superior and well-adapted seed sources (provenances) that match local site conditions. This ensures adaptability, better growth rates, and stress resistance.
- **Heritability of Growth Traits**: Traits such as height, diameter growth, biomass accumulation, and wood quality should exhibit high heritability, enabling reliable selection and consistent field performance.
- **Pest and Disease Resistance**: Genetically improved seedlings often exhibit enhanced resistance or tolerance to common pests and pathogens, reducing mortality and the need for chemical interventions.
- **Climate Adaptability**: Selection for traits that confer tolerance to abiotic stresses, such as drought, heat, and frost, ensures survival under changing climatic conditions.

4.3.2 Physiological Traits

- **Vigorous Root Development**: Well-developed and fibrous root systems enhance water and nutrient uptake, anchorage, and resistance to transplant shock.
- **High Photosynthetic Efficiency**: Seedlings with robust photosynthetic activity tend to grow faster and accumulate biomass more rapidly, supporting higher carbon capture.
- **Optimal Water Use Efficiency (WUE)**: Seedlings with high WUE maintain growth under water-limited conditions, making them suitable for semi-arid and degraded landscapes.
- **Nutrient Uptake Efficiency**: The efficient absorption and utilization of essential nutrients (N, P, K, and micronutrients) support strong early growth and health.
- **Stress Tolerance Mechanisms**: Traits such as osmotic adjustment, antioxidant enzyme activity (e.g., Superoxide Dismutase (SOD), Catalase (CAT), Peroxidase (POD)), and stomatal regulation enable seedlings to withstand drought, salinity, and temperature extremes (Singh, 2024d).

4.4 Sources of Quality Planting Materials (QPMs)

The effectiveness of QPM depends on its origin and the methods used for its propagation. Ensuring genetic superiority, physiological vigor, and adaptability begins with selecting appropriate sources. Sourcing QPM from scientifically managed and certified systems for forest and tree planting initiatives significantly improves survival rates, growth performance, and resilience. Thus, ensuring long-term ecological and economic benefits that align with sustainability and climate goals. The following are the key sources of QPM.

4.4.1 Seed Orchards

Seed orchards are established with genetically superior trees selected for desirable traits such as fast growth, high timber yield, pest resistance, and climate adaptability.

- Clonal and seedling seed orchards produce high-quality, tested seeds under controlled pollination conditions.
- These orchards ensure genetic purity, provenance identity, and trait stability, making them a reliable source for large-scale afforestation and reforestation programs.

4.4.2 Clonal Propagation

Clonal propagation involves the vegetative reproduction of elite trees (e.g., stem cuttings, grafting, layering) to produce genetically identical plants.

- It ensures uniformity in growth and performance, making it ideal for plantations that require consistent output.
- Clonal techniques are widely used in fast-growing species like *Eucalyptus*, *Populus*, and *Casuarina*.
- Clones selected for drought tolerance, biomass yield, or disease resistance are especially valuable in stressed or degraded ecosystems.

4.4.3 Tissue Culture (Micropropagation)

Tissue culture enables the mass production of disease-free and genetically uniform planting material from a small amount of parent tissue.

- It is beneficial for rare, slow-growing, **or** medicinally valuable species where seeds availability are limited.
- Micropropagation also supports the conservation of threatened species and the rapid dissemination of improved genotypes.

4.4.4 Certified Nurseries

Certified nurseries adhere to established standards for seed source, propagation technique, plant health, and nursery hygiene.

- The government or research institutions often accredit these nurseries to produce QPM that is traceable, healthy, and site-appropriate.

- Certified nurseries play a vital role in bridging the gap between research outputs (e.g., improved germplasm) and end-users like farmers, forest departments, and communities.

4.5 Significance of QPM in Enhancing Forest and Tree Cover

Notably, QPM plays a pivotal role in enhancing forest and tree cover by improving the success rate, productivity, and ecological impact of afforestation, reforestation, and restoration efforts. As countries strive to meet ambitious climate and biodiversity targets, including carbon neutrality, land degradation neutrality, and increased forest cover, the use of QPM ensures that tree-planting efforts are not only quantitatively extensive but also qualitatively effective.

One of the most significant contribution of QPM is its ability to improve seedling survival rates, especially in marginal and degraded lands. High-quality seedlings, derived from genetically superior and well-adapted sources, are more vigorous and resilient to biotic (pests and diseases) and abiotic (drought, poor soil, temperature extremes) stresses (Kumar et al., 2020). This leads to better establishment and growth of plantations, thereby reducing the need for replanting and improving long-term forest stability.

QPM also contributes to accelerated biomass accumulation and enhanced carbon sequestration. Trees grown from QPM tend to have better form, faster growth, and higher wood volume, all of which translate into increased carbon capture over time. This is especially important in the context of climate change mitigation, where the capacity of forests to act as carbon sinks is directly tied to the quality and health of the trees.

Beyond ecological benefits, QPM enhances the economic value of plantations. High-yielding, disease-resistant planting stock yields increased production of timber, fuelwood, fruits, and other non-timber forest products (NTFPs), thereby enhancing the economic returns of forestry and agroforestry systems. This not only boosts income for local communities and farmers but also reduces pressure on natural forests by making planted forests more productive and sustainable (Singh et al., 2018).

Besides, QPM plays a crucial role in promoting species diversity and ecological restoration. The selection of native and site-specific species with high survival rates and ecological compatibility enables QPM to contribute to the restoration of degraded ecosystems, the revival of biodiversity, and the re-establishment of critical ecological functions, such as soil conservation, water regulation, and pollinator support (Joshi et al., 2020).

The significance of QPM in enhancing forest and tree cover extends beyond the physical act of planting. It ensures that forest development is grounded in science, sustainability, and resilience, which are the key elements for achieving long-term environmental goals. The integration of QPM into national forestry strategies and restoration programs is thus essential for ensuring that every planted tree makes a meaningful contribution to ecological health, climate action, and human well-being.

4.6 QPM and Carbon Sequestration: Scientific Perspectives

The role of QPM in carbon sequestration is based on its ability to enhance the growth, biomass accumulation, and physiological performance of tree species. From a scientific perspective, the use of genetically improved and site-adapted planting material directly influences the rate at which trees absorb atmospheric carbon dioxide and convert it into biomass through photosynthesis. This makes QPM an essential component in maximizing the carbon sink potential of afforestation and reforestation programs (Singh, 2024c).

Carbon sequestration in trees primarily occurs through the fixation of atmospheric CO_2 in plant tissues, including stems, branches, leaves, and roots. QPM accelerates this process by promoting traits such as faster growth rates, higher wood density, improved canopy development, and better root architecture. These attributes contribute to higher Net Primary Productivity (NPP), which reflects the amount of carbon captured and retained by the plant system after accounting for respiration losses (Jangra et al., 2024).

Scientific studies have shown that trees derived from QPM may sequester 20–40% more carbon over their lifetime compared to those grown from unimproved or low-quality material. For instance, clonal plantations of fast-growing species, such as *Eucalyptus* sp., *Tectona grandis* (Teak), and *Populus* sp., have demonstrated significantly higher carbon storage potential under controlled trials and field conditions. These improvements are attributed to the enhanced photosynthetic efficiency, increased nutrient uptake, and improved stress tolerance resulting from the use of high-quality planting stock.

Moreover, QPM contributes to long-term soil carbon sequestration by promoting the development of more profound and more extensive root systems, which deposit organic matter into the soil. This below-ground carbon pool is significant for maintaining ecosystem stability and regulating the long-term climate. Enhanced litterfall and reduced mortality in QPM plantations further contribute to the accumulation of organic carbon in the forest floor and soil.

Advanced modeling tools and carbon accounting frameworks, such as the IPCC Guidelines, COMAP, and CO_2FIX, support the quantification of carbon sequestration in plantations established using the QPM. These tools confirm that plantations established with high-QPM offer greater predictability and reliability in long-term carbon storage, making them more suitable for carbon credit and payment for ecosystem services (PES) schemes under REDD+ and voluntary carbon markets.

In essence, QPM not only ensures ecological resilience and productivity but also strengthens the scientific and economic case for tree planting as a climate mitigation strategy. Its integration into reforestation programs, carbon offset initiatives, and landscape restoration efforts may significantly improve the effectiveness of nature-based solutions in achieving national and global climate targets.

4.7 QPM Impact on Degraded Land Restoration and Livelihood Improvement

Degraded lands characterized by poor soil fertility, reduced vegetation cover, and low productivity pose significant challenges to ecological sustainability and rural livelihoods. Restoring these lands is a critical priority for countries seeking to enhance climate resilience, biodiversity, and land productivity. In this context, QPM emerges as a transformative approach, offering ecological and socioeconomic benefits through improved plantation outcomes and community empowerment.

The advantages of QPM in degraded land restoration are its ability to thrive in harsh and stress-prone environments. High-QPM are selected for traits such as drought tolerance, disease resistance, and adaptability to degraded soils. This ensures better establishment, survival, and growth on lands where conventional seedlings often fail. The use of QPM leads to faster canopy closure, improved soil stabilization, enhanced microclimatic conditions, and a reduction in erosion, which are key indicators of successful restoration.

Moreover, plantations established with QPM contribute to soil fertility recovery by promoting litterfall, root biomass accumulation, and microbial activity. Species with nitrogen-fixing abilities or deep root systems should be prioritized in QPM-based restoration, improving nutrient cycling and water retention. This ecological recovery creates a positive feedback loop that supports the re-establishment of native flora and fauna, ultimately leading to the regeneration of ecosystem functions and services (Joshi et al., 2024).

From a livelihood perspective, QPM-based restoration projects have proven to generate higher yields of timber, fuelwood, fruits, fodder, and non-timber forest products (NTFPs). These resources directly contribute to household income, food security, and energy needs, particularly for rural and marginalized communities. High-yield plantations provide greater returns on investment and reduce the dependency on natural forests, promoting sustainable resource use and forest conservation.

In addition, QPM creates new opportunities in employment and enterprise development. Nursery development, seed collection, plantation management, and value-added processing of forest products all become viable livelihood activities. Community-based restoration projects using QPM have also been integrated into carbon credit programs, such as REDD+ and voluntary carbon markets, allowing local stakeholders to benefit financially from their conservation efforts.

It has been reported that QPM-driven reforestation on degraded lands significantly improves ecological indicators while enhancing rural incomes. For example, in India, the use of clonal *Eucalyptus* sp. and Teak on degraded farm and forest lands has led to rapid biomass accumulation and commercial returns. Similarly, community forestry programs in Nepal, utilizing improved planting stock, have enhanced forest health and household incomes (Ruhil et al., 2024).

QPM serves as a dual-purpose intervention, restoring ecological integrity while uplifting local communities. Investing in high-quality, site-appropriate planting material and linking it with inclusive restoration programs helps governments and organizations address land degradation, enhance ecosystem services, and improve the well-being of vulnerable populations in a sustainable and climate-resilient manner.

4.8 Institutional and Policy Frameworks

The successful development, distribution, and adoption of QPM requires a robust institutional and policy framework that supports quality assurance, research and development, capacity building, and stakeholder collaboration. Without such a framework, the benefits of QPM, including enhanced forest productivity, carbon sequestration, and support for livelihoods, remain underutilized. Therefore, strengthening the governance mechanisms that regulate and promote QPM is essential for achieving national reforestation targets and climate goals (Kumar & Singh, 2020).

At the institutional level, forestry research institutes, seed certification agencies, state forest departments, and agricultural universities play a crucial role in selecting, genetically improving, testing, and propagating high-quality planting material. Institutions such as the Indian Council of Forestry Research and Education (ICFRE), the Forest Research Institute (FRI), and its sister institutes play a crucial role in this endeavor. Establishing coordinated networks between research institutions and nurseries ensures that scientifically developed QPM reaches the field in a timely and cost-effective manner (Singh, 2024a).

The implementation of seed certification systems and accreditation protocols is crucial in maintaining the genetic quality, traceability, and authenticity of planting stock. National guidelines for seed standards, such as those developed under India's National Forest Policy and Agroforestry Policy, mandate the certification of seed sources and the registration of nurseries. These frameworks help eliminate the use of substandard material and promote trust among users, including farmers, forest managers, and private plantation developers.

Policy support for QPM is increasingly embedded in national and sub-national strategies related to climate change mitigation, land restoration, and rural development. Initiatives like the National Afforestation Programme (NAP) (https://www.nitiforstates.gov.in), Green India Mission (GIM) (https://moef.gov.in), National Bamboo Mission (NBM) (https://nbm.da.gov.in), and the National Agroforestry Policy (2014) (https://www.nitiforstates.gov.in) emphasize the use of QPM for large-scale planting interventions. Internationally, frameworks such as the UN Decade on Ecosystem Restoration (2021–2030) and REDD+ promote the use of high-quality, genetically diverse, and site-appropriate planting material to ensure long-term restoration success and climate benefits.

In addition to public institutions, public–private partnerships (PPPs) and community participation are crucial in scaling the adoption of QPM. Private-sector nurseries, seed companies, and farmer cooperatives can complement government efforts by expanding QPM production and ensuring last-mile delivery. Engaging communities, particularly through Joint Forest Management (JFM) and Community Forest User Groups (CFUGs), can foster local stewardship, promote indigenous species, and ensure that planting materials reflect local needs and preferences.

Capacity building and extension services are also necessary to disseminate best practices on QPM use, nursery management, and field establishment. Training programs for nursery workers, forest officials, and community members are crucial for enhancing awareness, technical skills, and compliance with certification standards. Establishing demonstration plots, knowledge-sharing platforms, and digital tools can further promote transparency and uptake of these practices.

Therefore, a supportive institutional and policy framework is foundational for the success of QPM-based forestry and restoration programs. Integrating QPM into national climate, forest, and development strategies and ensuring its scientific, economic, and social backing can significantly improve the quality and impact of countries' forest and tree cover enhancement efforts.

4.9 Strategies for Scaling up QPM Adoption

To realize the potential of QPM in enhancing forest and tree cover, restoring degraded landscapes, and achieving climate and livelihood goals, strategic interventions are required at multiple levels. Scaling up QPM adoption requires an integrated approach that combines science, policy, finance, and community engagement to achieve effective implementation. Embedding QPM in national forestry strategies and climate action plans may ensure that tree planting translates into meaningful environmental, economic, and social outcomes. The following key strategies support the large-scale adoption and mainstreaming of QPM.

4.9.1 Strengthening Research and Development (R&D)

Investment in tree improvement programs is crucial for developing genetically superior, site-specific, and climate-resilient planting materials. Research institutions should focus on both native and commercially important species, prioritizing traits such as rapid growth, pest resistance, drought tolerance, and wood quality. Collaboration among the forestry, agriculture, and biotechnology sectors can accelerate innovation in this field.

4.9.2 Expanding Certified Nurseries and Seed Production Zones

Establishing and upgrading regional and community-based nurseries with modern facilities and trained personnel can ensure a consistent supply of certified QPM. Creating seed production zones (SPZs) and seed orchards, as well as promoting clonal propagation techniques, will further enhance the genetic quality and availability of seeds.

4.9.3 Developing Robust Certification and Quality Control Systems

Implementing a national certification and accreditation framework for QPM, encompassing seed source registration, nursery standards, and traceability, is crucial. This system should be underpinned by periodic audits and capacity-building initiatives for seed inspectors and nursery managers to ensure compliance and transparency.

4.9.4 Ensuring Policy and Financial Support

Policies should mandate the use of QPM in government-supported afforestation, agroforestry, and restoration programs. Financial incentives, such as subsidies, credit schemes, and payment for ecosystem services (PES), should be extended to smallholder farmers and communities adopting QPM. Integrating QPM with schemes like the Green India Mission, Compensatory Afforestation, and REDD+ may amplify this impact.

4.9.5 Building Capacity and Awareness

Large-scale training and extension programs are needed to educate stakeholders, including nursery operators, farmers, forest officials, and community groups, on the benefits, handling, and field management of QPM. Demonstration plots, farmer field schools, mobile apps, and knowledge platforms can facilitate knowledge exchange and adoption.

4.9.6 Promoting Public–Private and Community Partnerships

Partnerships with private seed companies, NGOs, and farmer-producer organizations (FPOs) may expand QPM production and outreach. Encouraging community participation through Joint Forest Management (JFM), Van Panchayats, and self-help groups localizes production, improves ownership, and aligns species selection with local needs.

4.9.7 Monitoring, Evaluation, and Impact Assessment

Establishing monitoring systems to track survival rates, growth performance, and carbon sequestration benefits of QPM plantations helps refine practices and build evidence for scaling. Impact data can support policy advocacy and attract climate finance and donor support.

4.10 Integrating QPM into Climate and Sustainability Goals

The integration of QPM into climate and sustainability agendas is critical for realizing national and global commitments toward carbon neutrality, ecosystem restoration, and inclusive development. Integrating QPM into climate and sustainability strategies strengthens the ecological, economic, and social outcomes of afforestation and restoration programs. As nations strive for a carbon–neutral and resilient future, QPM emerges as a foundational tool for aligning local actions with global environmental and development commitments. QPM not only enhances the success of reforestation and afforestation efforts but also contributes directly to achieving key SDGs and targets under the Paris Agreement and the UN Decade on Ecosystem Restoration (2021–2030).

4.10.1 Contribution to Climate Mitigation and Carbon Neutrality

QPM plays a pivotal role in enhancing carbon sequestration through the cultivation of faster-growing, resilient, and high-biomass trees. The establishment of carbon-rich forest systems with genetically superior planting material enhances the capacity of terrestrial ecosystems to capture and store atmospheric CO_2, a fundamental requirement for achieving long-term climate neutrality goals under Nationally Determined Contributions (NDCs).

4.10.2 Alignment with Sustainable Development Goals (SDGs)

- QPM-driven forest and agroforestry initiatives contribute directly to multiple SDGs:
- (a) SDG 13 (Climate Action): through climate-resilient ecosystems and carbon sinks.
- (b) SDG 15 (Life on Land): via biodiversity conservation, land restoration, and ecosystem services.
- (c) SDG 1 (No Poverty) & SDG 2 (Zero Hunger): by enhancing livelihoods and food security from diversified forest products.
- (d) SDG 8 (Decent Work and Economic Growth): through employment generation in nurseries, plantations, and value chains.

4.10.3 Integration into National and Sub-national Climate Policies

Many countries, including India, have integrated QPM into national programs such as the Green India Mission, the National Afforestation Programme, and the REDD+ frameworks. Embedding QPM into forestry and agroforestry schemes ensures that restoration efforts are more effective, measurable, and scalable. Aligning QPM strategies with climate finance mechanisms, such as the Green Climate Fund (GCF) and voluntary carbon markets, may also leverage additional resources.

4.10.4 Support for Nature-Based Solutions (NbS)

QPM enhances the effectiveness of nature-based solutions by increasing the survival, adaptability, and productivity of planted trees, particularly under changing climatic conditions. QPM is a key pillar of NbS, contributing to biodiversity enhancement, water regulation, and disaster risk reduction.

4.10.5 *Biodiversity and Landscape Resilience*

Integrating QPM with site-specific and native species selections helps maintain ecological balance, supports habitat restoration, and enhances landscape resilience against climate-induced stresses, including drought, pests, and wildfires. QPM may serve as a bridge between ecological integrity and economic utility when designed with sustainability and diversity in mind.

4.10.6 *Monitoring and Reporting on Global Environmental Commitments*

High-quality planting material improves the accuracy and effectiveness of monitoring frameworks related to forest cover expansion, biomass accumulation, and carbon accounting. This supports countries in meeting their international reporting obligations under the UNFCCC, CBD, and Land Degradation Neutrality (LDN) frameworks.

4.11 QPM in Climate-Smart Forestry and Agroforestry Models

The integration of QPM into climate-smart forestry and agroforestry models is a critical pathway to enhance resilience, productivity, and sustainability in the face of climate change. Climate-smart approaches emphasize maximizing carbon sequestration, adapting to climate variability, and improving the livelihoods of rural and forest-dependent communities. QPM serves as a foundational input in these systems by ensuring that planted trees are genetically superior, physiologically robust, and adapted to local environmental conditions.

In climate-smart forestry, the use of QPM significantly boosts carbon sequestration potential by enabling faster growth, greater biomass accumulation, and improved survival of planted trees. Genetically improved seedlings possess traits such as drought tolerance, pest and disease resistance, and better adaptability to temperature fluctuations, which are essential for withstanding climate-induced stresses. Furthermore, QPM enhances the success rate of afforestation and reforestation efforts, particularly in degraded and marginal lands, by ensuring uniformity and vigor in the early establishment phase. This reduces the risk of plantation failure and enhances the ecological stability of forest ecosystems.

Agroforestry systems also benefit substantially from the use of QPM. High-quality seedlings of timber, fruit, fodder, and fuelwood species contribute to higher yields, shorter rotation periods, and better-quality produce. These outcomes translate into increased and diversified income for smallholders, thus enhancing rural livelihoods

(Barwa & Singh, 2024). In addition, trees with strong root systems derived from high-quality genetic stock improve soil structure, promote water retention, and increase organic matter, thereby contributing to land restoration and sustainable soil management. Such agroforestry models act as buffers against climate variability, offering shade, windbreaks, and microclimate regulation, which are vital for maintaining agricultural productivity in changing climates.

From a policy and programmatic standpoint, QPM is integral to national and subnational climate-resilient land-use strategies. Its importance is recognized in initiatives such as India's National Agroforestry Policy, the Green India Mission, and REDD+ programs, where high-quality planting stock is essential for achieving measurable outcomes in terms of carbon sequestration, biodiversity conservation, and socioeconomic upliftment. Furthermore, plantations established using QPM are better suited for monitoring, reporting, and verification (MRV) processes, making them eligible for support from climate finance mechanisms such as the Green Climate Fund and voluntary carbon markets (Garg et al., 2016).

Embedding QPM into climate-smart forestry and agroforestry enables countries to align local land-use practices with global climate and sustainability goals. These models offer a triple-win solution, mitigating greenhouse gas emissions, enhancing resilience to climate shocks, and improving livelihoods, making QPM a cornerstone of practical, scalable, and inclusive climate action. Ultimately, fostering the widespread adoption of QPM in tree-based systems will significantly contribute to building climate-resilient landscapes and achieving a carbon–neutral, sustainable future.

4.12 Challenges and Limitations

QPM can significantly enhance forest cover, carbon sequestration, and ecological restoration; however, several challenges hinder its large-scale adoption and effectiveness. Further, the limited availability and accessibility of QPM remain significant constraints. Many regions, especially remote or under-resourced areas, lack adequate infrastructure for producing, certifying, and distributing high-QPM. Most nurseries still rely on untested or locally sourced seeds, which results in poor germination, inconsistent growth, and increased vulnerability to pests and diseases (Singh & Kumar, 2022).

Inadequate seed source selection and certification systems pose a critical hurdle. In many countries, there is an absence of a well-established seed certification and accreditation framework. Without verified provenance and quality standards, there is a risk of using genetically inferior or non-site-specific material that may underperform or disrupt local ecosystems. This gap also limits the potential for scaling up restoration and plantation programs with predictable outcomes.

Technical and knowledge barriers affect both producers and end-users. Smallholder farmers, local communities, and even field-level forestry staff often lack the necessary training to identify, handle, and manage QPM effectively. Poor nursery practices, a lack of knowledge about species-site matching, and inadequate post-planting care can significantly reduce the benefits of QPM, resulting in high mortality rates and suboptimal growth.

Institutional fragmentation and weak coordination among research bodies, government agencies, and private nurseries hinder the development and dissemination of QPM. Tree improvement programs are often underfunded, and research findings may not reach field-level practitioners in a usable format. Furthermore, overlapping mandates between forest and agricultural departments can delay policy implementation and resource allocation for QPM initiatives.

Socio-economic limitations, including cost and equity issues, also restrict adoption. Producing or purchasing QPM is often more expensive than using locally available seed material. For resource-poor farmers and community groups, this cost can be prohibitive without external support. In addition, the benefits of using QPM, such as higher yields or carbon income, are long-term, whereas the costs are immediate, creating a disincentive for widespread adoption.

Ecological risks and monoculture tendencies must also be taken into consideration. Overemphasis on a few fast-growing or commercially valuable species in QPM programs can lead to monocultures, which reduce biodiversity, alter soil properties, and increase susceptibility to pests and diseases. Lack of attention to native and diverse species undermines the ecological objectives of reforestation and restoration.

Hence, addressing these challenges requires a multi-pronged approach that strengthens institutional frameworks, builds technical capacity, promotes equitable access, and ensures ecological safeguards. Overcoming these limitations is critical to unlocking the full potential of QPM as a tool for sustainable forestry, climate action, and rural development (Singh et al., 2024).

4.13 Conclusions and Way Forward

Enhancing forest and tree cover through the deployment of QPMs represents a strategic intervention for achieving carbon neutrality, restoring degraded ecosystems, and advancing SDGs. The use of genetically superior and physiologically robust seedlings ensures better survival rates, faster growth, and increased biomass accumulation, thereby strengthening carbon sinks and improving ecological resilience. Moreover, QPM significantly contributes to the productivity and profitability of forestry and agroforestry systems, creating livelihood opportunities for rural communities while also supporting climate mitigation and adaptation efforts. The evidence underscores that QPM is not merely a technical input but a transformative enabler of climate-smart land-use practices. From seed orchards and clonal propagation to tissue culture and certified nurseries, the availability and accessibility of high-QPMs must be scaled up to meet the growing demand for restoration, afforestation, and agroforestry initiatives. However, realizing the full potential of QPM requires a multi-pronged approach that includes strengthening institutional and policy frameworks, building nursery infrastructure, ensuring quality assurance through certification systems, and promoting partnerships among public, private, and community sectors. Additionally, integrating QPM into national climate action plans, sustainable forestry strategies, and rural development programs is essential. Targeted investments in research, capacity building, and decentralized production systems are critical to ensure that quality planting material reaches smallholders and community forestry groups across diverse landscapes. Furthermore, aligning QPM-based initiatives with carbon financing mechanisms and ecosystem service markets can enhance their financial sustainability and scale. Therefore, prioritizing QPM in forestry and agroforestry is a pragmatic and high-impact strategy for countries aiming to combat climate change, restore ecosystems, and improve livelihoods. Adopting a holistic and inclusive approach that bridges science, policy, and practice may pave the way for a more resilient, equitable, and carbon–neutral future.

References

Alessandri, S., Cabrer, A. M. R., Martìn, M. A., Mattioni, C., Pereira-Lorenzo, S., & Dondini, L. (2022). Genetic characterization of Italian and Spanish wild and domesticated chestnut trees. *Scientia Horticulturae, 295*, Article 110882.

Arunachalam, A., Handa, A. K., Dev, I., Kaushal, R., Panwar, P., & Dhyani, S. K. (2020). Alternate land use and agroforestry systems for resource conservation and enhanced productivity in the hills. *Indian J Hill Farming, 33*(1), 176–193.

Baliuckienė, A., & Baliuckas, V. (2006). Genetic variability of silver birch (*Betula pendula* L.) wood hardness in progeny testing at juvenile age. *Baltic Forestry, 12*(2), 134–140.

Bangarwa, K. S. (2002). Plus tree selection and progeny testing for establishment of first generation seed orchard in *Dalbergia sissoo* Roxb. *Indian Journal of Agroforestry, 4*(2). https://epubs.icar.org.in/index.php/IJA/article/view/102482

Barwa, S., & Singh, H. (2024). Climate-induced changes in forest ecosystems of Indian Himalayan region: A meta-analysis. In *Urban forests, climate change and environmental pollution: Physio-biochemical and molecular perspectives to enhance urban resilience* (pp. 839–866). Springer.

Bertin, T., Ann, D., Zacharie, T., Ebenezar, A., & Alain, T. (2012). Enhancing farmers access to quality planting materials through community-based seed and seedling systems: Experiences from the Western Highlands of Cameroon. *Middle East Journal of Science Research, 12*(4), 455–463. https://doi.org/10.5829/idosi.mejsr.2012.12.4.1625

Bisognin, D. A., Lencina, K. H., Greff, H. P., Tonetto, T., & Gazzana, D. (2022). Progeny evaluation and early selection for plant height in *Acacia mearnsii* improve genetic gains. *Crop Breeding and Applied Biotechnology, 22*(4), Article e432322413. https://doi.org/10.1590/198470332022v22n4a48

Bolibok, L., Andrzejczyk, T., Szeligowski, H., & Liziniewicz, M. (2021). New methods of oak planting require modification of tending prescriptions under high browsing pressure—A case study from North-Eastern Poland. *Forest Ecology and Management, 497*, Article 119449. https://doi.org/10.1016/j.foreco.2021.119449

Buiteveld, J., Vendramin, G. G., Leonardi, S., Kamer, K., & Geburek, T. (2007). Genetic diversity and differentiation in European beech (Fagus sylvatica L.) stands varying in management history. *Forest Ecology and Management, 247*(1–3), 98–106. https://doi.org/10.1016/j.foreco.2007.04.018

Chen, S., Zhang, Y., Zhang, T., Zhan, D., Pang, Z., Zhao, J., & Zhang, J. (2022). Comparative transcriptomic, anatomical and phytohormone analyses provide new insights into hormone-mediated tetraploid dwarfing in hybrid sweetgum (*Liquidambar styraciflua* × *L. formosana*). *Frontiers in Plant Science, 13*, 924044. https://doi.org/10.3389/fpls.2022.924044

Chen, Z. Q., Baison, J., Pan, J., Karlsson, B., Andersson, B., Westin, J., & Wu, H. X. (2018). Accuracy of genomic selection for growth and wood quality traits in two control-pollinated progeny trials using exome capture as the genotyping platform in Norway spruce. *BMC Genomics, 19*, 1–16. https://doi.org/10.1186/s12864-018-5256-y

Cornelius, J. P., Sotelo Montes, C., Ugarte-Guerra, L. J., & Weber, J. C. (2011). The effectiveness of phenotypic selection in natural populations: A case study from the Peruvian Amazon. *Silvae Genetica, 60*(5), 205. https://doi.org/10.1515/sg-2011-0027

David, S., & Sperling, L. (1999). Improving technology delivery mechanisms: Lessons from bean seed systems research in Eastern and Central Africa. *Agriculture and Human Values, 16*(4), 381–388.

Furtini, I. V., Ramalho, M. A. P., Abad, J. I. M., & Aguiar, A. M. (2012). Effect of different progeny test strategies in the performance of eucalypt clones. *Silvae Genetica, 61*(3), 116. https://doi.org/10.1515/sg-2012-0014

Garg, A., Pal, D., Singh, H., & Pandey, D. C. (2016). A comparative study of NDBI, NDISI and NDII for extraction of urban impervious surface of Dehradun [Uttarakhand, India] using Landsat 8 imagery. *International Conference on Emerging Trends in Communication Technologies (ETCT), 2016*, 1–5.

Haapanen, M., Hynynen, J., Ruotsalainen, S., Siipilehto, J., & Kilpeläinen, M. L. (2016). Realised and projected gains in growth, quality and simulated yield of genetically improved Scots pine in southern Finland. *European Journal of Forest Research, 135*, 997–1009. https://doi.org/10.1007/s10342-016-0989-0

Hardner, C. M., Healey, A. L., Downes, G., Herberling, M., & Gore, P. L. (2016). Improving prediction accuracy and selection of open-pollinated seed-lots in *Eucalyptus dunnii* Maiden using a multivariate mixed model approach. *Annals of Forest Science, 73*, 1035–1046. https://doi.org/10.1007/s13595-016-0587-9

Jangra, A., Kumar, K., Maikhuri, S., Bhandari, M. S., Pandey, S., Singh, H., & Barthwal, S. (2024). Unveiling stress-adapted endophytic bacteria: Characterizing plant growth-promoting traits and assessing cross-inoculation effects on *Populus deltoides* under abiotic stress. *Plant Physiology and Biochemistry, 210*, Article 108610.

Joshi, R., Sharma, B., Singh, H., Dhakal, N., Ayer, S., & Maraseni, T. (2024). Poplar plantation as an agroforestry approach: Economic benefits and its role in carbon sequestration in North India. *Journal of Resources and Ecology, 15*(4), 880–888.

Joshi, R., Singh, H., Chhetri, R., Poudel, S. R., & Rijal, S. (2021). Carbon sequestration potential of community forests: A comparative analysis of soil organic carbon stock in community managed forests of Far-Western Nepal. *Eurasian Journal of Soil Science, 10*(2), 96–104.

Joshi, R., Singh, H., Chhetri, R., & Yadav, K. (2020). Assessment of carbon sequestration potential in degraded and non-degraded community forests in Terai region of Nepal. *Journal of Forest and Environmental Science, 36*(2), 113–121.

Kellison, R. C., Dinus, R. J., Fins, L., Ching, K. K., Krugman, S. L., & Winieski, J. A. (1982). Forest tree improvement in the People's Republic of China. *Journal of Forestry, 80*(10), a0002-641.

Kumar, M., & Singh, H. (2020). Agroforestry as a nature-based solution for reducing community dependence on forests to safeguard forests in rainfed areas of India. In *Nature-based solutions for resilient ecosystems and societies* (pp. 289–306). Springer.

Kumar, M. S., Parameswari, N., Chin, C. F., Baharum, Z., Olalekan, K. K., & Aini, A. N. (2016). Selection and screening of superior genotypes for quality planting stock based on vegetative growth performance of some selected 12-year-old Acacia species. *Open Journal of Forestry, 6*(4), 217–229. https://doi.org/10.4236/ojf.2016.64018

Kumar, A., Dwivedi, G. K., Tewari, S., Sah, V. K., Singh, H., Kumar, P., Kumar, N., & Kaushal, R. (2020). Soil organic carbon pools under *Terminalia chebula* Retz. Based agroforestry system in Himalayan foothills, India. *Current Science, 118*(7), 1098–1103.

Lee, D. J. (2007). Achievements in forest tree genetic improvement in Australia and New Zealand 2: Development of Corymbia species and hybrids for plantations in Eastern Australia. *Australian Forestry, 70*(1), 11–16. https://doi.org/10.1080/00049158.2007.10676256

Maredia, M. K., Howard, J. A., Boughton, D., Naseem, A., Wanzala, M. N., & Kajisa, K. (1999). Increasing seed system efficiency in Africa: Concepts, strategies and issues.

Montes, C. S., & Weber, J. C. (2009). Genetic variation in wood density and correlations with tree growth in *Prosopis africana* from Burkina Faso and Niger. *Annals of Forest Science, 66*(7), 1–9. https://doi.org/10.1051/forest/2009060

Ozment, L. (2023a). *Increasing suitable ranges of woody biofuel species increase and interactive effects of genotype and environment in hybrid sweetgum plantations.* https://hdl.handle.net/20.500.14446/344285

Ozment, L. (2023b). *Increasing suitable ranges of woody biofuel species and interactive effects of genotype and environment in hybrid sweetgum plantations* (Master's thesis, Oklahoma State University).

Palanisamy, K., & Subramanian, K. (2001). Vegetative propagation of mature teak trees (*Tectona grandis* L.). *Silvae Genetica, 50*(5–6), 188–190.

Ramalho, M. A. P., Santos, H. G., & Souza, T. D. S. (2022). Eucalyptus breeding programs: A proposal for the use of inbred progênies. *Cerne, 28*, Article e103049. https://doi.org/10.1590/01047760202228013049

Rohrbach, D. D., Saadan, H. M., Monyo, E. S., Kiriwaggulu, J. A. B., Mtenga, K., & Mwaisela, F. (2002) Comparative study of three community seed supply strategies in Tanzania. Project Report. International Crops Research Institute for the Semi-Arid Tropics, Bulawayo, Zimbabawe.

Ruhil, T., Singh, H., & Barthwal, S. (2024). Long-term physiological monitoring for detecting climate change impacts on forest ecosystems. In *Forests and climate change: Biological perspectives on impact, adaptation, and mitigation strategies* (pp. 817–834). Springer.

Sato, A. S., Freitas, M. L. M., Lima, I. L. D., Zimback, L., Toniato, M. T. Z., & Sebbenn, A. M. (2010) Genetic variation among and within provenances and progenies of *Corymbia maculata* (Hook.) KD Hill and LAS Johnson, in Pederneiras, SP. Cerne, 16, 60–67. https://doi.org/10.1590/01047760201622022152

Singh, H. (2024a). Climate change and forests: Biological perspectives. In *Forests and climate change: Biological perspectives on impact, adaptation, and mitigation strategies* (pp. 1–17). Springer.

Singh, H. (2024b). Exploring adaptive modulation in plant functional traits and its impact on the productivity of *Acacia auriculiformis* under CO_2-enriched environment. *Industrial Crops and Products, 210*, Article 118186.

Singh, H. (2024c). *Forests and climate change: Biological perspectives on impact, adaptation, and mitigation strategies.* Springer Nature.

Singh, H. (2024d). *Urban forests, climate change and environmental pollution: Physio-biochemical and molecular perspectives to enhance urban resilience.* Springer Nature.

Singh, H. (2025). The role of plant functional traits in mitigating urban climate and environmental pollution. In *Plant Functional Traits* (pp. 297–308). Elsevier.

Singh, H., & Kumar, M. (2022). Climate change and its impact on Indian Himalayan forests: Current status and research needs. In *Climate change: Impacts, responses and sustainability in the Indian Himalaya* (pp. 223–242). Springer.

Singh, H., Kumar, N., & Kumar, A. (2024). Enhancing resource use efficiency in crops through plant functional traits. In *Plant functional traits for improving productivity* (pp. 97–117). Springer.

Singh, H., Verma, A., & Kumar, M. (2018). Can elevated CO_2 alter the growth dynamics, morphology, physiology and biochemistry of the forestry tree species? *Climate change and agriculture* (pp. 288–305). Discovery Publishing House.

Stackpole, D. J., Vaillancourt, R. E., de Aguigar, M., & Potts, B. M. (2010). Age trends in genetic parameters for growth and wood density in *Eucalyptus globulus*. *Tree Genetics & Genomes, 6*, 179–193. https://doi.org/10.1007/s11295-009-0239-4

Suontama, M., Klápště, J., Telfer, E., Graham, N., Stovold, T., Low, C., & Dungey, H. (2019). Efficiency of genomic prediction across two *Eucalyptus nitens* seed orchards with different selection histories. *Heredity, 122*(3), 370–379. https://doi.org/10.1038/s41437-018-0119-5

Swain, T. L., Verryn, S. D., & Laing, M. D. (2013). A comparison of the effect of genetic improvement, seed source and seedling seed orchard variables on progeny growth in Eucalyptus nitens in South Africa. *Tree Genetics & Genomes, 9*, 767–778. https://doi.org/10.1007/s11295-013-0593-0

Torres-Dini, D., Sebbenn, A. M., de Aguiar, A. V., Vargas, A., Rachid-Casnati, C., & Resquín, F. (2024). Progeny selection and genetic diversity in a *Pinus taeda* clonal seed orchard. *Forests, 15*(10), 1682. https://doi.org/10.3390/f15101682

Uleh, M., & Fagbemi, T. (2017) Progeny trial of *Prosopis africana* in Benue State, Nigeria. *Journal of Research in Forestry, Wildlife and Environment, 9*(1), 61–66. https://www.ajol.info/index.php/jrfwe/article/view/155247

Weber, J. C., Montes, C. S., Kalinganire, A., Abasse, T., & Larwanou, M. (2015). Genetic variation and clines in growth and survival of *Prosopis africana* from Burkina Faso and Niger: Comparing results and conclusions from a nursery test and a long-term field test in Niger. *Euphytica, 205*, 809–821. https://doi.org/10.1007/s10681-015-1413-4

[illegible]

Chapter 5
Wood as a Sustainable Energy Source: Unlocking Its Potential for a Greener Future

Shailendra Kumar and Shikhar Shukla

Abstract Wood and bio-based fuels are increasingly recognized for their role as renewable, versatile, and climate-friendly energy sources. This chapter examines the utilization of wood fuel, focusing on its carbon footprint, calorific value, and potential as a sustainable alternative to fossil fuels. Many nations are adopting bio-based fuels such as wood, biodiesel, and black liquor to meet renewable energy targets and reduce greenhouse gas emissions. In developing countries, wood remains a primary source of energy for heating and cooking due to its accessibility and affordability. When harvested and utilized sustainably, wood fuel is considered carbon neutral, contributing to climate change mitigation. The chapter also highlights India's growing emphasis on biomass-based fuels in thermal power plants, which can foster rural economic development and support local livelihoods. Besides, this chapter discusses key aspects, including the carbon implications of wood combustion, methods for estimating its energy output, and strategies to enhance its calorific value. It is inferred that responsibly managed wood resources offer a promising pathway to balance growing energy demands with environmental conservation goals, positioning wood as a practical and sustainable alternative to non-renewable energy sources.

Keywords Wood fuel · Renewable · Biodiesel · Carbon neutral · Calorific value · Conservation

5.1 Introduction

Throughout human history, wood and other bio-based fuels have played a fundamental role in meeting energy demands. From prehistoric times, when wood was the primary source of heat and illumination, to its critical use during the Industrial Revolution for powering steam engines, the contribution of wood as a fuel has been well-recognized (Ahorsu et al., 2018). Even in the modern era, the growing

S. Kumar (✉) · S. Shukla
Forest Products Division, ICFRE-Forest Research Institute, Dehradun, India
e-mail: kumarsro@icfre.org

H. Singh (ed.), *Forestry for a Carbon-Neutral and Sustainable Future*, Sustainability Solutions, https://doi.org/10.1007/978-3-032-06427-1_5

urgency to address climate change and reduce dependence on fossil fuels has renewed global interest in bio-based energy alternatives (Intergovernmental Panel on Climate Change, 2021; International Renewable Energy Agency, 2020).

Wood and other biomass-derived fuels offer significant potential as sustainable, low-carbon energy sources (Kumar et al., 2021). They are particularly relevant in regions where access to conventional energy infrastructure is limited or economically unfeasible. Wood, in particular, remains an indispensable energy resource in many parts of the world due to its wide availability, renewability, and notable calorific value. From domestic cooking and heating to decentralized electricity generation, wood-based energy systems provide practical, accessible solutions. Hardwoods such as oak and sal are favored for their higher energy density and extended burn duration, while softwoods like pine and spruce offer ease of ignition and handling (Singh et al., 2018). Additionally, the local sourcing of wood reduces dependence on long, carbon-intensive supply chains, contributing to energy security at the community level. Technological advancements such as pelletization and improved combustion methods have further enhanced wood's energy efficiency and environmental performance, enabling cleaner and more effective energy generation, even in resource-constrained settings (Singh et al., 2024).

Given its biological origin, wood energy holds unique significance in the global transition toward carbon neutrality. When harvested and utilized sustainably, wood can act as a carbon–neutral fuel, as the carbon dioxide released during combustion is offset by the carbon absorbed during tree growth. This characteristic positions wood as a crucial component of renewable energy strategies aimed at mitigating greenhouse gas emissions. Therefore, understanding wood's calorific value, a fundamental indicator of its energy potential, is essential for optimizing its use as an efficient, climate-resilient fuel source (Singh, 2024c). This chapter explores the national and global perspectives on wood fuel utilization, examines its carbon implications, and evaluates its role within the broader context of sustainable energy and climate change mitigation. Analyzing the scientific principles and real-world applications may provide a comprehensive overview of wood's role as a renewable energy resource and its contribution to achieving global climate goals.

5.2 Global Scenario

Industrialized nations are increasingly turning to bio-based fuels, such as wood pellets and biodiesel, to meet their renewable energy targets and reduce greenhouse gas emissions. For example, countries in the European Union have embraced wood pellets as a key component of their biomass energy strategies. Countries such as Germany, Sweden, and Denmark have invested heavily in biomass-based district heating systems that utilize wood pellets as a primary fuel source (Bioenetrgy Europe, 2020; Olofsson et al., 2013). These pellets are valued for their uniform size, low moisture content, and high energy density, making them efficient for both domestic and industrial energy applications. The EU's Renewable Energy Directive (RED) has been instrumental in fostering biomass adoption by establishing ambitious renewable energy benchmarks, driving significant growth in wood pellet usage (Singh & Kumar, 2022).

Notably, nations such as Italy and Austria have embraced pellet stoves and boilers for residential heating, driven by their efficiency and suitability for localized heating needs (IEA Bioenergy Annual Report, 2020). Meanwhile, the Netherlands and Belgium rely heavily on significant imports of wood pellets to enable biomass co-firing in large-scale energy plants, a crucial component of their efforts to transition away from coal. Concurrently, the United States and Canada have emerged as major exporters of wood pellets, leveraging their vast forest resources and advanced pelletization industries to meet growing global demand (FAO, 2021).

In developing countries across Asia, Africa, and Latin America, wood remains a primary source of energy for cooking and heating. For millions of households, particularly in rural areas, firewood and charcoal are indispensable due to their availability and affordability. In regions such as Sub-Saharan Africa, over 80% of households rely on wood fuels for their daily energy needs, underscoring their crucial role in maintaining basic living conditions (Singh, 2024d). In Asia, countries such as India and Nepal heavily rely on firewood collected from nearby forests, often by women and children, for cooking traditional meals. Similarly, in Latin America, rural communities in countries such as Honduras and Guatemala rely on wood as a primary source of fuel for both heating and preparing staple foods (Singh, 2025).

5.3 Indian Scenario

India has been at the forefront of integrating renewable energy sources into its energy mix, and wood, along with biofuels, plays a crucial role in this effort. The country's energy needs are substantial, given its population size and economic growth trajectory, and renewable sources, such as wood and biofuels, are essential for reducing dependence on fossil fuels. In rural India, wood remains a significant source of energy, especially for cooking and heating. Despite the push for modern energy solutions, firewood and charcoal are widely used due to their availability and affordability. Fuel wood is the primary source of domestic and commercial energy in rural and semi-urban areas, with a higher usage rate in rural areas. It is primarily used by women for cooking and is collected from forests, farmlands, homes, and common land. According to the NSSO (2012), the proportion of people who use firewood is higher in rural areas (67.3%) and lower in urban and semi-urban areas (14%). According to the 2011 Census, solid biomass—such as firewood, agricultural residue, and cow dung—is the primary source of fuel for cooking in 65.9% of Indian households. Fuel wood, which accounts for more than 93% of GHG emissions from India's forestry industry, is the primary energy source for residential and commercial usage in rural and semi-urban areas (Sharma, 2017). According to the IndiaState of the Forest Report (2018)), 85,000 tonnes of firewood are used in India annually, a problem that needs to be addressed. Reducing reliance on forest-based fuelwood burning is necessary to preserve the nation's forest carbon resource. Studies estimate that nearly 49% of rural households still rely on solid biomass, such as wood, for cooking. Because the incomplete combustion of biomass results in a variety of harmful pollutants, using traditional biomass for cooking on a basic traditional cookstove contributes to indoor air pollution. This reliance is driven by factors such as accessibility, cost-effectiveness, and cultural significance. Wood is locally available in forests, agroforestry plantations, and agricultural residues. In regions where access to LPG or electricity is limited, wood serves as a cost-effective alternative. Traditional cooking practices, such as the use of chulhas (clay stoves), continue to favor wood as a primary fuel source (Singh, 2024b).

Moreover, the Indian government has issued directives regarding the use of biomass as fuel in thermal power plants. In June 2023, the Ministry of Power revised its biomass policy, mandating a 5% co-firing of biomass with coal in thermal power plants starting from the fiscal year 2024–25. This requirement is set to increase to 7% from the fiscal year 2025–26 (PIB, 2023). The policy primarily focuses on the utilization of agro-residue-based biomass pellets, such as stubble, straw, stalks, and husks from various crops (Joshi et al., 2021). While the policy emphasizes the use of agricultural residues, it also permits the utilization of other biomass sources, including bamboo by-products, horticultural waste, and certain types of wood-derived biomass (PIB, 2023). That means India needs to bring more and more land under afforestation to cater to the growing demand for firewood. This would also stimulate rural livelihood and economic growth (MS Swaminathan Research Foundation, 2010).

5.4 Carbon Implications of Wood as a Fuel

Wood is considered a renewable fuel when harvested sustainably. However, its carbon implications depend on factors such as forest management practices and combustion efficiency. Wood, when used as a fuel, has lower carbon implications compared to coal, diesel, or gasoline when used sustainably and efficiently. It is not entirely free of environmental concerns, especially in areas of unsustainable forestry or inefficient combustion. To minimize its carbon footprint, efforts should focus on improving combustion technologies, promoting sustainable forestry practices, and reducing deforestation (Kumar et al., 2020).

On the other hand, wood is often described as carbon–neutral because trees absorb carbon dioxide (CO_2) during growth through the process of photosynthesis. When wood is burned, it releases CO_2, which is ideally reabsorbed by new trees if forests are replanted, completing a natural carbon cycle. However, several variables, including time, the accessibility and availability of wood feedstock, and the energy used for production, harvesting, shipping, processing, and other value chain activities, influence the net greenhouse gas emissions associated with generating wood energy (Joshi et al., 2020). It is challenging to calculate the net emissions associated with wood energy, and further research is ongoing to determine the specific circumstances under which it is carbon neutral. However, when wood is produced sustainably, the net carbon emissions from wood energy are generally thought to be lower than those from fossil fuels (Jangra et al., 2024).

On combustion, wood releases approximately 1.8 kg of CO_2 per kg of dry wood (FAO, 2021), which is comparable to some fossil fuels on a per-unit-energy basis. However, the net emissions are reduced if the wood is sourced from sustainably managed forests. Inefficient wood combustion (e.g., open fires) releases methane (CH_4) and fine particulates, both of which have severe climate and health impacts. Improved combustion technologies, such as modern wood stoves or biomass boilers, reduce these emissions. Table 5.1 presents different types of fuels and their calorific values (Singh et al., 2023).

Figure 5.1 presents a comparison of various types of fuels, along with raw wood as fuel (15 MJ/kg for dry and wet wood). It can be seen from Fig. 5.1 that the energy density of the wood is relatively lower compared to the other fuels. For example, it is 2 times lower than charcoal, 3.2 times lower than LPG, and 8.74 times lower than hydrogen. Thus, it is evident that the relative energy per unit weight in wood is lower compared to modern industrial fuels.

Figure 5.2 illustrates the amount of energy (in MJ) released per kilogram of CO_2 emitted. It is evident from the figure that the calorific value of wood, both in its dry and wet forms, is comparable to that of conventional fuels such as coal. When considering the average calorific value of dry wood (approximately 19.5 MJ/kg), this equates to about 11.3 MJ of energy per kilogram of CO_2 emitted. However, it is essential to note that, unlike fossil fuels, the carbon released from burning wood can be offset through the carbon uptake by trees during growth. Therefore, when harvested and managed sustainably, wood fuel has the potential to be carbon neutral, making it a more environmentally responsible energy source.

5.5 Estimation of the Available Heat from Wood Fuel

The calorific value (CV) of wood is a measure of the total energy released as heat when a specific quantity of wood undergoes complete combustion. This parameter is critical in determining the efficiency of wood as a fuel source for various applications, including residential heating, industrial energy production, and renewable energy initiatives (Joshi et al., 2024). Typically expressed in kilojoules per kilogram (kJ/kg) or megajoules per kilogram (MJ/kg), the calorific value of wood is influenced by several factors, including moisture content, wood density, and chemical composition. Shmulsky and Jone have described the terminology related to wood fuel energy, including higher heating value (HHV), gross heating value (GHV), and combustion efficiency (CE).

Table 5.1 Calorific values and carbon emissions for wood and alternative fuels (Basu, 2010; Demirbas, 2008; FAO, 2016; NREL, 2022)

Fuel type	Calorific value (MJ kg^{-1})	CO_2 emissions (kg kg^{-1} of fuel)	Remarks
Wood (dry)	16–20	1.65–1.80	It can be carbon–neutral if sustainably sourced; inefficient combustion increases methane and particulate emissions
Wood (wet)	8–12	1.20–1.50	Lower energy output due to high moisture content results in inefficient combustion and higher emissions per megajoule (MJ)
Charcoal	28–33	2.75–3.00	Higher emissions per kilogram due to higher carbon density; often linked to inefficient and unsustainable production methods
Coal (anthracite)	25–30	2.40–2.60	High carbon content is a significant contributor to global CO_2 emissions
Coal (bituminous)	24–28	2.30–2.50	Widely used for power generation, it produces higher CO_2 emissions per kg than wood
Liquefied Petroleum Gas (LPG)	46–50	3.00–3.20	Lower CO_2 emissions than coal or diesel; widely used for cooking and heating
Ethanol	27	1.90–2.00	Renewable fuel but lower energy density; emits less CO_2 than gasoline
Diesel	42–46	3.20–3.30	High carbon emissions, widely used in transportation and industry
Petrol (gasoline)	42–44	2.90–3.10	Similar to diesel, it is a significant source of transport-related CO_2 emissions
Biodiesel	37–41	2.50–2.80	Lower carbon emissions compared to fossil diesel, a renewable source of energy

(continued)

Table 5.1 (continued)

Fuel type	Calorific value (MJ kg^{-1})	CO_2 emissions (kg kg^{-1} of fuel)	Remarks
Agricultural residues	12–18	1.50–1.70	Includes rice husk and sawdust; emissions depend on specific residue and moisture content
Pellets (wood/ biomass)	18–20	1.70–1.80	Uniform size and low moisture content improve combustion efficiency compared to raw wood
Biogas	20–25	1.00–1.20 (per m3)	A mixture of methane and CO_2 is cleaner than fossil fuels, but emissions depend on methane leakage and combustion
Peat	10–15	1.60–1.90	High moisture and carbon content result in moderate CO_2 emissions, not widely utilized in industrialized nations
Hydrogen (compressed)	120–142	0	It emits only water vapor when burned; production emissions depend on the energy source (e.g., renewable vs. fossil)

5.5.1 *Higher Heating Value (HHV)*

HHV is the total heat produced by complete combustion under regulated conditions. Due to losses from vaporizing water in the fuel and other process losses, the actual heat that can be recovered in conventional burners is believed to be significantly less than the Higher Heating Value (HHV).

5.5.2 *Gross Heating Value (GHV)*

The usable potential of the heat energy after correcting for the heat decrease caused by the fuel's moisture content (MC) is known as the gross heating value. However, the MC% term used in the fuel wood and paper/pulp industry has a different meaning compared to the furniture, handicrafts, and panel industries.

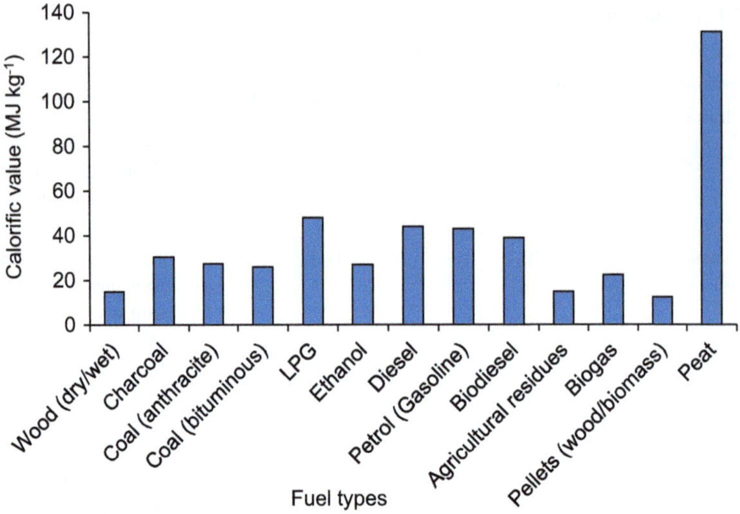

Fig. 5.1 Calorific value of one kilogram of fuel

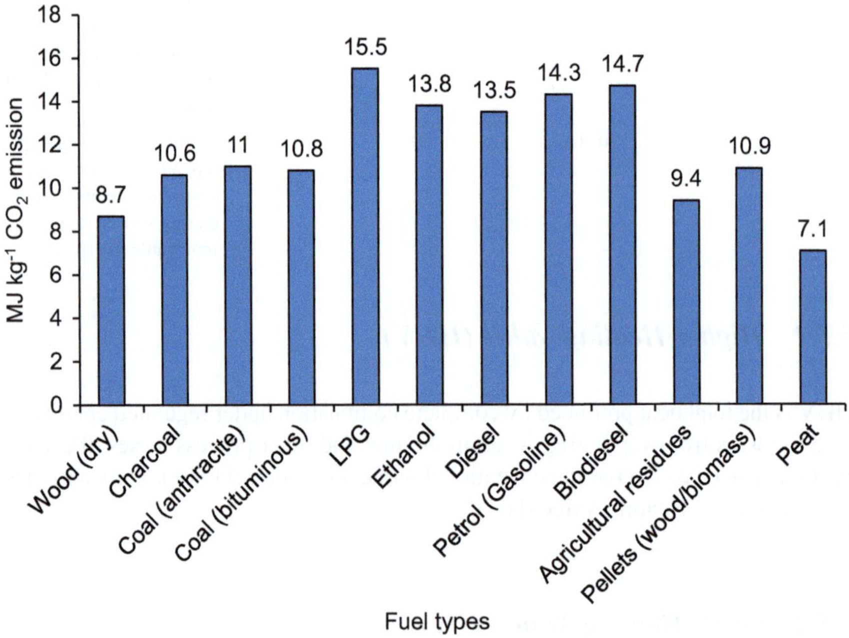

Fig. 5.2 Energy produced on emission of 1 kg CO_2

5.5.3 Moisture Content (MC%)

For furniture, handicrafts, panel wood, etc., the moisture content is the ratio of the weight of water to the weight of oven-dried wood (i.e., the weight of the wood when there is no water in it), on an oven-dry weight basis. It is often expressed in the form of percent (%). For the fuel wood/paper and pulp industry, the MC is the ratio of the weight of water to the weight of gross wood (i.e., the weight of both wood and water) on a wet weight basis. Both of the above MC% are interconvertible using the following equation:

$$MCtext\%_{wet\ weight\ basis} = \frac{MC\%_{\ od\ weight\ basis}}{100 + MC\%_{\ od\ weight\ basis}} \tag{5.1}$$

Thus, GHV can be calculated using HHV and MC% (wet weight basis) using the following equation:

$$GHV = HHV - \left(HHVxMC\%_{\ wet\ weight\ basis}\right). \tag{5.2}$$

5.5.4 Combustion Efficiency

The effectiveness with which a fuel's energy content is transformed into useful energy—usually heat or mechanical power—is measured by its combustion efficiency. It has a significant impact on the performance of the boiler, heater, and engine. For example, conventional wood stoves have a combustion efficiency of approximately 50% (Li et al., 2016).

For example, the HHV of 6-year-old wood of E. hybrid is found to be 19.5 MJ/kg (Kumar et al., 2010). Assuming air-dried E. hybrid wood (25% MC wet basis) is used in a traditional wood stove (50% combustion efficiency), GHV of 1 kg wood will be 14.625 MJ/kg, and the total available heat will be 7.31 MJ.

5.6 Factors Responsible for the Calorific Value of Wood

5.6.1 Moisture Content

Freshly cut wood (green wood) can have a moisture content as high as 50–60%, which significantly reduces its calorific value. A significant portion of the energy released during combustion is consumed by evaporating the water, reducing the usable energy output. For example, wood with a 50% moisture content may have an effective calorific value of approximately 8–10 MJ/kg, whereas kiln-dried wood with a moisture content of 10–15% can have a calorific value of up to 18–21 MJ/kg.

5.6.2 *Wood Density*

The density of wood directly impacts its calorific value due to the amount of combustible material present per unit volume.

5.6.2.1 High-Density Woods

Dense hardwoods (e.g., oak, sal) have higher calorific values because they contain more lignin and cellulose, which are energy-rich components.These woods burn slower and release heat more steadily, making them ideal for long-duration heating.Hardwoods typically have a calorific value of 18–21 megajoules per kilogram (MJ/kg).

5.6.2.2 Low-Density Woods

Softwoods (e.g., pine, spruce) have a lower density, resulting in a slightly lower calorific value (17–20 MJ/kg). However, they ignite more easily, making them suitable for quick-burning applications.

5.6.3 *Chemical Composition*

The proportions of cellulose, hemicelluloses, lignin, and extractives in wood influence its energy potential.

5.6.3.1 Lignin

Lignin has the highest calorific value among the three primary components, providing more energy per unit mass. Woods with higher lignin content, such as hardwoods, exhibit higher energy efficiency during combustion.

5.6.3.2 Cellulose and Hemicellulose

These components have moderate calorific values. Their combustion provides a steady release of energy, but less than that of lignin.

5.6.3.3 Extractives and Volatiles

Resins, oils, and other extractives can slightly increase the calorific value, especially in softwoods that contain natural resins. For example,resinous woodslike pine.

5.7 Biofuels as Renewable Energy

Although there are various methods for producing liquid fuels from wood, the three most viable ones are methanol, ethanol, and diesel fuel. Wood can also be used to make ethyl-tert-butyl ether and methyl-tert-butyl ether, two other significant fuels. Isopropyl alcohol, sec-butyl alcohol, tert-butyl alcohol, and mixed alcohol are further potential choices. Wood-based gasoline or diesel fuel can be produced in various ways (Zerbe, 1991).

India's biofuel sector has experienced rapid growth in recent years, driven by policy support and innovation. India has been actively advancing its biofuel sector to enhance energy security and reduce greenhouse gas emissions. The nation has set ambitious targets, including achieving a 20% ethanol blend in gasoline by 2025–26 and a 5% biodiesel blend by 2030 (IEA, 2024). In 2024, India's ethanol production is projected to reach 6.35 billion liters, a slight decrease from the previous year due to anticipated declines in sugarcane output and rice grain availability (USDA Foreign Agricultural Service, 2024). To meet its ethanol production goals, India has increased the procurement price of ethanol derived from corn, leading to a shift from being a net corn exporter to an importer. This strategic move underscores India's commitment to diversifying its biofuel feedstock and reducing dependence on traditional fossil fuels. Biofuels, including bioethanol, biodiesel, and biogas, are produced from agricultural residues, sugarcane, oilseeds, and organic waste (Ruhil et al., 2024).

5.8 Calorific Value Enhancement

Wood energy is the energy produced from wood or products obtained from wood, typically through combustion processes, and used to heat, cook, or create electricity. Wood and compounds made from wood that are utilized for energy purposes, whether in solid, liquid, or gaseous form, are also referred to as "wood energy." Charcoal, fuelwood (also known as firewood), and wood pellets (also referred to as briquettes or chips) made from wood or wood waste are examples of solid wood fuel. Unprocessed woody biomass obtained from tree stems, branches, or other parts is known as fuelwood. Wood wastes (such as sawdust and wood shavings) from the timber harvesting or wood-processing industries are occasionally included in this term to provide energy (Kumar & Singh, 2020).

Wood can be processed and treated through several methods to enhance its calorific value, energy efficiency, and usability as a fuel source. One of the most common methods is drying, also known as seasoning, which involves reducing the moisture content of wood. High moisture levels significantly reduce the effective calorific value, as a substantial amount of energy is consumed to evaporate the water during combustion. Drying can be achieved through air drying, where wood is stored in ventilated spaces for several months, or through kiln drying, which utilizes controlled heat to accelerate moisture removal (Cloutier & Bell, 2000). Kiln drying, in particular, reduces the moisture content to below 20%, resulting in a calorific value of approximately 16–21 MJ/kg, compared to the much lower energy potential of green wood. The throughput of the pelletizing machine can be significantly increased by preheating the biomass, which reduces the energy required per unit weight of the biomass pellets produced (Aqa & Bhattacharya, 1992).

*Biomass densification*is the process of transforming low-density biomass materials into high-density, compact forms. The goal is to enhance the handling, storage, transportation, and combustion qualities of biomass. For biomass densification, Klass (1998) and Bhattacharya et al. (1989) divided the process into three categories: roll briquetting, pelletizing, and extrusion. In an extruder, a screw or piston moves the raw material through a die and compresses it to create cylindrical or other-shaped compacts.

Pelletization involves compressing sawdust and wood residues into dense, standardized pellets. These pellets have low moisture content, typically around 8–10%, and a calorific value of 16–19 MJ/kg (Kazi & Sadiq, 2012). The process includes grinding the raw material, drying it, and then compressing it under high pressure to create uniform pellets. A pelletizer consists of one or two rollers and a perforated, hardened steel die. Densified pellets are produced by forcing the feedstock through the perforations as the die and rollers roll. The most popular machine type for pelletizing wood is a roller-and-die pelletizer. This method not only enhances the energy density of wood but also simplifies transportation and storage, making it suitable for residential stoves, industrial boilers, and co-firing in power plants (Demirbas, 2004).

Biomass resources are compacted into larger, solid blocks known as *briquettes* during the briquetting process. Briquettes are frequently used for cooking and in industrial settings. The feedstock is compressed into pillow-shaped briquettes in a briquetting roller press by being squeezed between two rollers that move in opposite directions. This process increases the energy density of wood, facilitating transportation and storage. Briquettes typically have a calorific value of 15–20 MJ/kg and are widely used in rural areas for domestic heating and small-scale industrial energy needs (Afsal et al., 2020).

Torrefaction is yet another advanced process used to improve wood's energy characteristics. Fuels produced by torrefaction are high in energy, low in moisture, weatherproof, and readily crushed or densified. In this mild pyrolysis technique, wood is heated to temperatures between 200 and 300 °C in a low-oxygen environment (Bian et al., 2016). This process removes volatile compounds, reduces the moisture content, and increases the carbon concentration, thereby increasing the calorific value to approximately 20–23 MJ/kg (Goh & Zainal, 2010). The resulting torrefied wood is hydrophobic, more brittle, and has a higher energy density, making it a suitable substitute for coal in biomass power plants. It is used as a direct substitute for coal in power plants, reducing greenhouse gas emissions. It is used as a direct substitute for coal in power plants, reducing greenhouse gas emissions. During torrefaction, volatile organic compounds (VOCs) are released and can be captured to produce heat or energy, making the process more energy-efficient.

Pyrolysis, a more intensive thermal process, decomposes wood in the absence of oxygen to produce charcoal, bio-oil, and syngas (Mohan et al., 2006). Depending on the process parameters, slow pyrolysis focuses on producing charcoal with a calorific value of 27–33 MJ/kg. In contrast, fast pyrolysis generates bio-oil that can be used as a renewable liquid fuel (Bridgwater, 2012). Charcoal is a solid carbon-rich material with a high calorific value. It is commonly used in cooking, heating, and metallurgy (e.g., in smelting iron). Due to its energy density and long combustion duration, it is an efficient alternative to firewood and coal. Charcoal is also used as an industrial reductant in metal extraction processes and as an adsorbent in water purification. Bio-oil, produced during fast pyrolysis, can be used as a renewable substitute for petroleum in heating applications and as a raw material for biofuel production. It is also gaining interest as a feedstock for chemicals and bioplastics (Singh, 2024a). Pyrolysis is particularly valuable for producing high-energy materials that are easily transportable and have diverse industrial applications, including cooking and metallurgy.

Gasification is another significant method, where wood is converted into *syngas*(a mixture of hydrogen, carbon monoxide, and methane) by heating it in a controlled, low-oxygen environment. This *syngas* has a calorific value ranging from 12 to 20 MJ/m^3 and serves as a clean-burning fuel for electricity generation and combined heat and power (CHP) systems. Gasification maximizes energy extraction from wood while minimizing waste (Escalante & Hernandez, 2017). It can also serve as a precursor for hydrogen production, which has applications in fuel cells and green energy systems. The *ash* produced during gasification is rich in minerals and can be used as a soil amendment in agriculture or as a raw material in cement and brick manufacturing.

Since hydrogen has an oxidation rate that is roughly four orders of magnitude higher than that of simple partially oxidized one-carbon compounds—which, in turn, have the same oxidation advantage over saturated hydrocarbons under the same conditions—it is the most active fuel that is frequently used in fuel cells (Wyman et al., 1993). Many biomass feedstocks, including wood chips from short-rotation forestry plantations, agricultural or forest wastes, and municipal solid waste, can be thermochemically gasified to produce hydrogen. It was found that the wood-fired molten carbonate fuel cell was highly effective and appropriate for small-scale power generation. For these systems, electrical efficiencies of $25 \pm 28\%$ are typical (McIlveen-Wright et al., 2000).

Lastly, *carbonization*, also known as charcoal production, involves heating wood in a low-oxygen environment to remove volatile compounds and produce carbon-rich charcoal (Demirbas, 2009). With a calorific value of 27–33 MJ/kg, charcoal is highly energy-dense and widely used in cooking, metallurgy, and other high-temperature applications. Carbonization produces charcoal as the primary product, along with wood vinegar and tar as secondary by-products. Also known as pyroligneous acid,

Table 5.2 Calorific values of improved wood fuels (Domingos et al., 2020; Lunguleasa et al., 2019)

Method	Calorific value (MJ/kg)	Key advantages	Applications
Drying (seasoning)	16–21	Increases energy efficiency by reducing moisture	Firewood for domestic heating
Pelletization	16–19	Standardized size, high energy density	Stoves, boilers, co-firing plants
Torrefaction	20–23	High energy density, hydrophobic, easy transport	Coal substitute in power plants
Pyrolysis	27–33 (Charcoal)	Produces high-energy charcoal, bio-oil, and syngas	Cooking, industrial energy, biofuels
Gasification	12–20 (Syngas)	Produces clean-burning syngas	Power generation, CHP systems
Briquetting	15–20	Reduces waste, improves energy density	Rural heating, small-scale industries
Carbonization	27–33	Produces high-energy charcoal	Cooking, metallurgy

wood vinegar is a liquid by-product rich in organic acids and phenolic compounds. It is used as a natural pesticide, soil conditioner, and deodorant. Tar is a viscous liquid used in waterproofing, road construction, and as a precursor for industrial chemicals.

These methods significantly enhance the energy efficiency and usability of wood, making it a highly adaptable and renewable source of energy suitable for a wide range of applications. Table 5.2 presents various types of improved wood fuels.

From the Table 5.2, it is evident that the calorific value of the raw wood can be improved by up to two times using suitable methods.

5.9 Black Liquor as a Fuel: A Sustainable Energy Source

Black liquor is a byproduct of the kraft pulping process, widely used in the paper and pulp industry. It consists of lignin, hemicellulose, and other organic substances dissolved from wood during the production of pulp, along with inorganic chemicals such as sodium hydroxide and sodium sulfide. Traditionally considered waste, black liquor is now recognized as a valuable biofuel due to its significant energy content and potential for sustainable energy generation.

The calorific value of black liquor ranges between 13 and 16 MJ/kg (dry solids), depending on its composition and moisture content. This makes it a viable alternative to fossil fuels for industrial applications. Burning black liquor also has environmental advantages. As a biomass-based fuel, its CO_2 emission is considered carbon–neutral because the carbon released during combustion is offset by the carbon absorbed by the trees used for paper production. However, the combustion process emits other pollutants, such as sulfur compounds and particulate matter, which require advanced scrubbing technologies to minimize environmental impact.

Innovative technologies, such as the gasification of black liquor, are being developed to improve its efficiency and sustainability. Gasification can produce syngas (a mixture of hydrogen and carbon monoxide), which can be used as a clean fuel or feedstock for chemical production. This approach reduces emissions and increases energy recovery compared to conventional combustion.

Experts advocate that a kraft bio-refinery operation would enable the optimal use of wood and its constituents for pulp, biofuels, biochemicals, and bio-power production. A significant amount of polysaccharides could be supplied to the biofuels sector by pre-extraction of certain hemicelluloses from kraft wood chips before pulping. Creation of pre-extraction methods that are optimized to produce a lingo-cellulosic stream for pulp production and a hemicellulose stream for biofuel production (Laitila et al., 2010).

Thus, black liquor is a promising biofuel that enhances the energy efficiency and sustainability of the pulp and paper industry. With advancements in recovery and gasification technologies, it has the potential to contribute significantly to renewable energy systems while reducing reliance on fossil fuels.

5.10 Future Directions

The future of wood as a sustainable energy source lies in addressing the dual challenge of maximizing its energy potential while ensuring ecological sustainability and climate resilience. Advancements in biomass processing technologies, coupled with improved forest resource management, will be pivotal in enhancing the role of wood within global and regional renewable energy frameworks (Barwa & Singh, 2024).

There is a critical need for the large-scale adoption of advanced densification techniques, such as pelletization, briquetting, and torrefaction, to improve the energy density, handling efficiency, and combustion performance of wood-based fuels. These processes not only enhance calorific value but also reduce logistical costs and storage limitations, making wood fuels viable for both decentralized and industrial-scale applications. Future research must focus on optimizing these processes for a broader range of wood species and biomass residues, including underutilized agroforestry by-products and forest waste.

Sustainable forest management (SFM) practices must be prioritized to ensure the long-term availability of wood resources without compromising forest carbon stocks or biodiversity. Integration of precision forestry techniques, remote sensing, and GIS-based monitoring can enable data-driven management of biomass resources, ensuring alignment with carbon neutrality goals.

There is significant potential for integrating wood-based fuels into hybrid renewable energy systems, particularly for rural electrification and decentralized power generation. Coupling biomass gasification or pyrolysis with solar or wind energy offers a reliable and low-carbon energy solution for off-grid communities, particularly in developing nations with limited access to modern energy infrastructure. Moreover, advancements in thermochemical conversion technologies, such as pyrolysis, gasification, and bio-refinery concepts, must be accelerated to diversify the range of high-energy, low-emission fuels derived from wood. The development of bio-oil, syngas, and value-added chemicals from wood feedstock presents opportunities to substitute for fossil fuels across various industrial sectors, aligning with circular economy principles.

Finally, policy frameworks need to evolve to support the responsible scaling of wood energy, including regulatory incentives for sustainable biomass production, financial mechanisms for technology deployment, and stringent standards to mitigate environmental externalities such as deforestation, air pollution, and land-use conflicts. Given India's policy push for biomass co-firing in thermal power plants and the global emphasis on reducing GHG emissions, coordinated efforts between policymakers, industries, and research institutions are essential to leverage wood energy in achieving national and international climate targets (Garg et al., 2016).

5.11 Conclusion

The global use of wood and bio-based fuels highlights their critical role as renewable, adaptable, and regionally accessible energy sources. These fuels play a significant role in meeting energy demands and reducing carbon emissions, particularly in areas with abundant biomass and limited access to modern energy infrastructure. However, their long-term viability depends on sustainable resource management and the integration of advanced technologies to mitigate associated environmental and socioeconomic challenges. While wood may offer a moderate calorific value compared to fossil fuels, its combination of renewability, low net carbon emissions, and compatibility with evolving energy systems positions it as a valuable component of the global renewable energy portfolio. When sourced and utilized responsibly, wood provides an effective and environmentally sound alternative to fossil fuels, helping to bridge the gap between growing energy needs and climate goals while promoting environmental sustainability and achieving a carbon–neutral future.

References

Afsal, A., David, R., Baiju, V., Suhail, N. M., Parvathy, U., & Rakhi, R. B. (2020). Experimental investigations on combustion characteristics of fuel briquettes made from vegetable market waste and saw dust. Materials Today: Proceedings, 33, 3826–3831.

Ahorsu, R., Medina, F., & Constantí, M. (2018). Significance and challenges of biomass as a suitable feedstock for bioenergy and biochemical production: A review. *Energies, 11*(12), 3366. https://doi.org/10.3390/en11123366

Aqa S, Bhattacharya S.C. (1992). Densication of preheated sawdust for energy conservation. *Energy 17*(6):575–8.

Barwa, S., & Singh, H. (2024). Climate-induced changes in forest ecosystems of Indian Himalayan Region: A meta-analysis. In *Urban forests, climate change and environmental pollution: Physio-biochemical and molecular perspectives to enhance urban resilience* (pp. 839–866). Springer.

Basu, P. (2010). *Biomass gasification and pyrolysis: Practical design and theory.* Academic Press.

Bioenergy Europe. (2020). *Bioheat: statistical report 2020.* Brussels, Belgium:Bioenergy Europe.

Bhattacharya, S. C., Sett, S., & Shrestha, R. M. (1989). State of the art for biomass densification. *Energy Sources, 11*, 161–182.

Bian, L., Liu, S., & Wei, Y. (2016). Torrefaction of wood and biomass for energy applications. *Renewable and Sustainable Energy Reviews, 64*, 417–433.

Bridgwater, A. V. (2012). Review of fast pyrolysis of biomass and product upgrading. *Biomass and Bioenergy, 38*, 68–94.

Cloutier, A., & Le Bel, L. (2000). Drying of wood: A review of the methods and applications. *Wood and Fiber Science, 32*(4), 399–412.

Demirbas, A. (2004). Combustion of biomass in a fixed bed: Modeling and performance. *Energy Sources, 26*(1), 17–25.

Demirbas, A. (2008). *Biofuels, solar and wind as renewable energy systems: Benefits and risks.* Springer.

Demirbas, A. (2009). Charcoal production from biomass. *Energy Sources, 31*(12), 1131–1142.

Domingos, I., Ayata, U., Ferreira, J., Cruz-Lopes, L., Sen, A., Sahin, S., & Esteves, B. (2020). Calorific power improvement of wood by heat treatment and its relation to chemical composition. *Energies, 13*(20), 5322. https://doi.org/10.3390/en13205322

Escalante, H., & Hernández, J. (2017). Wood gasification for energy production: An overview. *Renewable and Sustainable Energy Reviews, 69*, 323–336.

Food and Agriculture Organization (FAO). (2008). *Forestry Paper 162: Forests and energy.* FAO, Rome.

Food and agriculture organization of the United Nations. (2021). *Statistical pocketbook 2021: World Food and Agriculture.* Rome: FAO.

Food and Agriculture Organization (FAO). (2016). *Assessing woodfuel supply and demand in displacement settings.* FAO Corporate Document Repository.

Garg, A., Pal, D., Singh, H., & Pandey, D. C. (2016). A comparative study of NDBI, NDISI and NDII for extraction of urban impervious surface of Dehradun [Uttarakhand, India] using Landsat 8 imagery. In *2016 International Conference on Emerging Trends in Communication Technologies (ETCT)* (pp. 1–5).

Goh, C. S., & Zainal, Z. A. (2010). A review of torrefaction for biomass energy production. *Energy Conversion and Management, 51*(10), 2139–2146.

India state of Rorest Report 2017 (2018). Dehradun: Ministry oF Environment, Forest & Climate Change, Government of India. Forest Survey of India, Dehradun, India.

International Energy Agency (IEA). (2024). *Biofuel consumption in the accelerated case in India, 2015 to 2028.* IEA.

Intergovernmental Panel on Climate Change (IPCC). (2021). *Climate change 2021: The physical science basis.* Contribution of Working Group I to the Sixth Assessment Report. Cambridge University Press.

International Renewable Energy Agency (IRENA). (2020). *Renewable energy statistics 2020.* IRENA, Abu Dhabi.

IEA Bioenergy. (2020). *IEA bioenergy annual report 2020.* Paris: International Energy Agency Technology Collaboration Programme on Bioenergy. https://www.ieabioenergy.com/wp-content/uploads/2021/04/IEAB-Annual-Report-2020.pdf

Jangra, A., Kumar, K., Maikhuri, S., Bhandari, M. S., Pandey, S., Singh, H., & Barthwal, S. (2024). Unveiling stress-adapted endophytic bacteria: Characterizing plant growth-promoting traits and assessing cross-inoculation effects on *Populus deltoides* under abiotic stress. *Plant Physiology and Biochemistry, 210*, Article 108610.

Joshi, R., Singh, H., Chhetri, R., & Yadav, K. (2020). Assessment of carbon sequestration potential in degraded and non-degraded community forests in Terai region of Nepal. *Journal of Forest and Environmental Science, 36*(2), 113–121.

Joshi, R., Singh, H., Chhetri, R., Poudel, S. R., & Rijal, S. (2021). Carbon sequestration potential of community forests: A comparative analysis of soil organic carbon stock in community managed forests of Far-Western Nepal. *Eurasian Journal of Soil Science, 10*(2), 96–104.

Joshi, R., Sharma, B., Singh, H., Dhakal, N., Ayer, S., & Maraseni, T. (2024). Poplar plantation as an agroforestry approach: Economic benefits and its role in carbon sequestration in North India. *Journal of Resources and Ecology, 15*(4), 880–888.

Kazi, S. N., & Sadiq, M. (2012). Wood pelletization as an alternative energy source: Process and characteristics. *Renewable and Sustainable Energy Reviews, 16*(1), 137–146.

Klass, D. L. (1998). *Biomass for renewable energy, fuels, and chemicals.* Academic Press.

Kumar, M., & Singh, H. (2020). Agroforestry as a nature-based solution for reducing community dependence on forests to safeguard forests in Rainfed areas of India. In *Nature-based solutions for resilient ecosystems and societies* (pp. 289–306). Springer.

Kumar, R., Pandey, K. K., Chandrashekar, N., & Mohan, S. (2010). Effect of tree-age on calorific value and other fuel properties of Eucalyptus hybrid. *Journal of Forestry Research, 21*, 514–516.

Kumar, A., Dwivedi, G. K., Tewari, S., Sah, V. K., Singh, H., Kumar, P., Kumar, N., & Kaushal, R. (2020). Soil organic carbon pools under *Terminalia chebula* Retz. based agroforestry system in Himalayan foothills, India. *Current Science, 118*(7), 1098–1103.

Kumar, N., Singh, H., & Kumar, A. (Eds.). (2021). *Renewable energy and green technology: Principles and practices.* CRC Press.

Laitila, J., Heikkilä, J., & Anttila, P. (2010). Harvesting alternatives, accumulation, and procurement cost of small-diameter thinning wood for fuel in Central Finland.

Li, Q., Jiang, J., Cai, S., Zhou, W., Wang, S., Duan, L., & Hao, J. (2016). Gaseous ammonia emissions from coal and biomass combustion in household stoves with different combustion efficiencies. *Environmental Science & Technology Letters, 3*(3), 98–103.

Lunguleasa, A., Ayrilmis, N., Spirchez, C., & Croitoru, C. (2019). Increasing the calorific properties of sawdust waste from pellets by torrefaction. *BioResources, 14*(4), 7821–7839.

McIlveen-Wright, D. R., Williams, B. C., & McMullan, J. T. (2000). Wood gasification integrated with fuel cells. *Renewable Energy, 19*(1–2), 223–228.

M. S. Swaminathan Research Foundation. (2010). *Twentieth annual report 2009–2010.* Chennai: MSSRF. Retrieved fromhttps://www.yumpu.com/en/document/view/10563381/twentieth-annual-report-2009-2010-m-s-swaminathan.

Mohan, D., Pittman, C. U., & Steele, P. H. (2006). Pyrolysis of wood/biomass for bio-oil: A critical review. *Energy & Fuels, 20*(3), 848–889.

National Renewable Energy Laboratory (NREL). (2022). *Studies on hydrogen's calorific values and its usage.*

NSSO (2012).*Energy sources of Indian households for cooking and lighting: NSS report No. 542 (2009-10).* Ministry of Statistics & Programme Implementation, Government of India. Retrieved from https://mospi.gov.in/sites/default/files/publication_reports/nss_Report-542.pdf

Olofsson, D., Bellqvist, D., Karlsson, J., & Johansson, M. (2013). Optimising the operation of a district heating system. *Chemical Engineering Transactions, 35*, 637–642. https://doi.org/10.3303/CET1335106

Press Information Bureau. (2023, August 2). *Revised biomass policy mandates 5% biomass co-firing in thermal power plants from FY 2024–25: Union Minister for Power and New & Renewable Energy.* PIB, New Delhi.

Ruhil, T., Singh, H., & Barthwal, S. (2024). Long-term physiological monitoring for detecting climate change impacts on forest ecosystems. In *Forests and climate change: Biological perspectives on impact, adaptation, and mitigation strategies* (pp. 817–834). Springer.

Sharma, J. V. (2017). *Roadmap for achieving additional 2.5–3 billion tons eCO_2 sequestration from forestry sector by 2030.* Policy Brief, TERI, New Delhi.

Singh, H. (2024a). Climate change and forests: Biological perspectives. In *Forests and climate change: Biological perspectives on impact, adaptation, and mitigation strategies* (pp. 1–17). Springer.

Singh, H. (2024b). Exploring adaptive modulation in plant functional traits and its impact on the productivity of *Acacia auriculiformis* under CO_2-enriched environment. *Industrial Crops and Products, 210*, 118186.

Singh, H. (2024c). *Forests and climate change: Biological perspectives on impact, adaptation, and mitigation strategies*. Springer Nature.

Singh, H. (2024d). *Urban forests, climate change and environmental pollution: Physio-biochemical and molecular perspectives to enhance urban resilience*. Springer Nature.

Singh, H. (2025). The role of plant functional traits in mitigating urban climate and environmental pollution. In *Plant functional traits* (pp. 297–308). Elsevier.

Singh, H., & Kumar, M. (2022). Climate change and its impact on Indian Himalayan forests: Current status and research needs. In *Climate change: Impacts, responses and sustainability in the Indian Himalaya* (pp. 223–242). Springer.

Singh, H., Kumar, N., & Kumar, A. (2024). Enhancing resource use efficiency in crops through plant functional traits. In *Plant functional traits for improving productivity* (pp. 97–117). Springer.

Singh, H., Verma, A., & Kumar, M. (2018). *Can elevated CO_2 alter the growth dynamics, morphology, physiology and biochemistry of the forestry tree species? Climate change and agriculture* (pp. 288–305). Discovery Publishing House.

Singh, H., Kumar, N., Singh, R., & Kumar, M. (2023). Assessing the climate change impact on the habitat suitability of the range-restricted bird species (*Catreus wallichii*) in the Indian Himalayan ecosystem. *Environmental Science and Pollution Research, 30*(57), 121224–121235.

U.S. Department of Agriculture (USDA). (2024). *Foreign agricultural services: Biofuels annual report of India.* Report No. IN2024-0024, New Delhi.

Wyman, C. E., Bain, R. L., Hinman, N. D., & Stevens, D. J. (1993). Ethanol and methanol from cellulosic biomass. In *Renewable energy: Sources for fuel and electricity* (pp. 865–923). Island Press.

Zerbe, J. I. (1991). *Liquid fuels from wood—Ethanol, methanol*, diesel. *World Resource Review, 3*(4), 406–414.

Chapter 6
Climate-Smart Forestry: Strategies for Resilience and Sustainability

Vivek Chauhan, Rupali Singh, Ankit Chamoli, Alok Kumar Singh, Pooja, Tara Chand, and Hukum Singh

Abstract Climate-smart forestry (CSF) is an emerging and integrative framework designed to enhance the resilience of forests and forest-dependent communities to climate change while also contributing to climate change mitigation and achieving the Sustainable Development Goals (SDGs). CSF encompasses adaptive forest management, improved carbon sequestration, biodiversity conservation, technological innovation, and inclusive, participatory governance. This chapter examines both the conceptual foundations and practical applications of CSF, drawing on successful global case studies, analyzing policy integration mechanisms, and assessing its ecological benefits and socio-economic implications. A comprehensive assessment of current strategies, enabling technologies, implementation barriers, and future research directions provides insights for embedding CSF into national forest policies and climate action agendas. Notably, the analysis highlights that CSF surpasses conventional forestry practices, embodying a transformative, systems-based approach crucial for harmonizing ecological sustainability with socio-economic development in an era of accelerating climate change.

Keywords Climate change · Carbon sequestration · Forestry interventions · Mitigation · Nature-based solution · Sustainable development goals

V. Chauhan · R. Singh (✉) · A. Chamoli · T. Chand · H. Singh
ICFRE-Forest Research Institute, Dehradun, Uttarakhand, India
e-mail: srupali333@gmail.com

A. K. Singh
Rajiv Gandhi South Campus, Banaras Hindu University, Mirzapur, India

Pooja
Department of Silviculture and Agroforestry, College of Forestry, Dr. Y.S. Parmar University of Horticulture and Forestry, Nauni-Solan, Himachal Pradesh, India

H. Singh (ed.), *Forestry for a Carbon-Neutral and Sustainable Future*, Sustainability Solutions, https://doi.org/10.1007/978-3-032-06427-1_6

6.1 Introduction

The relationship between forests and climate is deeply reciprocal and increasingly significant. Forest ecosystems provide critical services, such as carbon sequestration, biodiversity conservation, water regulation, and livelihood support for over 1.6 billion people worldwide (FAO, 2020). However, these systems are under unprecedented stress from climate-induced hazards, including rising temperatures, erratic precipitation, extended droughts, pest outbreaks, and wildfires. According to the Intergovernmental Panel on Climate Change (IPCC, 2022), without enhanced mitigation and adaptation efforts, forest ecosystems risk reaching tipping points that compromise their functionality and resilience.

Forests have acted as substantial carbon sinks, absorbing about 30% of anthropogenic CO_2 emissions annually (Pan et al., 2011). Humans and natural ecosystems constantly interact with the climate, impacting all facets of life, including employment, diet, health, the economy, and overall well-being. Since the beginning of the Industrial Revolution, the excessive use of fossil fuels and changes in land use have altered weather patterns, with effects that frequently exceed the resilience limits of ecosystems (Singh et al., 2024). It is, therefore, recognized as one of the most critical problems facing the contemporary world. Since the recognition of climate change as a global political concern over fifty years ago, efforts to mitigate its effects have gradually intensified. The 1992 United Nations Conference on Environment and Development (UNCED), commonly referred to as the Earth Summit, led to the adoption of the United Nations Framework Convention on Climate Change (UNFCCC) and the Kyoto Protocol, signed in 1997, as well as the Paris Agreement, established in 2015. The Paris Agreement established ambitious goals to keep the average global temperature increase below 2 °C above pre-industrial levels and ideally below 1.5 °C. In keeping with these initiatives, the European Union has taken the lead in addressing climate change with the Green Deal, aiming to become climate-neutral by 2050 (Singh & Kumar, 2022). Nature-based solutions have recently garnered increased attention from the scientific community and policymakers as efforts to develop policies and plans for climate change adaptation and mitigation have intensified. To address societal and environmental issues while considering environmental, social, and economic factors and their interlinkages, these solutions provide a comprehensive framework of integrated, ecosystem-based approaches (Singh, 2021).

Climate-smart forestry (CSF) emerges as a next-generation approach that integrates climate mitigation and adaptation into forest policy and practice. It builds upon traditional sustainable forest management (SFM) principles, enhancing them through a climate-centric lens. CSF prioritizes ecosystem resilience, proactive planning, inclusive governance, and innovation. It addresses the triad of increasing carbon storage, enhancing ecosystem adaptability, and ensuring socio-economic sustainability.

The objective of this chapter is to offer a comprehensive examination of the concept, strategies, and implementation of CSF. It critically analyzes how CSF differs from and complements conventional forestry approaches, assesses its potential to deliver climate goals, and outlines a roadmap for its broader adoption (Kumar & Singh, 2020).

6.2 Understanding Climate Change and Forestry

The increasing frequency of climate-induced disturbances such as wildfires, droughts, floods, and pest outbreaks disrupts forest growth, species composition, and regenerative capacity. Warming temperatures cause range shifts for numerous tree species, leading to ecosystem imbalances and biodiversity loss. For instance, boreal forests are experiencing northward shifts, while tropical forests suffer from reduced rainfall and higher evapotranspiration rates (Allen et al., 2010).

Simultaneously, forests offer significant opportunities for climate mitigation by acting as carbon sinks. Tropical forests alone hold approximately 250 billion tons of carbon in their biomass (Baccini et al., 2012). However, when forests are degraded or deforested, they transition from sinks to sources of emissions. Land-use change, particularly deforestation in the tropics, accounts for approximately 11% of global greenhouse gas emissions (IPCC, 2019).

The interplay between climate and forestry is not unidirectional. Forests also influence local and regional climates through their effects on albedo, moisture recycling, and cloud formation. Their role in regulating water availability downstream highlights their importance in adaptation strategies. Therefore, integrating climate science into forest policy is critical for sustaining these ecosystems under changing conditions (Singh et al., 2018).

6.3 Concept and Principles of Climate-Smart Forestry

Climate-smart forestry (CSF) refers to forest management practices that achieve three interlinked goals: (1) increase the ability of forests to adapt to changing climatic conditions; (2) reduce or remove greenhouse gas emissions; and (3) maintain or enhance forest-based ecosystem services and livelihoods (Nabuurs et al., 2017). It is not a replacement for sustainable forest management (SFM) but a complementary approach that brings a climate-centric focus. Therefore, Climate-smart forestry is sustainable, adaptive forest management and governance that protects and enhances the potential of forests to adapt to and mitigate climate change.

6.3.1 Principles of CSF

6.3.1.1 Resilience and Adaptation

CSF enhances the resilience of forest ecosystems by diversifying species, promoting genetic variety, and implementing silvicultural systems that mimic natural processes (Puettmann et al., 2009).

6.3.1.2 Mitigation Through Sequestration

Activities such as afforestation, reforestation, and improved forest management sequester carbon in both biomass and soil.

6.3.1.3 Ecosystem-Based Management

Recognizing forests as dynamic systems with interdependent components, CSF adopts a holistic management approach integrating biodiversity, hydrology, and socio-economic needs (Messier et al., 2015).

6.3.1.4 Stakeholder Engagement

The active participation of local communities ensures that management strategies align with local needs and knowledge systems, thereby enhancing legitimacy and success.

6.3.1.5 Innovation and Technology Integration

CSF utilizes tools such as remote sensing, AI-driven modeling, and climate forecasts to facilitate data-informed decisions (Seidl et al., 2016) (Table 6.1 and Fig. 6.1).

6.4 Strategies for Enhancing Forest Resilience

Enhancing forest resilience is a cornerstone of Climate-Smart Forestry (CSF), particularly as climate change brings about more frequent and severe disturbances, including droughts, pest outbreaks, wildfires, and storms (Singh, 2024d). Resilient forests are better equipped to recover from these stressors while maintaining vital ecosystem functions. The following strategies are critical for bolstering forest resilience.

Table 6.1 The four I's for climate-smart forestry

I's	Description
Innovations	Development of resource-efficient bio-based products and business models to replace fossil-intensive and non-renewable materials, reducing emissions. It also includes innovations to measure, monitor, and manage forest ecosystem services and biodiversity
Institutions	Policy measures, such as public procurement, carbon pricing, standardization, and certification of bio-products, can create markets for bio-based solutions. Public–private partnerships are essential to invest in forest ecosystem services
Infrastructures	Mobilization and processing of sustainably harvested forest biomass into new wood-based products. Development of new and better-adapted forests as biological infrastructures
Investments	Investments in new business models, products, and technologies to create new value chains and markets. These value chains will enhance forest management, mitigate the impacts of, and adapt to, climate change while conserving biodiversity and providing valuable ecosystem services

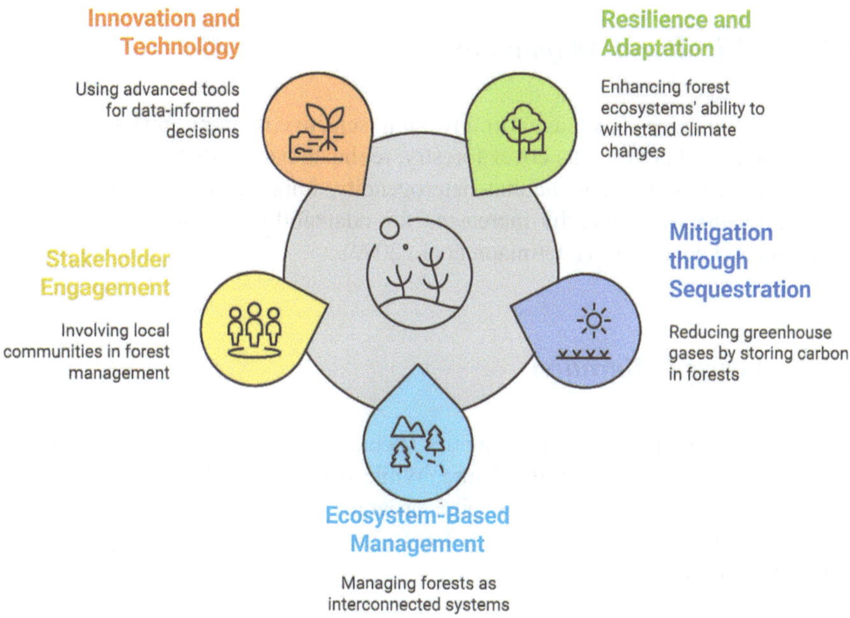

Fig. 6.1 Principles of CSF

6.4.1 Promoting Tree Species Diversity

Greater species diversity within forest stands contributes to increased ecosystem stability, productivity, and carbon storage. Diverse forests are more resistant to pests and diseases and better able to recover from climatic disturbances. Mixed-species plantations, in particular, demonstrate superior performance compared to monocultures in terms of growth, survival, and climate adaptation (Liang et al., 2016).

6.4.2 Landscape Connectivity

Ecological corridors are vital for maintaining biodiversity by facilitating the migration of species and the exchange of genes in response to shifting climate zones. Fragmented landscapes hinder these movements, elevating the risk of local extinctions and ecosystem service breakdown (Haddad et al., 2015). Creating connected forest patches through green belts and buffer zones enhances resilience at a broader landscape scale.

6.4.3 Silvicultural Adaptations

Silvicultural practices, such as those involving selective thinning, mixed-age stand development, and continuous cover forestry, replicate natural disturbance patterns. These approaches maintain habitat heterogeneity, improve light conditions, and promote regeneration, thereby increasing the adaptability of forests to changing environmental conditions (Puettmann et al., 2009).

6.4.4 Assisted Migration

Assisted migration involves relocating tree species or genotypes to regions expected to be climatically suitable in the future. While it remains a debated practice due to ecological and ethical concerns, pilot programs in North America and Europe suggest their potential to preserve key forest functions in the face of rapid climate shifts (Williams & Dumroese, 2013).

6.4.5 Agroforestry Integration

Integrating trees into agricultural systems enhances ecosystem resilience and supports diversified livelihoods. Agroforestry practices enhance soil health, reduce soil erosion, and improve water retention. They also provide shade, wind protection, and habitat continuity, contributing to resilience across rural landscapes (Raj et al., 2022).

6.5 Implementing Climate-Smart Forestry Practices

The practical implementation of CSF blends traditional ecological knowledge with modern science and robust policy frameworks. Effective CSF practices include.

6.5.1 Reforestation and Afforestation

Reforestation involves replanting forests on degraded or deforested land, while afforestation refers to planting trees on lands that were not previously forested. Both strategies are critical for enhancing carbon sinks, restoring biodiversity, and improving watershed functions. Selecting native and climate-resilient species ensures long-term ecological stability and adaptability (Singh, 2024b). In Ethiopia, the Green Legacy Initiative has mobilized millions of people to plant over 20 billion trees. Meanwhile, in Nepal, community-based reforestation programs have restored vast areas of degraded forest, resulting in measurable increases in forest cover and local income generation (FAO, 2013).

6.5.2 Natural Regeneration Support

Supporting natural regeneration allows ecosystems to recover through the natural processes of seed dispersal, germination, and growth. This approach often yields higher biodiversity, improved species adaptation, and lower costs compared to plantation forestry. Protecting natural regeneration areas from grazing, fire, and encroachment is key to success. Moreover, enhancing seed sources and soil conditions can accelerate the recovery process. Natural regeneration aligns well with conservation goals and can be especially effective in areas with remaining seed banks and minimal disturbance (Singh, 2024c).

6.5.3 Adaptive Harvesting

Adaptive harvesting incorporates ecological feedback into timber harvesting practices. It ensures that extraction rates align with forest regeneration capacity and ecological resilience. Techniques such as reduced-impact logging, extended rotation cycles, and selective cutting maintain canopy cover and minimize habitat fragmentation. In tropical forests, this approach has helped sustain timber yields while protecting ecosystem integrity. Decision-making is often informed by forest inventory data and growth models to avoid overexploitation and promote sustainability (Sodhi, 2011).

6.5.4 Fire Management

Wildfires are a natural part of many ecosystems, but their frequency and intensity have increased due to climate change and land-use pressures. Fire management under CSF includes the use of prescribed burns to reduce fuel loads and prevent uncontrolled wildfires. Creating fire breaks, thinning overgrown forests, and promoting fire-resilient species are also effective practices. Fire management is context-specific—what works in Mediterranean or Australian landscapes may differ significantly from what works in boreal forests. Community education, early warning systems, and fire response plans are crucial components (Moreira et al., 2020).

6.5.5 Community Involvement

Local communities are indispensable partners in CSF. Their knowledge of local landscapes, species, and resource management practices enhances the effectiveness and legitimacy of interventions. Community Forest User Groups (CFUGs), particularly in Nepal, exemplify successful models in which communities collectively manage forests under established legal frameworks (Singh, 2025). These groups often operate under REDD+ schemes, receiving financial compensation for verified emissions reductions and reinvesting in sustainable development. Participatory planning, benefit-sharing mechanisms, and capacity building are key to long-term community engagement and resilience.

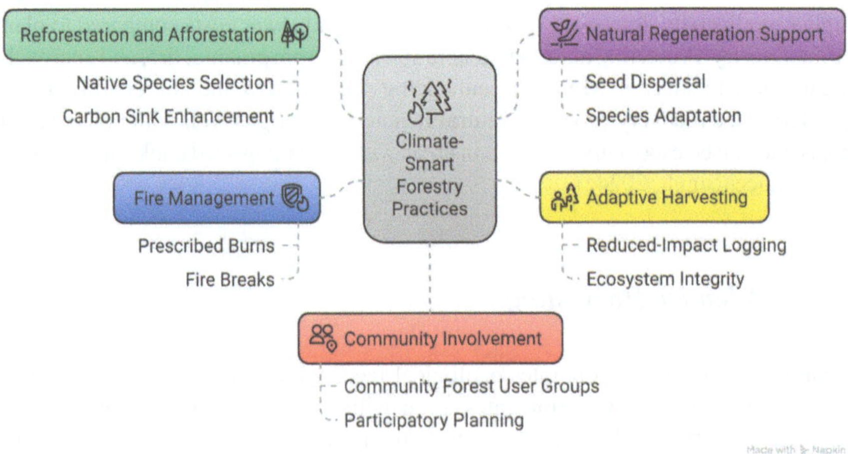

6.6 Mitigation Potential of Climate-Smart Forestry

Climate-smart forestry (CSF) presents significant opportunities to mitigate climate change through both direct and indirect pathways. Directly, CSF enhances forest carbon sinks by promoting sustainable management practices, afforestation, reforestation, and forest restoration, thereby increasing carbon sequestration and storage (Singh et al., 2023b). Indirectly, CSF contributes to reducing greenhouse gas (GHG) emissions from deforestation and forest degradation and promotes the substitution of wood-based products and bioenergy for more carbon-intensive materials and fossil fuels. Collectively, these strategies position CSF as a critical component in supporting global efforts to achieve climate mitigation targets, including those outlined in the Paris Agreement (Kumar et al., 2020).

6.6.1 Carbon Sequestration

Forests are vital carbon sinks, capturing atmospheric carbon dioxide and storing it in vegetation, litter, deadwood, and soil. CSF practices such as promoting rapid biomass growth, minimizing disturbance, and maintaining healthy forest structures improve the sequestration potential of both natural and managed forests. Afforestation, reforestation, and the protection of existing forests are core interventions that significantly enhance carbon storage capacity. Forests with high biodiversity and appropriate species composition also tend to be more resilient, maintaining consistent carbon uptake even under environmental stress (Pan et al., 2011).

6.6.2 Substitution Effects

The use of sustainably sourced timber and forest products can offset emissions from materials with higher embedded carbon, such as steel, aluminum, and concrete. For example, using engineered wood in construction reduces emissions associated with cement production. Additionally, wood products continue to store carbon throughout their lifespan, thereby further extending their climate benefits. Encouraging the use of wood in buildings, furniture, and textiles aligns with circular economy principles and enhances carbon retention beyond forest ecosystems (Sathre & O'Connor, 2010).

6.6.3 REDD+ Integration

Reducing Emissions from Deforestation and Forest Degradation (REDD+) is a global mechanism that provides financial compensation to countries for verified reductions in forest-related emissions. Integrating CSF practices within REDD+ initiatives strengthens the co-benefits of biodiversity conservation, community livelihoods, and climate mitigation. Brazil and Indonesia, for instance, have successfully secured REDD+ funding by implementing strategies that curb illegal deforestation and enhance sustainable forest governance. By aligning national forest management with REDD+ goals, countries can unlock international climate finance and improve their environmental outcomes.

6.6.4 Biochar and Soil Carbon

Biochar is a carbon-rich byproduct derived from the pyrolysis of organic biomass. When applied to soils, biochar enhances nutrient retention, boosts water-holding capacity, and remains stable for centuries, thereby sequestering carbon over the long term. CSF can incorporate biochar into land management by utilizing forest residues and invasive species for its production. This creates a closed-loop system in which waste biomass is transformed into a tool for soil enhancement and carbon sequestration. Emerging research suggests that biochar has the potential to contribute to both productivity and resilience under climate stress (Lehmann et al., 2006).

6.6.5 Emission Reduction Accounting

Credible monitoring, reporting, and verification (MRV) frameworks are crucial for accurately quantifying the mitigation benefits of CSF. Technologies such as remote sensing, satellite imagery, and forest inventory assessments enable the tracking of changes in biomass, forest cover, and carbon stocks over time. Advanced modeling tools, such as CO2FIX and CBM-CFS3, simulate carbon flows across forest systems under various scenarios, supporting strategic planning and fulfilling international reporting obligations. Transparent carbon accounting ensures that CSF practices are effectively integrated into national climate policies and global carbon markets (Kurz et al., 2009).

6.7 Monitoring and Evaluation of Climate-Smart Forestry

Monitoring and Evaluation (M&E) is a cornerstone of successful Climate-Smart Forestry implementation. It ensures transparency, facilitates adaptive management, validates carbon mitigation outcomes, and informs policymaking. A comprehensive M&E system blends technological innovation with grassroots participation to provide data that is both scientifically rigorous and contextually relevant (Jangra et al., 2024).

Effective CSF monitoring systems assess multiple dimensions of forest performance to ensure that environmental, social, and economic goals are met. The key indicators include forest health and productivity, which are typically assessed through metrics such as canopy cover, tree growth rates, and the presence of diseases or dieback. Monitoring biodiversity and species composition is also essential, involving the tracking of flora and fauna diversity to understand ecological balance and resilience (Joshi et al., 2021). Carbon sequestration levels are measured by estimating carbon stocks and changes in biomass both above and below ground, providing insight into the forest's role as a carbon sink. Socioeconomic outcomes are another key metric, examining the benefits that forests provide to local communities in terms of income, employment, and access to resources. Finally, climate risk reduction indicators assess the resilience of forests to environmental stresses, such as droughts, floods, wildfires, and pest outbreaks.

6.7.1 Technological Tools

Modern technological tools have revolutionized the monitoring, evaluation, and management of forest ecosystems, particularly in the context of climate change, biodiversity conservation, and sustainable forest management. These tools not only enhance the reach, accuracy, and efficiency of data collection and analysis but also facilitate evidence-based decision-making by integrating complex ecological and socio-economic datasets (Joshi et al., 2024).

6.7.2 Remote Sensing and Geographic Information Systems (GIS)

Remote Sensing (RS) involves the acquisition of forest data using satellite or aerial sensors that capture multi-temporal, multispectral, and high-resolution imagery. These tools provide (1) large-scale monitoring of forest cover, fragmentation, afforestation, and deforestation patterns, (2) real-time detection of forest fires, illegal encroachments, and pest outbreaks, and Temporal change analysis, enabling the tracking of seasonal and long-term ecological dynamics.

Geographic Information Systems (GIS) complement Remote Sensing (RS) by enabling advanced spatial analysis, integration, and visualization of multi-dimensional data. GIS facilitates the overlay of diverse data layers, including ecological variables (e.g., NDVI, canopy cover), socio-economic indicators (e.g., population density, land tenure), and administrative boundaries (e.g., forest divisions), to generate spatially explicit and policy-relevant insights. For example, Huang et al. (2018) demonstrated the utility of integrating MODIS-based forest disturbance detection with socio-political boundaries, thereby informing targeted and sustainable forest management interventions. This synergistic use of GIS and RS enhances decision-making processes in forest monitoring, planning, and policy formulation. GIS tools, such as ArcGIS, QGIS, and Google Earth Engine (GEE), support operations ranging from land classification to habitat modeling and carbon stock estimation (Singh, 2024a).

6.7.3 Drones and Unmanned Aerial Vehicles (UAVs)

Drones have become indispensable for site-specific forest monitoring, particularly in inaccessible or rugged terrains such as the Himalayas. Their applications include (1) high-resolution aerial surveys for canopy health, seedling regeneration, and slope stability; (2) 3D modeling and photogrammetry, which allow volumetric measurements of biomass and tree heights; (3) rapid post-disaster assessment following landslides, floods, or fires, and (4) monitoring forest encroachment, illegal logging, and the effectiveness of conservation interventions.

Compared to satellite-based remote sensing, UAVs provide finer spatial detail (e.g., sub-meter resolution), which is crucial for micro-level interventions such as restoration site selection or agroforestry assessments (Garg et al., 2016).

6.7.4 Forest Inventory and Carbon Modeling Tools

Monitoring forest biomass and estimating carbon sequestration potential is central to climate mitigation strategies. Advanced forest inventory and carbon simulation models include:

- CO2FIX: A model developed by Wageningen University that simulates carbon dynamics in forest ecosystems, plantations, and agroforestry systems over time. It allows the integration of growth rates, rotation cycles, harvesting, and management interventions to estimate long-term carbon stocks in vegetation, soil, and wood products.
- CBM-CFS3 (Carbon Budget Model of the Canadian Forest Sector): A highly detailed, forest-stand-based model used for national carbon accounting. It can project carbon fluxes under different land-use, land-cover change, and disturbance scenarios (Kurz et al., 2009).

These models are aligned with IPCC Tier 2 and Tier 3 methodologies, supporting countries in fulfilling their REDD+ and NDC (Nationally Determined Contributions) commitments by simulating future scenarios, calculating emission reductions, and providing scientifically robust data for carbon financing.

Characteristic	Remote Sensing & GIS	Drones & UAVs	Forest Inventory & Carbon Modeling	Integrative Digital Platforms & AI
Data Acquisition	Large-scale imagery	High-resolution aerial surveys	Simulates carbon dynamics	Automates species identification
Monitoring	Forest cover and fragmentation	Canopy health and encroachment	Biomass and carbon sequestration	Deforestation hotspots
Analysis	Spatial analysis and visualization	3D modeling and photogrammetry	Projects carbon fluxes	Detects anomalies in forest health
Applications	Sustainable forest policies	Post-disaster assessment	REDD+ and NDC commitments	Participatory forest monitoring

Fig. 6.2 Monitoring and evaluation of steps taken for CSF implementation using advanced tools

6.7.5 Integrative Digital Platforms and AI Tools

Emerging technologies now include AI-driven platforms, mobile apps, and machine learning algorithms that automate the identification of tree species using image recognition (e.g., PlantNet, LeafSnap). These techniques may help to predict deforestation hotspots using neural networks trained on land-use change datasets and process big data from weather stations, field surveys, and satellite feeds to detect anomalies in forest health (e.g., drought stress and phenological shifts).

Integrative platforms, such as SEPAL (System for Earth Observation Data Access, Processing, and Analysis for Land Monitoring) by FAO and Open Foris tools, are now being widely used in developing countries for participatory forest monitoring and capacity building (Fig. 6.2).

6.7.6 Community-Based Monitoring

Involving local communities in monitoring activities enhances trust, builds capacity, and ensures the timely collection of data. Local monitors can track tree growth, illegal activities, fire outbreaks, and biodiversity indicators. In countries like Nepal and Mexico, participatory monitoring programs have become integral to REDD+ implementation and forest governance, proving both cost-effective and socially empowering.

6.7.7 *Challenges*

Despite its importance, monitoring and evaluation (M&E) in Climate-Smart Forestry faces several significant challenges. One of the foremost issues is the technical capacity gap, as many forest agencies and local communities lack access to advanced tools, technologies, and the training needed to utilize them effectively. In addition, inconsistent methodologies and data standards across different regions complicate efforts to compare, aggregate, and interpret monitoring results, limiting the effectiveness of national and global assessments (Joshi et al., 2020).

Funding and data continuity also remain persistent challenges. Long-term monitoring and evaluation (M&E) efforts require sustained financial investment; however, many initiatives are constrained by short-term project cycles and limited donor support. Without consistent funding, valuable data collection efforts risk being fragmented or abandoned altogether. Moreover, integrating national monitoring and evaluation (M&E) systems with international frameworks, such as REDD+, presents further complexities. This includes aligning metrics, reporting formats, and verification protocols, all of which require high levels of coordination and technical expertise.

Therefore, addressing these challenges requires coordinated investments in training programs, digital and technical infrastructure, and the development of standardized yet adaptable monitoring and evaluation protocols. Strengthening institutional capacity and fostering collaborative networks across governmental, academic, and civil society organizations will be crucial for building robust and effective monitoring and evaluation systems that can support the long-term success of CSF.

6.8 Policy and Governance for Climate-Smart Forestry

Effective policy and governance structures form the institutional bedrock of Climate-Smart Forestry (CSF). These frameworks must promote transparency, ensure equitable benefit sharing, enable private sector engagement, and foster inter-agency coordination. A robust governance system is critical for creating an enabling environment that supports sustainable forest management, mobilizes resources, and promotes long-term commitment from all stakeholders.

6.8.1 *Enabling Policies*

Key policies that support CSF include legal instruments and national strategies that empower local actors and incentivize sustainable practices. The Forest Rights Act (India, 2006) is a landmark piece of legislation that recognizes the rights of indigenous and forest-dependent communities to access, manage, and benefit from forest resources. This legal empowerment has enabled community-based forestry to flourish in many parts of India. REDD+ strategies provide a performance-based framework that offers countries financial rewards for verified reductions in deforestation and forest degradation (Singh et al., 2023; Raj et al., 2022). For example, Indonesia and Guyana have successfully accessed international climate finance by integrating REDD+ into their national forest management systems. Nationally Determined Contributions (NDCs) under the Paris Agreement are increasingly incorporating Climate-Smart Forest (CSF) strategies, reflecting the growing recognition of forests' role in climate mitigation and adaptation.

6.8.2 Institutional Coordination

Siloed governance often impedes the effectiveness of CSF. Coordinated action among forestry, agriculture, water, and climate change departments is essential to ensure that land-use planning is coherent and mutually reinforcing. Inter-ministerial bodies, such as Kenya's National Climate Change Council and Costa Rica's National Forestry Financing Fund (FONAFIFO), serve as effective platforms for harmonizing climate and forestry objectives. These institutions ensure that sectoral strategies are aligned and that resources are allocated efficiently across competing land-use demands (Fig. 6.3).

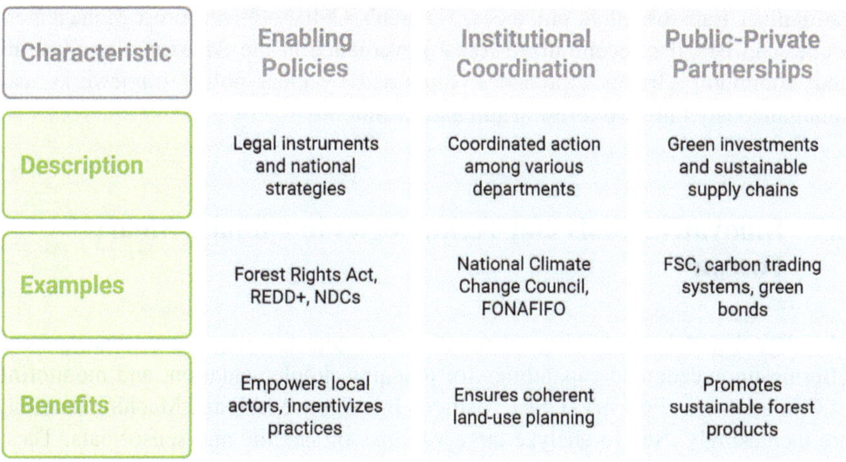

Fig. 6.3 CSF policy and governance elements

6.8.3 Public–Private Partnerships (PPPs)

The private sector plays a crucial role in scaling CSF through green investments, sustainable supply chains, and the commercialization of ecosystem services. Certification schemes, such as the Forest Stewardship Council (FSC), promote the production and sale of sustainable forest products, facilitating access to international markets. Carbon trading systems, both voluntary and compliance-based, offer additional incentives for forest conservation and restoration (Ruhil et al., 2024). In countries such as Peru and the Democratic Republic of the Congo, public–private partnerships have facilitated the development of carbon offset projects that benefit both biodiversity and the local communities' livelihoods. Furthermore, innovative financing mechanisms such as green bonds and climate funds are increasingly being tapped to support forestry initiatives.

Despite growing recognition and investment, governance challenges persist. Weak enforcement of forestry laws undermines efforts to curb illegal logging and encroachment. Corruption and lack of transparency in forest governance institutions can erode public trust and hinder progress. Land tenure conflicts remain a significant barrier, especially where customary land rights are not formally recognized. Political instability, particularly in conflict-prone regions, disrupts long-term forest planning and resource management (Barwa & Singh, 2024). However, successful examples from countries like Nepal—where community forestry programs are backed by legal recognition and devolved governance structures—demonstrate that with supportive legislation and local empowerment, Community-Scale Forests (CSF) can deliver transformative outcomes. Uganda's Collaborative Forest Management model and Brazil's decentralized forest governance in the Amazon also illustrate how community-driven approaches, supported by clear policy frameworks, can simultaneously improve forest health and livelihoods.

6.9 Innovative Tools and Technologies in Climate-Smart Forestry

The digital and biotechnological revolution is transforming forest management, offering unprecedented capabilities for planning, implementation, and monitoring of Climate-Smart Forestry (CSF). Artificial Intelligence (AI) and Machine Learning are increasingly used to analyze large volumes of satellite and sensor data. These technologies enable the prediction of wildfire risks based on climate and vegetation patterns, the automated analysis of wildlife movements using camera trap data, and the early detection of forest diseases or illegal logging through satellite change detection algorithms.

The Internet of Things (IoT) enhances on-the-ground monitoring by deploying soil sensors and microclimate monitors that provide real-time data on critical parameters, such as soil moisture, nutrient levels, air temperature, humidity, and tree growth rates.

This facilitates precision silviculture by allowing foresters to respond swiftly and effectively to changing environmental conditions, thereby improving forest health and productivity.

Decision Support Systems (DSS) integrate diverse data sources, including ecological models, socioeconomic information, and climate projections, to support evidence-based policymaking. These platforms help simulate and evaluate various land management strategies, enabling decision-makers to prioritize interventions that yield the greatest ecological and economic benefits.

Blockchain technology plays a vital role in enhancing transparency and accountability in forest-related transactions. It plays a crucial role in tracing the origin of timber products from source to market, managing carbon credits in both voluntary and compliance markets, and verifying the equitable distribution of REDD+ benefits to local communities. The immutable nature of blockchain records helps prevent fraud and builds trust among stakeholders (Fig. 6.4).

Bioinformatics and genomic tools represent a frontier in climate-adaptive forest management. Hence, identifying tree genotypes with favorable traits such as drought resistance, pest tolerance, and high carbon sequestration potential may support assisted breeding and species selection for reforestation and restoration projects. Genomic insights help tailor forest interventions to future climatic conditions, ensuring long-term resilience and productivity.

While these tools offer remarkable potential, their effective deployment requires adaptation to local contexts, robust digital infrastructure, and comprehensive digital literacy initiatives. Ensuring equitable access to these technologies is critical, particularly for marginalized and forest-dependent communities, to avoid exacerbating existing inequalities in forest governance and climate resilience.

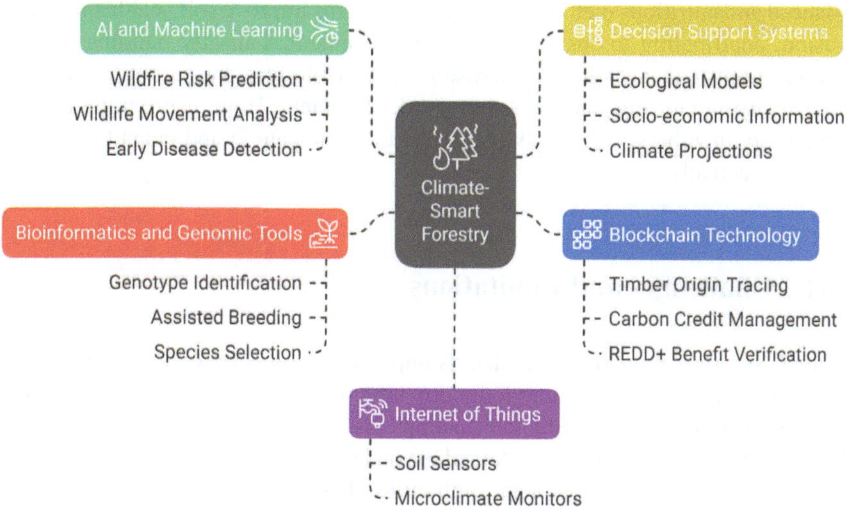

Fig. 6.4 Innovative tools for CSF

6.10 Socio-Economic Implications of Climate-Smart Forestry

Climate-smart forestry (CSF) has wide-ranging socio-economic implications that are crucial for the sustainability of forest landscapes and the well-being of the communities that depend on them. One of the most significant contributions of CSF is its potential to diversify rural livelihoods. Many forest communities rely heavily on a narrow range of forest or agricultural products, making them vulnerable to economic and climatic shocks. CSF encourages the sustainable harvesting of non-timber forest products (NTFPs) such as medicinal herbs, fruits, resins, and bamboo, which not only reduces the dependency on timber but also generates additional income. This diversification enhances economic resilience and offers communities a broader range of livelihood opportunities.

Another major aspect of CSF is its integration with carbon market mechanisms. Programs such as REDD+ (Reducing Emissions from Deforestation and Forest Degradation) offer monetary incentives to communities and forest managers for conserving forests and enhancing carbon stocks. These payments can significantly boost local economies and create new job opportunities in forest monitoring, management, and verification processes. Furthermore, CSF promotes gender equality and inclusiveness by encouraging the active involvement of women and marginalized groups in decision-making and benefit-sharing. For example, community forestry programs in Nepal have demonstrated the value of including women in leadership roles, resulting in improved forest governance and social equity.

Food and nutritional security is another critical benefit. Agroforestry practices, a key component of CSF, integrate the cultivation of trees with food crops and livestock. This integration not only increases the productivity of land but also ensures a more balanced and nutritious diet. Moreover, these systems improve soil health and reduce erosion, contributing to sustainable land management. Capacity building through CSF programs enhances local expertise in sustainable forestry, carbon accounting, biodiversity monitoring, and eco-enterprise development. Training programs provide community members with the skills needed to participate in and benefit from CSF initiatives actively.

6.11 Challenges and Limitations

While CSF offers numerous benefits, its implementation is fraught with challenges that must be addressed to ensure long-term effectiveness. A significant limitation is the lack of adequate and sustained funding. Many CSF projects are initiated with donor funding that may not be sustained beyond the initial phases, leading to disruptions in implementation. Financial limitations also restrict access to modern tools and technologies, which are crucial for effective monitoring, evaluation, and adaptive management (Larson et al., 2010).

Policy incoherence presents another serious barrier. In many countries, forestry, agriculture, and climate change policies are fragmented and sometimes conflicting. This lack of integration reduces the effectiveness of CSF interventions and complicates inter-agency coordination. Furthermore, land tenure insecurity undermines CSF initiatives (Nabuurs et al., 2017). Communities without secure land rights are less likely to invest in long-term forest conservation or restoration efforts. In such cases, unclear ownership can lead to disputes, illegal logging, and degradation.

The unpredictable impacts of climate change itself add another layer of complexity. As climate variables change rapidly, forests may experience increased stress from pests, diseases, and extreme weather events, making it difficult to plan and implement long-term forestry strategies. Additionally, there is a growing concern over instances of greenwashing, where the superficial adoption of CSF principles masks environmentally harmful activities, such as establishing monoculture plantations under the guise of reforestation.

6.12 Future Directions and Research Priorities

To scale up climate-smart forestry and maximize its benefits, it is crucial to focus on targeted research and innovation. One primary area for research is the identification of climate-resilient tree species. This involves long-term ecological trials to evaluate species performance under various climate scenarios. Selecting species with traits such as drought resistance, pest resilience, and fast growth can enhance forest productivity and stability.

In addition to biological research, integrated performance metrics that combine carbon sequestration, biodiversity conservation, and social equity need to be developed. Such metrics will help in evaluating the multifaceted outcomes of CSF projects and inform adaptive management strategies. Besides, research is required on the socio-economic impacts of CSF. Longitudinal studies that track changes in income, migration, gender roles, and community cohesion will provide valuable insights for designing inclusive policies.

Decision-support tools, such as simulation models and geographic information systems (GIS), can aid in forecasting the outcomes of various forest management strategies. These tools can support policymakers and practitioners in choosing the most effective approaches under different climate and socio-economic scenarios. Moreover, there is a need to establish innovation platforms where researchers, policymakers, and communities can collaborate to co-develop and test new approaches tailored to local contexts.

6.13 Conclusion

Climate-smart forestry (CSF) represents a transformative and forward-looking strategy that addresses the pressing challenges of climate change through sustainable forest management. Integrating climate adaptation, mitigation, and sustainable development into the CSF may provide a comprehensive framework for combating forest degradation while enhancing the resilience and well-being of forest-dependent communities. CSF strength lies in bridging ecological integrity with socio-economic empowerment, making it a robust model for inclusive, adaptive natural resource governance. The effective implementation of CSF requires addressing key operational and structural barriers. Ensuring predictable financing, harmonizing policies across sectors, securing land tenure, and fostering institutional capacity are critical to sustaining CSF initiatives. Moreover, the meaningful integration of traditional knowledge with scientific innovation may yield context-specific, culturally sensitive solutions that strengthen local engagement. Active community participation in planning, decision-making, and benefit-sharing processes enhances legitimacy, ownership, and long-term success. There is a need to embed CSF within national development strategies and global climate action frameworks. This calls for multi-stakeholder collaboration and a commitment to continuous learning and adaptive management. Therefore, it is inferred that CSF may contribute significantly to achieving climate targets, conserving biodiversity, and promoting resilient and equitable rural development.

References

Baccini, A., et al. (2012). Estimated carbon dioxide emissions from tropical deforestation improved by carbon-density maps. *Nature Climate Change, 2*(3), 182–185.

Barwa, S., & Singh, H. (2024). Climate-induced changes in forest ecosystems of Indian Himalayan region: A meta-analysis. In *Urban forests, climate change and environmental pollution: Physio-biochemical and molecular perspectives to enhance urban resilience* (pp. 839–866). Springer.

FAO. (2013). *Climate-smart agriculture sourcebook*. Food and Agriculture Organization of the United Nations.

FAO. (2020). *The State of the World's Forests 2020: Forests, biodiversity and people*. Rome: Food and Agriculture Organization of the United Nations. https://openknowledge.fao.org/server/api/core/bitstreams/dfb12960-44ee-4ddc-95f7-bec93fbb141e/content

Garg, A., Pal, D., Singh, H., & Pandey, D. C. (2016). A comparative study of NDBI, NDISI and NDII for extraction of urban impervious surface of Dehradun [Uttarakhand, India] using Landsat 8 imagery. In *2016 International Conference on Emerging Trends in Communication Technologies (ETCT)* (pp. 1–5).

Haddad, N. M., et al. (2015). Habitat fragmentation and its lasting impact on Earth's ecosystems. *Science Advances, 1*(2), Article e1500052.

Huang, Y., et al. (2018). Impacts of species richness on productivity in a large-scale subtropical forest experiment. *Science, 362*(6410), 80–83.

IPCC. (2022). *Climate Change 2022: Impacts, adaptation and vulnerability*. Contribution of Working Group II to the Sixth Assessment Report.

Jangra, A., Kumar, K., Maikhuri, S., Bhandari, M. S., Pandey, S., Singh, H., & Barthwal, S. (2024). Unveiling stress-adapted endophytic bacteria: Characterizing plant growth-promoting traits and assessing cross-inoculation effects on *Populus deltoides* under abiotic stress. *Plant Physiology and Biochemistry, 210*, Article 108610.

Joshi, R., Singh, H., Chhetri, R., & Yadav, K. (2020). Assessment of carbon sequestration potential in degraded and non-degraded community forests in Terai region of Nepal. *Journal of Forest and Environmental Science, 36*(2), 113–121.

Joshi, R., Singh, H., Chhetri, R., Poudel, S. R., & Rijal, S. (2021). Carbon sequestration potential of community forests: A comparative analysis of soil organic carbon stock in community managed forests of Far-Western Nepal. *Eurasian Journal of Soil Science, 10*(2), 96–104.

Joshi, R., Sharma, B., Singh, H., Dhakal, N., Ayer, S., & Maraseni, T. (2024). Poplar plantation as an agroforestry approach: Economic benefits and its role in carbon sequestration in North India. *Journal of Resources and Ecology, 15*(4), 880–888.

Kumar, M., & Singh, H. (2020). Agroforestry as a nature-based solution for reducing community dependence on forests to safeguard forests in Rainfed areas of India. In *Nature-based solutions for resilient ecosystems and societies* (pp. 289–306). Springer.

Kumar, A., Dwivedi, G. K., Tewari, S., Sah, V. K., Singh, H., Kumar, P., Kumar, N., & Kaushal, R. (2020). Soil organic carbon pools under *Terminalia chebula* Retz. based agroforestry system in Himalayan foothills, India. *Current Science, 118*(7), 1098–1103.

Kurz, W. A., et al. (2009). CBM-CFS3: A model of carbon-dynamics in forest ecosystems. *Ecological Modelling, 220*(4), 480–504.

Larson, A. M., et al. (2010). Tenure rights and REDD+: Lessons from the field. *Forest Governance and Policy, 21*(2), 1–7.

Lehmann, J., et al. (2006). Bio-char sequestration in terrestrial ecosystems—A review. *Mitigation and Adaptation Strategies for Global Change, 11*(2), 403–427.

Liang, J., et al. (2016). Positive biodiversity-productivity relationship predominant in global forests. *Science, 354*(6309), aaf8957.

Messier, C., Puettmann, K., & Coates, D. (2015). *Managing forests as complex adaptive systems: Building resilience to the challenge of global change*. Routledge.

Moreira, F., et al. (2020). Wildfire management in Mediterranean-type regions: Policy and practice gaps. *Environmental Science & Policy, 112*, 41–50.

Nabuurs, G. J., et al. (2017). By 2050 the mitigation effects of EU forests could nearly double through climate-smart forestry. *Forests*, 8(484).

Pan, Y., et al. (2011). A large and persistent carbon sink in the world's forests. *Science, 333*(6045), 988–993.

Puettmann, K. J., Coates, K. D., & Messier, C. (2009). *A critique of silviculture: Managing for complexity*. Island Press.

Raj, A., et al. (2022). Agroforestry a model for ecological sustainability. In *Natural resources conservation and advances for sustainability* (pp. 289–307).

Ruhil, T., Singh, H., & Barthwal, S. (2024). Long-term physiological monitoring for detecting climate change impacts on forest ecosystems. In *Forests and climate change: Biological perspectives on impact, adaptation, and mitigation strategies* (pp. 817–834). Springer.

Sathre, R., & O'Connor, J. (2010). Meta-analysis of greenhouse gas displacement factors of wood product substitution. *Environmental Science & Policy, 13*(2), 104–114.

Seidl, R., et al. (2016). Forest disturbances under climate change. *Nature Climate Change, 7*(6), 395–402.

Singh, H. (2024a). Climate change and forests: Biological perspectives. In *Forests and climate change: Biological perspectives on impact, adaptation, and mitigation strategies* (pp. 1–17). Springer.

Singh, H. (2024b). Exploring adaptive modulation in plant functional traits and its impact on the productivity of *Acacia auriculiformis* under CO_2-enriched environment. *Industrial Crops and Products, 210*, Article 118186.

Singh, H. (2024c). *Forests and climate change: Biological perspectives on impact, adaptation, and mitigation strategies*. Springer Nature.

Singh, H. (2024d). *Urban forests, climate change and environmental pollution: Physio-biochemical and molecular perspectives to enhance urban resilience*. Springer Nature.

Singh, H. (2025). The role of plant functional traits in mitigating urban climate and environmental pollution. In *Plant functional traits* (pp. 297–308). Elsevier.

Sodhi, N. S. (2011). *Southeast Asian biodiversity in crisis*. Cambridge University Press.

Singh, H., & Kumar, M. (2022). Climate change and its impact on Indian Himalayan forests: Current status and research needs. In *Climate change: Impacts, responses and sustainability in the Indian Himalaya* (pp. 223–242). Springer.

Singh, H., Kumar, N., & Kumar, A. (2024). Enhancing resource use efficiency in crops through plant functional traits. In *Plant functional traits for improving productivity* (pp. 97–117). Springer.

Singh, H., Verma, A., & Kumar, M. (2018). *Can elevated CO_2 alter the growth dynamics, morphology, physiology and biochemistry of the forestry tree species? Climate Change and Agriculture* (pp. 288–305). Discovery Publishing House.

Singh, H., Kumar, N., Singh, R., & Kumar, M. (2023a). Assessing the climate change impact on the habitat suitability of the range-restricted bird species (*Catreus wallichii*) in the Indian Himalayan ecosystem. *Environmental Science and Pollution Research, 30*(57), 121224–121235.

Singh, M., Kumar, P., Singh, H., Kumar, A. K. A., Kumar, R., & Kumar, R. (2023b). *Neolamarckia cadamba*: A comprehensive review on its physiological, ecological, phytochemical and pharmacological perspectives. *Ecology Environment and Conservation, 29*, 241–250.

Williams, M. I., & Dumroese, R. K. (2013). Preparing for climate change: Assisted migration and reforestation. *Journal of Forestry, 111*(4), 287–297.

Chapter 7
Agroforestry: A Key Contributor to Carbon Neutrality

Rupali Singh, Vivek Chauhan, Sangram B. Chavan, Anita Tomar, Pooja, Alok Kumar Singh, and Hukum Singh

Abstract Agroforestry, which integrates trees with crops and livestock, offers a robust approach to mitigating climate change, enhancing ecosystem services, and promoting sustainable development. This chapter examines the contribution of agroforestry to climate change mitigation, specifically through its carbon sequestration capabilities, ecological advantages, and socioeconomic benefits. As global temperatures continue to rise, addressing climate change requires strategies that effectively balance environmental, economic, and social goals. The chapter examines the mechanisms behind carbon sequestration in agroforestry systems, emphasizing the storage of carbon in both aboveground biomass and soil. Drawing on case studies from diverse geographical regions, it highlights the substantial carbon storage potential of agroforestry, with tropical systems sequestering carbon at rates ranging from 12 to 228 Mg C ha^{-1}. In India, it is estimated that agroforestry practices could sequester up to 2,400 million tonnes of carbon across 96 million hectares, demonstrating the potential of this practice for mitigating climate change. Beyond carbon sequestration, agroforestry systems enhance ecosystem resilience by improving soil fertility, regulating water cycles, and conserving biodiversity. This chapter also examines the socio-economic impacts of agroforestry, including its role in enhancing food security, alleviating poverty, and improving rural livelihoods. Integrating trees into agricultural practices may provide sustainable alternatives to traditional farming methods, thereby mitigating deforestation, land degradation, and greenhouse gas emissions. Furthermore, the chapter reviews policies like India's National Agroforestry Policy (2014), which supports agroforestry as a key tool for addressing environmental

R. Singh (✉) · V. Chauhan · A. Tomar · H. Singh
ICFRE-Forest Research Institute, Dehradun, Uttarakhand, India
e-mail: srupali333@gmail.com

R. Singh · S. B. Chavan
ICAR-National Institute of Abiotic Stress Management, Baramati, Maharashtra, India

Pooja
Department of Silviculture and Agroforestry, College of Forestry, Dr. Y.S. Parmar University of Horticulture and Forestry, Nauni, Solan, Himachal Pradesh, India

A. K. Singh
Rajiv Gandhi South Campus, Banaras Hindu University, Mirzapur, India

H. Singh (ed.), *Forestry for a Carbon-Neutral and Sustainable Future*, Sustainability Solutions, https://doi.org/10.1007/978-3-032-06427-1_7

challenges and achieving sustainable development objectives. Fostering climate-resilient landscapes, agroforestry contributes to the global goal of carbon neutrality and strengthens both ecological and economic resilience. The incorporation of traditional agroforestry knowledge alongside modern practices amplifies its effectiveness in mitigating climate change and supporting long-term sustainability.

Keywords Agroforestry · Climate change mitigation · Carbon sequestration · Ecosystem services · Biodiversity conservation · Sustainable development · Carbon neutrality · Policy frameworks

7.1 Introduction

The rising concentrations of greenhouse gases (GHGs), including carbon dioxide (CO_2), methane (CH_4), and nitrous oxide (N_2O), have led to global challenges such as increasing temperatures, rising sea levels, and more frequent extreme weather events impacting ecosystems, biodiversity, human health, food security, and socioeconomic development (Singh et al., 2025). Anthropogenic activities, notably the burning of fossil fuels, are the primary drivers of these changes. (Pörtner et al, 2022) If global temperatures exceed 2 °C above pre-industrial levels, scientists warn of severe and irreversible consequences for natural systems and human societies (Stern, 2007; Solomon et al., 2007). To combat these challenges, the Paris Agreement aims to limit warming to below 2 °C, with a preference for 1.5 °C, by achieving net-zero emissions by mid-century. Achieving carbon neutrality involves balancing greenhouse gas (GHG) emissions with removal mechanisms, such as carbon capture and storage technologies, as well as nature-based solutions, including agroforestry (Jose, 2009). Agroforestry, an ancient practice combining trees and shrubs with crops or livestock, has emerged as a key strategy for mitigating climate change and achieving carbon neutrality. Traditional systems, like the Taungya in northeastern India and agro-pastoral systems in Rajasthan, have historically provided essential resources while promoting ecological stability (Nair, 1993). Modern agroforestry builds on these foundations, offering diverse ecosystem services, including soil conservation, hydrological regulation, carbon sequestration, and biodiversity enhancement (Kay, 2018; Pandey, 2007). Fast-growing tree species, such as poplar (Populus deltoides) and eucalyptus (Eucalyptus spp.), which are extensively cultivated in the Indo-Gangetic Plains, demonstrate the commercial potential of agroforestry while supporting carbon storage and land restoration efforts. Beyond carbon sequestration, agroforestry systems enhance soil fertility, regulate water cycles, and provide habitats, contributing to overall climate resilience and biodiversity conservation (Jose, 2009). In India, where 2.4% of the global land supports 18% of the world's human population and 15% of livestock, agroforestry offers a sustainable solution to growing resource pressures. With the population expected to reach 1.6 billion by 2050, agroforestry aligns with the National Agroforestry Policy, which aims to increase tree cover to 33% of the geographical area (Dhyani et al., 2009). Over the decades,

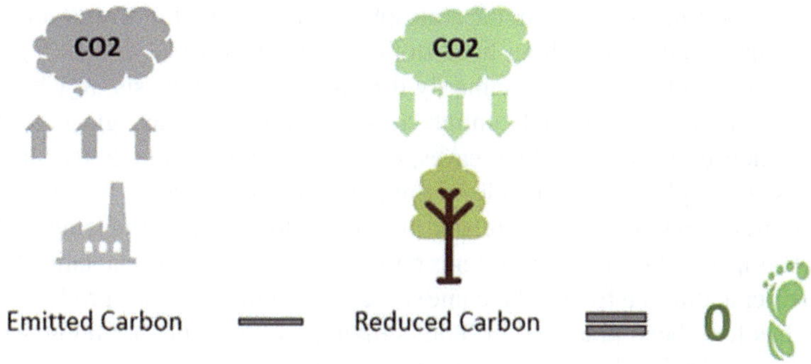

Fig. 7.1 Carbon neutrality concept (Al-Khaffaf et al., 2024)

research has highlighted the multifaceted benefits of agroforestry, including food security, poverty alleviation, and environmental protection, thereby solidifying its role as a cornerstone in climate mitigation and sustainable development (Dhyani et al., 2009). Agroforestry plays a crucial role in mitigating climate change by enhancing carbon sequestration in both biomass and soils, creating sustainable systems that integrate traditional knowledge with modern practices. These systems not only support livelihoods but also foster ecological resilience and balance. As the world faces the pressing challenge of climate change, agroforestry emerges as a key solution in the global push towards carbon neutrality and sustainable development. Achieving carbon neutrality, also known as net zero emissions, signifies a state where an equal amount of carbon removal or offsetting measures balances human-induced greenhouse gas (GHG) emissions, as demonstrated in Fig. 7.1. However, it is a complex endeavour requiring transformative changes across sectors, including energy, transportation, industry, agriculture, forestry, and waste management. It involves adopting low-carbon technologies, implementing sustainable practices, and making systemic shifts in production and consumption patterns. Agroforestry's potential to enhance carbon stocks complements these broader strategies, underscoring its importance in creating climate-resilient landscapes that align with the Sustainable Development Goals (Geels et al., 2017).

7.2 Carbon Sequestration

Climate change has emerged as a critical and global issue, demanding an urgent and coordinated response. In India, the temperature has increased by 0.60 °C over the past century, and the projected impacts of further warming could exacerbate fluctuations in crop yields, threatening food security. Studies indicate that medium-term climate change (2010–2039) could lead to yield reductions of 4.5–9%, depending on the scale and distribution of warming. According to the United Nations Framework Convention on Climate Change (UNFCCC), carbon sequestration refers to the process of capturing carbon from the atmosphere and storing it in a reservoir, which involves transferring CO_2 to long-lived storage pools (UNFCCC, 2013) (Fig. 7.2).

The carbon cycle in trees is primarily driven by photosynthesis, in which trees use carbon dioxide, water, and sunlight to produce oxygen and glucose, storing carbon in their biomass, including leaves, roots, and stems (Gorte, 2009). In agroecosystems, trees play a crucial role in mitigating climate change by sequestering carbon, buffering against weather-related production losses, and enhancing resilience to climate impacts. Agroforestry systems not only provide climate adaptation benefits but also offer additional food and income sources through tree-based products. Agroforestry has long been practiced in India, utilizing systems that are region-specific and influenced by local climate conditions (Garg et al., 2016). These systems are increasingly recognized as a viable option for mitigating climate change, as they provide multiple environmental benefits, including food security, enhanced land tenure, improved farm income, and biodiversity restoration. Agroforestry systems also maintain watershed hydrology and contribute to soil conservation (Yadava, 2010). Singh and Pandey (2011) emphasize that agroforestry's appeal as a carbon sequestration tool lies in several factors: (i) it captures carbon in vegetation and soils; (ii) it reduces the need for destructive agricultural practices like shifting cultivation; (iii) wood products from agroforestry can substitute unsustainably harvested products from natural forests; (iv) it reduces the economic incentives to exploit natural forests further; and (v) high-quality fodder species in agroforestry systems help reduce methane emissions from ruminants while also sequestering carbon. Several studies have demonstrated that agroforestry systems possess substantial carbon sequestration potential, with tropical systems sequestering between 12 and 228 Mg C ha^{-1}, yielding a mean value of 95 Mg C ha^{-1} (Pandey, 2007). Based on global estimates

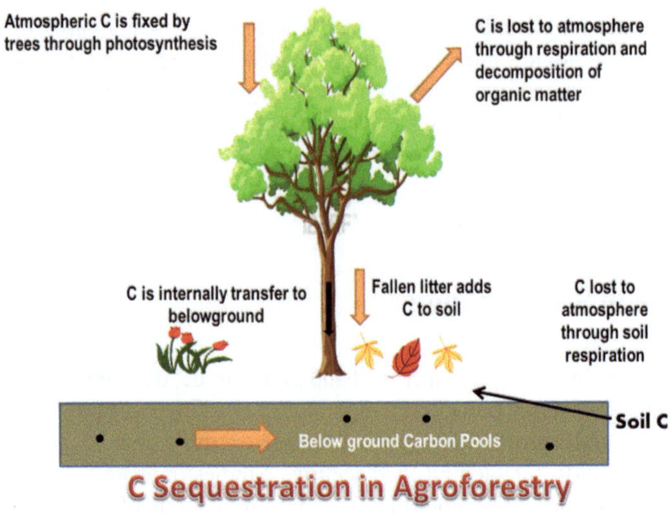

Fig. 7.2 Process of carbon sequestration through agroforestry systems

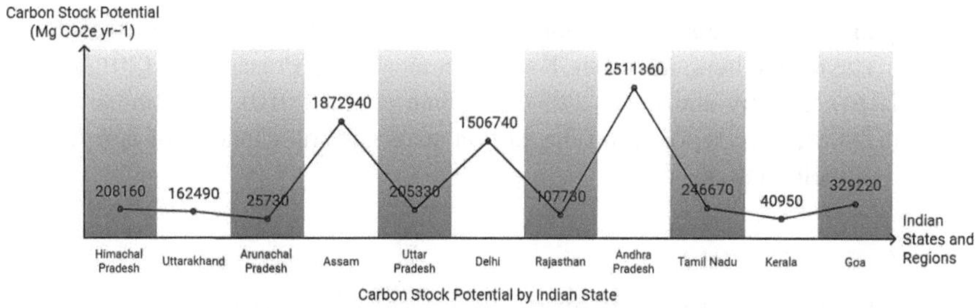

Fig. 7.3 State-wise C sequestration potential of agroforestry systems in India

of the area suitable for agroforestry, it is projected that 1.1–1.2 Pg of carbon could be sequestered in terrestrial ecosystems over the next 50 years (Albrecht & Kandji, 2003). In India, the carbon sequestration potential of agroforestry is significant, with estimates suggesting it could store around 2400 million tonnes of carbon over the country's 96 million hectares of agroforestry land (Sathaye & Ravindranath, 1998). Studies also indicate that agroforestry systems in Uttar Pradesh can sequester 19.56 Mg C ha^{-1} $year^{-1}$, and in Rajasthan's desert regions, they store between 23 and 47 Mg C ha^{-1} $year^{-1}$. (Fig. 7.3 state-wise C sequestration potential of agroforestry systems in India).

Carbon sequestration in agroforestry includes both above-ground biomass (wood, fuelwood) and below-ground components (roots, soil microbes). Agroforestry systems, particularly those that integrate trees and crops, are more effective than monoculture systems in enhancing soil fertility, increasing biomass production, and storing carbon (Nair et al., 2009a). The transition from grasslands to agroforestry systems has been shown to increase carbon sequestration by up to 36.94%, with the highest carbon stock observed in Agri-horticulture systems at 38.11 Mg C ha^{-1} (Kumara et al., 2023). Agroforestry systems provide significant opportunities for carbon storage, with estimates suggesting that over the next 50 years, they could sequester up to 2 Pg of carbon, offering a valuable strategy for mitigating and adapting to climate change. The effectiveness of carbon sequestration in agroforestry systems varies depending on the type of system, species composition, and geographic factors. These systems can also help reduce deforestation rates in tropical and subtropical regions while providing valuable ecosystem services to rural communities (Murthy et al., 2013).

The potential for agroforestry to contribute to climate change mitigation and adaptation has led to policy support, as seen in the introduction of the National Agroforestry Policy in India in 2014. This policy aims to promote research, extension services, and the production and marketing of agroforestry products. Moreover, the widespread adoption of agroforestry can enhance ecosystem services at various scales, from individual farms to entire agricultural landscapes (Panwar et al., 2022). However, studies such as by Kumara et al. (2023) highlight the influence of factors like geographical location, climate, soil conditions, and agronomic practices on the effectiveness of carbon sequestration in agroforestry systems. Table 7.1 presents the varying carbon sequestration potential of agroforestry systems across different regions. The researchers, including Kaushal et al. (2014), Prasad et al. (2012), Rizvi et al. (2014), Swamy and Puri (2005), Chauhan et al., (2010), and Ramnewaj and Dhyani (2008), have reported on the carbon sequestration capacity of agroforestry systems in their studies.

7.2.1 Carbon Sequestration and Carbon Neutrality Potential of Agroforestry Systems

Carbon sequestration refers to the net removal of CO_2 from the atmosphere through photosynthesis and its subsequent storage in long-term carbon reservoirs. These reservoirs include aboveground plant biomass, roots, and soil microbes, as well as stable organic and inorganic carbon in soils, deeper underground deposits, and long-lasting biomass-derived products such as timber (Singh & Kumar, 2022). Agroforestry systems are widely recognized for their higher carbon sequestration potential compared to grasslands or traditional agricultural fields. This is primarily due to the incorporation of trees into agricultural landscapes, which enhances carbon storage both aboveground and belowground (Panwar et al., 2022).

The dual benefits of increased carbon sequestration and reduced emissions make agroforestry a critical strategy for achieving net carbon neutrality, a goal emphasized by many nations under the Glasgow Agreement (COP-26). Several studies have evaluated carbon sequestration and emissions across various land-use systems. The CAB Abstracts database (http://www.cabi.org) lists 266 publications focused on agroforestry and carbon sequestration, with the majority of contributions emerging in the last 15 years. These assessments often combine data on aboveground time-averaged carbon stocks (typically 50% of the system's carbon stock at its maximum age or rotation duration) and soil carbon values.

7.2.2 Aboveground (Vegetation) Carbon Sequestration

Globally, forests are estimated to store up to 80% of all aboveground carbon and 40% of all belowground terrestrial carbon, including soils, litter, and roots. Assessing the biomass accumulated in forest ecosystems is vital for understanding their productivity and sustainability (Khasanah et al., 2015). Studies indicate that aboveground carbon sequestration potential (CSP) varies significantly based on vegetation characteristics. For example, 46–52% of the branch dry weight and 32% of the leaf dry weight contribute to the carbon content (Mohammadi et al., 2017).

In one study, roadside trees were found to have a total aboveground biomass of 154.53 metric tonnes, carbon stock of 72.63 metric tonnes, and carbon equivalent of

Table 7.1 Reported carbon sequestration potential (Mg C ha^{-1} yr^{-1}) of various agroforestry systems in India

Location	Agroforestry system	Tree species	No. of tree per hectare	Age (year)	CSP (Mg C ha^{-1} yr^{-1})	References
Uttarakhand	Agrisilviculture	*D. hamiltonii*	1000	7	15.91	Kaushal et al. (2014)
Himachal Pradesh	Agrihorticulture	Fruit trees	69	-	12.15	Goswami et al. (2013)
Khammam, Andhra Pradesh	Agrisilviculture	*L. leucocephala*	4444	4	14.42	Prasad et al. (2012)
			10,000	4	15.51	
Uttarakhand	Agrisilviculture	*P. deltoids*	500	8	12.02	Singh and Lodhiyal (2009)
SBS Nagar, Punjab	Agrisilviculture	*P. deltoids*	740	7	9.40	Chauhan et al. (2010)
Dehradun, Uttarakhand	Silviculture	*E. tereticornis*	2500	3.5	4.40	Dhyani et al. (1996)
			2777*	2.5	5.90	
Kurukkhetra, Haryana	Silvipasture	*A. nilotica*	1250	7	2.81	Kaur et al. (2002)
		D. sissoo	1250	7	5.37	
		P. juliflora	1250	7	6.50	
Chandigarh	Agrisilviculture	*L. leucocephala*	10,666	6	10.48	Mittal and Singh (1989)
Tripura	Silviculture	*T. grandis*	444	20	3.32	Negi et al. (1990)
		G. arborea	452	20	3.95	
Tarai Region Uttarakhand	Silviculture	*T. grandis*	570	10	3.74	Negi et al. (1995)
			500	20	2.25	
			494	30	2.87	
Jhansi, Uttar Pradesh	Agrisilviculture	A. *procera*	312	7	3.70	Ramnewaj and Dhyani (2008)
Jhansi, Uttar Pradesh	Agrisilviculture	*A. pendula*	1666	5.3	0.43	Rai et al. (2002)
Jhansi, Uttar Pradesh	Silviculture	A. *procera*	312	10	1.79	Rai et al. (2002)
		A. amara	312	10	1.00	
		A. pendula	312	10	0.95	
		D. sissoo	312	10	2.55	
		D. cinerea	312	10	1.05	
		E. officinalis	312	10	1.55	
		H. binata	312	10	0.58	
		M. azaderach	312	10	0.49	

(continued)

Table 7.1 (continued)

Location	Agroforestry system	Tree species	No. of tree per hectare	Age (year)	CSP (Mg C ha^{-1} yr^{-1})	References
Hyderabad, Andhra Pradesh	Silviculture	*L. leucocephala*	2500	9	10.32	Rao et al. (2000)
		E. camaldulensis	2500	9	8.01	
		D. sissoo	2500	9	11.47	
		A. lebbeck	625	9	0.62	
		A. albida	1111	9	0.82	
		A. tortilis	1111	9	0.39	
		A. auriculiformis	2500	9	8.64	
Hyderabad, Andhra Pradesh	Agrisilviculture	*L. leucocephala*	11,111	4	2.77	Rao et al. (1991)
			6666	4	1.90	
Raipur Chhattisgarh	Agrisilviculture	*G. arborea*	592	5	3.23	Swamy and Puri (2005)
Coimbatore Tamilnadu	Agrisilviculture	*C. equisetifolia*	833	4	1.57	Viswanath et al. (2004)
Kerala	Home garden	Mixed tree spp.	667	71	1.60	Saha et al. (2009)

266.55 metric tonnes, highlighting their significant role in mitigating climate change through carbon storage (Adekanmbi et al., 2023).

As shown in Table 7.2, agroforestry systems exhibit substantial variability in their carbon sequestration potential (CSP). This variability is influenced by factors such as site characteristics, land-use types, species composition, stand age, and management practices. Agroforests in arid, semi-arid, and degraded landscapes generally exhibit a lower carbon sequestration potential (CSP) compared to those in fertile, wetter regions. Similarly, tropical agroforestry systems typically have higher CSP than temperate systems. Sub-humid tropical regions, which practice intensive continuous cropping and short-term fallow systems, tend to have lower CSP in vegetation compared to the slash-and-burn systems commonly found in humid tropical areas (Kumar & Kunhamu, 2021).

7.2.3 Belowground (Soil) Carbon Sequestration

Soils play a crucial role in the global carbon cycle (Lal, 2008). Significant regional variations in soil carbon sequestration have been observed in forests across China (Bin et al., 2022). For instance, forest soils in provinces such as Jiangxi, Hunan,

Table 7.2 Aboveground carbon sequestration by different agroforestry systems

Sl.no	Agroforestry/Land-use system	Mean vegetation carbon	Source
1	Fodder Trees of the Lower and Middle Ouémé Valley, Benin	0.21–54.17 Mg ha^{-1} y^{-1}	Kollie et al. (2024)
2	Coconut-based Intercropping System	0.037–0.056 Mg ha^{-1} y^{-1}	Namitha et al. (2022)
3	Agroforestry at Different Altitudes in the Garhwal Himalayas	353.48–373.23 t ha^{-1}	Singh et al. (2024a)
4	Agroforestry Food Crop System on Saobi Island, Madura	11.59–14.97 t ha^{-1}	Mandasari et al. (2022)
5	Indigenous Agroforestry Systems in Silte Wereda, Ethiopia	1.28–7 Mg ha^{-1}	Semere et al. (2024)
6	Coffee Agroforests in the Western Highlands of Guatemala	74.0–259.0 Mg C ha^{-1}	Schmitt-Harsh et al. (2012)
7	Sesbania Alley Cropping in Rainfed Systems	1.72 Mg ha^{-1} y^{-1}	Palsaniya et al. (2023)
8	Tropical Seagrass Meadows in Indonesia	1.6–7.4 Mt C y^{-1}	Wahyudi et al. (2020)
9	Urban Afforestation in Prato Municipality, Italy	33.1 kt CO_2 yr^{-1}	Brilli et al. (2022)
10	Mangroves in Sukol River, Philippines	10,187.05 Mg ha^{-1}	Quitain (2021)
11	Prominent Agroforestry Systems in North-Western Himalaya, India	66.55 t ha^{-1} to 34.87 t ha^{-1}	Saleem et al. (2023)
12	Araucaria angustifolia Agroforestry in Brazil	50 to 140 Mg ha^{-1}	Zinn et al. (2024)

Zhejiang, Fujian, Anhui, Shanxi, Shaanxi, Guangxi, and Liaoning acted as carbon sources, releasing approximately 25.507 Tg C annually. In contrast, the soils in 22 other provinces served as carbon sinks, with an average sequestration rate of 103.300 Tg C per year.

The global soil carbon pool is estimated at 2,300 Pg (1 pentagram = 10^{15} g or 1 billion tonnes), which is roughly three times larger than the atmospheric carbon pool of 770 Pg and 3.8 times greater than the vegetation carbon pool of 610 Pg. A reduction of 1 Pg in the soil carbon pool corresponds to an increase of approximately 0.47 ppm in atmospheric CO_2. Hence, even minor changes in the soil carbon pool can have a significant impact on the global carbon budget. Historical CO_2 emissions

Table 7.3 Belowground carbon sequestration by different agroforestry systems

Sl.no	Agroforestry/Land-use system	Mean vegetation carbon	Source
1	Caragana korshinskii Kom Plantations, Loess Plateau	20.52 Mg ha^{-1}	Deng et al. (2017)
2	Mangroves in Eastern Niger Delta	732,595.71 ± 55.64 Mg CO_2	Drexler et al. (2024)
3	Shelterbelt Trees in Canada	20.8 g C kg^{-1}	Mayrinck et al. (2019)
4	Commercial Willow Plantation	0.07 to 0.99 Mg ha^{-1} y^{-1}	Niksa et al. (2020)
5	Tankawati Natural Forest in Bangladesh	36.26 to 522.24 kg·ha^{-1}	Ullah and Al-Amin (2012)
6	Carbon Storage in Old Hedgerows, Nigeria	43.23 Mg ha^{-1}	Nwankwo et al. (2023)
7	Typical Steppe of Nei Monggol in North China	277.35 to 7307.59 g m^{-2}	Yan et al. (2008)

from terrestrial ecosystems are estimated at 136–55 Pg, with soils contributing 78–12 Pg of these emissions. Table 7.3 provides an overview of the average potential for belowground carbon sequestration across various agroforestry systems based on multiple studies Singh et al. (2024b).

The interaction between vegetation and soil microorganisms is fundamental for long-term carbon storage. Beyond its capacity for carbon sequestration, agroforestry systems deliver a wide array of environmental and socioeconomic benefits. According to the IPCC report (Solomon et al., 2007), agroforestry enhances biodiversity conservation and ecosystem resilience. Economically, it provides diversified income sources for farmers through the production of timber, fruits, and crops while also improving farm productivity. Environmentally, agroforestry systems enhance soil fertility, improve water retention, and regulate the microclimate, contributing to their value as a sustainable land-use practice.

7.2.4 Tree Species Selection and Silvicultural Management

The "native versus exotic" species debate, along with differences in growth rates among tree species, remains a significant biological issue in the context of carbon sequestration within agroforestry systems (Oren et al., 2001). Much of this discussion stems from studies on carbon sequestration in tropical tree plantations, where carbon sequestration is often conflated with carbon stock. This concept may not fully capture the dynamics of carbon flux (Joshi et al., 2024). Although tropical tree plantations occupy only around 5% of tropical forests, their potential is growing, as these plantations are expected to expand in the future, with many species promoted for plantations being integrated into agroforestry systems (Solomon et al., 2007). The question remains whether native species, which are thought to be better suited to local conditions, will outperform exotic species in such systems (Fig. 7.4). The idea that afforestation can be a cost-effective solution for sequestering CO_2 emissions is also under scrutiny (Xu et al., 2023). Research in loblolly pine (*Pinus taeda*) forests in North Carolina, USA, revealed that while these trees initially showed accelerated growth, they eventually exhibited reduced growth rates, absorbing less carbon than anticipated (Schlesinger & Lichter, 2001). Experiments on Pinus taeda trees exposed to high atmospheric CO_2 showed no significant increase in biomass carbon at a nutrient-deficient site, and only a temporary boost at a nutrient-rich site, which stabilized after three years. A synergistic effect was observed when nutrients were added, especially at poorer sites. These findings underscore that soil fertility and nitrogen deposition have a significant influence on carbon sequestration outcomes.

Additionally, a study on the decomposition of leaves and roots in pine forest plots exposed to elevated CO_2 found that while litter production increased, decomposition rates also accelerated, resulting in the release of carbon back into the atmosphere (Koutika et al., 2021). This suggests that planting trees alone may not fully replace the need for reducing greenhouse gas emissions. Another area of concern is the variation in wood quality and carbon accumulation rates among species (Singh, 2024a).

Mixed-species plantings, particularly those involving nitrogen-fixing tropical species alongside commercial timber trees, have been shown to produce more aboveground biomass than monoculture stands. These mixed systems also offer enhanced resistance to pests and diseases. Recent research has indicated that integrating trees into vineyard agroforestry systems can improve arthropod-based pest management strategies (Favor et al., 2024). Silvicultural practices such as stand density and rotation periods further influence biomass production and carbon sequestration. High-density stands tend to sequester more carbon than lower-density stands, although this may conflict with management objectives, such as timber production. Thus, balancing

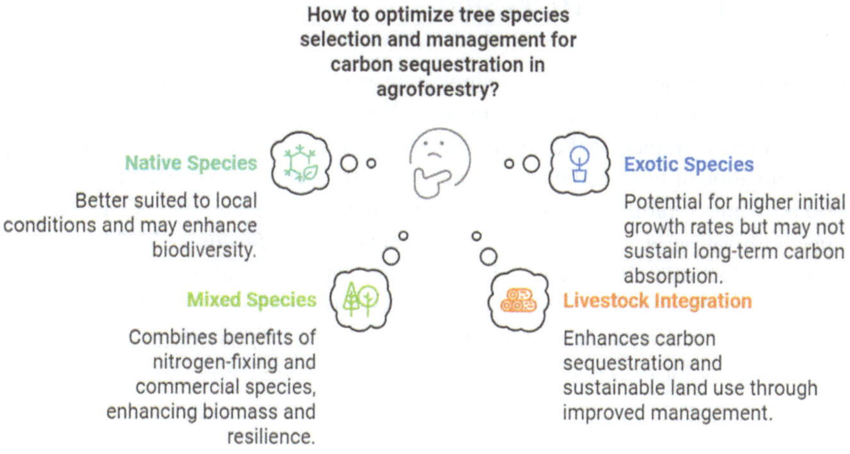

Fig. 7.4 Tree species selection and management for carbon sequestration in agroforestry

ecological services, such as carbon sequestration, with economic outputs, like timber, requires careful management of stand density (Kumar & Singh, 2020). Agroforestry implementation varies significantly across regions, adapting to local environmental and socioeconomic conditions. Successful case studies from both developed and developing countries highlight the potential for agroforestry to meet diverse regional needs. However, challenges remain, including high initial costs, long waiting periods for returns, and the need for specialized knowledge. Institutional barriers and policy gaps also hinder broader adoption, while limited land availability and competing land-use demands are significant obstacles (Kumar & Singh, 2020).

Integrating livestock into agroforestry systems provides additional opportunities for carbon sequestration and sustainable land use. Bussoni et al. (2019) demonstrated that silvo-pastoral systems can enhance soil carbon storage through improved manure distribution and grazing management. The choice of livestock breeds—whether native or exotic—can significantly affect system efficiency. Native breeds, which are often better adapted to local conditions, may promote more sustainable carbon sequestration and biodiversity conservation (Sánchez, 2005). Moreover, alley-cropping systems with nitrogen-fixing trees have been shown to optimize crop yields while maintaining substantial carbon sequestration benefits (Garrett et al., 2021). Certain agroforestry configurations naturally suppress pest populations by enhancing predator diversity, potentially reducing the need for chemical pesticides, a particularly beneficial outcome during the transition from monocropping systems (Favor et al., 2024; Pumariño et al., 2015).

7.3 Few Case Studies Across the Glob

7.3.1 MPT-Based Black Pepper-Based Production System (AICROAF-KAU, Kerala)

Black pepper (Piper nigrum) is a significant spice crop grown in Kerala's home gardens, traditionally supported by trees like Erythrina, Macaranga, and Acacia. The pepper vines are trained on these trees starting in the second year, with annual pruning and application of farmyard manure and cow dung. The climate of Kerala, with over 2000 mm annual rainfall and temperatures ranging from 38 to 19.5 °C, is ideal for pepper cultivation. Among support trees, ***Acacia auriculiformis*** yielded the highest pepper production (1.98 tons/ha). The average cost of cultivation is Rs. 44,160 per hectare. Farmers earned approximately Rs. 158,000 per hectare with Acacia and Rs. 141,600 with Artocarpus heterophyllus. This agroforestry practice proves profitable and sustainable for both marginal and large-scale farmers (Joshi et al., 2024).

7.3.2 *Integration of* Prosopis Cineraria *and* Tecomella Undulata *in Agroforestry Systems for Improved Agricultural Production in Arid Zones (ICFRE Technology; Bahuguna, 2011).*

In response to the growing demand for fodder and fuelwood due to increasing human and livestock populations, this case study explores the benefits of integrating *Prosopis cineraria* (khejri) and *Tecomella undulata* (rohida) into agricultural systems in arid regions. The region faces poor forest cover, which makes it crucial to adopt agroforestry practices that can diversify and intensify traditional agricultural systems (Singh, 2024c). Initiated in 1991, this study examined the impact of varying densities of *P. cineraria* (1666, 833, and 417 trees per hectare) and their interaction with different crops (Barwa & Singh, 2024). The tree densities were gradually reduced over the years to optimize crop yields, with thinning carried out in 1995 and 2002 to achieve densities of 417, 278, and 208 trees per hectare. Crops such as *Vigna mungo* (mungbean) and *Pennisetum glaucum* (pearl millet) were grown under rainfed conditions during the monsoon season. The results indicated that optimal tree density (417 trees per hectare at 6–7 years, 278 trees per hectare at 6–7 years, and 208 trees per hectare at 10 years) significantly improved crop yields by 10–15% compared to sole cropping systems. *P. cineraria* also provided valuable fruits (350–1040 g per tree) and biomass, including 0.85 tons of leaf fodder per hectare per year at 12 years of age (Joshi et al., 2020). The study highlighted that optimum tree density not only increased agricultural production but also offered additional biomass, demonstrating the bio-economic benefits of agroforestry. Moreover, *P. cineraria* was less competitive with crops than other tree species, making it an ideal choice for integration into farming systems. This approach has been successfully adopted in Western Rajasthan under the Marugauchar Yojana, showcasing the potential of agroforestry in arid regions for both improved agricultural productivity and sustainable land use (Singh, 2024d).

7.3.3 *Alder-Based Large Cardamom Agroforestry in Sikkim—Enhancing Ecological and Economic Sustainability*

The traditional large cardamom agroforestry system in Sikkim, historically practiced by the Lepcha people, has evolved to include the use of *Alnus nepalensis* (alder) as a shade tree. This system, initially rooted in the collection of cardamom from natural forests, has transitioned to include monocultures of *Alnus nepalensis*, a N2-fixing actinorhizal tree, alongside other native species such as *Schima wallichii*, *Engelhardtia acerfolia*, and Ficus species, thereby contributing to the region's tree biodiversity. In response to the growing demand for food and ecosystem services, this agroforestry practice plays a vital role in sustainable land management, particularly in mountainous areas where biodiversity conservation is crucial (Singh, 2025). The large cardamom agroforestry system has been shown to enhance nutrient cycling, increase soil fertility, reduce soil erosion, conserve water and biodiversity, and serve as a significant carbon sink. Notably, it stores 3.5 times more carbon than rainfed agriculture, offering substantial mitigation potential for atmospheric carbon. Moreover, this system boosts farm income and provides aesthetic and cultural value to local communities. The cardamom-based agroforestry system is an efficient model with a cost–benefit ratio exceeding 13:1, making it 5.7 times more profitable than conventional rainfed agriculture (Jangra et al., 2024). The system is primarily self-sustaining, requiring minimal external inputs, with fuelwood for curing cardamom capsules sourced from the shade trees within the system. The annual production of woody biomass in cardamom agroforestry ranges from 4.5 to 5.5 tons per hectare, more than sufficient to meet fuelwood requirements. The contribution of large cardamom to Sikkim's state income highlights the economic significance of this agroforestry practice, with a benefit–cost ratio exceeding six from the fourth year onward. This system exemplifies the potential of agroforestry to support both ecological sustainability and economic viability in mountain regions (Singh, 2025).

7.4 Economics of Agroforestry Systems in India

Numerous studies across India highlight that agroforestry can be more profitable than traditional agriculture or forestry in specific areas. A comprehensive analysis of 24 agroforestry systems from different agroclimatic regions revealed economic viability, with Internal Rate of Return (IRR) ranging from 25 to 68% and Benefit-Cost (B:C) ratios between 1.01 and 4.17 (Planning Commission, 2001). In regions such as Haryana, Punjab, and Western Uttar Pradesh, poplar-based agroforestry systems, where poplar trees are planted on agricultural fields and along field boundaries, have proven highly profitable. These systems, with a rotation period of 6–8 years, generate significantly higher economic returns than sole crops (Singh et al., 2023). A study by Ram Newaj and Rai (2005) on aonla-based agroforestry in marginal, rainfed lands showed a B:C ratio of 3.28 (discounted to 2.61), indicating strong profitability. However, most economic analyses of agroforestry have focused on input–output costs in material terms, often neglecting the temporal and spatial costs of goods and services. Future research should include more comprehensive analyses. Collaboration with wood-based industries is crucial for promoting agroforestry models, particularly those based on species like poplar, which have been successful in northern India. While species like Eucalyptus and Casuarina have been widely used, no universal agroforestry models exist for other regions (Tirkey et al., 2024). Therefore, there is significant potential to promote diverse agroforestry systems nationwide through partnerships among industries, banks, research institutes, and farmers. Although ICFRE and ICAR have developed several agroforestry models, these should be extended to farmers' fields to enhance livelihoods (Singh, 2024b).

7.5 Policy Recommendations for Advancing Agroforestry

To successfully scale agroforestry, coordinated policy actions and research initiatives are crucial. The development of standardized carbon measurement protocols is essential to facilitate global carbon credit systems and enhance the credibility of agroforestry's role in climate mitigation. Several key policy actions, as illustrated in Fig. 7.5, can support the adoption and implementation of agroforestry.

7.5.1 Financial Incentives and Support

Governments should develop financial mechanisms such as grants, subsidies, and low-interest loans to offset the initial costs of establishing agroforestry systems. Providing direct financial incentives for farmers who adopt agroforestry practices can encourage broader participation, particularly in marginal areas where initial investment may otherwise be prohibitive (Cechin et al., 2021; Franzel et al., 2001).

7.5.2 Strengthening Research and Extension Services

Research into agroforestry practices, including optimizing tree-crop-livestock combinations and enhancing ecosystem services, should be prioritized. Strengthening extension services will ensure farmers receive the knowledge and tools necessary for successful agroforestry implementation. Additionally, integrating agroforestry into agricultural education can foster future adoption by creating a skilled workforce capable of supporting agroforestry practices (Ntawuruhunga et al., 2023; Tranchina et al., 2024).

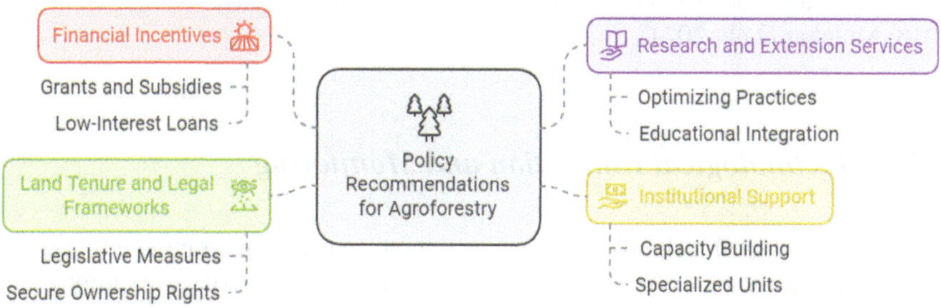

Fig. 7.5 Policy recommendations for agroforestry

7.5.3 Institutional Support and Capacity Building

Creating specialized agroforestry units within agricultural ministries, as recommended by several studies (Alavalapati et al., 2005; Akamani & Holzmueller, 2017), can help integrate agroforestry into national policy frameworks. Strengthening institutional capacities at both local and national levels will help address barriers to implementation. Capacity-building initiatives, such as training programs for farmers and extension workers, are critical to increasing awareness and technical knowledge about agroforestry (Zinngrebe et al., 2020; Katic, 2021).

7.5.4 Addressing Land Tenure and Legal Frameworks

Land tenure insecurity is a significant barrier to the adoption of agroforestry. Policymakers should implement legislative measures that provide clear land ownership and user rights to smallholder farmers, ensuring their secure tenure. Establishing secure land tenure will encourage long-term investment in agroforestry practices and help build confidence among farmers in committing to agroforestry systems (Franzel et al., 2001; Jahan et al., 2022).

7.5.5 Policy Integration and Coordination

There is often a misalignment between agricultural and forestry policies, which can hinder the promotion of agroforestry. To overcome this, agroforestry should be integrated into both national and international climate strategies, ensuring that policies in these sectors align with broader environmental goals (Joshi et al., 2021). Coordination across government sectors is crucial for developing coherent strategies that support agroforestry practices and enhance climate resilience (Rosenstock et al., 2018; Malakar et al., 2024).

7.5.6 Technological Innovation and Monitoring

Incorporating digital technologies, such as remote sensing, geographic information systems (GIS), and blockchain, can enhance the efficiency of agroforestry implementation and carbon credit verification through transparent carbon monitoring. These technologies will enable better monitoring and planning, allowing for more accurate measurement of agroforestry's carbon sequestration potential and ensuring greater transparency in carbon credit systems (Azlan et al., 2024; Nair et al. 2009b).

7.5.7 Gender-Responsive Policies

Ensuring that agroforestry policies are gender-sensitive can enhance adoption rates and improve project outcomes. Addressing the specific needs and roles of women in agroforestry systems can increase the overall effectiveness of implementation, as women often play a central role in farming and resource management (Kiptot, 2015).

Therefore, developing integrated frameworks that combine technology, financial support, institutional capacity, and gender-sensitive policies may help governments effectively address the barriers to agroforestry adoption and create an enabling environment for scaling agroforestry systems. These efforts will not only contribute to climate mitigation but also provide socio-economic benefits, enhancing the resilience of agricultural landscapes globally (Kumar et al., 2020).

7.6 Conclusion

Agroforestry stands as a transformative strategy for achieving global net-zero goals, offering a remarkable synergy of climate mitigation and adaptation benefits. Integrating trees with crop and livestock systems enhances carbon sequestration significantly compared to traditional agricultural practices, redefining agricultural carbon management. These systems effectively store carbon in above-ground biomass, root networks, and soil organic matter through their multi-layered structure, benefiting from the interplay between woody and non-woody elements. Although the theoretical basis for agroforestry's enhanced carbon sequestration is well established, current evaluations often focus on estimating carbon stocks with limited methodological precision under diverse conditions. The challenges in accurately quantifying agroforestry's carbon benefits underscore the need for refined measurement approaches to capture its full potential across varied landscapes. Despite these challenges, the adaptability of agroforestry across different geographical and socio-economic contexts reinforces its status as a globally viable solution for combating climate change. In addition to its environmental benefits, agroforestry is one of the most cost-effective nature-based approaches to achieving carbon neutrality. Beyond carbon sequestration, it delivers a range of co-benefits, including biodiversity conservation, enhanced ecosystem services, improved agricultural productivity, and strengthened rural livelihoods. Critical factors, such as standardized measurement protocols, supportive policies, accessible financing, technical expertise, and accurate land-use assessment, must be addressed to unlock the full potential of agroforestry in reaching net-zero targets. Agroforestry offers a sustainable approach to achieving carbon neutrality while simultaneously enhancing agricultural resilience and ecosystem health when implemented effectively with adequate support. The establishment of robust monitoring and evaluation systems will be essential to leverage its cost-effective advantages. As the global community intensifies its efforts to mitigate climate change, agroforestry emerges as an indispensable solution that requires enhanced focus, investment, and policy integration at local, national, and international levels. There is a need for continued research to deepen the understanding of its carbon sequestration dynamics, which will further strengthen its role in climate strategies.

References

Adekanmbi, D. I., Yevide, I. A., Koko, K. I. E. D., et al. (2023). Quantification of above-ground biomass and carbon sequestration potential of roadside trees in the plateau department of Benin Republic. *Journal of Geoscience and Environment Protection, 11*(9), 20–27. https://doi.org/10.4236/gep.2023.119002

Akamani, K., & Holzmueller, E. J. (2017). Socioeconomic and policy considerations in the adoption of agroforestry systems: An ecosystem-based adaptive governance approach. *Agroforestry*, 833–855. https://doi.org/10.1007/978-981-10-7650-3_35

Al Khaffaf, I., Tamimi, A., & Ahmed, V. (2024). Pathways to carbon neutrality: a review of strategies and technologies across sectors. *Energies, 17*(23), 6129.

Alavalapati, J. R., Mercer, D. E., Kant, S., & Lehrer, E. (2005). A framework for institutional analysis of agroforestry systems. In *Valuing agroforestry systems: Methods and applications* (pp. 279–302). https://doi.org/10.1007/1-4020-2413-4_15

Albrecht, A., & Kandji, S. T. (2003). Carbon sequestration in tropical agroforestry systems. *Agriculture, Ecosystems & Environment, 99*, 15–27.

Azlan, Z. H. Z., Junaini, S. N., & Bolhassan, N. A. (2024). Evidence of the potential benefits of digital technology integration in Asian agronomy and forestry: A systematic review. *Agricultural Systems, 217*, Article 103947. https://doi.org/10.1016/j.agsy.2024.103947

Bahuguna, V. K. (2011). Forestry in the service of nation: ICFRE technologies: A comprehensive account of R&D of Indian council of forestry research and education. Indian Council of Forestry Research & Education, Dehradun (p. 768).

Barwa, S., & Singh, H. (2024). Climate-induced changes in forest ecosystems of Indian Himalayan region: A meta-analysis. In *Urban forests, climate change and environmental pollution: physio-biochemical and molecular perspectives to enhance urban resilience* (pp. 839–866). Springer.

Bin, W., Moucheng, L., & Zhichun, Z. (2022). Preliminary estimation of soil carbon sequestration of China's forests during 1999–2008. *Journal of Resources and Ecology, 13*(1). https://doi.org/10.5814/j.issn.1674-764x.2022.01.002

Brilli, L., Carotenuto, F., Chiesi, M., Fiorillo, E., Genesio, L., Magno, R., Morabito, M., Nardino, M., Zaldei, A., & Gioli, B. (2022). An integrated approach to estimate how much urban afforestation can contribute to move towards carbon neutrality. *Social Science Research Network*. https://doi.org/10.2139/ssrn.4096154

Bussoni, A., Alvarez, J., Cubbage, F., Ferreira, G., & Picasso, V. (2019). Diverse strategies for integration of forestry and livestock production. *Agroforestry Systems, 93*(1), 333–344. https://doi.org/10.1007/s10457-017-0092-7

Cechin, A., da Silva Araújo, V., & Amand, L. (2021). Exploring the synergy between community supported agriculture and agroforestry: Institutional innovation from smallholders in a Brazilian rural settlement. *Journal of Rural Studies, 81*, 246–258. https://doi.org/10.1016/j.jrurstud.2020.10.031

Chauhan, S. K., Sharma, S. C., Chauhan, R., Gupta, N., & Srivastava, R. (2010). Accounting poplar and wheat productivity for carbon sequestration in Agri silviculture system. *Indian Forester, 136*(9), 1174–1182.

Deng, L., Han, Q. S., Zhang, C., Tang, Z. S., & Shangguan, Z. P. (2017). Above-ground and below-ground ecosystem biomass accumulation and carbon sequestration with Caragana korshinskii Kom plantation development. *Land Degradation & Development, 28*(3), 906–917. https://doi.org/10.1002/ldr.2642

Dhyani, S. K., Chauhan, D. S., Kushwaha, D. K., Lepcha, S. (1996). Sericulture-based agroforestry systems for hilly areas of north-east India. *Agroforestry Systems, 34*, 247–258.

Dhyani, S. K., Karemulla, K., Ajit, & Handa, A. K. (2009). Agroforestry potential and scope for development across agroclimatic zones of India. *Indian Journal of Forestry, 32*(2), 181–190.

Drexler, S., Thiessen, E., & Don, A. (2024). Carbon storage in old hedgerows: The importance of below-ground biomass. *GCB Bioenergy, 16*(1), Article e13112. https://doi.org/10.1111/gcbb.13112

Favor, K., Gold, M., Halsey, S., Hall, M., & Vallone, R. (2024). Agroforestry for enhanced arthropod pest management in vineyards. *Agroforestry Systems, 98*(1), 213–227. https://doi.org/10.1007/s10457-023-00900-9

Franzel, S., Coe, R., Cooper, P., Place, F., & Scherr, S. J. (2001). Assessing the adoption potential of agroforestry practices in sub-Saharan Africa. *Agricultural Systems, 69*(1–2), 37–62. https://doi.org/10.1016/S0308-521X(01)00017-8

Garg, A., Pal, D., Singh, H., & Pandey, D. C. (2016). A comparative study of NDBI, NDISI and NDII for extraction of urban impervious surface of Dehradun [Uttarakhand, India] using Landsat 8 imagery. In *2016 International Conference on Emerging Trends in Communication Technologies (ETCT)* (pp. 1–5).

Garrett, H. E., Wolz, K. J., Walter, W. D., Godsey, L. D., & McGraw, R. L. (2021). Alley cropping practices. In *North American agroforestry* (pp. 163–204). https://doi.org/10.1002/9780891183785.ch7. (13 Dec 2021)

Geels, F. W., Sovacool, B. K., Schwanen, T., & Sorrell, S. (2017). The socio-technical dynamics of low-carbon transitions. *Joule, 1*(3), 463–479.

Gorte, R. W. (2009). *Carbon sequestration in forests*. DIANE Publishing.

Goswami, S., Verma, K. S., & Kaushal, R. (2013). Biomass and carbon sequestration in different agroforestry systems of a Western Himalayan watershed. *Biological Agriculture & Horticulture*. https://doi.org/10.1080/01448765.2013.855990

Jahan, H., Rahman, M. W., Islam, M. S., Rezwan-Al-Ramim, A., Tuhin, M. M.-U.-J., & Hossain, M. E. (2022). Adoption of agroforestry practices in Bangladesh as a climate change mitigation option: Investment, drivers, and SWOT analysis perspectives. *Environmental Challenges, 7*, 100509. https://doi.org/10.1016/j.envc.2022.100509

Jangra, A., Kumar, K., Maikhuri, S., Bhandari, M. S., Pandey, S., Singh, H., & Barthwal, S. (2024). Unveiling stress-adapted endophytic bacteria: Characterizing plant growth-promoting traits and assessing cross-inoculation effects on *Populus deltoides* under abiotic stress. *Plant Physiology and Biochemistry, 210*, Article 108610.

Jose, S. (2009). *Agroforestry for ecosystem services and environmental benefits: An overview* (pp. 1–10). Springer.

Joshi, R., Sharma, B., Singh, H., Dhakal, N., Ayer, S., & Maraseni, T. (2024). Poplar plantation as an agroforestry approach: Economic benefits and its role in carbon sequestration in North India. *Journal of Resources and Ecology, 15*(4), 880–888.

Joshi, R., Singh, H., Chhetri, R., Poudel, S. R., & Rijal, S. (2021). Carbon sequestration potential of community forests: A comparative analysis of soil organic carbon stock in community managed forests of Far-Western Nepal. *Eurasian Journal of Soil Science, 10*(2), 96–104.

Joshi, R., Singh, H., Chhetri, R., & Yadav, K. (2020). Assessment of carbon sequestration potential in degraded and non-degraded community forests in Terai region of Nepal. *Journal of Forest and Environmental Science, 36*(2), 113–121.

Katic, P. (2021). Climate change governance in agroforestry systems: A systematic review. *Journal of the British Academy, 9s10*(s10), 7–20. https://doi.org/10.5871/jba/009s10.007

Kaur, B., Gupta, S. R., Singh, G. (2002). Carbon storage and nitrogen cycling in silvopastoral systems on a sodic soil in northwestern India. *Agroforestry Systems, 54*, 21–29.

Kaushal, R., Tewari, S. K., Banik, R. L., Chaturvedi, S. (2014). Growth, biomass production and soil properties under different bamboo species. In: *ISTS-IUFRO Conference on Sustainable Resource Management for Climate Change mitigation and social security at Chandigarh India* (13–14 March 2014)

Kay, S. (2018). *Assessment of ecosystem services provided by agroforestry systems at the landscape scale* (Doctoral dissertation, University of Zurich).

Khasanah, N. M., van Noordwijk, M., & Ningsih, H. (2015). Aboveground carbon stocks in oil palm plantations and the threshold for carbon-neutral vegetation conversion on mineral soils. *Cogent Environmental Science, 1*(1), 1119964.

Kiptot, E. (2015). Gender roles, responsibilities, and spaces: Implications for agroforestry research and development in Africa. *International Forestry Review, 17*(4), 11–21. https://doi.org/10.1505/146554815816086426

Kollie, A. Y., Fangninou, F. F., Teka, O., Aboh, A. B., Honvou, S. H. S., & Mumbi, A. W. (2024). Diversity and carbon storage in fodder trees of the lower and middle Ouémé valley, Benin: Ecological and pastoral sustainability. *European Journal of Theoretical and Applied Sciences, 2*(3), 947–960. https://doi.org/10.59324/ejtas.2024.2(3).74

Koutika, L. S., Taba, K., Ndongo, M., & Kaonga, M. (2021). Nitrogen-fixing trees increase organic carbon sequestration in forest and agroforestry ecosystems in the Congo basin. *Regional Environmental Change, 21*(4), 109.

Kumar, B. M., & Kunhamu, T. K. (2021). Carbon sequestration potential of agroforestry systems in India: A synthesis. *Agroforestry and Ecosystem Services* (pp. 389–430). Springer.

Kumar, A., Dwivedi, G. K., Tewari, S., Sah, V. K., Singh, H., Kumar, P., Kumar, N., & Kaushal, R. (2020). Soil organic carbon pools under *Terminalia chebula* Retz. Based agroforestry system in Himalayan foothills, India. *Current Science, 118*(7), 1098–1103.

Kumar, M., & Singh, H. (2020). Agroforestry as a nature-based solution for reducing community dependence on forests to safeguard forests in rainfed areas of India. In *Nature-based solutions for resilient ecosystems and societies* (pp. 289–306). Springer.

Kumara, K., Pal, S., Chand, P., & Kandpal, A. (2023). Carbon sequestration potential of agroforestry systems in Indian agricultural landscape: A meta-analysis. *Ecosystem Services, 62*, Article 101537.

Lal, R. (2008). Soil carbon stocks under present and future climate with specific reference to European ecoregions. *Nutrient Cycling in Agroecosystems, 81*(2), 113–127. https://doi.org/10.1007/s10705-007-9147-x

Malakar, A., Kumar, R., Prasad, S. R., & Malakar, A. (2024). Agroforestry for carbon neutrality: An effective pathway to net zero. *Journal of Advances in Biology & Biotechnology, 27*(11), 591–602.

Mandasari, P. A., Adim, M., Aisjah, S., Supriyadi, S., & Murniyanto, E. (2022). The capacity of agroforestry and food crop system for C Stock and sequestration (case study on Saobi Island madura). In *IOP Conference Series: Earth and Environmental Science: Earth Environ Sci IOP Conference Series* (Vol. 1005, no. 1). https://doi.org/10.1088/1755-1315/1005/1/012011

Mayrinck, R. C., Laroque, C. P., Amichev, B. Y., & Van Rees, K. (2019). Above- and belowground carbon sequestration in shelterbelt trees in Canada: A review. *Forests, 10*(10), 922. https://doi.org/10.3390/f10100922

Mittal, S. P., & Singh, P. (1989). Intercropping field crops between rows of *Leucaena leucocephala* under rainfed conditions in northern India. *Agroforestry System, 8*(2), 165–172.

Mohammadi, Z., Mohammadi Limaei, S., Lohmander, P., & Olsson, L. (2017). Estimating the aboveground carbon sequestration and its economic value: Case study: Iranian Caspian Forests. *Journal of Forest Research, 63*(11), 511–518.

Murthy, I. K., Gupta, M., Tomar, S., Munsi, M., Tiwari, R., Hegde, G. T., & Ravindranath, N. H. (2013). Carbon sequestration potential of agroforestry systems in India. *Journal of Earth Science Climate Change, 4*(1), 1–7.

Nair, P. K. R., Kumar, B. M., & Nair, V. D. (2009a). Agroforestry as a strategy for carbon sequestration. *Journal of Plant Nutrition and Soil Science, 172*, 10–23.

Nair, P. K. R. (1993). State-of-the-art of agroforestry research and education. *Agroforestry Systems, 23*, 95–119.

Nair, P. K. R., Nair, V. D., Kumar, B. M., & Haile, S. G. (2009). Soil carbon sequestration in tropical agroforestry systems: A feasibility appraisal. *Environmental Science and Policy, 12*(8), 1099–1111. https://doi.org/10.1016/j.envsci.2009.01.010

Negi, J. D. S., Bahuguna, V. K., & Sharma, D. C. (1990). Biomass production and distribution of nutrients in 2 years old teak (*Tectona grandis*) and gamar (*Gmelina arborea*) plantation in Tripura. *Indian Forester, 116*(9), 681–686.

Negi, M. S., Tandon, V. N., & Rawat, H. S. (1995). Biomass and nutrient distribution in young teak (*Tectona grandis*) plantation in Tarai Region of Uttar Pradesh. *Indian Forester, 121*(6), 455–463.

Newaj, R., & Rai, P. (2005). Aonla-based agroforestry system: A source of higher income under rainfed conditions. *Indian Farming, 55*(9), 24–27.

Niksa, D., Krzyżaniak, M., & Stolarski, M. J. (2020). The estimation of above- and below-ground biomass residues and carbon sequestration potential in soil on commercial willow Plantation. In M. Wróbel, M. Jewiarz, & A. Szlęk (Eds.), *Renewable energy sources: Engineering, technology, innovation* (pp. 257–266). Springer proceedings in energy. Springer. https://doi.org/10.1007/978-3-030-13888-2_25

Ntawuruhunga, D., Ngowi, E. E., Mangi, H. O., Salanga, R. J., & Shikuku, K. M. (2023). Climate-smart agroforestry systems and practices: A systematic review of what works, what doesn't work, and why. *Forest Policy and Economics, 150*, 102937. https://doi.org/10.1016/j.forpol.2023.102937

Nwankwo, C., Tse, A. C., Nwankwoala, H. O., Giadom, F. D., & Acra, E. J. (2023). Below ground carbon stock and carbon sequestration potentials of mangrove sediments in Eastern Niger Delta, Nigeria: Implication for climate change. *Scientific African, 22*, Article e01898. https://doi.org/10.1016/j.sciaf.2023.e01898

Oren, R., Ellsworth, D. S., Johnsen, K. H., Phillips, N., Ewers, B. E., Maier, C., Schäfer, K. V., McCarthy, H., Hendrey, G., McNulty, S. G., & Katul, G. G. (2001). Soil fertility limits carbon sequestration by forest ecosystems in a CO2-enriched atmosphere. *Nature, 411*(6836), 469–472. https://doi.org/10.1038/35078064. PubMed: 11373677.

Palsaniya, D. R., Kumar, T. K., Chaudhary, M., & Choudhary, M. (2023). Effect of reduced tillage and mulching on soil health in Sesbania Alley cropping based rainfed food–Fodder systems. *Archives of Agronomy and Soil Science, 69*(10), 1750–1764. https://doi.org/10.1080/03650340.2022.2111025

Pandey, D. N. (2007). Multifunctional agroforestry systems in India. *Current Science, 92*(4), 455–463.

Panwar, P., Mahalingappa, D. G., Kaushal, R., Bhardwaj, D. R., Chakravarty, S., Shukla, G., Thakur, N. S., Chavan, S. B., Pal, S., & Nayak, B. G. (2022). Biomass production and carbon sequestration potential of different agroforestry systems in India: a critical review. *Forests*, 13, 1274.

Planning Commission. (2001). Report of the task force on greening India for livelihood security and sustainable development (p. 254). Planning Commission, GoI, New Delhi.

Pörtner, H. O., Roberts, D. C., Adams, H., Adler, C., Aldunce, P., Ali, E., Begum, R. A., Betts, R., Kerr, R. B., Biesbroek, R., Climate Change.(2022). *Impacts, adaptation and vulnerability*. Cambridge University Press.

Prasad, J. V. N. S., Srinivas, K., Rao, C. S., Ramesh, C., Venkatravamma, K., & Venkateswarlu, B. (2012). Biomass productivity and carbon stocks of farm forestry and agroforestry systems of leucaena and eucalyptus in Andhra Pradesh, India. *Current Science, 103*(5), 536–540.

Pumariño, L., Sileshi, G. W., Gripenberg, S., Kaartinen, R., Barrios, E., Muchane, M. N., Midega, C., & Jonsson, M. (2015). Effects of agroforestry on pest, disease and weed control: A metaanalysis. *Basic and Applied Ecology, 16*(7), 573–582. https://doi.org/10.1016/j.baae.2015.08.006

Quitain, R. A. (2021). Describing the greenhouse gas reduction capacity of mangroves by carbon stock assessment using allometric data in Sukol River, Bongabong, Oriental Mindoro, Philippines. *International Multidisciplinary Research Journal, 3*(4), 139–147. https://doi.org/10.54476/iimrj215

Rai, A. K., Solanki, K. R., & Rai, P. (2002). Performance of *Anogeissus pendula* genotypes under Agri silviculture system. *Indian Jornal of Agroforestry, 4*(1), 71–77.

Ramachandran Nair, P. K., Mohan Kumar, B., & Nair, V. D. (2009). Agroforestry as a strategy for carbon sequestration. *Journal of Plant Nutrition and Soil Science, 172*(1), 10–23.

Ramnewaj, & Dhyani, S. K. (2008). Agroforestry for carbon sequestration: Scope and present status. *Indian Journal of Agroforestry* 10:1–9.

Rao, L. G. G., Joseph, B., & Sreemannarayana, B. (2000). Growth and biomass production of some important multipurpose tree species on rainfed sandy loam soils. *Indian Forester, 126*(7), 772–781.

Rao, M. R., Ong, C. K., Pathak, P., & Sharma, M. M. (1991). Productivity of annual cropping and agroforestry systems on a shallow Alfisol in semi-arid India. *Agroforestry System, 15*, 51–63.

Rizvi, R. H., Dhyani, S. K., Newaj, R., Karmakar, P. S., Saxena, A. (2014). Mapping agroforestry area in India through remote sensing and preliminary estimates. *Indian Farming, 63*(11), 62–64.

Rosenstock, T. S., Wilkes, A., Jallo, C., Namoi, N., Bulusu, M., Suber, M., Bernard, F., & Mboi, D. (2018). Making trees count: Measurement, reporting and verification of agroforestry under the UNFCCC [CCAFS working paper]. (30 Nov 2018).

Saha, S., Nair, P. K. R., Nair, V. D., & Kumar, B. M. (2009). Soil carbon stocks in relation to plant diversity of home gardens in Kerala, India. *Agroforestry System, 76*, 53–65.

Saleem, I., Mugloo, J. A., Pala, N. A., Bhat, G. M., Masoodi, T. H., Mughal, A. H., Baba, A. A., & Mehraj, B. (2023). Biomass production, carbon Stock and sequestration potential of prominent agroforestry systems in North-Western Himalaya, India. *Frontiers in Forests and Global Change, 6*. https://doi.org/10.3389/ffgc.2023.1192382

Sánchez, L. (2005). Indigenous breeds and silvopastoral systems. In Silvopastoralism and sustainable land management. In *Proceedings of an International Congress on Silvopastoralism and Sustainable Management Held in Lugo, Spain, April 2004* (pp. 231–235). CABI Publishing. https://doi.org/10.1079/9781845930011.0231

Sathaye, J. A., & Ravindranath, N. H. (1998). Climate change mitigation in the energy and forestry sectors of developing countries. *Annual Review of Environment Resources, 23*, 387–437.

Schlesinger, W. H., & Lichter, J. (2001). Limited carbon storage in soil and litter of experimental forest plots under increased atmospheric CO2. *Nature, 411*(6836), 466–469. https://doi.org/10.1038/35078060. PubMed: 11373676.

Schmitt-Harsh, M., Evans, T. P., Castellanos, E., & Randolph, J. C. (2012). Carbon stocks in coffee agroforests and mixed dry tropical forests in the Western Highlands of Guatemala. *Agroforestry Systems, 86*(2), 141–157. https://doi.org/10.1007/s10457-012-9549-x

Semere, M., Tadesse, C., Abebe, T., Cherinet, A., & Gebreyesus, M. (2024). Comparative carbon stock potential of indigenous agroforestry systems in silte wereda, Southern Ethiopia. *East African Journal of Forestry and Agroforestry*, *7*(1), 134–145. https://doi.org/10.37284/eajfa.7.1.1868

Singh, H. (2024a). Climate change and forests: biological perspectives. In *Forests and climate change: biological perspectives on impact, adaptation, and mitigation strategies* (pp. 1–17). Springer.

Singh, H. (2024b). Exploring adaptive modulation in plant functional traits and its impact on the productivity of *Acacia auriculiformis* under CO_2-enriched environment. *Industrial Crops and Products, 210*, Article 118186.

Singh, H. (2024c). *Forests and climate change: Biological perspectives on impact, adaptation, and mitigation strategies*. Springer Nature.

Singh, H. (2024d). *Urban forests, climate change and environmental pollution: Physio-biochemical and molecular perspectives to enhance urban resilience*. Springer Nature.

Singh, H. (2025). The role of plant functional traits in mitigating urban climate and environmental pollution. In *Plant functional traits* (pp. 297–308). Elsevier.

Singh, H., & Kumar, M. (2022). Climate change and its impact on Indian Himalayan forests: current status and research needs. In *Climate change: impacts, responses and sustainability in the Indian Himalaya* (pp. 223–242). Springer.

Singh, H., Kumar, N., & Kumar, A. (2024). Enhancing resource use efficiency in crops through plant functional traits. In *Plant functional traits for improving productivity* (pp. 97–117). Springer.

Singh, H., Kumar, N., Singh, R., & Kumar, M. (2023). Assessing the climate change impact on the habitat suitability of the range-restricted bird species (*Catreus wallichii*) in the Indian Himalayan ecosystem. *Environmental Science and Pollution Research, 30*(57), 121224–121235.

Singh, P., & Lodhiyal, L. S. (2009). Biomass and carbon allocation in 8-year-old poplar (*Populus deltoides* Marsh) plantation in Tarai agroforestry systems of central Himalaya, India. *New York Science Journal, 2*, 49–53.

Singh, N., Riyal, M. K., Singh, B., Khanduri, V. P., Rawat, D., Singh, C., Pinto, M. M. S. C., & Kumar, M. (2024). Carbon sequestration potential of agroforestry versus adjoining forests at different altitudes in the Garhwal Himalayas. *Atmosphere, 15*(3). https://doi.org/10.3390/atmos15030313

Singh, V. S., & Pandey, D. N. (2011). Multifunctional agroforestry systems in India: Science-based policy options. RSPCB Occasional Paper No. 4/2011. Rajasthan State Pollution Control Board.

Singh, R., Prakash, P., Singh, A. K., Shukla, S., Sharma, P., Chauhan, V., & Singh, B. K. (2025). Altitudinal variation in carbon stocks of *Quercus leucotrichophora* forests in the Western Himalayas: implications for climate change mitigation. *Trees, Forests and People,* 101059.

Solomon, S., Qin, D., Manning, M., Chen, Z., Marquis, M., Averyt, K., & Miller, H. (2007). IPCC fourth assessment report (AR4). *Climate change, 374.*

Stern, N. H. (2007). *The economics of climate change: the stern review.* Cambridge University Press.

Swamy, S. L., & Puri, S. (2005). Biomass production and C-sequestration of *Gmelina arborea* in plantation and agroforestry system in India. *Agroforestry System, 64*(3), 181–195.

Tirkey, A. I. P., Kumar, R., Prasad, R. S., & Malakar, A. (2024). Agroforestry for carbon neutrality: an effective pathway to net zero. *Journal of Advances in Biology & Biotechnology, 27*(11), 591–602.

Tranchina, M., Reubens, B., Frey, M., Mele, M., & Mantino, A. (2024). What challenges impede the adoption of agroforestry practices? A global perspective through a systematic literature review. *Agroforestry Systems, 98*(6), 1817–1837. https://doi.org/10.1007/s10457-024-00993-w

Ullah, M. R., & Al-Amin, M. (2012). Above- and below-ground carbon stock estimation in a natural forest of Bangladesh. *Journal of Forest Science, 58*(8), 372–379. https://doi.org/10.17221/103/2011-JFS

UNFCCC, (2013). United Nations farmework convention on climate change. https://cdm.unfccc.int/methodologies/index.html

Viswanath, S., Peddappaiah, R. S., Subramoniam, V., Manivachakam, P., & George, M. (2004). Management of *Casuarina equisetifolia* in wide-row intercropping systems for enhanced productivity. *Indian Journal of Agroforestry, 6*(2), 19–25.

Wahyudi, A. A. J., Rahmawati, S., Irawan, A., Hadiyanto, H., Prayudha, B., Hafizt, M., Afdal, A., Adi, N. S., Rustam, A., Hernawan, U. E., Rahayu, Y. P., & Kiswara, W. (2020). Assessing carbon stock and sequestration of the tropical seagrass meadows in Indonesia. *Ocean Science Journal, 55*, 85–97. https://doi.org/10.1007/s12601-020-0003-0

Xu, X., Wu, H., Yue, J., Tang, S., & Cheng, W. (2023). Effects of snow cover on carbon dioxide emissions and their δ13C values of temperate forest soils with and without litter. *Forests, 14*(7). https://doi.org/10.3390/f14071384

Yadav, G. S., Kandpal, B. K., Das, A., Babu, S., Mohapatra, K. P., Devi, A. G., Chandra, P., Singh, R., & Barman, K. K. (2021). Impact of 28-year-old agroforestry systems on soil carbon dynamics in Eastern Himalayas. *Journal of Environmental Management, 283*, Article 111978.

Yadava, A. K. (2010). Carbon Sequestration A: Under exploited environmental benefits of *Tari* Agroforestry systems. *Report and Opinion, 2*(11), 35–41.

Yan, Y. C., Tang, H. P., Chang, R. Y., & Liu, L. (2008). Variation of below-ground carbon sequestration under long term cultivation and grazing in the typical steppe of Nei Monggol in North China. *Huan Jing ke Xue= Huanjing Kexue*, *29*(5), 1388–1393.

Zinn, Y. L., Cardoso, R., & Silva, C. A. (2024). Soil organic carbon sequestration under Araucaria angustifolia plantations but not under exotic tree species on a mountain range. *Revista Brasileira de Ciência do Solo*, *48*. https://doi.org/10.36783/18069657rbcs20230146

Zinngrebe, Y., Borasino, E., Chiputwa, B., Dobie, P., Garcia, E., Gassner, A., Kihumuro, P., Komarudin, H., Liswanti, N., Makui, P., Plieninger, T., Winter, E., & Hauck, J. (2020). Agroforestry governance for operationalising the landscape approach: Connecting conservation and farming actors. *Sustainability Science*, *15*(5), 1417–1434. https://doi.org/10.1007/s11625-020-00840-8

Chapter 8
Emerald Canopies in Concrete Jungles: Harnessing Urban Forests for Climate Action and Environmental Sustainability

Bhavya Thapa, Agnish Kumar Das, Diksha Verma, and Bhoomika Ghale

Abstract The Industrial Revolution brought exponential economic growth and opportunities, catalyzing rapid urbanization. While this has helped humanity to make significant progress, it has also confronted us with the caveats of environmental degradation, such as the emergence of urban heat islands and the surge in human ailments, which are greatly intensified under climate change scenarios. With demographic trends indicating rapid growth in urban centers, there is an urgent need to redesign our cities to foster sustainable urban landscapes. The growing focus on urban forestry and its capacity to address contemporary environmental challenges has garnered significant interest. A science-backed approach to managing and developing urban forests can offer a hyper-local solution to address critical environmental challenges and harness the flow of ecosystem services in urban landscapes. This capacity of urban forests to act as a first line of defense in localized pollution abatement and climate change mitigation positions them as a pivotal tool for achieving the global objectives of carbon neutrality. Urban forest's ability to enhance resilience aligns seamlessly with environmental sustainability goals. Optimizing ecosystem services from urban forests amidst challenges like elevated temperatures, water scarcity, and a heavy pollution load is imperative for the anthropogenic cause. This requires a scientific screening of tree species that exhibit positive physiological responses and adaptations to various abiotic stressors, accompanied by effective management strategies for urban planning under different climatic and socio-economic scenarios. Urban forests

B. Thapa (✉)
Dr YS Parmar University of Horticulture and Forestry, Nauni, Himachal Pradesh, India
e-mail: bhavyathapa18@gmail.com

A. K. Das
Faculty of Forestry and Wood Sciences, Czech University of Life Sciences, Prague, Czech Republic
e-mail: kumar_das@fld.czu.cz

D. Verma
Forest Research Institute Deemed to be University, Dehradun, India

B. Ghale
ISRO-Indian Institute of Remote Sensing, Hathibarkala, Dehradun, Uttarakhand, India

H. Singh (ed.), *Forestry for a Carbon-Neutral and Sustainable Future*, Sustainability Solutions, https://doi.org/10.1007/978-3-032-06427-1_8

are recognized as a potential, cost-effective, nature-based solution that also ensures enhanced accessibility to green spaces for all, improving the livability of cities by many folds. These green sanctuaries prophesied to play a key role in helping develop a future-proof and sustainable urban landscape demand a multi-pronged comprehension. This chapter examines the origins of this concept, its global evolution, and the prevailing discussions within the field of urban forests to inform future researchers and urban planners in developing a sustainable and resilient urban landscape.

Keywords Urban forestry · Green belts · Sponge cities · Urban biodiversity · Ecological infrastructure · Carbon neutrality

8.1 Introduction

Urbanization and explosive global population growth are two defining demographic trends of the modern era (Brown, 2013). Despite cities covering less than 2% of the Earth's surface, they are responsible for a disproportionately large share of emissions (Brown, 2013). The evidence of this urban sprawl's environmental and socio-economic unsustainability has necessitated innovative solutions for sustainable urban models that can respond to the demands for basic ecosystem services.

Urban forestry has emerged as a vital tool for bridging the gap between nature and cities, providing a sustained flow of essential ecosystem services within urban environments. Urban forestry, being a broad concept in itself, is a multi-managerial system that includes the sustained planning, planting, protection, maintenance, and care of trees, forests, green space, and related resources such as watersheds, wildlife habitats, landscape design, and outdoor recreational opportunities in and around cities and communities for economic, environmental, social, and public health benefits for people (Deneke, 1993). Urban forests encompass the significant presence of trees thriving within urban environments. These trees, located in parks, along roads, highways, yards, and gardens, provide critical ecological services, including climate amelioration, water-soil cycle regulation, disaster risk reduction, biodiversity conservation, and climate change mitigation, thereby enhancing the livability of urban spaces (Singh et al., 2024). Figure 8.1 illustrates the visual representation of the most commonly associated keywords generated from a Scopus database on "Urban Forestry," highlighting the primary themes and concepts in the field. The association analysis of keywords, based on research in urban forestry over the past 13 years (2010–2023), was considered to generate the network. Five major clusters of keywords emerged in the analysis. Among them, keywords such as "air pollution", "population density", "remote sensing", "vegetation", "cities", and "urban areas" exhibit robust linkages with "*urban forestry*". The figure summarizes the broad concept of interdisciplinary research collaboration for a comprehensive discussion on urban forestry (Singh, 2024d).

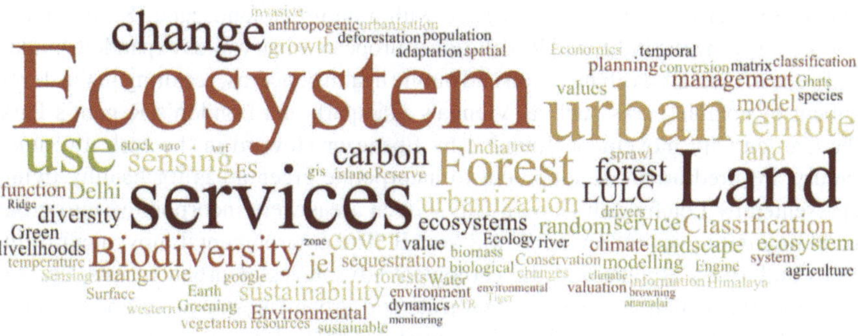

Fig. 8.1 Co-occurrence network of keywords in Urban Forestry research (The size of the word frame highlights the frequency of occurrence in the publication, and the width of the link highlights the strength between interrelated words)

8.1.1 The Growth and Development of Urban Forests: A Historical Perspective

The concept and idea of urban forests have gained attraction in recent years, both as an independent domain of research and as a matter of scientific and intellectual interest. However, the idea of urban forests and their potential impact on metropolitan life and society has lingered around the fringes of the mainstream for quite some time. Historically, urban forests emerged predominantly in the Western Hemisphere because this region experienced the early waves of the Industrial Revolution, which consequently spurred the initial phases of urbanization. The concept and development of urban forests have differed in the Western Hemisphere, particularly in Europe and the United States of America (Konijnendijk, 1997).

Urban forest patches can be historically attributed to many processes, including the remnants of former hunting grounds, large estates, spiritual sites, and abandoned lands (Lawrence, 1997). Despite the existence of these forest patches within an urban mosaic, the rapid increase in urban population following the rise of industrialization in the USA, along with the living conditions of the working class, prompted architects and planners in the 1850s to champion the creation of urban parks (Johnson et al., 2021). These urban parks had an English pastoral style, incorporating forested areas and expansive lawns for recreation (Cranz, 1982). Later, the emergence of zoning laws in the U.S. and suburbanization following World War II led to the creation of green spaces, which were both natural and designed, with aesthetic and private and public ownership (Johnson et al., 2021). Only with the growth of technological platforms in the mid-twentieth century did scientific planning and management of urban forests begin in the U.S.

Contrary to the development of urban forests and urban forestry in the U.S., the attention and application of scientific thinking to urban forests in Europe until the last decade of the twentieth century primarily focused on forests under a pervasive urban influence rather than on any greenspace within an urban landscape (Konijnendijk, 1997). Following the Industrial Revolution, European urban woods expanded dramatically due to increased urbanization and public health concerns, resulting in the development of notable parks such as Vienna's Stadtpark and London's Regent's Park. The Garden City movement, founded by Ebenezer Howard in the early twentieth century, inspired urban development to incorporate green areas for healthy living, represented by Letchworth Garden City. Green areas were incorporated into urban rebuilding initiatives following World War II. The environmental movement of the 1960s and 1970s emphasized the ecological advantages of urban forests, encouraging measures such as green belt projects and urban forestry programs in places such as Copenhagen and Berlin ('Konijnendijk, 2003). Recently, sustainable urban development has been a key focus in the twenty-first century, with green urban areas promoted by the EU Biodiversity Strategy and the European Green Capital Award (European Commission, 2023). Modern attempts include technological integration for forest maintenance and community engagement, as seen in Berlin's urban forest (Larondelle & Hasse, 2017) and Milan's Bosco Verticale (Giacomello, 2015).

Currently, urban forests refer to any green infrastructure within an urban space that contributes positively to ecological functioning, encompassing all trees within the cityscape's boundaries (Konijnendijk, 2005). Over the last three decades, the discourse, research, and policies on urban forests and urban forestry have expanded beyond the Western Hemisphere, driven by the global surge in urbanization (Singh & Kumar, 2022).

8.1.2 Where is the Discourse on Urban Forests Centered?

Research on urban forests has primarily focused on the Western Hemisphere. This is not appalling because the idea of urban forests developed in the west. A search in the Web of Knowledge database using the keyword "urban forest" revealed that the United States of America is the leading country in terms of publications on this topic, with China ranking as a relatively distant second. This finding aligns with the rate of urbanization in the countries mentioned above (Du & Xia, 2018). However, the same research reveals certain alarming facts, the most prominent of which is the severe underrepresentation of the countries from the Global South. In the list of the top 25 countries by publications on urban forests, the Global South is represented by only China, India, and Brazil, in descending order of the number of publications (Fig. 8.2).

The Global South is part of the world where most of the planet's urban growth has occurred over the last seven decades (Randolph & Storper, 2023). As urbanization accelerates in the Global South, the implementation and management of urban forests become crucial for long-term city planning, air quality improvement, and people's quality of life. Without enough study and documentation from these locations, personalized initiatives that address distinct local concerns and harness indigenous knowledge are unavailable. As a result, the Global South's engagement is crucial

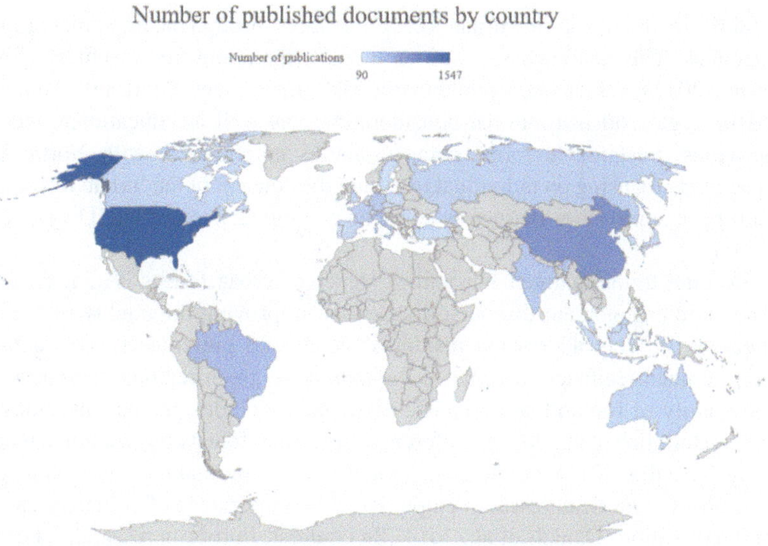

Fig. 8.2 A map showing the top 25 countries with the most published documents in the Web of Knowledge database. The search term "Urban forest" was used to compile all relevant published documents in the database

in establishing inclusive, effective, and context-specific urban forestry practices that promote the health and sustainability of the rapidly urbanizing environments in these emerging areas.

The rise of interest in ecosystem services provided by urban forests has been the major propellor for the existing discourse in research on urban forests. Although the discourse has a negligible contribution from the Global South, it is not unidimensional but multi-faceted. A sincere attempt must be made to understand this, and based on this understanding, the future pathway must be determined.

8.1.3 Discourses in Research on Urban Forests

Urban forestry, the academic discipline focused on the study of urban forests, is a complex and multifaceted field. It encompasses various influencing factors, including political, economic, social, and administrative aspects, as well as ecological and physiological considerations related to plants. Consequently, the discourse on urban forests is diverse and multidisciplinary. Krajter Ostoić and Konijnendijk van den Bosch (2015) explored the scientific discourses on urban forestry and identified six key areas that have found space in urban forestry research.

First and foremost is the discourse on the management of urban forests. The primary goal of managerial discourse in urban forest management is to produce healthy, resilient, and safe urban forests and trees through the implementation of best practices. This encompasses various themes, including tree inventory (Sreetheran et al., 2011), forest management, tree risk assessment (Martinez-Trinidad et al., 2010), legal and institutional considerations, as well as educational and research activities. Sustainable forest management is emphasized, with North American specialists focusing on individual trees and the concept of the 'urban forest,' whereas European approaches frequently adopt an ecological perspective (Dwyer & Nowak, 2003).

Natural to the maintenance and existence of urban forests is civic participation. The civic engagement discourse focuses on improving the quality of life in urban areas by actively addressing residents' needs and preferences. This entails gathering citizen feedback to enhance the quality of green regions, thereby improving the quality of life and welfare. The discussion includes people's attitudes towards trees (Heimlich et al., 2008), preferences for urban forest constructions, recreational usage (Gentin, 2011), public engagement in decision-making, and volunteer activities. North American research emphasizes the importance of volunteer engagement and motivation (Daniels et al., 2014). In contrast, European research focuses on the healing effects of green areas and social inclusion (Roe & Aspinall, 2011). Questionnaire surveys are the most commonly used research methodologies; however, additional qualitative approaches are necessary to gain a deeper understanding of human-urban forest interactions.

Ecosystem services (ES) have been a central focus of interest for scientists and policymakers in urban forests. The primary narrative of this discourse centers on urban woods as sources of diverse ecosystem services (ES), with advantages that outweigh the costs of their maintenance. The primary objective is to explain the costs of providing services while gaining public support and funding, especially in light of declining public funding. A major difficulty is quantifying and valuing these benefits, which are generally public goods, utilizing techniques such as i-Tree (Forest Service, 2021) and non-market valuation methodologies. This expert-driven discourse has gained traction in North America, with a focus on services such as carbon sequestration (Zheng et al., 2013), air pollutant absorption (Wang & Lin, 2012), property value impacts (Saphores & Li, 2012), and energy savings. The discussion highlights the importance of cost–benefit analysis in demonstrating the efficacy and efficiency of urban forest management (McPherson & Simpson, 2002).

In addition to focusing on the benefits of urban forests for humans, increasing attention has been paid to their contribution to biodiversity. The biodiversity discourse emphasizes that urban areas may both support and contribute to biodiversity protection. This rhetoric is common in Europe and North America, with research highlighting the negative impacts of urbanization, including habitat fragmentation, biodiversity loss, and changes in species composition (Hutyra et al., 2011). Forest age structure, forest regeneration, and habitat connectivity are crucial for developing green networks to conserve biodiversity. The discussion highlights the importance of mitigating the impacts of urban sprawl, residential development, and other human activities on urban forests (Schmitt & Suffling, 2006). Researchers primarily employ quantitative approaches, such as field measurements, spatial analysis, remote sensing, and biodiversity indices, to provide evidence and answers for sustainable urban forest management.

The related discourses surrounding urban planning and green infrastructure were also identified by Krajter Ostoić and Konijnendijk van den Bosch (2015). Urban planning and green infrastructure (GI) discourses both emphasize the integration of green areas into urban surroundings to achieve sustainable cities. The urban planning discourse argues for preserving existing green spaces while adding new ones as part of a complete approach to urban planning and strategic planning for urban forests and green spaces. The GI discourse, which first emerged in the United States in the mid-1990s, focuses on the optimal placement and connection of green areas to effectively supply ecosystem services (Ahern, 2007). It covers planning, ecology, recreation, and institutional issues, focusing on the network structure and connectedness of green spaces. Both discourses emphasize the value of green areas in urban sustainability and the necessity for integrated planning to maximize their advantages (Krajter Ostoić & Konijnendijk van den Bosch, 2015).

While the existing discourses primarily encompass most issues related to urban forests, some gaps warrant addressing and can be explored by researchers in the future. The underrepresentation of the specific socioeconomic dynamics of urban centers in the global south prevents us from delving into the deeper aspects of these discourses that must be discussed (Kumar & Singh, 2020). Despite overall social tendencies towards civic involvement in natural resource governance, research on citizens' active participation in decision-making, hands-on activities, and collaborative procedures or partnerships in urban forestry is limited (Krajter Ostoić & Konijnendijk van den Bosch, 2015). Addressing this gap will enable policymakers and decision-makers to recognize the benefits of partnerships and understand what constitutes an effective partnership, particularly in the context of green infrastructure. Furthermore, utilizing current technology, such as remote sensing, can aid in designing and managing urban forests, providing a more holistic approach to urban environment planning (Singh et al., 2018a, 2018b).

As cities grow and adapt, closing these gaps is critical to establishing sustainable urban ecosystems that can withstand future challenges. This leads us to another essential component of urban forestry and its ability to mitigate and adapt to the effects of climate change. Understanding the relationship between urbanization and climate dynamics is critical, given the increasing environmental stresses. This prompts us to investigate the relationship between climate change and urban forests, particularly in light of the rising urbanization and its associated consequences.

8.2 Climate Change and Urban Forests

8.2.1 Urban Expansion and Climate Dynamics

The challenges posed by climate change occupy a central position in many current policy frameworks and initiatives, underscoring its significance for our era. Urban cities have garnered growing attention from the scientific community regarding urban climate and the multidimensional impacts of urbanization (Argüeso et al., 2014). Urban centers, envisioned as hubs of human creativity and progress, have consequently distanced themselves from the rhythms of nature (Singh, 2024c). Unlike natural landscapes, which are shaped over time, cities lack inherent resilience, presenting significant concerns regarding potential economic and social repercussions in the event of a system failure. Rapid urbanization, characterized by high population densities and high consumption patterns, intensifies emission levels through increased demands for energy, infrastructure, and transportation. Cities worldwide are grappling with unprecedented heat stress events (Kotharkar et al., 2021), shifting surface temperatures, and their consequential impacts on heat and cold patterns (Lee & Livermore, 2010), compounded by alarming levels of air pollutants such as SOx, NOx, O3, VOCs, and particulate matter (Campbell et al., 2018). Instances of intense heat are predicted to increase in frequency (i.e., the number of hot days and nights), duration (heatwaves), and intensity (hotter days and nights) (Monteiro et al., 2022). Jacinto et al. (2012) investigated the multifaceted effects of climate change on ecosystem services, including water supply and demand, as well as the increasing incidence and severity of droughts in urban areas. Kumar (2021) discussed a study on the exposure and vulnerability of 1,860 cities to natural disasters. They emphasized the urgent need for metropolitan planners and decision-makers to strengthen resilience and adaptation strategies to address the imminent challenges posed by potential natural disasters. Such evidence strongly suggests that profound transformations are occurring in urban spaces under climate change scenarios. The temperature response to urbanization is primarily driven by the compounding effects of two factors: reduced surface evapotranspiration and the thermal properties of urban structures (Gaur & Singh, 2024). Together, these factors alter the partitioning of surface energy fluxes and the daily cycles of surface fluxes (Argüeso et al., 2014).

The distinctive biophysical features of urban spaces characterized by diminished green spaces, intricate geometries, and impervious surfaces, create a microclimate of its own that tends to trap heat at the surface, contributing to modification of fundamental processes notably altering albedo, reduced transpiration rates and perturbing latent heat flux dynamics (Zeleňáková et al., 2015). This results in relative warming of urban areas compared to their rural surroundings, also known as the Urban Heat Island (UHI) effect, first coined by Howard in 1833 (Chapman et al., 2017). Interestingly, the 'sky view' factor is also a key contributor to the UHI effect. It traps air and delays the release of heat to the outer atmosphere (Argüeso et al., 2014). UHI has garnered the focus of researchers, given its role in exacerbating the vulnerability of urban dwellers to the adverse impacts of climate change (Chapman et al.,

2017). Such extremes in climatic conditions likely affect the vitality and performance of species, altering the composition of urban forests (Esperon-Rodriguez et al., 2022b). The asynchronous relationship between impending climate change and the time plant species are required to respond (also known as macro-climatic debt) suggests that most urban forest species are already vulnerable to climate change (Weissert et al., 2014). Notably, 50% of the urban forest species currently surpass their climatic tolerance for mean annual temperature. By 2050, this percentage is projected to escalate, posing a heightened risk to the resilience of urban ecosystems (Esperon-Rodriguez et al., 2022b). While the impacts of climate change on tree growth have been extensively studied in forests, less is known about how the combined effects of climate change and the urban heat island affect the vitality and growth of urban plants (Zhang & Brack, 2021). In such a scenario, the size, composition, and structural attributes of urban forests are crucial in determining how they can mitigate climate effects and provide ecosystem services (Esperon-Rodriguez et al., 2022a). Current species selection during urban planning is primarily based on past and present climate conditions, without accounting for future climate change, and on management considerations that prioritize characteristics such as canopy size or aesthetics (Ordóñez & Duinker, 2015). Considering future climate change in urban forest species selection as a prospective strategy should become a priority in cities worldwide (Esperon-Rodriguez et al., 2022b). Hence, contemporary urban planning and the selection of tree species are crucial for ensuring a successful climate mitigation strategy for the future (Zhang & Brack, 2021).

8.2.2 Climate Variability and Challenges for Urban Forests

Urban forests endure significant stresses from elevated temperatures, limited water availability, concretization, and vehicular emissions. Researchers have explored the responses of urban trees to these environmental pressures, yielding crucial insights into their resilience and adaptive behaviors (Singh, 2025). While higher temperatures can enhance the rates of photosynthesis, provided other resources are available, it is reported that carbon losses due to increased respiration may also occur. Therefore, it is crucial to determine whether enhanced photosynthesis will offset the increased carbon losses from respiration (Saugier & Roy, 2001). Higher air temperatures also affect soil temperatures, negatively impacting root functions and nutrient availability. It impairs root growth and function, making it more difficult for plants to absorb essential nutrients (Kjelgren & Montague, 1998).

The increased evaporation of moisture exacerbates water stress, promoting soil compaction. This compaction reduces soil aeration and limits root penetration, creating a challenging environment for tree growth and health (Kjelgren & Montague, 1998). With alterations in precipitation patterns, low-lying areas will be particularly susceptible to flooding and waterlogging, leading to localized declines in tree growth and increased mortality. Trees exhibit varying degrees of tolerance to flooding, with some species able to withstand brief periods of less than one week (Lorenz et al.,

2010). Increased temperatures may also result in increased survivability of pests and diseases (Herms & Mattson, 1992).

Pests often have short life cycles and high reproductive potential compared to long-lived, slowly adapting tree species. This disparity makes trees particularly vulnerable to rapidly evolving pests and diseases (Herms & Mattson, 1992). Environmental changes may also alter plant's palatability to pests. Broadmeadow and Jackson (2000) reported that a surge in CO_2 in the atmosphere can make some species more palatable due to the increased production of soluble carbohydrates within phloem tissue. Such shifts in resource allocation within plants may affect their growth and defence mechanisms (Broadmeadow & Jackson, 2000).

8.2.3 Climate Variability and Urban Forests: Physiological and Biochemical Insights

Air pollution is one of the most significant challenges trees face in urban environments. When exposed to pollutants, plants function as scavengers, attracting, gathering, and combining pollutants that impinge on leaf surfaces, primarily by adapting to environmental conditions (Gupta et al., 2011). As air pollution increases over time, trees exhibit noticeable changes depending on their sensitivity levels. Changes are observed at various levels, including morphological (leaf area, number, thickness, stomatal frequency), phenological (change in phenophases), physiological (carbon assimilation, stomatal conductance, water use efficiency), and biochemical (change in protein, enzyme, sugars, etc.) (Singh et al., 2017). Some of these changes have been observed to drive empirical inferences about how trees respond to different levels of air pollution.

The most widely employed method for evaluating their tolerance is the Air Pollution Tolerance Index (APTI), which considers four key plant traits: leaf pH, relative water content (RWC), chlorophyll concentration, and ascorbic acid concentration. The combined value of these four parameters gives an idea of a plant's response to pollution (Singh et al., 2017). It reflects plants' inherent potential to withstand stress and determines their resistance and susceptibility to pollution, which varies by species based on their capacity to endure pollution without external injuries. Besides APTI, the Anticipated performance index (API) is another parameter that can be used to select species for urban forestry. API considers the morphological and socio-economic status of the plant species along with its APTI (Singh et al., 2020). Ascorbic acid, an antioxidant found in growing parts of plants, influences a tree's ability to counter adverse conditions. It is a key parameter tested to determine plants' response to pollution stress due to its role in metabolism and defence mechanisms (Singh et al., 2020). Its activity is- pH-dependent and is typically higher at elevated pH levels. Leaf pH content indicates air pollution, with plants exposed to pollution tending to have lower pH due to increased acidity from the adsorption of SO_2 and

NOx (Rai, 2020). Total chlorophyll reflects the level of photosynthetic activity in plants, indicating biomass growth and development.

Highly stable under normal conditions, chlorophyll undergoes various photochemical reactions (oxidation, reduction, and reversible bleaching) when exposed to unfavourable conditions (Singh et al., 2020). Such situations cause abnormal functioning of stomata, resulting in a malfunction of gaseous exchange and leading to a rise in leaf temperature. This results in decreased stomatal numbers and reduced CO_2 availability, leading to impaired photosynthetic activity and carbon fixation. Relative water content (RWC) measures the water in leaves compared to their fully turgid state. Low RWC causes early leaf senescence (Rai, 2020). It influences water use efficiency, which regulates the concentration of key metabolic processes such as proline accumulation and soluble sugars (Singh et al., 2020). Many studies have provided fascinating insights into the response of trees to environmental discrepancies under heavy pollution loads. Due to the deposition of dust particles and particulate matter on leaves, a study on Lagerstroemia speciosa trees in northern India recorded reduced CO_2 assimilation, transpiration, and stomatal conductance (Singh et al., 2017; Singh, 2024b). Bojarczuk et al. (2002) reported a 31% reduction in the carbon assimilation rate of *Betula pendula* plants in a polluted environment compared to plants growing in unpolluted conditions. Such studies are crucial for understanding how and which trees perform best in contaminated environments. Observing plant changes at both biochemical and physiological levels is vital for predicting their future performance and selecting resilient species that contribute to climate change mitigation. These researches boost urban resilience and play a crucial role in climate change mitigation and adaptation (Singh, 2024a).

8.3 New Frontiers and Advances

8.3.1 Breakthroughs in Urban Forest Initiatives: The Latest Developments

Global steps to bring forests into the cities

A program launched by the Food and Agriculture Organization (FAO) recognized *Tree Cities of the World* (TCW) to showcase forest-based solutions that contribute to building cities as sustainable and resilient models to a greater extent. This initiative aims to support the First World Forum on Urban Forests (WFUF) organized by FAO and its affiliates (Joshi et al., 2021). Additionally, the urban ecosystem is a key focus area of the UN Decade on Ecosystem Restoration (2021–2030), highlighting the mitigative approach to climate change impacts and the importance of reconnecting with nature by integrating urban planning at the landscape level.

Community stewardship

The synergy between local community engagement and land stewardship in urban fringe forests offers numerous benefits, as community engagement is pivotal in preserving urban forests. Rebuilding the connection between local people and nature holds immense value. There are several possibilities to experience nature and its dynamics in the surrounding area through the range of species assemblages, structural characteristics, and the proximity of growing urban forests to urban dwellers. In research, perception-based assessments have gained interest in urban forestry, particularly in developing countries (Kumar et al., 2020).

Recognition of UF as a nature-based solution

Over the past few years, urban metropolitan areas have been experiencing the impact of climate anomalies, underscoring the need for urban forestry in the era of sustainability. Urban forestry is widely accepted as a nature-based solution that contributes to climate action. Among the 17 Sustainable Development Goals (SDGs) of the United Nations, Sustainable Cities and Communities (SDG 11) can be achieved through sustainable urban forestry, making cities resilient to climate-related issues such as the urban heat island effect, irregular weather patterns, and an increased risk of disastrous floods and landslides (Singh, 2024c). The significance of urban forests as a cost-effective approach under nature-based solutions (NBS) was deliberated jointly by UNECE and FAO during the UNECE 2022 Regional Forum on Sustainable Development. Additionally, a 3-30-300 rule was envisioned and proposed by the Nature-Based Solution Institutes (NBSI), emphasizing the importance of urban forests for improved health and human well-being. The numbers represent the performance criteria of individual access to urban green spaces: at least '3' mature trees should be visible from the house, not less than '30' percent of tree canopy in the neighborhood and a maximum of '300' meters of distance to the nearest urban green parks (Konijnendijk, 2023).

8.3.2 Geospatial Approaches in Urban Forestry

In urban forestry, geospatial techniques are vital for assessing, managing, and planning green spaces within cities (Barwa & Singh, 2024). Using satellite imagery, UAVs, LiDAR sensors, and machine learning algorithms, these tools play a crucial role in analyzing urban forest trends and ecosystem services, thereby enhancing decision-making in urban ecosystems. They enable comprehensive assessments of urban expansion, vegetation loss, environmental factors, land use, and land cover (LULC) changes, which are essential for sustainable urban planning in rapidly urbanizing areas (Tiwari et al., 2024). For example, Bae and Ryu (2015) investigated the spatiotemporal variations in soil organic carbon (SOC) within an urban park in Seoul, examining how changes in land use and land cover (LULC) influenced the SOC. The effects of climate change, including rising temperatures and extreme weather events, further underscore the importance of geospatial technology in assessing urban forest vulnerability and adaptive capacity (Salleh et al., 2023). Ren et al. (2013) investigated how different features of urban parks influenced the intensity of park cool islands in China using multi-source satellite imagery.

The use of Remote Sensing (RS) and Geographic Information Systems (GIS) are revolutionizing the study and management of urban forestry (Fig. 8.3) by providing innovative solutions to the multifaceted challenges faced by urban planners and environmental managers. These technologies offer comprehensive, data-driven approaches to understanding and managing urban forests, addressing environmental degradation, urban heat islands (UHIs), pollution, and equitable access to green spaces (Joshi et al., 2020). Here's how RS-GIS addresses these current challenges.

Mapping and Monitoring Urban Forests

Traditional urban forest monitoring methods are often labor-intensive, time-consuming, and limited in scope. However, high-resolution satellite imagery and aerial photography enable detailed, up-to-date, and large-scale mapping of tree canopy cover, species distribution, and green space availability (Neyns & Canters, 2022). Multispectral and hyperspectral remote sensing technologies can detect subtle changes in vegetation health, such as variations in leaf colour, moisture content, and

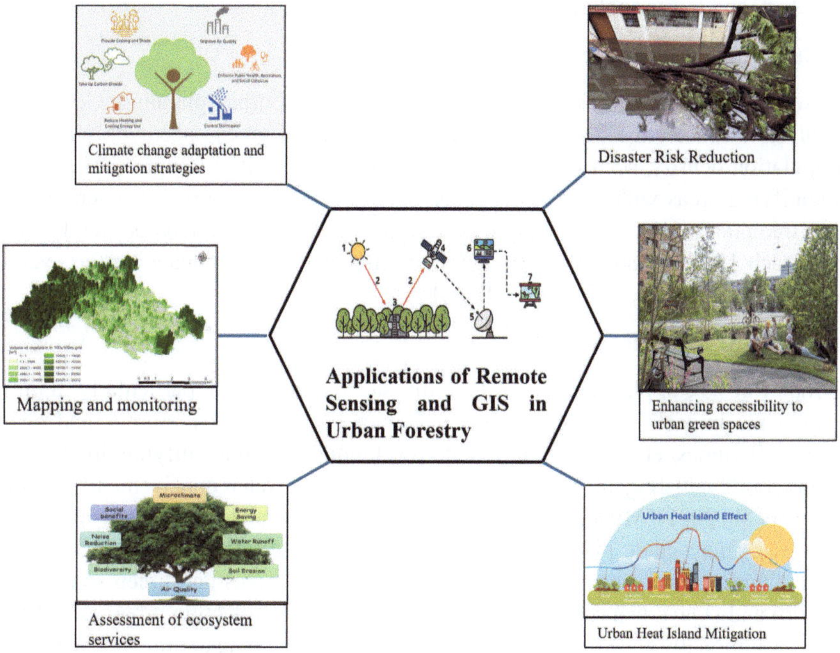

Fig. 8.3 A pictorial representation of the emerging applications of remote sensing and GIS in the field of urban forestry (Image sources: Pace (2020); Lange and Rodrigues (2021); Zięba-Kulawik et al. (2021); Climate Adaptation Knowledge Exchange (2023))

chlorophyll concentration, identifying stressed or diseased trees early for prompt intervention (Qamar et al., 2022). Additionally, GIS integrates these health assessments with other spatial data, such as soil quality and pollution levels, to provide a comprehensive understanding of urban forest health (Cârlan et al., 2020).

Assessing Ecosystem Services

Geospatial techniques are crucial in quantifying the ecosystem services of urban forests (García-Pardo et al., 2022). Integrating spatial data with ecological models can help researchers estimate the benefits of urban forests in terms of carbon sequestration, improved air quality, temperature regulation, and stormwater management (Bottalico et al., 2016; Cetin et al., 2023; Lefrançois, 2015). This approach maximizes the contribution of urban forests in mitigating the Urban Heat Island (UHI) effect, reducing pollution, and enhancing urban biodiversity (Müller et al., 2018).

UHI Mitigation

Thermal remote sensing maps land surface temperatures in urban areas, identifying hotspots and areas most affected by the Urban Heat Island (UHI) effect. Correlating temperature data with tree canopy cover may help urban planners prioritize greening initiatives in the areas with the highest temperatures (Ren et al., 2018; Zhang et al., 2017). This targeted approach helps reduce ambient temperatures, enhance thermal comfort, and lower cooling energy consumption.

Enhancing Green Space Accessibility

Ensuring equitable access to green spaces is crucial in urban planning. Geospatial analysis evaluates the distribution and accessibility of green spaces in relation to population density, socio-economic factors, and accessibility (Gupta et al., 2016). Identifying areas with insufficient green spaces would help urban planners prioritize tree planting and green space development in underserved neighborhoods. This approach promotes social equity and environmental justice, enhancing the overall quality of life in cities (Wolch et al., 2014).

Disaster Risk Reduction

Urban forests can play a crucial role in disaster risk reduction, particularly in mitigating the impacts of extreme weather events, such as floods, storms, and heatwaves (Twumasi et al., 2019). Geospatial techniques aid in identifying flood-prone areas and evaluating the role of vegetation in stormwater management (Kadaverugu et al., 2021). Modeling the interaction between urban forests and hydrological processes may guide planners in designing blue-green infrastructure (BGI) solutions that enhance urban resilience (Sörensen, 2018). Trees and green spaces can absorb and slow down runoff, reducing flood risks and protecting urban infrastructure. Additionally, they provide shade and cool the air, helping to mitigate the effects of heatwaves (Ribeiro et al., 2021).

Climate Change Adaptation and Mitigation Strategies

The escalating threats of climate change to urban ecosystems underscore the increasing reliance on RS-GIS technologies. These tools analyze historical and current climate data alongside spatial information on urban forests, enabling planners to model future climate scenarios and assess their impacts on urban green spaces (Barron et al., 2023; Menconi et al., 2024). This holistic approach facilitates the selection of resilient tree species, the design of climate-adaptive green infrastructure, and the development of sustainable urban forestry strategies to enhance city resilience (Kaur et al., 2024).

Future Scenarios and Planning

Geospatial techniques are crucial for scenario planning in urban development, as they simulate various growth scenarios and their impacts on green spaces to prioritize sustainability and resilience (Mansourihanis et al., 2023). These models integrate climate projections, land-use changes, and socioeconomic trends to forecast urban forest needs (Barron et al., 2023). This forward-looking approach ensures urban forests are integrated into long-term planning for sustainable cities. Future RS-GIS opportunities include the integration of IoT sensors for real-time monitoring of tree health and environmental conditions, as well as advancements in AI and machine learning for more precise remote sensing analysis (Rathore et al., 2016).

Cities worldwide are increasingly utilizing RS-GIS technologies to enhance the management of urban forests. For instance, New York City's TreesCount initiative utilizes high-resolution aerial imagery and LiDAR data for street tree mapping and inventorying, guiding strategies for maintenance and planting (Johnson et al., 2018). Similarly, the Green View Index utilizes Google Street View images to assess tree cover and inform urban greening initiatives (Sánchez & Labib, 2024).

In India, the forest cover of major metropolitan cities, including Ahmedabad, Bangalore, Chennai, Delhi, Hyderabad, Kolkata, and Mumbai, was mapped and included in the Indian State Forest Report (Forest Survey of India, 2021). In 2020, urban forestry was promoted through the launch of *Nagar Van Yojana* (NVY) (Fig. 8.4), which engaged stakeholders at various levels, including educational institutions, NGOs, local bodies, and local communities (Garg et al., 2016).

8.4 Research Needs and Prospects

Urban forests provide habitat to more trees than peri-urban areas (Blood et al., 2016). Further research is necessary to fully comprehend the dynamics of regeneration in fragmented and urban forest environments, despite studies indicating that urban forest breaches and regeneration exhibit similar patterns and issues to those found in non-urban forests (Massad et al., 2019). Urban forests also require ongoing maintenance; however, their execution is frequently impeded by a lack of time, funding, and participation from individual property owners and municipal administrations (Jangra et al.,

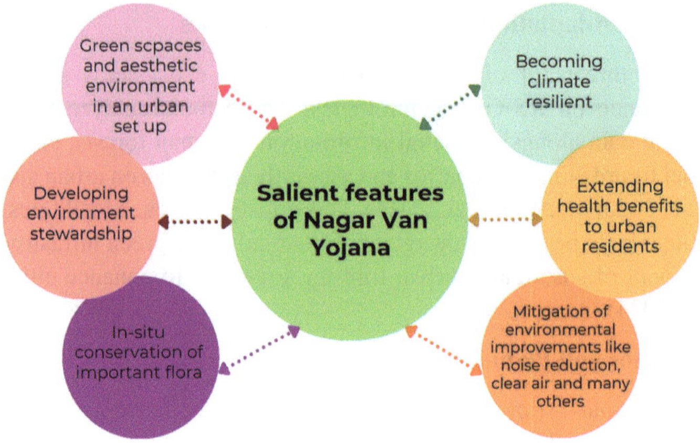

Fig. 8.4 A graphic showcasing the features of Nagar Van Yojana

2024). Future research on creating urban forest management systems to optimize advantages for urban environmental quality and human well-being (Escobedo et al., 2006) will benefit from developing an urban forestry database of urban floras (Li et al., 2006). The collection will be particularly valuable for ongoing and regular monitoring programs. The research is being discussed at the local level, aided by ample literature emphasizing local forest structure (Nowak et al., 2013) and tree design at the local scale (Nowak et al., 2013). The contribution overlooks the conflicting impacts of urban forests at various scales (Vos et al., 2013) due to spatial heterogeneity. Thereby providing a generalization that is not entirely accurate. Urban foresters can streamline the services of urban ecosystems by developing nested approaches (Jeanjean et al., 2015). The regulations and legal framework often do not effectively acknowledge urban forests. There is a drawback to the policy framework, which lacks an understanding of the diverse aspects of scale dynamics and patterns of ecosystem service (ES) functionality. For this reason, there is a need to incorporate the urban local's perception of various urban forest ES (Velarde et al., 2007). Local communities must be involved in the upkeep of urban forests, as community-based initiatives can enhance sustainable management techniques. Implementing conservation approaches, such as zoning, green corridors, and ecologically oriented urban architecture, that consider the particular constraints urban forests face. This service is vital when opportunities to enjoy nature are dwindling (Soga & Gaston, 2016), and beneficial responses to people's willingness to conserve biodiversity have been anticipated (Soga et al., 2016). Citizen's awareness of the importance of urban forest ecosystem services has increased over the past years. Despite this, it remains unclear how we can manage the urban forest to provide citizens with an optimal combination of ecosystem services (ESs) (Singh et al., 2023).

Additionally, we remain uncertain about the type of urban forest ecosystem services (ESs) that citizens prefer (Jang-Hwan et al., 2020). Incorporating stewardship principles such as enhanced motives, sustainable harvesting, inspection, and responsive management; establishing laws and policies at the local and state levels that support them; formulating strategies to make them more resilient to climate change; and initiating outreach and education campaigns to raise awareness of the importance of urban fringe forests and promote stewardship (Joshi et al., 2024). Despite facing pervasive issues, including urban densification, pollution, and climatic variation, urban forests have been withstanding the far-reaching repercussions of these globally addressed issues and contribute to maintaining the urban ecosystem's sustainability. The importance of urban nature in creating habitable surroundings for urban inhabitants should be incorporated into biodiversity conservation goals, particularly in urban environments (Dearborn & Kark, 2010). Therefore, when discussing the contribution of new urban forests to the development of urban green infrastructure, several socially and environmentally based aims must be considered.

8.5 Conclusion

The discussions around urban forests are poised to gain further momentum, as they are pivotal in building climate-resilient cities. Understanding species-specific responses and performance is pertinent to developing a robust knowledge base of urban forests as major carbon sinks. Current research, coupled with precise modeling, advanced geospatial tools, and quantification, can substantially aid in addressing urban environmental challenges. This approach will enable us to provide well-informed recommendations for future research and policies that enhance urban greenery and promote carbon neutrality.

References

Ahern, J. (2007). Green infrastructure for cities: The spatial dimension. In *Cities of the future: Towards integrated sustainable water and landscape management* (Vol. 13, pp. 267–283).

Argüeso, D., Evans, J. P., Fita, L., & Bormann, K. J. (2014). Temperature response to future urbanization and climate change. *Climate Dynamics, 42*, 2183–2199. https://doi.org/10.1007/s00382-013-1789-6

Bae, J., & Ryu, Y. (2015). Land use and land cover changes explain spatial and temporal variations of the soil organic carbon stocks in a constructed urban park. *Landscape and Urban Planning, 136*, 57–67.

Barron, S., Rugel, E., Cheng, Z., et al. (2023). Achieving the urban tree trifecta: Scenario modelling for salubrious, resilient, and diverse urban forests in Densifying Cities. *Arboriculture & Urban Forestry (AUF), 49*, 313–328.

Barwa, S., & Singh, H. (2024). Climate-induced changes in forest ecosystems of Indian Himalayan region: A meta-analysis. In *Urban forests, climate change and environmental pollution: Physio-biochemical and molecular perspectives to enhance urban resilience* (pp. 839–866). Springer.

Blood, A., Starr, G., Escobedo, F., et al. (2016). How do urban forests compare? Tree diversity in urban and Periurban forests of the Southeastern U.S. *Forests, 7*, 120.

Bojarczuk, K., Karolewski, P., Oleksyn, J., et al. (2002). Effect of polluted soil and fertilisation on growth and physiology of silver birch (Betula pendula Roth.) Seedlings

Bottalico, F., Chirici, G., Giannetti, F., et al. (2016). Air pollution removal by green infrastructures and urban forests in the city of Florence. *Agriculture and Agricultural Science Procedia, 8*, 243–251.

Broadmeadow, M. S. J., & Jackson, S. B. (2000). Growth responses of Quercus petraea, Fraxinus excelsior and Pinus sylvestris to elevated carbon dioxide, ozone and water supply. *New Phytologist, 146*, 437–451. https://doi.org/10.1046/j.1469-8137.2000.00665.x

Brown, L. R. (2013). *Eco-economy: Building an economy for the earth*. Routledge.

Campbell, P., Zhang, Y., Yan, F., et al. (2018). Impacts of transportation sector emissions on future U.S. air quality in a changing climate. Part II: Air quality projections and the interplay between emissions and climate change. *Environmental Pollution, 238*, 918–930. https://doi.org/10.1016/j.envpol.2018.03.016

Cârlan, I., Mihai, B.-A., Nistor, C., & Große-Stoltenberg, A. (2020). Identifying urban vegetation stress factors based on open access remote sensing imagery and field observations. *Ecological Informatics, 55*, Article 101032.

Cetin, M., Adiguzel, F., & Zeren Cetin, I. (2023). Determination of the effect of urban forests and other green areas on surface temperature in Antalya. In: *Concepts and applications of remote sensing in forestry* (pp 319–336). Springer

Chapman, S., Watson, J. E. M., Salazar, A., et al. (2017). The impact of urbanization and climate change on urban temperatures: A systematic review. *Landscape Ecology, 32*, 1921–1935.

Climate Adaptation Knowledge Exchange. (2023). Urban forest climate adaptation tool. https://www.cakex.org/tools/urban-forest-climate-adaptation-tool

Cranz, G. (1982). *The politics of park design: A history of urban parks in America*. The MIT Press.

Daniels, J. M., Robbins, A. S. T., Brinkley, W. R., et al. (2014). Toward estimating the value of stewardship volunteers: A cost-based valuation approach in King County, Washington, USA. *Urban Forestry & Urban Greening, 13*, 285–289.

Dearborn, D. C., & Kark, S. (2010). Motivations for conserving urban biodiversity. *Conservation Biology, 24*, 432–440.

Deneke, F. (1993). Urban forestry in North America: Towards a global ecosystem perspective. In: *Proceedings of the First Canadian Urban Forests Conference* (pp. 4–8).

Du, W. C., & Xia, X. H. (2018). How does urbanization affect GHG emissions? A cross-country panel threshold data analysis. *Applied Energy, 229*, 872–883.

Dwyer, J. F., & Nowak, D. J. (2003) Sustaining urban forests. *Journal of Arboriculture, 29*(1), 49–55.

Escobedo, F. J., Nowak, D. J., Wagner, J. E., et al. (2006). The socioeconomics and management of Santiago de Chile's public urban forests. *Urban Forestry & Urban Greening, 4*, 105–114.

Esperon-Rodriguez, M., Rymer, P. D., Power, S. A., et al. (2022a). Assessing climate risk to support urban forests in a changing climate. *Plants People Planet, 4*, 201–213. https://doi.org/10.1002/ppp3.10240

Esperon-Rodriguez, M., Tjoelker, M. G., Lenoir, J., et al. (2022b). Climate change increases global risk to urban forests. *Nature Clinical Practice Endocrinology & Metabolism, 12*, 950–955. https://doi.org/10.1038/s41558-022-01465-8

European Commission. (2023). The European green capital book. European Union (pp. 1–84). https://circabc.europa.eu/ui/group/c6e126de-5b8c-4cd7-8d36-a1978a2a63de/library/d041bd31-a918-4148-8231-36bb507e6c7c/details?download=true

Forest Service U. (2021). Eco User's Manual.

Forest Survey of India. (2021). Indian State of Forest Report.

García-Pardo, K. A., Moreno-Rangel, D., Domínguez-Amarillo, S., & García-Chávez, J. R. (2022). Remote sensing for the assessment of ecosystem services provided by urban vegetation: A review of the methods applied. *Urban Forestry & Urban Greening, 74*, Article 127636.

Garg, A., Pal, D., Singh, H., & Pandey, D. C. (2016). A comparative study of NDBI, NDISI and NDII for extraction of urban impervious surface of Dehradun [Uttarakhand, India] using Landsat 8 imagery. In *2016 International Conference on Emerging Trends in Communication Technologies (ETCT)* (pp. 1–5). tudy on the effective

Gaur, R., & Singh, H. (2024). Urban tree responses to climate change and environmental pollution. In *Urban forests, climate change and environmental pollution: Physio-biochemical and molecular perspectives to enhance urban resilience* (pp. 279–305). Cham: Springer Nature Switzerland.

Gentin, S. (2011). Outdoor recreation and ethnicity in Europe—A review. *Urban Forestry & Urban Greening, 10*, 153–161.

Giacomello, E. (2015). Case study: Bosco verticale, Milan: A new urban forest rises in Milan. *CTBUH Journal, 1*, 12–18.

Gupta, K., Roy, A., Luthra, K., & Maithani, S. (2016). GIS based analysis for assessing the accessibility at hierarchical levels of urban green spaces. *Urban Forestry & Urban Greening, 18*, 198–211.

Gupta, S., Nayek, S., & Bhattacharya, P. (2011). Effect of air-borne heavy metals on the biochemical signature of tree species in an industrial region, with an emphasis on anticipated performance index. *Chemistry and Ecology, 27*, 381–392. https://doi.org/10.1080/02757540.2011.561791

Heimlich, J., Sydnor, T. D., Bumgardner, M., O Brien, P. (2008). Attitudes of residents toward street trees on four streets in Toledo, Ohio, U.S. before removal of ash trees (Fraxinus spp.) from emerald ash borer (Agrllus planlpennis). *Arboriculture & Urban Forestry, 34*, 47.

Herms, D. A., Mattson, W. J. (1992) The dilemma of plants: To grow or defend.

Hutyra, L. R., Yoon, B., Hepinstall-Cymerman, J., & Alberti, M. (2011). Carbon consequences of land cover change and expansion of urban lands: A case study in the Seattle metropolitan region. *Landscape and Urban Planning, 103*, 83–93.

Jacinto, R., Cruz, M. J., & Santos, F. D. (2012). Water use scenarios as a tool for adaptation to climate change of a water supply company. *Drinking Water Engineering and Science Discussions, 5*, 265–289. https://doi.org/10.5194/dwesd-5-265-2012

Jang-Hwan, J., So-Hee, P., JaChoon, K., et al. (2020). Preferences for ecosystem services provided by urban forests in South Korea. *Forest Science and Technology, 16*, 86–103.

Jangra, A., Kumar, K., Maikhuri, S., Bhandari, M. S., Pandey, S., Singh, H., & Barthwal, S. (2024). Unveiling stress-adapted endophytic bacteria: Characterizing plant growth-promoting traits and assessing cross-inoculation effects on *Populus deltoides* under abiotic stress. *Plant Physiology and Biochemistry, 210*, 108610.

Jeanjean, A. P. R., Hinchliffe, G., McMullan, W. A., et al. (2015). A CFD study on the effectiveness of trees to disperse road traffic emissions at a city scale. *Atmos Environmental, 120*, 1–14

Johnson, L. R., Johnson, M. L., Aronson, M. F. J., et al. (2021). Conceptualizing social-ecological drivers of change in urban forest patches. *Urban Ecosystems, 24*, 633–648. https://doi.org/10.1007/s11252-020-00977-5

Johnson, M. L., Campbell, L. K., Svendsen, E. S., & Silva, P. (2018). Why count trees? Volunteer motivations and experiences with tree monitoring in New York City. *Arboriculture & Urban Forestry, 44*, 59–72.

Joshi, R., Sharma, B., Singh, H., Dhakal, N., Ayer, S., & Maraseni, T. (2024). Poplar plantation as an agroforestry approach: Economic benefits and its role in carbon sequestration in North India. *Journal of Resources and Ecology, 15*(4), 880–888.

Joshi, R., Singh, H., Chhetri, R., Poudel, S. R., & Rijal, S. (2021). Carbon sequestration potential of community forests: A comparative analysis of soil organic carbon stock in community managed forests of Far-Western Nepal. *Eurasian Journal of Soil Science, 10*(2), 96–104.

Joshi, R., Singh, H., Chhetri, R., & Yadav, K. (2020). Assessment of carbon sequestration potential in degraded and non-degraded community forests in Terai region of Nepal. *Journal of Forest and Environmental Science, 36*(2), 113–121.

Kadaverugu, A., Nageshwar Rao, C., & Viswanadh, G. K. (2021). Quantification of flood mitigation services by urban green spaces using in VEST model: A case study of Hyderabad city, India. *Modeling Earth Systems and Environment, 7*, 589–602.

Kaur, R., Hallett, R. A., & Strauss, N. (2024). Building urban forest resilience to sea level rise: A GIS-based climate adaptation tool for New York City. *Forests, 15*, 92.

Kjelgren, R., & Montague, T. (1998). Urban tree transpiration over turf and asphalt surfaces. *Atmospheric Environment, 32*, 35–41.

Konijnendijk. C. C. (1997) A short history of urban forestry in Europe.
Konijnendijk, C. C. (2003). A decade of urban forestry in Europe. *Forest Policy and Economics, 5*(2), 173–186.
Konijnendijk, C. C. (2005) New perspectives for urban forests: Introducing wild woodlands. In: *Wild urban woodlands: New perspectives for urban forestry*. (pp. 33–45). Springer
Konijnendijk, C. C. (2023). Evidence-based guidelines for greener, healthier, more resilient neighbourhoods: Introducing the 3–30–300 rule. *Journal of Forestry Research (Harbin), 34*, 821–830.
Kotharkar, R., Ghosh, A., Kotharkar, V. (2021). Estimating summertime heat stress in a tropical Indian city using Local Climate Zone (LCZ) framework. *Urban Climate, 36*. https://doi.org/10.1016/j.uclim.2021.100784
Krajter Ostoić, S., & Konijnendijk van den Bosch, C. C. (2015). Exploring global scientific discourses on urban forestry. *Urban Forestry & Urban Greening, 14*, 129–138.
Kumar, A., Dwivedi, G. K., Tewari, S., Sah, V. K., Singh, H., Kumar, P., Kumar, N., & Kaushal, R. (2020). Soil organic carbon pools under *Terminalia chebula* Retz. based agroforestry system in Himalayan foothills, India. *Current Science, 118*(7), 1098–1103.
Kumar, M., & Singh, H. (2020). Agroforestry as a nature-based solution for reducing community dependence on forests to safeguard forests in rainfed areas of India. In *Nature-based solutions for resilient ecosystems and societies* (pp. 289–306). Springer.
Lange, I. S. G., & Rodrigues, C. N. (2021). Urban green spaces: combining goals for sustainability and placemaking. *Europe Now-A Journal of Research & Art*
Larondelle, N., & Haase, D. (2017). Back to nature! Or not? Urban dwellers and their forest in berlin. *Urban Ecosystems, 20*, 1069–1079.
Lawrence, H. W. (1997). From private Allée to public shade tree: Historic roots of the urban forest
Lee, K. H., & Levermore, G. J. (2010). Weather data for future climate change for South Korean building design: Analysis for trends. *Architectural Science Review, 53*, 157–171. https://doi.org/10.3763/asre.2008.0055
Lefrançois, C. B. (2015). Designing effective stormwater management policies: the role of the urban forest and impervious cover in Vancouver, BC.
Li, W., Ouyang, Z., Meng, X., & Wang, X. (2006). Plant species composition in relation to green cover configuration and function of urban parks in Beijing, China. *Ecological Research, 21*, 221–237.
Lorenz, M., Clarke, N., Paoletti, E., et al. (2010). Air pollution impacts on forests in a changing climate. In: *Forests and society: responding to global drivers of change. International Union of Forest Research Organizations (IUFRO)* (pp. 55–75).
Mansourihanis, O., Maghsoodi Tilaki, M. J., Yousefian, S., & Zaroujtaghi, A. (2023) A computational geospatial approach to assessing land-use compatibility in urban planning. *Land (Basel), 12*, 2083.
Martinez-Trinidad, T., Watson, W. T., Arnold, M. A., et al. (2010). Comparing various techniques to measure tree vitality of live oaks. *Urban Forestry & Urban Greening, 9*, 199–203.
Massad, T. J., Williams, G. L., Wilson, M., et al. (2019). Regeneration dynamics in old-growth urban forest gaps. *Urban Forestry & Urban Greening, 43*, Article 126364.
McPherson, E. G., & Simpson, J. R. (2002). A comparison of municipal forest benefits and costs in Modesto and Santa Monica, California, USA. *Urban Forestry & Urban Greening, 1*, 61–74.
Menconi, M. E., Bonciarelli, L., & Grohmann, D. (2024). Assessment of the potential contribution of the urban green system to the carbon balance of cities. *Environments, 11*, 98.
Monteiro, A., Ankrah, J., Madureira, H., & Pacheco, M. O. (2022). Climate risk mitigation and adaptation concerns in urban areas: A systematic review of the impact of IPCC assessment reports. *Climate, 10*
Müller, A., Bøcher, P. K., Fischer, C., & Svenning, J.-C. (2018). 'Wild'in the city context: Do relative wild areas offer opportunities for urban biodiversity? *Landscape and Urban Planning, 170*, 256–265.

Neyns, R., & Canters, F. (2022). Mapping of urban vegetation with high-resolution remote sensing: A review. *Remote Sensing (Basel), 14*, 1031.

Nowak, D. J., Greenfield, E. J., Hoehn, R. E., & Lapoint, E. (2013). Carbon storage and sequestration by trees in urban and community areas of the United States. *Environmental Pollution, 178*, 229–236.

Ordóñez, C., & Duinker, P. N. (2015). Climate change vulnerability assessment of the urban forest in three Canadian cities. *Climate Change, 131*, 531–543. https://doi.org/10.1007/s10584-015-1394-2

Pace, R. (2020). Modeling ecosystem services of urban trees to improve air quality and microclimate. Technical University of Munich.

Qamar, F., Sharma, M. S., & Dobler, G. (2022). The impacts of air quality on vegetation health in dense urban environments: A ground-based hyperspectral imaging approach. *Remote Sensing (Basel), 14*, 3854.

Rai, P. K. (2020). Particulate matter tolerance of plants (APTI and API) in a biodiversity hotspot located in a tropical region: Implications for eco-control. *Particulate Science and Technology, 38*, 193–202. https://doi.org/10.1080/02726351.2018.1527800

Randolph, G. F., & Storper, M. (2023). Is urbanisation in the Global South fundamentally different? Comparative global urban analysis for the 21st century. *Urban Studies, 60*, 3–25.

Rathore, M. M., Ahmad, A., Paul, A., & Rho, S. (2016). Urban planning and building smart cities based on the internet of things using big data analytics. *Computer Networks, 101*, 63–80.

Ren, Z., He, X., Pu, R., & Zheng, H. (2018). The impact of urban forest structure and its spatial location on urban cool island intensity. *Urban Ecosystems, 21*, 863–874.

Ren, Z., He, X., Zheng, H., et al. (2013). Estimation of the relationship between urban park characteristics and park cool island intensity by remote sensing data and field measurement. *Forests, 4*, 868–886.

Ribeiro, A. P., Bollmann, H. A., De Oliveira, A., et al. (2021). The role of tree landscape to reduce effects of urban heat islands: a study in two Brazilian cities. *Trees*, 1–14

Roe, J., & Aspinall, P. (2011). The restorative outcomes of forest school and conventional school in young people with good and poor behaviour. *Urban Forestry & Urban Greening, 10*, 205–212.

Salleh, S. A., Abd. Latif, Z., Pardi, F., et al (2023) Conceptualising the citizen-driven urban forest framework to improve local climate condition: Geospatial data fusion and numerical simulation. In: *Concepts and applications of remote sensing in forestry* (pp. 337–353). Springer

Sánchez, I. A. V., & Labib, S. M. (2024). Accessing eye-level greenness visibility from open-source street view images: A methodological development and implementation in multi-city and multi-country contexts. *Sustainable Cities and Society, 103*, Article 105262.

Saphores, J.-D., & Li, W. (2012). Estimating the value of urban green areas: A hedonic pricing analysis of the single family housing market in Los Angeles, CA. *Landscape and Urban Planning, 104*, 373–387.

Saugier, B., & Roy, J. (2001) Estimations of global terrestrial productivity: Converging toward a single number?

Schmitt, D., & Suffling, R. (2006). Managing eastern North American woodlands in a cultural context. *Landscape and Urban Planning, 78*, 457–464.

Singh, H., Savita, Sharma, R., et al. (2017). Physiological functioning of *Lagerstroemia speciosa* L. under heavy roadside traffic: An approach to screen potential species for abatement of urban air pollution. *3 Biotech, 7*. https://doi.org/10.1007/s13205-017-0690-0

Singh, H., Yadav, M., Kumar, N., et al. (2020). Assessing adaptation and mitigation potential of roadside trees under the influence of vehicular emissions: A case study of *Grevillea robusta* and *Mangifera indica* planted in an urban city of India. *PLoS One, 15*. https://doi.org/10.1371/journal.pone.0227380

Singh, H. (2024a). Climate change and forests: Biological perspectives. In *Forests and climate change: Biological perspectives on impact, adaptation, and mitigation strategies* (pp. 1–17). Springer.

Singh, H. (2024b). Exploring adaptive modulation in plant functional traits and its impact on the productivity of *Acacia auriculiformis* under CO_2-enriched environment. *Industrial Crops and Products, 210*, Article 118186.

Singh, H. (2024c). *Forests and climate change: Biological perspectives on impact, adaptation, and mitigation strategies*. Springer Nature.

Singh, H. (2024d). *Urban forests, climate change and environmental pollution: Physio-biochemical and molecular perspectives to enhance urban resilience*. Springer Nature.

Singh, H. (2025). The role of plant functional traits in mitigating urban climate and environmental pollution. In *Plant functional traits* (pp. 297–308). Elsevier.

Singh, H., & Kumar, M. (2022). Climate change and its impact on Indian Himalayan forests: current status and research needs. In *Climate change: Impacts, responses and sustainability in the Indian Himalaya* (pp. 223–242). Springer.

Singh, H., Kumar, N., & Kumar, A. (2024). Enhancing resource use efficiency in crops through plant functional traits. In *Plant functional traits for improving productivity* (pp. 97–117). Springer.

Singh, H., Kumar, N., Singh, R., & Kumar, M. (2023). Assessing the climate change impact on the habitat suitability of the range-restricted bird species (*Catreus wallichii*) in the Indian Himalayan ecosystem. *Environmental Science and Pollution Research, 30*(57), 121224–121235.

Singh, H., Verma, A., & Kumar, M. (2018a). *Can elevated CO_2 alter the growth dynamics, morphology, physiology and biochemistry of the forestry tree species?* (pp. 288–305). Climate Change and Agriculture. Discovery Publishing House.

Singh, K. K., Gagné, S. A., & Meentemeyer, R. K. (2018b). Urban forests and human well-being

Soga, M., & Gaston, K. J. (2016). Extinction of experience: The loss of human–nature interactions. *Frontiers in Ecology and the Environment, 14*, 94–101.

Soga, M., Gaston, K. J., Yamaura, Y., et al. (2016). Both direct and vicarious experiences of nature affect children's willingness to conserve biodiversity. *International Journal of Environmental Research and Public Health, 13*, 529.

Sörensen, J. (2018). Urban, pluvial flooding: Blue-green infrastructure as a strategy for resilience

Sreetheran, M., Adnan, M., & Azuar, A. K. K. (2011). Street tree inventory and tree risk assessment of selected major roads in Kuala Lumpur, Malaysia. *Arboriculture & Urban Forestry (AUF), 37*, 226–235.

Tiwari, A. K., Pal, A., & Kanchan, R. (2024). Mapping and monitoring of land use/land cover transformation using geospatial techniques in Varanasi city development region, India. *Nature Environment & Pollution Technology, 23*.

Twumasi, N. Y. D., Shao, Z., & Orhan, A. (2019). Remote sensing and gis methods in urban disaster monitoring and management–an overview. *International Journal of Trend in Research and Development, 3*, 918–926.

Velarde, M. D., Fry, G., & Tveit, M. (2007). Health effects of viewing landscapes–Landscape types in environmental psychology. *Urban Forestry & Urban Greening, 6*, 199–212.

Vos, P. E. J., Maiheu, B., Vankerkom, J., & Janssen, S. (2013). Improving local air quality in cities: To tree or not to tree? *Environmental Pollution, 183*, 113–122.

Wang, Y.-C., & Lin, J.-C. (2012). Air quality enhancement zones in Taiwan: A carbon reduction benefit assessment. *Forest Policy and Economics, 23*, 40–45.

Weissert, L. F., Salmond, J. A., & Schwendenmann, L. (2014). A review of the current progress in quantifying the potential of urban forests to mitigate urban CO2 emissions. *Urban Climate, 8*, 100–125.

Wolch, J. R., Byrne, J., & Newell, J. P. (2014). Urban green space, public health, and environmental justice: The challenge of making cities 'just green enough.' *Landscape and Urban Planning, 125*, 234–244.

Zeleňáková, M., Purcz, P., Hlavatá, H., Blištan, P. (2015). Climate change in urban versus rural areas. Procedia Engineering. Elsevier Ltd. (pp. 1171–1180)

Zhang, B., & Brack, C. L. (2021). Urban forest responses to climate change: A case study in Canberra. *Urban Forestry & Urban Greening, 57*. https://doi.org/10.1016/j.ufug.2020.126910

Zhang, X., Wang, D., Hao, H., et al. (2017). Effects of land use/cover changes and urban forest configuration on urban heat islands in a loess hilly region: Case study based on Yan'an City, China. *International Journal of Environmental Research and Public Health, 14*, 840.

Zheng, D., Ducey, M. J., & Heath, L. S. (2013). Assessing net carbon sequestration on urban and community forests of northern New England, USA. *Urban Forestry & Urban Greening, 12*, 61–68.

Zięba-Kulawik, K., Skoczylas, K., Wężyk, P., et al. (2021). Monitoring of urban forests using 3D spatial indices based on LiDAR point clouds and voxel approach. *Urban Forestry & Urban Greening, 65*, Article 127324.

Chapter 9
Policy Frameworks and Global Initiatives for Carbon Neutrality

Manish Singh, Garima Mishra, and Hukum Singh

Abstract Achieving carbon neutrality is a pivotal objective in the global fight against climate change, requiring a strategic balance between greenhouse gas emissions and their effective removal. This chapter highlights the vital role of public policy in promoting sustainable forestry practices, particularly in regions facing complex land tenure and governance challenges. It examines the evolution of international climate agreements and their impact on shaping collective efforts to reduce global carbon emissions. Key strategies such as emissions trading, carbon pricing, and the integration of renewable energy are examined alongside the growing significance of innovative technologies, including carbon capture and storage. Furthermore, it critically evaluates existing carbon offset mechanisms, advocating for the establishment of robust governance structures and transparent valuation systems to enhance their credibility and effectiveness. It highlights the importance of accurate, transparent, and verifiable monitoring systems to inform policy planning and implementation. Additionally, it addresses the potential trade-offs between short-term timber yields and long-term carbon sequestration objectives within forest-based mitigation strategies. Notably, the chapter advocates for aligning decarbonization efforts with the United Nations Sustainable Development Goals (SDGs), emphasizing the need for inclusive global cooperation, technological innovation, and policy reform. It also presents a forward-looking perspective that calls for a unified global commitment to accelerate the transition toward a sustainable, resilient, and low-carbon future.

Keywords Carbon neutrality · Carbon offset · Environmental sustainability · Forest conservation

M. Singh (✉)
Centre for Environmental Assessment and Climate Change, G.B. Pant National Institute of Himalayan Environment, Kosi-Katarmal, Almora, Uttarakhand, India
e-mail: manish.s92@gmail.com

G. Mishra
Genetics and Tree Improvement Division, ICFRE-Forest Research Institute, Dehradun, Uttarakhand, India

H. Singh
ICFRE-Forest Research Institute, Dehradun, Uttarakhand, India

H. Singh (ed.), *Forestry for a Carbon-Neutral and Sustainable Future*, Sustainability Solutions, https://doi.org/10.1007/978-3-032-06427-1_9

9.1 Introduction

Forests play a major role in combating climate change, providing sustainable livelihoods, and ensuring the overall well-being of our planet (Raman et al., 2024). Due to their role in sequestering carbon, forests have emerged as a crucial component in achieving global carbon neutrality goals (Lamb et al., 2021). However, the financing and implementation of sustainable forest management practices have long been a challenge (Boscolo et al., 2010). Hence, a comprehensive policy framework that aligns with global initiatives, such as the United Nations' Sustainable Development Goals, is essential to achieving carbon neutrality (Goparaju et al., 2020; Raman et al., 2024).

The recognition of forests' multiple functions and values is a key aspect of the policy framework (Boscolo et al., 2010). Forests not only store carbon and regulate the climate but also provide vital habitats for biodiversity and support the livelihoods of Indigenous communities and small-scale farmers (Boscolo et al., 2010). Accordingly, the policy framework should foster an integrated approach to forest management, where conservation, restoration, and sustainable utilization are balanced and mutually reinforcing.

The policy framework should prioritize the active use of the UN Sustainable Development Goals as a guiding framework for action. This would involve aligning forestry-related policies and strategies with the relevant Sustainable Development Goals (SDGs), such as SDG 15 (Life on Land), SDG 13 (Climate Action), and SDG 1 (No Poverty), among others (Baumgartner, 2019). The management of forests for sustainability encompasses at least eight of the 17 UN Sustainable Development Goals (SDGs), with a primary focus on combating desertification, halting biodiversity loss, and reversing land degradation (Mahari et al., 2020). Moreover, one-to-one reforestation is insufficient to restore the atmospheric carbon dioxide levels to their original state if deforestation persists (Choi & Manousiouthakis, 2020). Direct air capture (DAC) has also emerged as a potential alternative to forestry, though its efficiency remains under scrutiny (Gambhir & Tavoni, 2019).

Governments and institutions should work towards fully adopting the UN SDGs and actively incorporating them into their forestry sector policies and initiatives. Moreover, the policy framework should address the financing challenges that have historically hindered sustainable forest management (Singh et al., 2024). Emerging initiatives that enhance forest carbon stocks and reduce greenhouse gas emissions may provide additional funding options for sustainable forestry (Bustamante et al., 2014). These include voluntary carbon markets and REDD+ programs. Importantly, the policy framework should ensure that these initiatives benefit local communities, small-scale farmers, and indigenous peoples (Boscolo et al., 2010).

Agroforestry, which integrates trees, crops, and livestock, has also been identified as a key driver in achieving the Sustainable Development Goals. This approach fosters synergy among technologies, landscapes, rights, and markets, ultimately contributing to the restoration of multifunctionality in forest landscapes (Goparaju et al., 2020).

Additionally, the policy framework should address the need for strengthened governance, transparency, and accountability in the forestry sector. This includes ensuring the rights and livelihoods of indigenous communities and small-scale farmers, as well as enhancing the transparency and accountability of decision-making processes (IPCC, 2021). Hence, it can be stated that formulating a comprehensive policy framework is essential for achieving the SDGs and ensuring the long-term sustainability of forests. This framework should be guided by the principles of integrated forest management, alignment with the UN Sustainable Development Goals (SDGs), innovative financing mechanisms, and strengthened governance and transparency (Fig. 9.1).

Moreover, carbon offset initiatives, such as renewable energy projects or energy efficiency programs that mitigate emissions, may help achieve carbon neutrality (Kotchen, 2009). The predominantly carbon-intensive chemical industry has begun adopting various strategies to achieve carbon neutrality. These strategies include carbon capture and utilization (CCU), biomass production and processing, and earning or purchasing carbon credits (Blaufelder et al., 2021; Gabrielli et al., 2020). Moreover, activities such as investments in projects like afforestation, energy conservation, and waste-to-energy conversion are also feasible to acquire carbon credits (Bleuel, 2024).

Several key strategies are necessary to achieve meaningful reductions in greenhouse gases. Since the burning of fossil fuels is the primary source of greenhouse gases, transitioning to renewable energy sources, such as solar, wind, and geothermal energy, is essential (Sims, 2014). Additionally, changes in forestry practices, such as adopting regenerative methods that improve soil health and reduce emissions, are critical for addressing potent greenhouse gases like methane and nitrous oxide (Ghosh et al., 2024). Sustainable management of forests and ecosystems, including

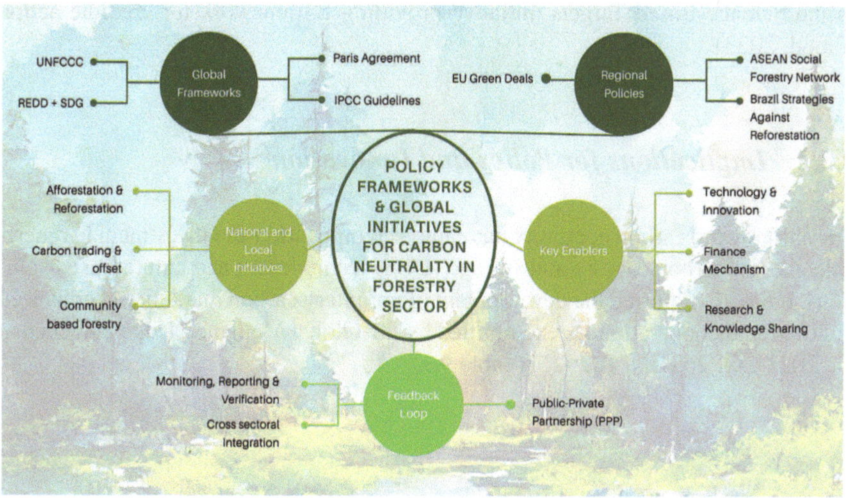

Fig. 9.1 Policy frameworks and global initiatives for carbon neutrality in forestry

wetlands, peatlands, and grasslands, is essential for maintaining carbon sinks and supporting biodiversity (Piabuo et al., 2018). Ultimately, establishing a policy environment that incorporates carbon pricing, emission caps, and incentives for clean energy transitions is crucial for driving large-scale mitigation efforts (UNEP, 2021).

The UN Sustainable Development Goals, the UNFCCC, and the Convention on Biological Diversity (CBD) emphasize the importance of integrating habitat conservation and restoration with climate change mitigation efforts (Duncanson et al., 2023). International collaboration, equitable policies, and advancements in renewable energy technology are crucial for achieving these goals and ensuring a sustainable future. This comprehensive approach helps reconcile carbon neutrality goals with practical implementation challenges, providing a more sustainable and effective transition toward reduced atmospheric carbon dioxide levels.

9.2 The Rise of Carbon Neutrality as a Global Target

9.2.1 Emergence and Adoption

The concept of carbon neutrality, or achieving net-zero emissions, has gained momentum since the mid-2010s as a cornerstone of climate action. To achieve net zero, the quantity of GHG released is balanced by an equal amount retrieved from the atmosphere (IPCC, 2018). Countries and corporations alike have embraced net-zero targets. As of 2023, over 130 nations, including major emitters such as China, the European Union, and the United States, have committed to achieving net-zero emissions by 2050 or earlier. Corporations, driven by consumer demand and investor pressure, have also committed to achieving carbon neutrality, with initiatives such as the Science-Based Targets initiative providing a framework for credible action (Flood, 2023).

9.2.2 Implications for Policy and Innovation

The shift toward carbon neutrality has catalyzed innovation and investment in renewable energy, carbon capture and storage (CCS) technologies and nature-based solutions. It has also spurred the development of voluntary carbon markets and enhanced international cooperation, as exemplified by the Global Methane Pledge launched at the 2021 COP26 (UNEP, 2021) (Table 9.1).

Table 9.1 illustrates key milestones in the global pursuit of carbon neutrality

Year	Milestone	Impact
2015	Ratification of the Paris Agreement	Universal commitment to emission cuts
2018	Special Report of the IPCC on 1.5 °C Global Warming	Highlighted urgency of net-zero
2021	Global Methane Pledge at COP26	Focus on potent GHG reductions
2023	First Global Stocktake under the Paris Agreement	Assessed progress, identified gaps

9.2.3 Carbon Accounting for Land Use, Land-Use Change, and Forestry (LULUCF)

Land use, land-use change, and forestry activities play crucial roles in the global carbon cycle, accounting for a significant portion of anthropogenic greenhouse gas emissions (IPCC, 2018). The IPCC has developed comprehensive guidelines to help countries and organizations accurately account for and report on these complex carbon dynamics. (Olorunfemi et al., 2020; Grassi et al., 2018).

The land use sector is a double-edged sword, with the potential to both emit and sequester carbon. As highlighted in the literature, natural climate solutions, such as improved land stewardship, can increase carbon sinks, whereas deforestation and land-use change can lead to significant carbon emissions (Singh & Kumar, 2022). Therefore, careful management of land resources is crucial for achieving global climate goals, as land use can influence future terrestrial climate-carbon cycle interactions (Yue et al., 2020).

Yue et al. (2020) have provided insights into the impact of land use on the interannual variability of the terrestrial carbon cycle. These findings underscore the need to prioritize mitigation strategies that reduce land-use emissions and enhance the climate resilience of carbon sinks, particularly in managed lands. Furthermore, Olorunfemi et al. (2020) emphasized the potential of agroforestry and sustainable forestry practices to sequester carbon, underscoring the importance of understanding land-use management practices and their temporal and spatial variability.

9.2.4 Evaluating the IPCC's Methodologies for Forest Carbon Measurement

The IPCC has been at the forefront of global efforts to mitigate climate change, providing authoritative guidance on greenhouse gas accounting and emission reduction strategies. As forests play a crucial role in the global carbon cycle, the IPCC's methods for measuring and modeling forest carbon stocks and sequestration have significant policy implications. However, recent research has highlighted potential limitations in the IPCC's approaches, suggesting the need for more granular and dynamic assessments of forest carbon dynamics (Cook-Patton et al., 2020).

It has been emphasized that the IPCC's default forest carbon accumulation rates may underestimate the actual potential of natural forest regrowth (Cook-Patton et al., 2020). Compiling over 13,000 field measurements, Cook-Patton et al. (2020) generated a high-resolution global map of potential aboveground carbon accumulation rates, revealing a variation of more than 100-fold across different climatic regions. This finding suggests that the IPCC's one-size-fits-all approach may fail to capture the nuanced dynamics of forest carbon sequestration, which may vary significantly depending on local environmental conditions.

Moreover, the IPCC's methodologies have been criticized for their reliance on coarse-resolution data and limited spatial coverage, hindering their applicability for local and regional-scale planning. In contrast, recent advancements in remote sensing and spatial modeling techniques have enabled the development of high-resolution forest carbon mapping approaches that can quantify both current stocks and future sequestration potential with greater accuracy and granularity (Kumar & Singh, 2020).

As policy-makers and land managers increasingly rely on forest-based climate mitigation strategies, the need for robust and dynamic forest carbon accounting systems has become paramount. Incorporation of the latest scientific insights and leveraging advanced geospatial tools strengthen the credibility and relevance of its guidance, ensuring that global efforts to harness the carbon sequestration potential of forests are grounded in the best available evidence (Singh et al., 2018).

9.3 Regional Policies

9.3.1 EU Forest Strategy for 2030

The European Union has recognized the importance of mapping and assessing ecosystems and their services to inform environmental policy decisions. The EU's Biodiversity Strategy has urged member states to delineate and evaluate the condition of ecosystems and their services within their national jurisdictions, with support from the European Commission. This initiative provides comprehensive knowledge based on ecosystems and their services in Europe, which is crucial for achieving the strategy's six targets and is relevant to various EU sectoral policies, including agriculture, fisheries, and maritime affairs (Gordeeva et al., 2022).

Quantitative geographic data on the demand for ecosystem services provide foundational metrics for evaluating future net gains or losses, which is crucial for assessing policy impacts and may facilitate the creation of financial instruments to fund ecosystem initiatives (Maes et al., 2012). As the European Union moves towards an ecosystem-based approach to environmental management, there is a growing emphasis on understanding the economic, ecological, hydrological, and other processes that influence ecosystem health and the delivery of essential services (Apitz et al., 2006).

The 2010 Biodiversity objective, established worldwide and in Europe, was a significant political commitment to enhance biodiversity conservation and management. However, achieving this target depends on the biodiversity indicators that are currently under development (Singh, 2024c). The successful execution in Europe requires extensive collaboration and integration of regulatory and technical information across EU member countries, agencies, and disciplines (Sara et al., 2015; Maes et al., 2012).

9.3.2 *Carbon Farming and Biodiversity Targets*

The EU's Forest Strategy for 2030 has set ambitious targets for carbon farming and biodiversity conservation. The strategy aims to enhance the EU's carbon sink capacity through sustainable forest management practices, including afforestation, reforestation, and improved soil management (Singh, 2024d). Additionally, the strategy aims to protect and restore forest ecosystems, reverse biodiversity loss, and ensure the long-term resilience of forest habitats. (Apitz et al., 2006; Maes et al., 2012).

To achieve these goals, the EU closely monitors and assesses the state of its forest ecosystems and the services they provide. Mapping and quantifying the delivery of ecosystem services, such as carbon sequestration and habitat provision, will be crucial for informing policy decisions and developing financial instruments to support investments in forest conservation and restoration (Maes et al., 2012).

9.3.3 *ASEAN Social Forestry Network*

The ASEAN Social Forestry Network is a regional platform that promotes collaboration and knowledge sharing among ASEAN member states on social forestry. The network aims to promote forest sustainability and the equitable sharing of benefits with local communities and indigenous peoples (Singh, 2025). The management of forests for sustainability encompasses at least eight of the 17 UN Sustainable Development Goals, including those related to combating desertification, halting biodiversity loss, and reversing land degradation. Substantial alterations are necessary to preserve the planet's rainforests, consequently safeguarding the global climate and biodiversity (Mahari et al., 2020). These worldwide solutions and mitigations require developed and emerging countries to adjust their policies and meet their high demand for hardwood. Governments and organizations should use non-profit tree-planting browsers wherever feasible to support forest conservation efforts (Mahari et al., 2020).

9.4 The Role of Forestry in Voluntary and Compliance Carbon Markets

The global effort to mitigate climate change has led to the development of various carbon markets, both voluntary and compliance-driven, in which forestry plays a significant role. In voluntary carbon markets, forest owners can gain recognition and financial compensation for their efforts in sustainable forest management and carbon sequestration. (Vickers et al., 2012) However, the complexities of these markets present challenges for forest owners, who may risk surrendering the potential benefits to more informed actors (Vickers et al., 2012).

Emerging initiatives to reduce greenhouse gas emissions associated with deforestation have opened up additional funding options for sustainable forest management. These initiatives, along with voluntary carbon markets, offer promising opportunities for Indigenous peoples, local communities, and small-scale farmers to participate in and benefit from the carbon economy (Boscolo et al., 2010).

While the potential benefits of integrating forestry into carbon markets are evident, significant barriers exist to widespread landowner participation. Long-term monitoring obligations, high initial project development costs, and legislative uncertainty can deter family forest landowners from participating in carbon offset projects (Kerchner & Keeton, 2015). Moreover, the success of such initiatives may be hindered by public policy challenges in frontier landscapes with unclear land tenure. (Griscom et al., 2017).

The relative success of Brazil's efforts to slow deforestation through a strong regulatory framework and supply chain interventions provides a promising model for other regions to emulate (Kerchner & Keeton, 2015; Vickers et al., 2012). Implementation of Payment for Ecosystem Services (PES). Schemes, such as those focused on forest carbon sequestration, can serve as a tool for financing sustainable small-scale forestry (Singh et al., 2023). However, the financial viability of these market-based approaches for small-scale forest owners remains a complex issue, with a range of factors to consider, including initial project costs, long-term monitoring obligations, and policy uncertainty (Singh, 2024b). The careful design and implementation of these initiatives, along with supportive public policies and capacity building for local stakeholders, will be crucial to ensuring that the benefits of forestry's integration into carbon markets are equitably distributed and contribute to the broader goals of climate change mitigation and sustainable development (Joshi et al., 2021).

9.5 Satellite Remote Sensing and GIS for Monitoring Deforestation

Satellite remote sensing and geographic information systems have become invaluable tools in the fight against deforestation, providing researchers and policymakers with a wealth of data and analytical capabilities to monitor and assess the extent and patterns of forest loss worldwide (Kumar et al., 2020). In a study of the Niger Delta region, researchers utilized satellite imagery and GIS applications to acquire spatial and temporal information about the region's wetlands ecosystem, identifying changes in the mangrove ecosystem over time (Elekwachi et al., 2021). Similarly, a study in the Oued Isser watershed in western Algeria utilized high-resolution Sentinel-2A satellite imagery and GIS software to map land use activities and vegetation cover, providing critical data for implementing the Revised Universal Soil Loss Equation model to assess water erosion (Joshi et al., 2020).

The integration of remote sensing and GIS technologies has proven particularly useful in monitoring and managing water resources in agricultural settings. Satellite measurements can be combined with in-situ data and GIS tools to estimate crop yields, evapotranspiration, and net groundwater use, providing valuable insights for irrigation management (Bastiaanssen et al., 2003). Despite the clear benefits of these technologies, their adoption by water managers and policymakers has been limited, underscoring the need for better communication and collaboration between the remote sensing community and end-users (Ruhil et al., 2024).

9.6 Bioengineering for Resilient Tree Species

Bioengineering has emerged as a promising field in addressing the pressing challenges faced by forests in the face of climate change and environmental degradation. The stabilization of slopes through the application of vegetation is known as soil bioengineering. One key aspect of bioengineering is the integration of ecological principles with engineering practices to assess, design, construct, and maintain living vegetation systems (Stokes et al., 2014) This approach has been particularly beneficial in the rehabilitation of severely degraded hillslopes, where the establishment of vegetation is often hindered by strong erosive forces, especially in dry climates and poor soil conditions (Stokes et al., 2014). The primary mechanisms through which vegetation contributes to slope stabilization are both hydrological and mechanical. The hydrological impact of vegetation involves reducing soil water content through processes such as transpiration, interception, and evapotranspiration, thereby enhancing the overall stability of the slope. Conversely, the mechanical reinforcement of soil through the extensive root systems of vegetation provides additional structural support, further improving the slope's resistance to erosion and failure (Mulyono et al., 2018; Stokes et al., 2014).

The effective implementation of soil bioengineering techniques requires a thorough understanding of the complex interplay between the hydromechanical characteristics of vegetation and the unique environmental conditions of the target site (Stokes et al., 2014). Consequently, the selection of appropriate plant species and the integration of engineering structures or soil bioengineering techniques become crucial in addressing these challenges (Jangra et al., 2024).

9.7 Community-Based Forestry Management

Community-based forestry management has emerged as a promising approach to sustainable forest conservation and management, offering an alternative to top-down, centralized models that have often failed to address the needs and concerns of local communities (Singh et al., 2023). This approach aims to empower local communities to participate in the decision-making, management, and utilization of forest resources to achieve both conservation and livelihood objectives simultaneously. (Arts & Koning, 2017; Zyambo, 2018).

However, the success of community-based forestry management has been mixed. While some initiatives have demonstrated positive outcomes in terms of forest conservation and community well-being, others have faced significant challenges. An example of successful community-based forestry management is in Nepal, where community forest management groups have been able to enhance natural, economic, and social capital concurrently. These groups have not only been able to harvest forest products sustainably but have also worked to increase the overall forest stock, demonstrating the ideals of strong sustainability (Joshi et al., 2024).

In Cameroon, the concept of community forestry has evolved, with researchers highlighting the importance of good governance as a critical condition for success. However, the proliferation of management conflicts in Cameroonian community forests has shown that the broad expectations of decentralization and community management have often not been met (Mbolo, 2013; Piabuo et al., 2018). Researchers have identified a range of factors that can contribute to the success or failure of community-based forestry management. These include the degree of community engagement and empowerment, the level of transparency and accountability in decision-making processes, the ability to resolve conflicts and manage resource use, and the extent to which local livelihoods and ecological sustainability are aligned (Singh, 2024a).

9.8 Policy Enforcement and Capacity Building

Strengthening community capacity and effective policy enforcement is key to overcoming the challenges faced in community-based forestry management (Mbolo, 2013; Piabuo et al., 2018). Providing communities with the necessary skills, resources, and institutional support to manage forest resources effectively can help ensure the long-term sustainability of these initiatives (Piabuo et al., 2018). The European Green Deal represents a transformative framework aimed at achieving carbon neutrality within the European Union (EU) by 2050. It is a bold and comprehensive strategy designed to address climate change, enhance resource efficiency, and promote sustainable economic growth. The Green Deal not only sets ambitious climate targets but also outlines specific policy mechanisms to achieve them (Barwa & Singh, 2024).

9.9 Carbon Taxation

While governments have not implemented a nationwide carbon tax, they have endorsed market-based mechanisms to reduce emissions. Voluntary carbon markets, supported by robust federal guidelines, play a pivotal role in complementing other policies. These markets enable businesses to offset emissions by investing in verified carbon reduction projects, such as reforestation and renewable energy initiatives (Sang et al., 2023).

Efforts to price carbon emissions have also been encouraged at the state and regional levels. Cap-and-trade programs, such as those in California and the Regional Greenhouse Gas Initiative (RGGI), have garnered federal support as models for broader adoption. These programs align with the administration's broader commitment to leveraging innovative market mechanisms to achieve climate goals (Rezai et al., 2018).

The Biden administration's federal policies underscore a commitment to integrating clean energy standards, renewable energy incentives, and carbon market principles into a cohesive strategy for addressing climate change. The United States aims to lead by example, both domestically and internationally, by fostering innovation, strengthening public–private partnerships, and prioritizing equity (Garg et al., 2016).

9.10 Developing Nations' Challenges

9.10.1 Balancing Economic Growth with Carbon Neutrality

Developing nations face a unique challenge in balancing the dual imperatives of economic growth and carbon neutrality. Many of these countries rely heavily on carbon-intensive industries and fossil fuels to drive development while grappling with limited financial and technological resources to transition toward sustainable practices. Striving for carbon neutrality often involves trade-offs that may hinder immediate economic goals, particularly in sectors such as energy, agriculture, forestry, and transportation (Budiana, 2024).

Rapid industrialization in developing nations contributes to greenhouse gas emissions, but it is also essential for poverty alleviation, infrastructure development, and job creation. Achieving carbon neutrality requires integrating sustainable practices without compromising the economic well-being of vulnerable populations. Strategies such as renewable energy deployment, energy efficiency improvements, and reforestation initiatives can help strike a balance between these objectives. However, such measures necessitate substantial investment and technological advancements that many developing nations cannot afford without external assistance (Rezai et al., 2018).

9.10.2 Financial and Technological Support from Global Institutions

Global institutions play a crucial role in supporting the climate transitions of developing nations by providing financial resources and facilitating technology transfer. Under the Paris Agreement, developed countries are obligated to mobilize $100 billion annually to assist developing nations in addressing the impacts of climate change. These funds aim to strengthen resilience, adapt to climate risks, and reduce emissions in line with the goals of the agreement (Budiana, 2024).

Technology transfer is another essential pillar of global support. Sharing advanced technologies such as solar panels, wind turbines, and carbon capture systems may enable developing nations to leapfrog to cleaner and more efficient energy systems. This process is supported through mechanisms established under the United Nations Framework Convention on Climate Change (UNFCCC), including the Technology Mechanism, which fosters innovation and the dissemination of climate-friendly technologies (Budiana, 2024). Additionally, multilateral development banks and regional initiatives offer concessional financing to reduce the cost burden of climate investments (Sorooshian, 2024).

9.10.3 Role of Climate Finance and Technology Transfer

Climate finance is a cornerstone of global efforts to support developing nations in their climate ambitions. It encompasses financial flows from developed to developing countries, enabling investments in projects that mitigate emissions, enhance climate resilience, and foster sustainable development. These funds come from public, private, and multilateral sources, with a growing emphasis on leveraging private-sector investments to bridge funding gaps (Budiana, 2024).

However, high financing costs and risk perceptions often deter private investments in green initiatives within developing nations. Addressing these challenges requires innovative financial instruments, such as green bonds, blended finance, and risk-sharing mechanisms, which can lower the barriers to entry for private investors (Rezai et al., 2018).

Technology transfer complements climate finance by providing developing nations with access to cutting-edge solutions that reduce emissions and improve efficiency. These technologies enable countries to meet their climate goals without compromising growth. For instance, clean energy technologies and sustainable agricultural practices can reduce dependence on fossil fuels while fostering economic productivity. Capacity-building programs further enhance the ability of developing nations to integrate and maintain these technologies effectively (Sorooshian, 2024).

Balancing economic growth with carbon neutrality is a big challenge for developing nations. However, with robust financial and technological support from global institutions, this transition will become more equitable and achievable. Climate finance and technology transfer remain indispensable tools in empowering these nations to make meaningful contributions to global climate goals while ensuring sustainable development for their populations.

9.11 Carbon Pricing Mechanisms

9.11.1 Carbon Taxes

Carbon taxes are one of the most straightforward approaches to internalizing the cost of carbon emissions. Assigning a clear monetary value to the carbon content of fossil fuels motivates enterprises and individuals to reduce emissions and adopt greener technologies. Notable case studies of successful implementation include Sweden and British Columbia, which demonstrate the effectiveness of this mechanism in reducing greenhouse gas emissions.

9.11.1.1 Sweden: Pioneering Carbon Taxation

Sweden introduced its carbon tax in 1991 as part of a broader environmental tax reform. Initially set at approximately €25 per ton of CO_2, the tax has progressively increased to €120 per ton, making it one of the highest in the world (Martinsson et al., 2024). This tax has been applied broadly, covering fuels used for heating and transportation, while industries subject to the European Union Emissions Trading System (EU ETS) received partial exemptions.

Sweden's carbon tax has significantly reduced emissions, particularly in the transportation and heating sectors. Between 1990 and 2015, emissions from Sweden's manufacturing sector decreased by 31%, with 18% of this reduction attributed directly to carbon taxation policies (Martinsson et al., 2024). Despite concerns about economic competitiveness, Sweden has maintained strong economic growth, showcasing the feasibility of integrating carbon taxes with economic development (Pan et al., 2024).

9.11.1.2 British Columbia: Revenue-Neutral Carbon Tax

British Columbia implemented a revenue-neutral carbon tax in 2008, starting at CAD 10 per ton of CO_2 and increasing to CAD 50 by 2022. The tax accounts for nearly 70% of the province's emissions, including those from fossil fuels used in transportation and heating (Pan et al., 2024). What sets British Columbia's carbon tax apart is its revenue-neutral design, all tax revenue is returned to the public through income tax reductions and rebates. This approach has mitigated public resistance while maintaining environmental effectiveness. A study found that between 2008 and 2015, per capita fossil fuel consumption in British Columbia decreased by 16%, compared to a 3% reduction in the rest of Canada (Emeka-Okoli et al., 2024). Significantly, the province's GDP grew in line with the national average, illustrating that well-designed carbon taxes can achieve emissions reductions without compromising economic growth (Pan et al., 2024).

Therefore, carbon taxes are a powerful tool for addressing climate change, as demonstrated by the experiences of Sweden and British Columbia (Emeka-Okoli et al., 2024). Carefully designing and implementing these taxes with considerations for equity, economic impact, and public acceptance may help governments achieve significant emissions reductions while maintaining economic stability (Martinsson et al., 2024).

9.12 Carbon Offset Markets

Carbon offset markets are a critical component of global climate action strategies. They enable entities to mitigate greenhouse gas emissions by investing in projects that reduce or remove carbon dioxide from the atmosphere. These markets are categorized into two primary types: voluntary markets and compliance markets, each with distinct objectives, participants, and regulatory frameworks (Ruhil et al., 2024).

9.12.1 Voluntary Carbon Markets

Voluntary carbon markets are designed for organizations or individuals who choose to offset their emissions beyond the requirements of regulations. These markets are typically driven by corporate sustainability goals, brand reputation, and consumer demand for environmentally responsible products (Preston, 1984). Participants in voluntary markets acquire carbon credits produced by initiatives such as renewable energy projects, forestry efforts, and methane collection systems (Montel et al., 2023). Voluntary markets offer flexibility but face challenges in ensuring the credibility and integrity of carbon credits. Concerns have arisen over credits of dubious quality that do not represent real or additional emissions reductions. As a result, third-party certification standards play a critical role in validating projects and ensuring transparency (Espenan, 2023; Henriksen et al., 2004).

9.12.2 Compliance Carbon Markets

Compliance carbon markets are regulated frameworks set up by international organizations or governments to meet mandatory targets for reducing emissions. Examples include EU ETS and India's proposed carbon trading mechanism (Jain, 2015;

Vashishtha and Vashishtha, 2024). These markets operate under a cap-and-trade model, where the total allowable emissions are capped, and entities trade allowances within this limit. Compliance markets are typically more structured and robust than voluntary markets, with strict monitoring, reporting, and verification (MRV) systems to track emissions. They are crucial for attaining national and international climate objectives, as delineated in the Paris Agreement (Bleuel, 2024).

9.13 Certification Standards

Certification standards are fundamental to the credibility of carbon offset markets. In both voluntary and compliance markets, these requirements imply that carbon credits signify authentic, quantifiable, and supplementary reductions in emissions. Prominent certification frameworks include the Verified Carbon Standard (VCS), Gold Standard, and the Climate Action Reserve (CAR). Key elements of certification include:

1. **Additionality**: Ensuring that the project's emissions reductions would not have occurred without the offset initiative.
2. **Permanence**: Verifying those emissions reductions are long-lasting and not easily reversed.
3. **Baseline Measurements**: Establishing a clear reference point for emissions reductions.

In compliance markets, certification is often embedded within the regulatory framework, ensuring alignment with legal and environmental standards. Voluntary markets rely on independent certifiers to establish trust and avoid greenwashing (Deloitte Consulting, 2009; Orozco, 2019) (Fig. 9.2).

9.14 Achieving Real Carbon Reductions

Carbon offset markets, when properly managed, can drive substantial reductions in emissions and channel investments into critical climate solutions. However, the effectiveness of these markets depends on addressing challenges such as double counting, low-quality credits, and a lack of transparency. Certification standards and robust governance frameworks play an essential role in ensuring that carbon offsets deliver tangible climate benefits (Santier, 2023).

For compliance markets, the structured approach and regulatory oversight provide a more reliable pathway to achieving national and global climate goals. Voluntary markets, on the other hand, offer an avenue for private sector innovation and engagement but require stringent standards to maintain integrity.

Carbon offset markets, encompassing both voluntary and compliance frameworks, are indispensable tools in global efforts to reduce emissions. Adhering to

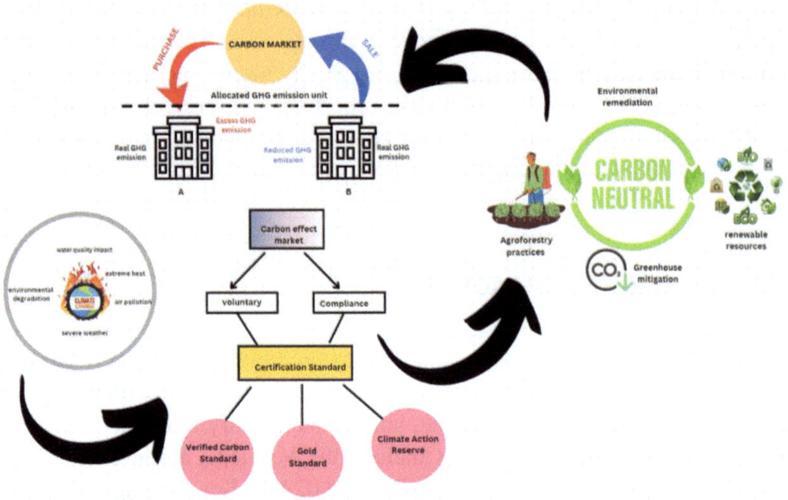

Fig. 9.2 Carbon market mechanism for achieving carbon neutrality and mitigating climate change impacts

rigorous certification standards and addressing existing challenges, these markets can enhance their credibility and maximize their contribution to achieving tangible and measurable reductions in carbon emissions.

9.15 Voluntary Carbon Markets and REDD+ Programs in Achieving Carbon Neutrality

The forestry sector has emerged as a crucial component in the global effort to mitigate climate change, with voluntary carbon markets and Reducing Emissions from Deforestation and Forest Degradation (REDD+) programs playing a significant role (Agrawal et al., 2011; Kim et al., 2021). Sustainable forest management, which encompasses initiatives to replenish forest carbon stores and mitigate greenhouse gas emissions associated with deforestation, is providing new funding options and opportunities for Indigenous communities, local farmers, and small-scale forestry enterprises (Boscolo et al., 2010).

Voluntary carbon markets, in particular, have gained traction as a means of incentivizing forest conservation and restoration. These markets provide a platform for individuals, corporations, and organizations to offset their carbon emissions by purchasing carbon credits generated through verified emission reduction activities, such as those implemented through REDD+ programs (Kim et al., 2021). However, the successful implementation of REDD+ and the effective utilization of voluntary carbon markets face several challenges. One key challenge is the issue of non-permanence, where deforestation and forest degradation may still occur due to various drivers, such as policy changes, shifts in governmental regimes, and economic advancements in REDD+ nations (Agrawal et al., 2011). To address this, REDD+ countries and their partners must establish reliable monitoring, reporting, and verification systems to precisely assess and represent the impact of REDD+ initiatives on climate change mitigation (Espenan, 2023; Henriksen et al., 2004). The inclusion of the forestry sector, particularly through initiatives such as REDD+ and voluntary carbon markets, is considered essential for achieving the objectives of the UNFCCC and the goal of carbon neutrality.

9.16 Policy Frameworks and Global Initiatives for Carbon Neutrality

Achieving global carbon neutrality is a pressing challenge that requires comprehensive policy frameworks and strategic initiatives across various sectors, including forestry. In India, the forestry sector plays a crucial role in the country's climate change mitigation efforts, with the potential to make a significant contribution to achieving carbon neutrality overall. (Duncanson et al., 2023; Griscom et al., 2017).

Incorporating forest carbon into regional climate mitigation strategies is crucial for achieving carbon neutrality. While many Indian states have established greenhouse gas reduction goals and framework documents, the integration of forest carbon accounting remains inconsistent. A uniform, efficient, and standardized forest carbon monitoring system would improve transparency and progress tracking to meet state, national, and international climate objectives (Lamb et al., 2021).

The development of high-resolution forest carbon models further enhances planning and decision-making processes, enabling the quantification of carbon stocks over large, policy-relevant areas with sufficient accuracy (Hurtt et al., 2024). Such advanced modeling approaches can significantly advance climate mitigation planning in the forestry sector.

Importantly, a systems approach is necessary to comprehensively assess the climate change mitigation options in the forestry sector, as the benefits of certain scenarios may be outweighed by increased carbon emissions in other parts of the system. Preserving forests, lengthening harvest cycles, and transitioning to durable goods have shown the most significant mitigation advantages across several decades (Griscom et al., 2017). India may learn from Brazil's successful fight against deforestation by adopting some of the country's policies, such as its robust regulatory framework, transparent and accurate federal monitoring, and supply chain interventions.

9.17 Green Climate Fund and Corporate Investments in Forestry-Based Carbon Offsets

Sustainable forestry management faces significant financing challenges but also unprecedented opportunities. Emerging initiatives, such as the Green Climate Fund and corporate investments in forestry-based carbon offsets, are offering new funding options for sustainable forestry (Boscolo et al., 2010). The world's mountain grasslands and shrublands store vast amounts of carbon, yet their conservation has been largely overlooked in climate policy discussions (Ward et al., 2014). Forests, too, have acquired new value as important carbon sinks, leading to the rise of forestry voluntary carbon markets where forest owners can gain financial compensation for sustainably managing their lands. (Vickers et al., 2012) However, forest owners have generally been slow to understand and capitalize on these markets, risking the potential benefits being captured by other, more informed actors. (Vickers et al., 2012).

To address the financing challenges in sustainable forestry management, national strategies are being developed that leverage a variety of funding sources. Indigenous peoples, local communities, and small-scale farmers are increasingly recognized as key players in implementing such financing initiatives (Boscolo et al., 2010). Overall, the future of global forest conservation will depend on the ability to mobilize diverse sources of finance, from public funds to private sector investment, in support of sustainable forestry practices.

9.18 Technological and Industrial Strategies to Achieve Carbon Neutrality

9.18.1 Decarbonization Technologies

Achieving global net-zero emissions by 2050 requires a transformative shift in technological and industrial strategies to reduce greenhouse gas emissions across sectors. Key decarbonization technologies include renewable energy and carbon capture and storage (CCS), which collectively play pivotal roles in mitigating climate change while supporting sustainable economic growth.

9.18.2 Renewable Energy

Renewable energy technologies, including solar, wind, and hydropower, are foundational to decarbonizing the global energy system (Kumar et al., 2021). The International Energy Agency (IEA) highlights that renewables will account for nearly 90% of electricity generation by 2050 under a net-zero scenario (Liang et al., 2023).

Solar Power: Solar photovoltaics (PV) have emerged as the fastest-growing renewable technology, driven by declining costs and widespread adoption. Innovations in PV materials, such as perovskites, are enhancing efficiency and scalability, while utility-scale solar projects contribute to decarbonizing electricity grids (Ahmed et al., 2025).

Wind Power: Wind energy, both onshore and offshore, is a critical component of renewable energy portfolios. Offshore wind projects, supported by advances in floating turbine technologies, are expected to make a significant contribution to supplying clean energy to densely populated coastal regions (Mitigation, 2011; Sims, 2014).

Hydropower: One of the most established renewable energy sources, hydropower provides reliable baseload energy. Modernizing existing hydropower infrastructure and integrating pumped storage systems are key strategies for maximizing its potential while addressing environmental concerns (Mitigation, 2011; Sims, 2014).

9.18.3 Carbon Capture and Storage (CCS)

Carbon Capture and Storage (CCS) is a crucial technology for decarbonizing sectors that are challenging to abate, such as the cement, steel, and chemicals industries. CO_2 emissions are captured at their source, transported, and stored underground in geological formations as part of CCS. According to the IEA, large-scale deployment of CCS could mitigate up to 15% of cumulative emissions reductions required by 2050 (Liang et al., 2023). Key advancements in CCS include:

Direct Air Capture (DAC): Technologies that extract CO_2 directly from the atmosphere, although still in the early stages of development, offer the potential for negative emissions.

CCS in Industrial Applications: Industrial-scale carbon capture and storage (CCS) projects, such as those implemented in oil and gas refineries, are demonstrating the technology's viability in reducing emissions from high-intensity processes (Terzić, 2024). International collaboration is crucial for addressing technology gaps in carbon capture and storage (CCS), including cost reduction, scalability, and infrastructure development. Partnerships between developed and developing nations can accelerate CCS deployment through shared knowledge and financing mechanisms (Terzić, 2024).

9.18.4 Decarbonization Through the Forestry Sector

The forestry sector holds immense potential in mitigating climate change through the deployment of decarbonization technologies. Careful consideration must also be given to the impacts of bioenergy, as the carbon cycle benefits depend on the timeframe, feedstocks, and alternative uses of biomass (Dugan et al., 2021). In addition, the forestry sector contributes to decarbonization through the production of sustainable, low-emission materials and biofuels. Agroforestry, forest plantations, and sustainable forestry practices should be given serious consideration as strategies to enhance carbon storage and reduce emissions (Dugan et al., 2021; Olorunfemi et al., 2020; Waring et al., 2020).

9.18.5 Energy Sector: Shifting from Fossil Fuels to Renewables

The energy sector is at the forefront of decarbonization efforts, with a global shift from fossil fuels to renewable energy sources, including wind, solar, and hydropower. Renewable energy deployment has accelerated due to declining costs, supportive policies, and technological advancements (Khan & Al-Ghamdi, 2023).

Renewable Electrification: In countries like India, initiatives for green electrification aim to decarbonize industrial energy use by increasing reliance on renewable energy sources. India's National Green Hydrogen Mission, for instance, emphasizes the role of renewable electricity in producing green hydrogen for industrial and transportation applications (Khan & Al-Ghamdi, 2023).

Phasing Out Fossil Fuels: Governments and industries worldwide are phasing out coal and natural gas in favour of clean energy solutions. This transition not only reduces emissions but also enhances energy security and economic resilience (Khan & Al-Ghamdi, 2023).

9.18.6 Industry: Green Steel, Cement, and Sustainable Supply Chains

The industrial sector accounts for a significant portion of global greenhouse gas emissions, primarily from energy-intensive processes, including steel and cement production. Transitioning to greener technologies and sustainable supply chains is essential for decarbonizing this sector.

Green Steel: The use of hydrogen in steel production as a substitute for coal in blast furnaces is a promising development. Green steel reduces emissions while meeting global demand for this critical material (Souza Filho et al., 2023).

Low-Carbon Cement: Innovations such as clinker substitution and carbon capture in cement production can significantly reduce emissions from this sector. Pilot projects worldwide demonstrate the viability of low-carbon cement technologies (Mohammed et al., 2023).

Sustainable Supply Chains: Establishing circular economies and improving resource efficiency are critical strategies for reducing industrial emissions. Companies are adopting green energy solutions, such as on-site solar and wind installations, to power manufacturing processes sustainably (Vertes et al., 2020).

9.18.7 Agriculture: Reducing Emissions Through Regenerative Farming and Reforestation

The agriculture sector, a major contributor to methane and nitrous oxide emissions, can achieve substantial reductions through regenerative farming practices and reforestation efforts.

Regenerative Farming: Practices such as no-till farming, use of cover crops, and crop rotation can enhance soil health, sequester carbon, and reduce the need for synthetic fertilizers. These methods align with sustainable agricultural goals while maintaining productivity (Ghosh et al., 2024).

Livestock Management: Strategies to reduce methane emissions include improving feed quality and implementing anaerobic digesters to capture methane from manure (Subbarao et al., 2023).

Reforestation: Large-scale reforestation and afforestation projects not only sequester carbon but also enhance biodiversity and improve local water cycles. These initiatives are integral to long-term strategies for offsetting emissions in agriculture.

Therefore, decarbonizing the energy, industry, and agriculture sectors is pivotal to achieving global climate goals. Adopting renewable energy, green industrial practices, and sustainable agricultural methods can facilitate a transition to low-carbon operations while maintaining economic growth and social equity. Collaborative efforts among governments, industries, and communities are crucial for scaling these innovations and meeting ambitious climate targets.

9.19 Challenges and Barriers to Achieving Carbon Neutrality

9.19.1 Political and Economic Obstacles

Transitioning to carbon neutrality is a multifaceted challenge influenced by political resistance and economic constraints. Despite widespread recognition of the importance of decarbonization, obstacles such as industry pushback, high transition costs, and inequality hinder global progress toward this goal.

9.19.2 Resistance from High-Carbon Industries

High-carbon industries, including fossil fuel producers, heavy manufacturing, and certain agricultural sectors, often resist decarbonization due to concerns over profitability and competitiveness. These industries, which play critical roles in national economies and employment, are wary of policies that impose stricter emissions regulations or increase operational costs (Kupzok & Nahm, 2024).

Economic Dependence: Many nations heavily reliant on fossil fuel exports or energy-intensive industries face significant economic risks from decarbonization. Resistance is amplified in countries where these industries form the backbone of economic stability and fiscal revenues (Tcvetkov, 2022).

Lobbying and Political Influence: Fossil fuel companies and other high-carbon sectors often wield significant political influence, lobbying against stringent climate policies. For example, global delays in implementing carbon pricing mechanisms have been attributed to such industry opposition (Espagne et al., 2021).

Job Security Concerns: The prospect of job losses in traditional energy sectors generates socio-political resistance, particularly in regions where alternatives like renewable energy have yet to establish strong footholds (Espagne et al., 2021; Tcvetkov, 2022).

Achieving carbon neutrality is one of the most pressing challenges of the twenty-first century. Addressing the multifaceted barriers, ranging from geopolitical tensions and economic disparities to technological gaps, requires coordinated, equitable, and innovative solutions, as shown in Table 9.2 (Table 9.3).

9.20 Conclusion

The transition to carbon neutrality stands as one of the most pressing and complex challenges of the twenty-first century, necessitating coordinated, cross-sectoral, and global efforts. There is an urgent need to adopt diverse and integrated mechanisms, strategies, and policy frameworks that aim to reduce greenhouse gas emissions and foster sustainable development pathways. Achieving net-zero emissions depends on a dynamic balance among technological innovation, effective policy interventions, and international cooperation encompassing the deployment of renewable energy systems, carbon pricing mechanisms, and the enforcement of global environmental agreements. Nonetheless, this transition is not without obstacles. Geopolitical tensions, economic disparities, and technological constraints pose significant barriers, demanding targeted, inclusive, and adaptive solutions to ensure equitable progress. A multifaceted approach that combines cutting-edge technologies, strengthens international collaboration, and prioritizes social and environmental equity is essential to overcoming these challenges. Achieving carbon neutrality goes beyond technical and economic goals, representing a moral responsibility integral to the broader pursuit of sustainable development. As countries and industries intensify

Table 9.2 Challenges and barriers to achieving carbon neutrality

Barrier type	Specific issues	Examples/impacts	Proposed solutions	References
Political resistance	Opposition from high-carbon industries, lack of global consensus	Fossil fuel lobbying is delaying the implementation of carbon pricing, resulting in a divergence in climate goals between developed and developing nations	Transparent policy frameworks, more vigorous enforcement, and incentives for cleaner alternatives	Blondeel et al. (2021), Kupzok and Nahm (2024)
Economic inequality	Uneven access to technology and finance, regressive impacts of carbon pricing	Developing nations are unable to invest in renewable infrastructure, resulting in a disproportionate impact on low-income households	Equitable climate finance, revenue recycling from carbon taxes to support vulnerable groups	Tcvetkov, (2022)
Technological gaps	Limited deployment of CCS, direct air capture, and scalable renewables	High costs and inefficiencies of CCS technologies; challenges in integrating renewable energy due to intermittency	Public–private partnerships for innovation, international collaboration on R&D, and adoption subsidies	Espagne et al. (2021), Souza Filho et al. (2023)
Trade and geopolitical issues	Disputes over carbon tariffs, supply chain disruptions, and competition over critical minerals	Trade barriers result from the EU's Carbon Border Adjustment Mechanism (CBAM), as well as supply chain bottlenecks in the production of lithium and cobalt for electric vehicle (EV) batteries	Establishing fair trade policies, diversifying mineral supply chains, and enhancing recycling initiatives	Tcvetkov (2022)

their efforts to reduce their carbon footprints, emphasis must be placed on building resilience, promoting social justice, and accelerating the global transition toward low-carbon economies. For instance, Brazil's regulatory success in reducing deforestation highlights the potential of governance and policy to drive change. Emerging opportunities to enhance forest carbon stocks and expand voluntary carbon markets further underscore the transformative potential of collective action. Ultimately, the path to a carbon–neutral future hinges on innovation, accountability, and global solidarity. The knowledge generated through such efforts holds significant implications for transformative change, offering a hopeful and actionable trajectory toward a sustainable and climate-resilient world.

Table 9.3 Strategic framework for advancing global climate cooperation and decarbonization

Strategy	Key actions	Implementation examples	Expected benefits	References
Strengthening global cooperation	Align international climate goals, ensure equitable climate finance	$100 billion annual climate finance commitment under the Paris Agreement	Equitable resource distribution accelerated global decarbonization	Benioff et al. (2010), Kupzok and Nahm (2024)
Investing in green innovation	Prioritize R&D for CCS, battery storage, and hydrogen production	U.S. National Renewable Energy Laboratory's advancements in green hydrogen technologies	Scalable and cost-effective carbon–neutral solutions	Ghiasi et al. (2023)
Promoting just transition policies	Support affected workers and regions with job retraining and economic diversification	EU's Just Transition Fund, U.S. Infrastructure Act initiatives for coal-dependent communities	Mitigated socioeconomic impacts of decarbonization, reduced inequality	Tcvetkov, 2022; Espagne et al. (2021)
Enhancing carbon markets	Strengthen MRV systems, ensure transparency in carbon trading, and implement robust certification standards to enhance accountability	Blockchain-based carbon markets, Verified Carbon Standard (VCS), and the EU ETS	Credible emissions reductions, enhanced private sector participation, and global market alignment	Blaufelder et al., 2021

References

Agrawal, A., Nepstad, D., & Chhatre, A. (2011). Reducing emissions from deforestation and forest degradation. *Annual Review of Environment and Resources, 36*(1), 373–396.

Ahmed, S., Ali, A., Ansari, J. A., Qadir, S. A., & Kumar, L. (2025). A comprehensive review of solar photovoltaic systems: scope, technologies, applications, progress, challenges and recommendations. *IEEE Access.*

Apitz, S. E., Elliott, M., Fountain, M., & Galloway, T. S. (2006). European environmental management: Moving to an ecosystem approach. *Integrated Environmental Assessment and Management: An International Journal, 2*(1), 80–85.

Arts, B., & De Koning, J. (2017). Community forest management: An assessment and explanation of its performance through QCA. *World Development, 96*, 315–325.

Barwa, S., & Singh, H. (2024). Climate-induced changes in forest ecosystems of Indian Himalayan Region: A meta-analysis. In *Urban forests, climate change and environmental pollution: Physio-biochemical and molecular perspectives to enhance urban resilience* (pp. 839–866). Springer.

Bastiaanssen, W., Mobin-ud-Din Ahmad, M. U. D. A., & Zubair Tahir, Z. T. (2003). Upscaling water productivity in irrigated agriculture using remote-sensing and GIS technologies. In *Water productivity in agriculture: Limits and opportunities for improvement* (pp. 289-300). CABI Publishing.

Baumgartner, R. J. (2019). Sustainable development goals and the forest sector—A complex relationship. *Forests, 10*(2), 152.

Benioff, R., de Coninck, H., Dhar, S., Hansen, U., McLaren, J., & Painuly, J. (2010). Strengthening clean energy technology cooperation under the UNFCCC: Steps toward implementation.

Blaufelder, C., Levy, C., Mannion, P., & Pinner, D. (2021). A blueprint for scaling voluntary carbon markets to meet the climate challenge. *McKinsey Report*.

Bleuel, S. (2024). *Including forest carbon credits in compliance emission markets* (Doctoral dissertation, Universität Ulm).

Blondeel, M., Bradshaw, M. J., Bridge, G., & Kuzemko, C. (2021). The geopolitics of energy system transformation: A review. *Geography Compass, 15*(7), Article e12580.

Boscolo, M., van Dijk, K., & Savenije, H. (2010). Financing sustainable small-scale forestry: Lessons from developing national forest financing strategies in Latin America. *Forests, 1*(4), 230. (Multidisciplinary Digital Publishing Institute).

Budiana, M. (2024). Climate change and international politics: cooperation or conflict. *Journal of Law, Social Science and Humanities, 2*(1), 128–139.

Bustamante, M., Robledo-Abad, C., Harper, R., Mbow, C., Ravindranat, N. H., Sperling, F., & Smith, P. (2014). Co-benefits, trade-offs, barriers and policies for greenhouse gas mitigation in the agriculture, forestry and other land use (AFOLU) sector. *Global Change Biology, 20*(10), 3270–3290.

Choi, S. H., & Manousiouthakis, V. I. (2020). On the carbon cycle impact of combustion of harvested plant biomass vs. fossil carbon resources. *Computers & Chemical Engineering, 140*, 106942.

Cook-Patton, S. C., Leavitt, S. M., Gibbs, D. A., Harris, N. L., Lister, K., Anderson-Teixeira, K. J., Griscom, B. W. (2020). Mapping carbon accumulation potential from global natural forest regrowth. *Nature, 585*(7826), 545. (Nature Portfolio).

Deloitte Consulting, L. L. P. (2009). Certification and accreditation of SOA implementations: Programmatic rules for the DoD.

Dugan, A. J., Lichstein, J. W., Steele, A., Metsaranta, J. M., Bick, S., & Hollinger, D. Y. (2021). Opportunities for forest sector emissions reductions: A state-level analysis. *Ecological Applications, 31*(5), Article e02327.

Duncanson, L., Liang, M., Leitold, V., Armston, J., Krishna Moorthy, S. M., Dubayah, R., & Zvoleff, A. (2023). The effectiveness of global protected areas for climate change mitigation. *Nature Communications, 14*(1), 2908.

Elekwachi, W., Phil-Eze, P. O., Etuk, E. A., Wizor, C. H., & Onyishi, C. J. (2021). Spatiotemporal characteristics of wetlands ecosystem in the Niger Delta Region. *Journal of Geoscience and Environment Protection, 9*(12), 244. (Scientific Research Publishing).

Emeka-Okoli, S., Otonnah, C. A., Nwankwo, T. C., & Nwankwo, E. E. (2024). Review of carbon pricing mechanisms: Effectiveness and policy implications. *International Journal of Applied Research in Social Sciences, 6*(3), 337–347.

Espagne, É., Godin, A., Magacho, G., Mantes, A., & Yilmaz, D. (2021). Developing countries' macroeconomic exposure to the low-carbon transition. *AFD Research Papers, 220*, 1–42.

Espenan, N. P. (2023). Improving voluntary carbon markets through standardization and blockchain technology. *Wyo. l. Rev., 23*, 141.

Flood, J. (2023). Net-zero transition for companies–hindering and supporting factors: The influence of SBTi, sustainability reporting, and ratings.

Gabrielli, P., Wegner, A., Sierra-Hernández, M. R., Beaudon, E., Davis, M., Barker, J. D., & Thompson, L. G. (2020). Early atmospheric contamination on the top of the Himalayas since the onset of the European Industrial Revolution. *Proceedings of the National Academy of Sciences, 117*(8), 3967–3973.

Gambhir, A., & Tavoni, M. (2019). Direct air carbon capture and sequestration: How it works and how it could contribute to climate-change mitigation. *One Earth, 1*(4), 405–409.

Garg, A., Pal, D., Singh, H., & Pandey, D. C. (2016). A comparative study of NDBI, NDISI and NDII for extraction of urban impervious surface of Dehradun [Uttarakhand, India] using Landsat 8 imagery. In *2016 International Conference on Emerging Trends in Communication Technologies (ETCT)* (pp. 1–5).

Ghiasi, M., Wang, Z., Mehrandezh, M., Jalilian, S., & Ghadimi, N. (2023). Evolution of smart grids towards the Internet of energy: Concept and essential components for deep decarbonisation. *IET Smart Grid, 6*(1), 86–102.

Ghosh, S., Saha, A., Kumar, S., & Pathania, S. (2024). Regenerative agriculture for climate change mitigation and food security. In *Regenerative agriculture for sustainable food systems* (pp. 83–133). Springer Nature Singapore.

Goparaju, L., Ahmad, F., Uddin, M., & Rizvi, J. (2020). Agroforestry: An effective multi-dimensional mechanism for achieving Sustainable Development Goals. *Ecological Questions, 31*(3), 63–71.

Gordeeva, E., Weber, N., & Wolfslehner, B. (2022). The new EU Forest strategy for 2030—An analysis of major interests. *Forests, 13*(9), 1503.

Grassi, G., House, J. I., Kurz, W. A., Cescatti, A., Houghton, R. A., Peters, G. P., Zaehle, S. (2018). Reconciling global-model estimates and country reporting of anthropogenic forest CO_2 sinks. *Nature Climate Change, 8*(10), 914. Nature Portfolio.

Griscom, B. W., Adams, J., Ellis, P. W., Houghton, R. A., Lomax, G., Miteva, D. A., & Fargione, J. (2017). Natural climate solutions. *Proceedings of the National Academy of Sciences, 114*(44), 11645. (National Academy of Sciences).

Henriksen, H. Z., Mahnke, V., & Hansen, J. M. (2004). Public e-procurement adoption: economic and political rationality. In *37th Annual Hawaii International Conference on System Sciences, 2004. Proceedings of the* (pp. 9-pp). IEEE. (Jan 2004)

Hurtt, G. C., Ma, L., Lamb, R., Campbell, E., Dubayah, R. O., Hansen, M., & Tang, H. (2024). Beyond MRV: Combining remote sensing and ecosystem modeling for geospatial monitoring and attribution of forest carbon fluxes over Maryland, USA. *Environmental Research Letters, 19*(12), Article 124058.

Intergovernmental Panel on Climate Change. (2021). *Climate change 2021: The physical science basis. Contribution of Working Group I to the Sixth Assessment Report of the Intergovernmental Panel on Climate Change* (V. Masson-Delmotte, P. Zhai, A. Pirani, S. L. Connors, C. Péan, S. Berger,... & B. Zhou, Eds.). Cambridge University Press. https://doi.org/10.1017/9781009157896

IPCC (Intergovernmental Panel on Climate Change). (2018). *Global Warming of 1.5° C. An IPCC Special Report on the impacts of global warming of 1.5° C above pre-industrial levels and related global greenhouse gas emission pathways, in the context of strengthening the global response to the threat of climate change, sustainable development, and efforts to eradicate poverty.* Geneva.

Jain, S. (2015). Globalization and its impact on consumer culture and competition policy. Available at SSRN 2777387.

Jangra, A., Kumar, K., Maikhuri, S., Bhandari, M. S., Pandey, S., Singh, H., & Barthwal, S. (2024). Unveiling stress-adapted endophytic bacteria: Characterizing plant growth-promoting traits and assessing cross-inoculation effects on *Populus deltoides* under abiotic stress. *Plant Physiology and Biochemistry, 210*, Article 108610.

Joshi, R., Sharma, B., Singh, H., Dhakal, N., Ayer, S., & Maraseni, T. (2024). Poplar plantation as an agroforestry approach: Economic benefits and its role in carbon sequestration in North India. *Journal of Resources and Ecology, 15*(4), 880–888.

Joshi, R., Singh, H., Chhetri, R., Poudel, S. R., & Rijal, S. (2021). Carbon sequestration potential of community forests: A comparative analysis of soil organic carbon stock in community managed forests of Far-Western Nepal. *Eurasian Journal of Soil Science, 10*(2), 96–104.

Joshi, R., Singh, H., Chhetri, R., & Yadav, K. (2020). Assessment of carbon sequestration potential in degraded and non-degraded community forests in Terai region of Nepal. *Journal of Forest and Environmental Science, 36*(2), 113–121.

Kerchner, C., & Keeton, W. S. (2015). California's regulatory forest carbon market: Viability for northeast landowners. *Forest Policy and Economics, 50,* 70. Elsevier BV.

Khan, M. I., & Al-Ghamdi, S. G. (2023). Hydrogen economy for sustainable development in GCC countries: A SWOT analysis considering current situation, challenges, and prospects. *International Journal of Hydrogen Energy, 48*(28), 10315–10344.

Kim, M., Kraxner, F., Forsell, N., Song, C., & Lee, W. K. (2021). Enhancing the provisioning of ecosystem services in South Korea under climate change: The benefits and pitfalls of current forest management strategies. *Regional Environmental Change, 21*(1), 6.

Kotchen, M. J. (2009). Offsetting green guilt. *Stanford Social Innovation Review, 7*(2), 26–31.

Kumar, A., Dwivedi, G. K., Tewari, S., Sah, V. K., Singh, H., Kumar, P., Kumar, N., & Kaushal, R. (2020). Soil organic carbon pools under *Terminalia chebula* Retz. based agroforestry system in Himalayan foothills, India. *Current Science, 118*(7), 1098–1103.

Kumar, M., & Singh, H. (2020). Agroforestry as a nature-based solution for reducing community dependence on forests to safeguard forests in rainfed areas of India. In Nature-based solutions for resilient ecosystems and societies (pp. 289–306). Springer.

Kumar, N., Singh, H., & Kumar, A. (Eds.). (2021). Renewable energy and green technology: principles and practices. CRC Press.

Kupzok, N., & Nahm, J. (2024). The Decarbonization bargain: How the decarbonizable sector shapes climate politics. *Perspectives on Politics*, 1–21.

Lamb, W. F., Wiedmann, T., Pongratz, J., Andrew, R., Crippa, M., Olivier, J. G., & Minx, J. (2021). A review of trends and drivers of greenhouse gas emissions by sector from 1990 to 2018. *Environmental Research Letters, 16*(7), Article 073005.

Liang, Y., Kleijn, R., & Van der Voet, E. (2023). Increase in demand for critical materials under IEA Net-Zero emission by 2050 scenario. *Applied Energy, 346*, Article 121400.

Maes, J., Egoh, B. N., Willemen, L., Liquete, C., Vihervaara, P., Schägner, J. P., & Bidoglio, G. (2012). Mapping ecosystem services for policy support and decision making in the European Union. *Ecosystem Services, 1*(1), 31. Elsevier BV.

Mahari, W. A. W., Azwar, E., Li, Y., Wang, Y., Peng, W., Ling, N., & Sonne, C. (2020). Deforestation of rainforests requires active use of UN's Sustainable Development Goals. *The Science of the Total Environment, 742,* 140681. Elsevier BV.

Martinsson, G., Sajtos, L., Strömberg, P., & Thomann, C. (2024). The effect of carbon pricing on firm emissions: Evidence from the swedish co2 tax. *The Review of Financial Studies, 37*(6), 1848–1886.

Mbolo, C. (2013). Analyzing the establishment of community forestry (CF) and its processes examples from the south west region of Cameroon.

Mitigation, C. C. (2011). IPCC special report on renewable energy sources and climate change mitigation. *Renewable Energy, 20*(11).

Mohammed, A. A., Nahazanan, H., Nasir, N. A. M., Huseien, G. F., & Saad, A. H. (2023). Calcium-based binders in concrete or soil stabilization: Challenges, problems, and calcined clay as partial replacement to produce low-carbon cement. *Materials, 16*(5), 2020.

Montel, S., Martin, J., & Crerand, A. (2023). Lessons Learned from Large Scale Hydrogen Production Project.

Mulyono, A., Subardja, A., Ekasari, I., Lailati, M., Sudirja, R., & Ningrum, W. (2018). The hydromechanics of vegetation for slope stabilization. *IOP Conference Series Earth and Environmental Science, 118,* 012038. IOP Publishing.

Olorunfemi, I. E., Fasinmirin, J. T., Olufayo, A. A., & Komolafe, A. A. (2020). Total carbon and nitrogen stocks under different land use/land cover types in the Southwestern region of Nigeria. *Geoderma Regional, 22,* e00320. Elsevier BV.

Orozco, D. (2019). A systems theory of compliance law. *University of Pennsylvania Journal of Business Law, 22,* 244.

Pan, J., Cross, J. L., Zou, X., & Zhang, B. (2024). To tax or to trade? A global review of carbon emissions reduction strategies. *Energy Strategy Reviews, 55*, Article 101508.

Piabuo, S. M., Foundjem-Tita, D., & Minang, P. A. (2018). Community forest governance in Cameroon. *Ecology and Society*, *23*(3).

Preston, L. M. (1984). Freedom, markets, and voluntary exchange. *American Political Science Review*, *78*(4), 959–970. https://doi.org/10.2307/1955801

Raman, R., Manalil, S., Dénes, D. L., & Nedungadi, P. (2024). The role of forestry sciences in combating climate change and advancing sustainable development goals. *Frontiers in Forests and Global Change, 7*, 1409667.

Rezai, A., Taylor, L., & Foley, D. (2018). Economic growth, income distribution, and climate change. *Ecological Economics, 146*, 164–172.

Ruhil, T., Singh, H., & Barthwal, S. (2024). Long-term physiological monitoring for detecting climate change impacts on forest ecosystems. In *Forests and climate change: Biological perspectives on impact, adaptation, and mitigation strategies* (pp. 817–834). Springer.

Sang, Z., Hamann, A., & Rweyongeza, D. (2023). Adapting reforestation programs to observed and projected climate change. *Mitigation and Adaptation Strategies for Global Change, 28*(2), 14.

Santier, T. (2023). *The next priorities of the Voluntary Carbon Market for mass adoption: The need for new technologies, carbon policy frameworks, and a meta registry* (Doctoral dissertation, Université Paris-Saclay).

Sims, R. E. (2014). Renewable energy and climate change mitigation: an overview of the IPCC Special Report. *Weather matters for energy*, 91–110.

Singh, H. (2024a). Climate change and forests: Biological perspectives. In *Forests and climate change: Biological perspectives on impact, adaptation, and mitigation strategies* (pp. 1–17). Springer.

Singh, H. (2024b). Exploring adaptive modulation in plant functional traits and its impact on the productivity of *Acacia auriculiformis* under CO_2-enriched environment. *Industrial Crops and Products, 210*, Article 118186.

Singh, H. (2024c). *Forests and climate change: Biological perspectives on impact, adaptation, and mitigation strategies*. Springer Nature.

Singh, H. (2024d). *Urban forests, climate change and environmental pollution: Physio-biochemical and molecular perspectives to enhance urban resilience*. Springer Nature.

Singh, H. (2025). The role of plant functional traits in mitigating urban climate and environmental pollution. In *Plant functional traits* (pp. 297–308). Elsevier.

Singh, H., & Kumar, M. (2022). Climate change and its impact on Indian Himalayan forests: current status and research needs. In *Climate change: Impacts, responses and sustainability in the Indian Himalaya* (pp. 223–242). Springer.

Singh, H., Kumar, N., & Kumar, A. (2024). Enhancing resource use efficiency in crops through plant functional traits. In *Plant functional traits for improving productivity* (pp. 97–117). Springer.

Singh, H., Kumar, N., Singh, R., & Kumar, M. (2023). Assessing the climate change impact on the habitat suitability of the range-restricted bird species (*Catreus wallichii*) in the Indian Himalayan ecosystem. *Environmental Science and Pollution Research, 30*(57), 121224–121235.

Singh, M., Kumar, P., Singh, H., Kumar, A. K. A., Kumar, R., & Kumar, R. (2023). *Neolamarckia cadamba*: a comprehensive review on its physiological, ecological, phytochemical and pharmacological perspectives. *Ecology Environment and Conservation, 29*, 241–250.

Singh, H., Verma, A., & Kumar, M. (2018). *Can elevated CO_2 alter the growth dynamics, morphology, physiology and biochemistry of the forestry tree species?* (pp. 288–305). Climate Change and Agriculture. Discovery Publishing House.

Sorooshian, S. (2024). The sustainable development goals of the united nations: A comparative midterm research review. *Journal of Cleaner Production*, 142272.

Souza Filho, I. R., Ma, Y., Raabe, D., & Springer, H. (2023). Fundamentals of green steel production: On the role of gas pressure during hydrogen reduction of iron ores. *JOM Journal of the Minerals Metals and Materials Society, 75*(7), 2274–2286.

Stokes, A., Douglas, G., Fourcaud, T., Giadrossich, F., Gillies, C. T., Hubble, T., & Walker, L. R. (2014). Ecological mitigation of hillslope instability: Ten key issues facing researchers and practitioners. *Plant and Soil, 377*(1), 1. (Springer Science+Business Media).

Subbarao, P. M., D'Silva, T. C., Adlak, K., Kumar, S., Chandra, R., & Vijay, V. K. (2023). Anaerobic digestion as a sustainable technology for efficiently utilizing biomass in the context of carbon neutrality and circular economy. *Environmental Research, 234,* Article 116286.

Tcvetkov, P. (2022). Engagement of resource-based economies in the fight against rising carbon emissions. *Energy Reports, 8*, 874–883.

Terzić, L. (2024). Assessing progress towards a circular carbon economy in the G20 countries. *Икономическа Мисъл, 3*, 281–322.

UNEP, U. (2021). *Emissions gap report 2020* (p. 112). UN.

Vashishtha, S., & Vashishtha, D. S. (2024). The evolution of the Indian carbon market: Challenges, opportunities, and policy frameworks. *Opportunities, and Policy Frameworks***.**

Vertes, A. A., Qureshi, N., Blaschek, H. P., & Yukawa, H. (Eds.). (2020). *Green energy to sustainability: Strategies for global industries.* John Wiley & Sons.

Vickers, B., Trines, E., & Pohnan, E. (2012). Community guidelines for accessing forestry voluntary carbon markets (Issue 1).

Ward, A., Dargusch, P., Thomas, S., Liu, Y., & Fulton, E. A. (2014). A global estimate of carbon stored in the world's mountain grasslands and shrublands, and the implications for climate policy. *Global Environmental Change, 28,* 14. Elsevier BV.

Waring, B., Neumann, M., Prentice, I. C., Adams, M., Smith, P., & Siegert, M. (2020). Forests and decarbonization–roles of natural and planted forests. *Frontiers in Forests and Global Change, 3*, Article 534891.

Yue, C., Ciais, P., Houghton, R. A., & Nassikas, A. A. (2020). Contribution of land use to the interannual variability of the land carbon cycle. *Nature Communications, 11*(1). (Nature Portfolio).

Zyambo, P. (2018). What is limiting success of community-based approach to conservation of natural resources in Southern Africa. *Journal of Ecology & Natural Resources*, 2(4).

Chapter 10
Community Engagement and Indigenous Knowledge in Forest Conservation: A Pathway for Achieving Carbon Neutrality in Nepal

Rajeev Joshi, Nabin Dhakal, Renuka Khamcha, Pratiksha Sharma, Aman Prabhakar, Tek Maraseni, and Han Zhang

Abstract Forests are crucial for Nepal to achieve its carbon neutrality goals; however, a disconnect often exists between national policies and the local, Indigenous knowledge systems that ensure sustainable forest management. While community-based forest management (CBFM) is recognized as a successful model, a comprehensive analysis of how the integration of community engagement and Indigenous knowledge contributes explicitly to carbon sequestration remains underexplored. This study aims to fill this gap by examining the synergistic role of these elements in enhancing forest conservation and advancing carbon neutrality in Nepal. Using the standard PRISMA framework, a systematic literature search was conducted across five academic databases: Web of Science, Scopus, Google Scholar, ResearchGate, and Sci-Hub, to ensure comprehensive coverage of relevant studies. Following a series of rigorous screening steps, 74 studies were identified as the final sample for in-depth analysis. Our findings demonstrate that community-managed forests significantly increase carbon stocks, with sequestration rates of 1–3 tC ha^{-1} annually, alongside notable improvements in forest cover and biodiversity. The study identifies that the success of CBFM is rooted in traditional practices, such as rotational grazing

R. Joshi (✉) · H. Zhang (✉)
College of Economics and Management, Northwest A&F University, Yangling District, Xianyang, Shaanxi, China
e-mail: joshi.rajeev20@gmail.com

H. Zhang
e-mail: hanzhang@nwafu.edu.cn

R. Joshi · N. Dhakal · R. Khamcha · P. Sharma · Aman Prabhakar
College of Natural Resource Management, Faculty of Forestry, Agriculture and Forestry University, Katari, Udayapur, Nepal

T. Maraseni
Institute for Life Sciences and the Environment, University of Southern Queensland, Toowoomba, QLD 4350, Australia

H. Singh (ed.), *Forestry for a Carbon-Neutral and Sustainable Future*, Sustainability Solutions, https://doi.org/10.1007/978-3-032-06427-1_10

and farm-forest integration. However, challenges such as the exclusion of marginalized groups from decision-making processes can undermine equity and effectiveness. We conclude that supportive policies which formally recognize community rights, integrate Indigenous knowledge into conservation planning, and promote equitable benefit-sharing are essential. Prioritizing this integrated approach not only accelerates progress toward carbon neutrality but also fosters social equity and climate resilience, offering a replicable model for other forest-dependent nations.

Keywords Climate change · Sustainable development · Indigenous knowledge · Forest conservation · Carbon sequestration · Nepal

10.1 Introduction

Climate change poses a significant threat to the Hindu Kush Himalayan region, with profound implications for countries such as Nepal, affecting its cryosphere, water resources, and ecosystem stability (Wester et al., 2019). Climate change is intensifying glacier retreat and increasing the risk of glacial lake outburst floods (GLOFs) across the Himalaya, presenting a severe threat to Nepal's communities and infrastructure (Nie et al., 2021; Veh et al., 2020). The Himalayas, which are known as the "Water Towers of Asia," are losing glaciers at an alarming rate, threatening the water supply for millions of people downstream. Unpredictable monsoons and protracted droughts reduce crop yields and exacerbate food insecurity in Nepal's agricultural sector, which employs more than 60% of the country's workforce (Khatri-Chhetri et al., 2020). Farming communities, many of whom depend on subsistence agriculture, bear a disproportionate burden due to climate-induced soil erosion, land degradation, and water scarcity (Gentle & Maraseni, 2021).

Furthermore, changing climate conditions disrupt habitats and endanger Nepal's unique biodiversity, including species such as the snow leopard, red panda, and Himalayan musk deer, which rely on fragile alpine ecosystems (Aryal et al., 2022; Singh & Yadav, 2024). Shifting temperature zones are forcing species to migrate to higher altitudes, reducing viable habitats and increasing the risk of extinction (Scheffers et al., 2016; Peel et al., 2017). Forest degradation and increased forest fires, fueled by rising temperatures and dry spells, threaten carbon stocks and diminish the forests' ability to act as carbon sinks (Thapa et al., 2025). These cascading impacts underscore the urgent need for sustainable strategies to mitigate and adapt to the evolving threats of climate change (Wester et al., 2023).

Achieving carbon neutrality is crucial for Nepal to safeguard its natural environment and build resilient communities (Pandey et al., 2024; Wu et al., 2022). Forests play a pivotal role in this endeavor, acting as vital carbon sinks that absorb and store atmospheric CO_2 (Joshi et al., 2023; Pandey et al., 2024). Community-based forest management (CBFM) systems, which empower local communities to oversee forest resources, play a central role in increasing carbon sequestration while enhancing biodiversity and local livelihoods (Mengie& Szemethy, 2025). Studies show that community-managed forests can store up to 160 t ha^{-1} of soil carbon, with annual sequestration rates of 1–3 t ha^{-1} (Pandey et al., 2014). The success of CBFM is rooted in the deep ecological knowledge of Indigenous communities, who practice sustainable forest management techniques such as rotational grazing, selective harvesting, and forest fire prevention, all of which contribute to maintaining healthy, carbon-rich ecosystems (Pandey et al., 2024; Poudyal et al., 2020). A sustainable future rooted in carbon neutrality ensures long-term ecological balance and economic stability. It limits climate change and improves public health by decreasing air pollutants, reducing respiratory diseases, and enhancing overall well-being (Shi et al., 2021). It fosters energy independence by promoting renewable energy sources, which is particularly beneficial for remote communities that rely on forest biomass for fuel.

Additionally, reforestation and afforestation efforts, supported by international mechanisms like REDD+, can incentivize communities to protect and restore forests, creating a positive feedback loop for conservation and carbon sequestration (Cadaman & Maraseni, 2012). Climate change presents a profound threat to Nepal, impacting its critical Himalayan ecosystems, agricultural-dependent economy, and unique biodiversity through mechanisms such as glacial retreat, unpredictable monsoons, and habitat disruption (Wester et al., 2023). In response, achieving carbon neutrality has become a national priority, with forests playing a pivotal role as significant carbon sinks (Pandey et al., 2024). The sustainable management of these forests is therefore essential not only for climate mitigation but also for supporting the livelihoods of the majority of the population, which depends on forest resources for energy, fodder, and construction materials (Maharjan, 2005a, b; Pan et al., 2011).

Nepal's innovative approach to this challenge is embodied in its system of community-based forest management (CBFM), particularly through Community Forest User Groups (CFUGs). This model empowers local communities to manage forest resources, demonstrating that local governance, when legally supported, can lead to sustainable outcomes (Ojha et al., 2023). The success of CBFM is deeply rooted in the integration of Indigenous knowledge, including practices such as rotational grazing and agroforestry, which have maintained forest health for generations (Poudyal et al., 2020). This alignment with Elinor Ostrom's (1990) principles for managing common-pool resources underscores the effectiveness of clear boundaries and collaborative decision-making. Furthermore, international mechanisms like REDD + have begun to recognize and incentivize this community-led model, offering financial avenues for conservation through carbon credits (Cadman & Maraseni, 2012).

However, a critical gap persists in the literature. While the individual components of CBFM, Indigenous knowledge, and policy frameworks like REDD+ are well-documented, there is a lack of synthesized analysis examining their synergistic interaction as an integrated pathway to carbon neutrality. Existing studies often treat community forestry either as a conservation strategy or a livelihood mechanism, failing to situate it within a holistic climate mitigation framework that captures both biophysical outcomes (e.g., carbon sequestration) and socio-institutional dimensions (e.g., equity, participation).

Therefore, the specific aim of this chapter is to bridge this gap by critically examining how the interplay between community engagement and Indigenous knowledge contributes to forest conservation and advances carbon neutrality goals in Nepal. To achieve this aim, the chapter is guided by the following objectives:

1. To analyze the historical evolution and current role of Indigenous knowledge systems in sustainable forest management.
2. To evaluate the effectiveness of community engagement strategies, particularly through CFUGs, in enhancing carbon sequestration and forest health.
3. To assess the policy and legal frameworks that enable or hinder the integration of community-based approaches into national carbon neutrality strategies.

The novelty of this chapter lies in its interdisciplinary integration of ecological science, Indigenous knowledge, and institutional frameworks. Highlighting the interaction between community engagement and traditional practices with modern policy mechanisms may advances a holistic framework. This approach provides new insights into how Nepal's community forestry model can inform global discourses on sustainable forest governance and climate resilience.

10.2 Methodology

This study employed a systematic literature review methodology, guided by the PRISMA (Preferred Reporting Items for Systematic Reviews and Meta-Analyses) framework (Page et al., 2021), to ensure a comprehensive, transparent, and reproducible analysis of the role of community engagement and Indigenous knowledge in forest conservation and carbon neutrality in Nepal (Fig. 10.1).

10.2.1 Search Strategy

A systematic literature search was conducted in June 2024 across five academic databases—Web of Science, Scopus, Google Scholar, ResearchGate, and Sci-Hub—to ensure comprehensive coverage of relevant peer-reviewed literature. The search strategy was built around three core conceptual domains: (1) the geographic context, using terms such as "Nepal" OR "Himalaya"; (2) the intervention, with keywords like "community forest" OR "CBFM" OR "CFUG" OR "community-based forest management"; and (3) the outcomes of interest, including "carbon sequest" OR "carbon stock" OR "carbon neutral" OR "indigenous knowledge" OR "traditional ecological knowledge". These terms were combined using Boolean operators (AND, OR) to create targeted search strings. For instance, a search executed in Scopus was structured as follows: (TITLE-ABS-KEY ("community forest" OR cfug) AND TITLE-ABS-KEY ("carbon sequest" OR "carbon neutral") AND TITLE-ABS-KEY (nepal)). This methodical approach aimed to capture the broadest possible

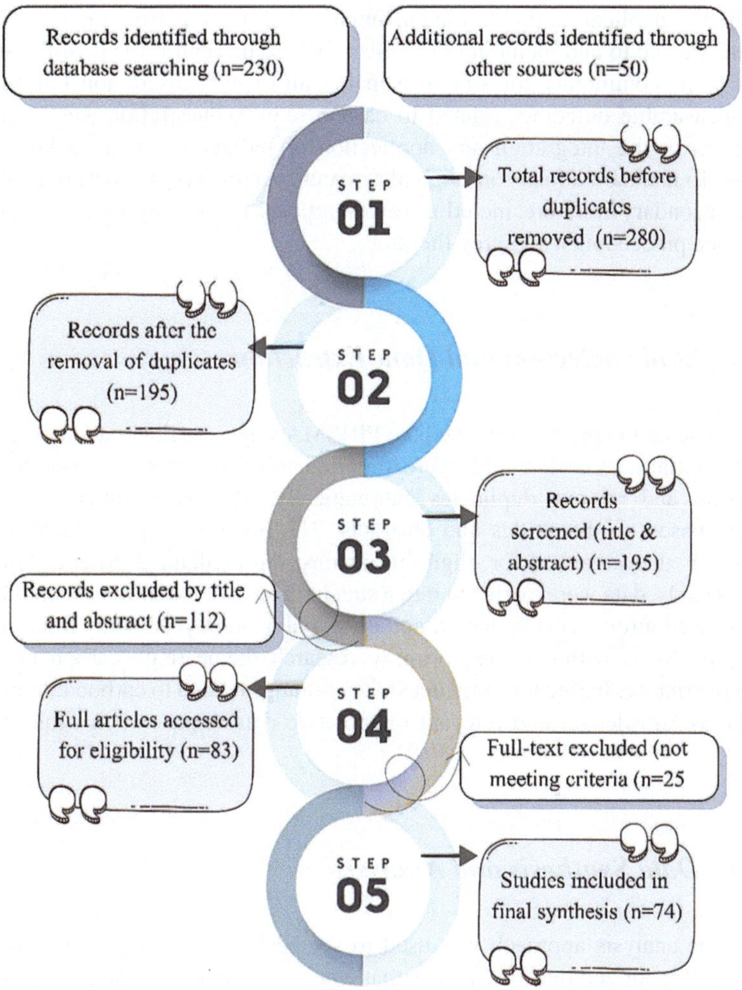

Fig. 10.1 Flowchart of the study and steps followed during the review process

set of studies examining the interlinkages between community forestry, indigenous knowledge, and carbon dynamics in Nepal.

10.2.2 Eligibility Criteria

The study selection was guided by predefined eligibility criteria to ensure the inclusion of relevant and high-quality evidence. Included studies were required to be primary research articles published in peer-reviewed journals and written in

English. The publication timeframe encompassed all records from the inception of the databases up to and including December 2023. The central focus of the studies needed to be community forestry in Nepal, with explicit discussion linking it to either measurable outcomes related to carbon (e.g., sequestration, stock, climate mitigation) or the integration and application of Indigenous or local knowledge systems. To maintain a focus on original research, exclusion criteria were applied to remove secondary literature, including review articles, meta-analyses, book chapters, conference proceedings, and grey literature.

10.2.3 Study Selection and Data Extraction

The study selection process followed the PRISMA stages, as illustrated in Fig. 10.1. After the initial search, all records were imported into Zotero, the reference manager, to identify and remove duplicates automatically. The remaining records were screened based on their titles and abstracts. The full text of potentially relevant articles was then assessed for eligibility against the predefined criteria. For each included study, data were extracted into a standardized Microsoft Excel spreadsheet. The extracted information included: author(s), publication year, study title, journal, geographic focus within Nepal, primary research theme (e.g., carbon measurement, governance, Indigenous practices), key findings related to carbon/community/Indigenous knowledge, and relevant quantitative data (e.g., carbon sequestration rates).

10.2.4 Data Synthesis and Analysis

A thematic analysis approach was used to synthesize the findings. The extracted data were categorized into key themes that emerged from the literature, such as: (1) Historical context of CBFM, (2) Role of Indigenous Knowledge, (3) Community Engagement Strategies, (4) Policy and Legal Frameworks, and (5) Carbon Sequestration Outcomes. Trends, consistencies, and gaps within each theme were identified and narratively summarized.

10.3 Results and Discussion

10.3.1 Historical Context

The history of community-based forest protection and management in Nepal's hills dates back to when local communities developed their indigenous systems for managing forests despite the absence of government institutional and policy support (Fig. 10.2). However, this changed significantly with the passing of the Private Forest Nationalisation Act of 1957, which nationalized all privately owned forests larger than 1.2 hectares in the hills and 3.5 hectares in the Terai (Tamrakar & Mohns, 2016). As a result, local communities gradually lost interest in the conservation, use, and management of forests, which led to a steady decline in forest health and an increase in degradation, particularly in the hills. Recognizing the need for local participation in forest management, the Nepalese government took significant steps toward community-based forestry in the 1960s. The Forest Act of 1961 was introduced, followed by an amendment in 1978 that specifically incorporated provisions for community forestry. The Master Plan for the Forestry Sector, established in 1989, reinforced the idea that community forestry should become a central focus of the country's forestry sector. The government further strengthened community forestry through the Forest Act of 1993 and the Forest Rules of 1995, which legally empowered local forest user groups to sell forest products, raise funds, and engage in income-generating activities through grants and donations. This historical shift toward local governance, supported by legislative backing, laid the groundwork for Nepal's success in community-driven conservation and carbon sequestration (Singh et al., 2018).

The historical background of international support for Nepal's forest sector is marked by significant donor involvement. Australia's engagement began in 1966, with the Nepal-Australia Community Resource Management Project starting in 1997 to develop sustainable community resource management systems in two districts (Singh, 2024c). In 1998, DANIDA launched the National Resource Management Sector Assistance Program (NARMSAP), which focused on building national capacity in natural resource management and continued the work of the Bank-supported Hill Community Forestry project, which ended in 1999. The UK's DFID has been supporting Nepal's forest sector since 1979 and is now planning a 10-year livelihood and forestry program to improve the livelihoods of poor rural communities. These initiatives represent a long-standing commitment to promoting sustainable forest management and supporting rural development in Nepal, along with other global environmental initiatives such as the Convention on Biological Diversity (CBD) and Reducing Emissions from Deforestation and Forest Degradation (REDD+), which also recognize community engagement as a means to conserve biodiversity and reduce carbon emissions (Agrawal & Angelsen, 2010). This historical foundation of international support has been crucial in shaping the policy landscape that enables community forestry to contribute to climate goals. The early investments by donors, including DFID, DANIDA, and Australia, in building community

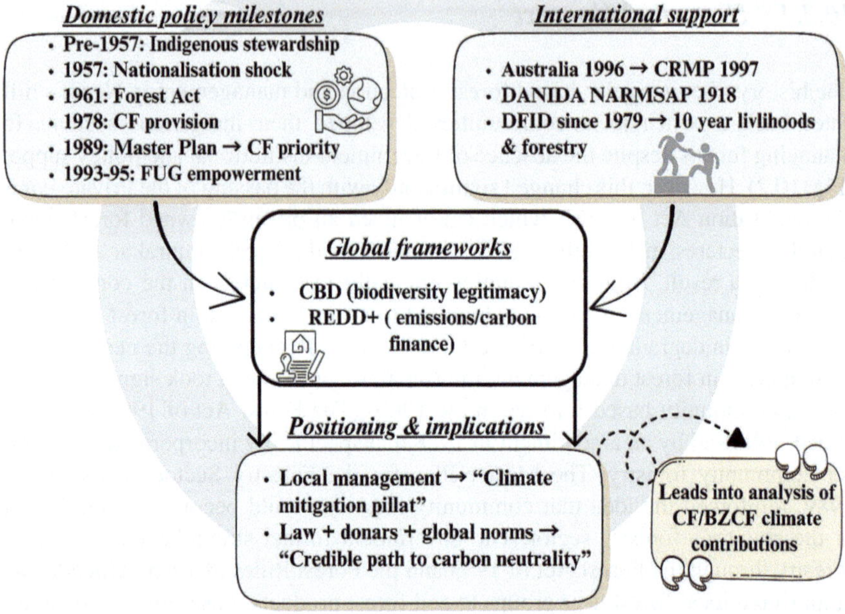

Fig. 10.2 Evolution of Nepal's Community Forestry strategy, illustrating the interplay of domestic policy milestones, international donor support, and global frameworks that established its role in climate mitigation

capacity and institutional frameworks provided the essential groundwork upon which later carbon-focused programs, such as REDD+, could be integrated. This evolution marks a significant shift from viewing community forestry solely as a strategy for local resource management and poverty alleviation to recognizing its critical role in national and global climate mitigation efforts. Therefore, understanding this historical trajectory of donor involvement is crucial to analyzing how Nepal's community forestry model is positioned to serve as a viable pathway for achieving carbon neutrality, the central focus of this chapter.

10.3.2 Role of Indigenous Knowledge

The recent transfer of the Kanchenjunga Conservation Area (CA) to community-based organizations marks a growing trend in participatory, community-driven forest management in Nepal. The community forestry program is regarded as a successful model for sustainable forest management, and its policy is considered one of the most progressive globally (DFID-Nepal, 2000). Community involvement in resource management refers to the sustainable use and management of natural resources by residents of a region, taking into account its ecological, social, and cultural context. Indigenous communities in Nepal's hills and mountains possess a wealth of Indigenous local knowledge (ILK), technologies, and practices that have been utilized for centuries to manage natural resources and address climate change adaptation and disaster risk reduction solutions.

In recent years, Indigenous communities, especially in rural areas of Nepal, have been using Indigenous and local knowledge (ILK) and practices to adapt to climatic and non-climatic changes (Chaudhary et al., 2017) where Indigenous communities in Nepal's hills and mountains possess a wealth of local knowledge passed down through generations. Modernization has frequently led to the disregard of ancient customs. Traditional traditions passed down through centuries can offer valuable insights for building sustainable practices in today's world. Indigenous knowledge and practices help them in adaptive planning and local resource management by (a) taking stock of the available forest ecosystem resources, (b) enhancing people's understanding of the nature of climatic risks, (c) identifying gaps in the adaptive capacity of individuals and institutions, and (c) reducing the cost of adaptation by using local knowledge and physical resources. Indigenous practices, such as the Farm-Forest Integration System, include crop, livestock, and forest (including grasslands) as interrelated production sub-systems (see Fig. 10.3).

The agricultural subsystem provides animal food in the form of crop byproducts (Singh, 2025). The livestock subsystem supplies the necessary draft force and manure to sustain the agricultural subsystem. Forests provide food, manure, and other benefits essential for a sustainable life. Before 1990, forests were government-owned, and villagers had to negotiate with officials to access forest products (Shrestha & Tiwari, 2006).The transition from state-controlled to community-based forest management in Nepal's mid-hills demonstrably improved both forest health and local livelihoods, as evidenced by specific case studies. Under the previous government-controlled

Fig. 10.3 Traditional farm-forest integration system practiced by indigenous communities in mid-hill Nepal. (*Source* Bikram Singh)

system, households faced severe restrictions that hindered their well-being and provided little incentive for conservation. For instance, allocations were limited to just 12 ft^3 of timber annually per household—an amount insufficient for basic needs—along with fixed quotas for fuelwood (216 *bhari*) and fodder (183 *bhari*). These constraints forced villagers to spend an average of 3.4 h daily collecting fuelwood and 2.4 h gathering fodder, with many families resorting to burning dung cakes due to shortages (Singh, 2024b).

The handover of forest management to Community Forest User Groups (CFUGs) in the 1990s marked a dramatic reversal. A prominent example is the Gadhanta-Banjhkateri Community Forest in Kaski district. Following its formation, the CFUG established clear operational plans with localized regulations for rotational harvesting, penalties for illegal grazing, and dedicated conservation areas (Gautam et al., 2004). This community-driven governance led to tangible ecological and socio-economic benefits: forest canopy cover increased from 45 to 75% within a decade, while the average timber harvest increased by more than 250% during the 1990–2000 period. Consequently, villagers could meet their fodder needs in just two hours a day, signifying a significant gain in efficiency and well-being, alongside improved housing conditions resulting from greater timber access (Singh, 2024b; Ojha et al., 2023). This case concretely demonstrates how the shift to CFUGs, by leveraging shared responsibilities and localized rules, simultaneously enhanced forest conditions and socio-economic outcomes, aligning with the historical precedent of Indigenous communities employing sustainable practices without formal government intervention (Acharya et al., 2008; Wiersum, 1997).

10.3.3 Community Engagement Strategies

Community engagement is central to forest conservation in Nepal, where Community Forest User Groups (CFUGs) manage forest areas, collect forest products, and determine market prices under the supervision of an elected committee (refer to Fig. 10.4).

These groups foster collective ownership, local leadership, and participatory decision-making, ensuring that conservation strategies are culturally appropriate and context-specific (Joshi et al., 2021). CFUGs have become powerful platforms for knowledge exchange and capacity building, where members share traditional practices, learn new techniques, and collaboratively address conservation challenges. The communities in the Khumbu and lower Mustang regions are found to be employing a co-management system of forest and pasture management, where a good integration of indigenous or traditional and modern institutional arrangements is observed in Mustang. Both conventional (Mukhiya) and modern Annapurna Conservation Area Project (ACAP) systems are operating in parallel. This is the reason why the Sagarmatha National Park (SNP) management has recognized the Shingi Nawa system of the Sherpa community and included the Nawas as an integral part of their Buffer Zone Committee staff (UNESCO, 2025).

Fig. 10.4 A Community Forest User Group (CFUG) and officials meeting held to discuss forest resource allocation in Kanchanpur district. (*Source* DFO, Kanchanpur)

In contrast, the ACAP has not recognized the Mukhiya system of the Thakali and Lho-ba communities. Integrating the Shingi Nawa system into a formal system of protected area management in the SNP is a good example of a synergy between indigenous and modern practices. This reinforces the earlier conclusion on this good practice (Bhatta, 2013; Stevens, 1993). The ACAP has been operating a parallel structure of Conservation Area Committees (CACs), Conservation Area Management Councils (CAMCs), and Forest Protection Committees (FPCs), which, besides alienating the indigenous peoples, has also resulted in duplication of efforts and less-than-satisfactory conservation achievements. In the Annapurna Conservation Area (ACA), CAMCs have been established as the leading local institution for planning, designing, and implementing conservation and development plans and programs in ACA villages. It plays a crucial role in conservation planning, policy formulation, and the management of forest resources. During the initial stages of conservation intervention, the legitimacy of CAMCs was based on a shared value system and collective cohesiveness, similar to a traditional authority structure (Baral, 2009). The conservation regulation endorsed by the government in 1996 gave CAMCs a legal designation and role. As foreseen in the ACA Operational Plan, income from fines and the issue of permits for resource harvesting (for example, timber) is deposited into a CAMC account and used for incentives and local development (Bajracharya et al., 2005). This participatory model has improved forest cover, biodiversity, and livelihoods, as communities gain easier access to resources and income through sustainable forest management (Joshi et al., 2020).

In Nepal, community forestry initiatives have shown quantifiable results in raising carbon stocks and forest cover. Studies from mixed broad-leaf forests showed that carbon sequestration rates were 3 t ha^{-1} yr^{-1}, whereas those in pine forests were 1 t ha^{-1} yr^{-1} (Shrestha & Devkota, 2016; Subedi et al., 2022). Alongside improvements in local livelihoods, there has been an increase in canopy cover and forest growing stock (Kanel & Niraula, 2004). Community-controlled forests account for approximately 15.5% of all forests worldwide, and this percentage is expected to continue increasing. Community forestry supports global efforts to achieve carbon neutrality by reducing deforestation, increasing biodiversity, and storing carbon (Sikor et al., 2013). Tropical forests, which store approximately 40% of the world's terrestrial carbon, play a crucial role in these initiatives. This strategy increases resistance to environmental changes and promotes sustainable ecosystem management. It is supported by the Farm-Forest Integration System, which enriches soil fertility by utilizing forest products such as manure and fodder, thereby establishing a self-sustaining cycle (Sharma et al., 2010). According to Bhatta (2004), community-managed mixed broadleaf forests had soil carbon levels of 150–160 t ha^{-1} and above-ground carbon stocks ranging from 55 to 67 t ha^{-1}. These results highlight the part that community management plays in sequestering carbon. The success of CFUGs lies in their ability to build trust and accountability, as local stakeholders directly experience the benefits of conservation efforts, thereby reinforcing their commitment to long-term stewardship (Jangra et al., 2024).

10.3.4 Policy and Legal Framework

The management of forest resources in Nepal is a cornerstone of rural livelihoods, particularly in the hilly regions where communities depend on forests for energy, fodder, construction materials, and various non-timber forest products. The historical context of forest management in Nepal reveals a complex evolution of policies and legal frameworks that have significantly influenced the current landscape of forest governance. The nationalization of forests in 1957, through the Private Forest Nationalization Act, marked a pivotal moment in the management of forests as a national resource. However, this initiative inadvertently stripped local communities of their traditional rights and access to forest resources, leading to widespread illegal harvesting and a disconnect between local needs and government policies. The subsequent introduction of the Panchayat Forest System in 1978 aimed to empower local governance in forest management; however, it was hindered by weak institutional support, inadequate training for local leaders, and a lack of genuine community involvement in decision-making processes.

In response to these historical challenges, the Forest Sector Master Plan of 1988 was developed to lay the groundwork for sustainable forest management and emphasize the importance of community participation. This plan recognized that local communities possess valuable traditional knowledge and practices that can contribute to effective forest management. The Forest Act of 1993 marked a significant shift in policy by formally recognizing the role of Community Forest User Groups (CFUGs) in managing forest resources. This act enabled local communities to establish Community Forest User Groups (CFUGs), which are empowered to manage and utilize forest resources in accordance with an operational plan that they develop and approve, with the assistance of the divisional forest officer. The

Forest Rule of 1995 further elaborated on the operational guidelines for these groups, detailing their formation, rights, responsibilities, and processes for resource allocation and management. CFUGs have become integral to forest management in Nepal, allowing local communities to collectively manage and utilize forest resources while promoting sustainable practices that align with local needs.

Despite these advancements, significant challenges remain in implementing community forest management. One of the most pressing issues is the exclusion of marginalized groups, notably lower caste communities like the Kami, from CFUGs and decision-making processes. This exclusion perpetuates inequalities in access to forest resources and undermines the potential benefits of community management. The historical context reveals that the socially marginalized Kami have often been overlooked in the distribution of forest benefits, which has hindered their ability to improve their well-being. To address these disparities, it is essential to implement policies that promote equitable access and ensure the participation of all community members in forest management. This can be achieved through targeted outreach and capacity-building initiatives to empower marginalized groups, ensuring that their voices are heard and their needs are met within the framework of CFUGs. Capacity-building initiatives are crucial to enhancing the skills and knowledge of CFUG members, enabling them to manage forests sustainably and effectively. Training programs should focus on sustainable forest management practices, conflict resolution, and the development of operational plans that reflect the diverse needs of the community (Singh et al., 2023b).

Additionally, fostering a sense of ownership and accountability among CFUG members is vital for the long-term success of community forest management. This can be achieved by promoting transparency in decision-making processes and ensuring that benefits derived from forest resources are equitably shared among community members. Monitoring and evaluation mechanisms must also be established to assess the effectiveness of community forest management and its impact on local livelihoods (Joshi et al., 2024). These mechanisms should include participatory approaches that involve community members in the assessment process, allowing them to provide feedback on the management practices and their outcomes. By fostering a culture of accountability and transparency, these mechanisms can help ensure that the benefits of forest resources are distributed fairly among all community members. Furthermore, regular assessments can identify areas for improvement and inform adaptive management strategies that respond to changing environmental and social conditions (Singh, 2024a).

Thus, policy and legal frameworks for forest resource management in Nepal must continue to evolve, addressing historical injustices and promoting sustainability. By strengthening the role of CFUGs and ensuring equitable access to resources, the well-being of all community members can be significantly improved. Integrating marginalized voices into the decision-making process, coupled with capacity-building initiatives and robust monitoring mechanisms, will enhance the effectiveness of community forest management. Ultimately, a more inclusive and participatory approach to forest governance will contribute to the sustainable development of Nepal's forest resources, ensuring that they continue supporting rural communities' livelihoods for generations to come. The alignment of national policies with international frameworks has further strengthened conservation outcomes, as community initiatives are supported by global funding and technical expertise. The policy environment in Nepal demonstrates that localized conservation efforts, when legally recognized and adequately supported, can make significant contributions to global sustainability goals.

10.3.5 Pathway to Achieve Carbon Neutrality Through Community Engagement

Achieving carbon neutrality in Nepal is a crucial and urgent goal, particularly given the country's vast forest resources, which are vital for carbon sequestration and climate regulation. Community engagement has emerged as a transformative strategy that enhances forest conservation efforts and empowers local populations to take an active role in managing their natural resources (Barwa & Singh, 2024). Research has consistently shown that community-based forest management initiatives can lead to more effective resource management by integrating local knowledge and practices that are often more attuned to the ecological dynamics of the area (Agrawal & Gupta, 2005). For instance, the Community Forestry Program in Nepal has successfully empowered local communities to manage forest resources sustainably, resulting in a remarkable increase in forest cover and improved carbon stocks. According to the Food and Agriculture Organization (FAO, 2016), community-managed forests in Nepal sequester approximately 1.6 million tons of carbon dioxide annually, underscoring the significant impact of local engagement in conservation efforts (as shown in Fig. 10.5).

Fig. 10.5 Enhanced canopy cover in a community forest of Eastern Nepal, achieved through assisted natural regeneration and participatory management by the Community Forest User Group (CFUG). (*Source* DFO, Triveni)

Indigenous knowledge is a cornerstone of effective community engagement in forest management. Local communities possess a wealth of traditional ecological knowledge developed over generations, encompassing sustainable land-use practices, species selection for reforestation, and methods for maintaining soil health. This Indigenous knowledge is particularly relevant in the context of climate change as it provides insights into adaptive strategies that can enhance resilience. A study by Berkes (2009) emphasizes the importance of integrating indigenous knowledge into formal conservation strategies, arguing that such integration can lead to more effective and culturally appropriate approaches to carbon management. For example, practices such as rotational grazing, agroforestry, and the use of native species for reforestation not only enhance biodiversity but also improve carbon sequestration capabilities (Singh et al., 2023a).

Successful case studies in Nepal illustrate the effectiveness of community engagement in achieving carbon neutrality. The management of community forests has demonstrated that when local communities are actively involved in decision-making processes, significant environmental benefits can be realized. For instance, establishing Community Forest User Groups (CFUGs) has led to increased forest cover, improved biodiversity, and enhanced livelihoods for local populations. The FAO (2016) reports that community-managed forests have substantially reduced deforestation rates, which is critical for maintaining carbon stocks and mitigating climate change. Furthermore, the active participation of local communities in monitoring forest health and biodiversity has proven to be an effective strategy for ensuring sustainable management practices.

Policymakers need to create an enabling environment that supports local participation to maximize the potential of community engagement in achieving carbon neutrality. This includes recognizing the rights of communities to manage their forest resources, providing training and capacity-building opportunities, and facilitating access to financial resources for sustainable practices. A study by Chhatre and Agrawal (2008) emphasizes the importance of policies that promote equitable benefit-sharing mechanisms, ensuring that local communities derive tangible benefits from their conservation efforts. Such policies can incentivize community participation and enhance the overall effectiveness of carbon management strategies. Moreover, integrating community engagement into national climate policies can significantly improve the effectiveness of carbon neutrality initiatives. Collaboration among government agencies, non-governmental organizations, and local communities can help Nepal establish a supportive environment for sustainable forest management that aligns with national and global climate objectives (Garg et al., 2016). This collaborative approach not only strengthens the capacity of local communities to engage in carbon accounting and monitoring efforts but also ensures that conservation strategies are culturally relevant and ecologically sound.

Thus, community engagement is critical to achieving carbon neutrality in Nepal. Harnessing local communities' knowledge and participation significantly enhances forest conservation efforts, increases carbon sequestration, and promotes sustainable development. Integrating indigenous knowledge and sustainable practices into formal conservation strategies is essential for creating effective and culturally relevant approaches to carbon management. Future policies should prioritize community engagement, ensuring that local voices are heard and that conservation efforts contribute to environmental sustainability and social equity. This holistic approach addresses the urgent need for carbon neutrality. It fosters resilient communities capable of adapting to the challenges posed by climate change, ultimately leading to a sustainable and equitable future for all.

10.4 Conclusion

Nepal's pathway toward carbon neutrality is strongly supported by its community-based forest management, particularly through Community Forest User Groups (CFUGs). These groups have demonstrated significant success in enhancing forest health, biodiversity, and carbon sequestration while simultaneously supporting local livelihoods. Indigenous knowledge and traditional practices enrich these efforts with adaptive, culturally grounded strategies for managing forest resources under changing climatic conditions. However, the transformative potential of community engagement can only be fully realized through supportive policies. Strengthening legal recognition of community rights, embedding equitable benefit-sharing mechanisms, and investing in capacity building are essential to ensure that local stewardship is sustained and scaled. Integrating community engagement into Nepal's national climate strategies not only enhances domestic resilience but also positions Nepal as a model for other countries with similar socio-ecological contexts. Much of the current evidence, however, is drawn from case studies and secondary literature, highlighting the need for longitudinal and quantitative research to assess the precise carbon sequestration potential of community-led practices. Future studies should also explore governance challenges, particularly those related to conflict resolution and cross-scale coordination, to further strengthen participatory models of conservation. Hence, harnessing the synergies between Indigenous knowledge, community participation, and supportive governance provides Nepal with a robust and transferable framework for pursuing carbon neutrality. This holistic approach not only advances environmental sustainability but also promotes social equity and resilience, offering lessons of global relevance in the collective fight against climate change.

Competing Interests The authors declare that they have no known competing financial interests or personal relationships that could have appeared to influence the work reported in this chapter.

Funding This work was supported by the Major Project of the National Social Science Foundation of China [grant number 24&ZD108].

References

Acharya, K. P., Dangi, R. B., & Tripathi, D. M. (2008). Community forestry in Nepal: Decentralized forest governance. *International Forestry Review, 10*(1), 136–147.

Agrawal, A., & Angelsen, A. (2010). *Using community forest management to achieve REDD+ goals*. World Bank Policy Research Working Paper Series.

Agrawal, A., & Gupta, K. (2005). Decentralization and participation: The governance of forests in Nepal. In *The role of local institutions in the management of natural resources* (pp. 1–20).

Aryal, K., Maraseni, T., & Apan, A. (2022). How much do we know about trade-offs in ecosystem services? A systematic review of empirical research observations. *Science of the Total Environment, 806*, Article 151229. https://doi.org/10.1016/j.scitotenv.2021.151229

Bajracharya, S. B., Furley, P. A., & Newton, A. C. (2005). Effectiveness of community involvement in delivering conservation benefits to the Annapurna Conservation Area, Nepal. *Environmental Conservation, 32*(3), 239–247.

Baral, N. (2009). Institutional resilience of community-based conservation to the Maoist insurgency in Nepal. *doctoral diss., Virginia Polytechnic Institute and State University, 2.*

Barwa, S., & Singh, H. (2024). Climate-induced changes in forest ecosystems of Indian Himalayan region: a meta-analysis. In *Urban forests, climate change and environmental pollution: Physio-biochemical and molecular perspectives to enhance urban resilience* (pp. 839–866). Springer.

Berkes, F. (2009). Evolution of co-management: Role of knowledge generation, bridging organizations and social learning. *Journal of Environmental Management, 90*(5), 1692–1702.

Bhatta, B. (2004). *Carbon stock in the forests of Phulchowki Watershed, Lalitpur*. Forest Research and Development Centre, Nepal.

Bhatta C. (2013). An evaluation of the Sagarmatha National Park Forestry Project (SNPFP). Khumbu, Nepal: A Community Stakeholder Approach, Unpublished M.Sc. Thesis. Ryerson University, Toronto, Canada.

Cadman, T., & Maraseni, T. (2012). The governance of REDD+: An institutional analysis in the Asia Pacific region and beyond. *Journal of Environmental Planning and Management, 55*(5), 617–635.

Chaudhary, R. P., Bhattarai, S. H., Basnet, G., Bhatta, K. P., Uprety, Y., Bhatta, L. D., & Sharma, U. R. (2017). *Traditional practice and knowledge of indigenous and local communities in Kailash Sacred Landscape*, Nepal.

Chhatre, A., & Agrawal, A. (2008). Trade-offs and synergies between carbon storage and livelihoods in forest commons. *Proceedings of the National Academy of Sciences, 105*(42), 16392–16397.

DFID-Nepal. (2000). *Community forestry programme in Nepal*. Department for International Development, Nepal.

Food and Agriculture Organization (FAO). (2016). State of the World's Forests 2016: Forests and Agriculture: Land-Use Challenges and Opportunities. FAO.

Garg, A., Pal, D., Singh, H., & Pandey, D. C. (2016). A comparative study of NDBI, NDISI and NDII for extraction of urban impervious surface of Dehradun [Uttarakhand, India] using Landsat 8 imagery. *International Conference on Emerging Trends in Communication Technologies (ETCT), 2016*, 1–5.

Gautam, A. P., Webb, E. L., & Shivakoti, G. P. (2004). Forest cover change, physiography, local economy, and institutions in a mountain watershed in Nepal. *Environmental Management, 33*(1), 48–61.

Gentle, P., & Maraseni, T. N. (2021). Climate change, poverty and livelihoods: adaptation practices by rural mountain communities in Nepal. *Environmental Science & Policy*, 115, 13–22. (Excellent for the second sentence, highlighting the burden on rural communities).

Jangra, A., Kumar, K., Maikhuri, S., Bhandari, M. S., Pandey, S., Singh, H., & Barthwal, S. (2024). Unveiling stress-adapted endophytic bacteria: Characterizing plant growth-promoting traits and assessing cross-inoculation effects on *Populus deltoides* under abiotic stress. *Plant Physiology and Biochemistry, 210*, Article 108610.

Joshi, R., Singh, H., Chhetri, R., & Yadav, K. (2020). Assessment of carbon sequestration potential in degraded and non-degraded community forests in Terai region of Nepal. *Journal of Forest and Environmental Science, 36*(2), 113–121.

Joshi, R., Singh, H., Chhetri, R., Poudel, S. R., & Rijal, S. (2021). Carbon sequestration potential of community forests: A comparative analysis of soil organic carbon stock in community managed forests of Far-Western Nepal. *Eurasian Journal of Soil Science, 10*(2), 96–104.

Joshi, R., Sharma, B., Singh, H., Dhakal, N., Ayer, S., & Maraseni, T. (2024). Poplar plantation as an agroforestry approach: Economic benefits and its role in carbon sequestration in North India. *Journal of Resources and Ecology, 15*(4), 880–888.

Joshi, R., Shrestha, T. K., Mishra, B., Gautam, J., Maharjan, B., & Gosai, K. R. (2023). Assessment of carbon sequestration in private forests across two different physiographic regions of Nepal: Implications for conservation and climate change mitigation. *Scientifica*, *2023*(1), 6599067. https://doi.org/10.1155/2023/6599067

Kanel, K. R., & Niraula, D. R. (2004). Can rural livelihoods be improved in Nepal through community forestry? *Forest Livelihood Briefs, 2*, 1–6.

Khatri-Chhetri, A., Regmi, P. P., Chanana, N., & Aggarwal, P. K. (2020). Potential of climate-smart agriculture in reducing women farmers' drudgery in high climatic risk areas. *Climatic Change, 158*(1), 29–42. (This study is specifically on Nepal and climate risks in agriculture).

Kumar, M., & Singh, H. (2020). Agroforestry as a nature-based solution for reducing community dependence on forests to safeguard forests in rainfed areas of India. In *Nature-based solutions for resilient ecosystems and societies* (pp. 289–306). Springer.

Kumar, A., Dwivedi, G. K., Tewari, S., Sah, V. K., Singh, H., Kumar, P., Kumar, N., & Kaushal, R. (2020). Soil organic carbon pools under *Terminalia chebula* Retz. based agroforestry system in Himalayan foothills, India. *Current Science, 118*(7), 1098–1103.

Maharjan, K. L. (2005a). Community forestry in Nepal: A review of the current status and future directions. *Journal of Forest and Livelihood, 4*(1), 1–10.

Maharjan, K. L. (2005b). Community participation in forest resource management in Nepal. *Journal of Mountain Science, 2*, 32–41.

Mengie, T., & Szemethy, L. (2025). The practice of community-based forest management in Northwest Ethiopia. *Land, 14*(7), 1407.

Nie, Y., Pritchard, H. D., Liu, Q., Hennig, T., Wang, W., Wang, X., & Liu, S. (2021). Glacial change and hydrological implications in the Himalaya and Karakoram. *Nature Reviews Earth & Environment, 2*(2), 91–106.

Ojha, H. R., et al. (2023). *Community-based forest management in Nepal: Policies, practices, and outcomes.*

Ostrom, E. (1990). *Governing the commons: The evolution of institutions for collective action.* Cambridge University Press.

Page, M. J., McKenzie, J. E., Bossuyt, P. M., Boutron, I., Hoffmann, T. C., Mulrow, C. D., Shamseer, L., Tetzlaff, J. M., Akl, E. A., Brennan, S. E., Chou, R., Glanville, J., Grimshaw, J. M., Hróbjartsson, A., Lalu, M. M., Li, T., Loder, E. W., Mayo-Wilson, E., McDonald, S., & Moher, D. (2021). The PRISMA 2020 statement: An updated guideline for reporting systematic reviews. *BMJ, 372*, Article n71. https://doi.org/10.1136/bmj.n71

Pan, Y., Birdsey, R. A., Fang, J., Houghton, R., Kauppi, P. E., Kurz, W. A., & Hayes, D. (2011). A large and persistent carbon sink in the world's forests. *Science, 333*(6045), 988–993.

Pandey, S. S., Maraseni, T. N., & Cockfield, G. (2014). Carbon stock dynamics in different vegetation dominated community forests under REDD+: A case from Nepal. *Forest Ecology and Management, 327*, 40–47.

Pandey, B., Bhandari, T. S., & Parajuli, R. (2024). Pathways to net-zero emissions in the land-use sector: A case study from Nepal. *Environmental Science & Policy, 151*, Article 103628. https://doi.org/10.1016/j.envsci.2024.103628

Peel, A. B., Smith, C. D., & Johnson, E. F. (2017). Shifting temperature zones and mountaintop habitat loss in montane species under climate change. *Global Change Biology, 23*(4), 1234–1245.

Poudyal, B. H., Maraseni, T., Cockfield, G., & Bhattarai, B. (2020). Recognition of historical contribution of indigenous peoples and local communities through benefit sharing plans (BSPs) in REDD+. *Environmental Science & Policy, 106*, 111–114.

Scheffers, B. R., De Meester, L., Bridge, T. C., Hoffmann, A. A., Pandolfi, J. M., Corlett, R. T., & Watson, J. E. (2016). The broad footprint of climate change from genes to biomes to people. *Science, 354*(6313), aaf7671.

Sharma, R. (2010). The role of farm-forest integration in sustainable livelihoods. *Mountain Research and Development, 30*(1), 59–67.

Sharma, D. K., Behera, U. K., & Mahapatra, S. (2010). *Integrated farming system: An approach for sustainable agriculture.* National Institute of Agricultural Extension Management (MANAGE). https://anrcm.org/wp-content/uploads/2021/11/01-DK-Sharma.pdf

Shi, X., Zheng, Y., Lei, Y., Xue, W., Yan, G., Liu, X., & Wang, J. (2021). Air quality benefits of achieving carbon neutrality in China. *Science of the Total Environment, 795*, Article 148784.

Shrestha, B. P., & Devkota, B. P. (2016). Carbon stocks in the oak and pine forests in Salyan District, Nepal. *Banko Janakari, 23*(2), 30–36. https://doi.org/10.3126/banko.v23i2.15478

Shrestha, B. B., & Tiwari, S. (2006). Indigenous knowledge and practices in climate change adaptation: A Nepalese perspective. *Environmental Management Journal, 3*(4), 45–51.

Sikor, T., et al. (2013). Forests and carbon: The importance of community rights. *Global Environmental Change, 23*(1), 40–45.

Singh, H. (2024a). Climate change and forests: Biological perspectives. In *Forests and climate change: Biological perspectives on impact, adaptation, and mitigation strategies* (pp. 1–17). Springer.

Singh, H. (2024b). Exploring adaptive modulation in plant functional traits and its impact on the productivity of *Acacia auriculiformis* under CO_2-enriched environment. *Industrial Crops and Products, 210*, Article 118186.

Singh, H. (2024c). *Forests and climate change: Biological perspectives on impact, adaptation, and mitigation strategies.* Springer Nature.

Singh, H. (2024d). *Urban forests, climate change and environmental pollution: Physio-biochemical and molecular perspectives to enhance urban resilience.* Springer Nature.

Singh, H. (2025). The role of plant functional traits in mitigating urban climate and environmental pollution. In *Plant functional traits* (pp. 297–308). Elsevier.

Singh, H., Verma, A., & Kumar, M. (2018). *Can elevated CO_2 alter the growth dynamics, morphology, physiology and biochemistry of the forestry tree species?* (pp. 288–305). Climate Change and Agriculture. Discovery Publishing House.

Singh, M., Kumar, P., Singh, H., Kumar, A. K. A., Kumar, R., & Kumar, R. (2023a). *Neolamarckia cadamba*: A comprehensive review on its physiological, ecological, phytochemical and pharmacological perspectives. *Ecology Environment and Conservation, 29*, 241–250.

Singh, H., Kumar, N., Singh, R., & Kumar, M. (2023b). Assessing the climate change impact on the habitat suitability of the range-restricted bird species (*Catreus wallichii*) in the Indian Himalayan ecosystem. *Environmental Science and Pollution Research, 30*(57), 121224–121235.

Singh, H., & Kumar, M. (2022). Climate change and its impact on Indian Himalayan forests: current status and research needs. In *Climate change: Impacts, responses and sustainability in the Indian Himalaya* (pp. 223–242). Springer.

Singh, S., & Yadav, A. (2024). Unveiling anthropogenic environmental burdens impacting ecosystem services in the Himalayas. In *The resilience of traditional knowledge systems for a sustainable future: A focus on agriculture and food practices in the Himalayas* (pp. 221–246). Cham: Springer Nature Switzerland.

Singh, H., Kumar, N., & Kumar, A. (2024). Enhancing resource use efficiency in crops through plant functional traits. In *Plant functional traits for improving productivity* (pp. 97–117). Springer.

Stevens, S. F. (1993). *Claiming the high ground: Sherppas subsistence and environmental change in the Highest Himalaya.* Berkeley, USA: University of California Press.

Subedi, B., Lamichhane, P., Magar, L. K., & Subedi, T. (2022). Aboveground carbon stocks and sequestration rates of forests under different management regimes in Churia region of Nepal. *Banko Janakari, 32*(1), 15–24. https://doi.org/10.3126/banko.v32i1.45442

Tamrakar, P. R., & Mohns, B. (2016). Forest restoration at the landscape level in Nepal. *Forest Landscape Restoration for Asia-Pacific Forests, 1990*(92), 103.

Thapa, S., Maraseni, T., Dhonju, H. K., Shakya, K., Shakya, B., Apan, A., & Banerjee, B. (2025). Predictive assessment of forest fire risk in the Hindu Kush Himalaya (HKH) region using HIWAT data integration. *Remote Sensing, 17*(13), 2255. https://doi.org/10.3390/rs17132255

UNESCO. (2025). *Dynamic climate leadership of the Sherpa people—Sagarmāthā National Park*. UNESCO. https://www.unesco.org/en/articles/dynamic-climate-leadership-sherpa-people-sagarmatha-national-park

Veh, G., Korup, O., & Walz, A. (2020). Hazard from Himalayan glacier lake outburst floods. *Proceedings of the National Academy of Sciences, 117*(2), 907–914.

Wester, P., Mishra, A., Mukherji, A., & Shrestha, A. B. (Eds.). (2019). *The Hindu Kush Himalaya assessment: Mountains, climate change*. Springer International Publishing.

Wester, P., Chaudhary, S., Chettri, N., Jackson, M., Maharjan, A., Nepal, S., & Steiner, J. F. (2023). Water, ice, society, and ecosystems in the Hindu Kush Himalaya: an outlook.

Wiersum, K. F. (1997). Indigenous exploitation and management of tropical forest resources: An evolutionary continuum in forest-people interactions. *Agriculture, Ecosystems & Environment, 63*(1), 1–16.

Wu, X., Tian, Z., & Guo, J. (2022). A review of the theoretical research and practical progress of carbon neutrality. *Sustainable Operations and Computers, 3*, 54–66. https://doi.org/10.1016/j.susoc.2021.12.004

Chapter 11
Forests as Catalysts for Socio-Economic Development: Unveiling Their Comprehensive Benefits

Prachi Upadhyay, Neetu Arya, and Thiyam Seityajit Singh

Abstract Forests are distributed globally, covering a total area of 4.1 billion hectares. Forests play a key role in regulating the Earth's climate by influencing physical, chemical, and biological processes. Forests are vital ecosystems that provide a wide array of direct and indirect benefits, making them indispensable to socio-economic development at both global and local levels. The forest acts as a provider of various ecological, social, economic, and aesthetic services. This chapter explores the multi-faceted contributions of forests, categorizing them into tropical, temperate, and boreal types, each offering unique ecosystem services and resources. Direct benefits, such as timber, non-timber forest products, and industry raw materials, are complemented by indirect benefits, including carbon sequestration, biodiversity conservation, and water regulation. National and global strategies, such as India's Joint Forest Management program and initiatives under the UN's REDD + framework, highlight the role of forests in driving economic growth, enhancing livelihoods, and promoting sustainability. Moreover, the chapter features case studies that examine the transformative impact of forests on socio-economic progress, ranging from community-based forestry models in Nepal to afforestation projects in Africa that combat desertification while creating employment opportunities. Besides, this chapter underscores the interplay between forest resources and development, emphasizing forests as catalysts for reducing poverty, achieving food security, and enhancing climate resilience. It advocates for innovative policies and sustainable management practices that balance forest conservation with socio-economic advancement, ensuring forests thrive as a cornerstone of global prosperity.

Keywords Forest · Forest types · Direct benefits · Indirect benefits · Socio-economic role · Climate change · Sustainable forest management

P. Upadhyay (✉) · N. Arya
Indian Institute of Forest Management, Bhopal, Madhya Pradesh, India
e-mail: prachiupadhyay093@gmail.com

Forest Research Institute Deemed to be University, Dehradun, Uttarakhand, India

T. S. Singh
Department of Anthropology, University of Delhi, Delhi, Delhi, India

H. Singh (ed.), *Forestry for a Carbon-Neutral and Sustainable Future*, Sustainability Solutions, https://doi.org/10.1007/978-3-032-06427-1_11

11.1 Introduction

Forests are distributed in both urban and rural environments, covering a total area of 4.1 billion hectares (Vose et al., 2018; FAO, 2024), which is nearly one-third of the global land cover (Le Gear et al., 2023). They influence the global climate through various physical, chemical, and biological processes, impacting the hydrological cycle, atmospheric composition, and planetary energetics. Forests, in particular, cover ~42 million km^2 in the form of tropical, temperate, and boreal forests, which account for around 30% of the land surface. This interaction between the forest and the atmosphere is complex and nonlinear and can impact the results of climate change (Bonan, 2008a, b). Globally, forests are distributed differently, and the top three countries with the most forest area are Russia, Brazil, and the United States of America (Singh & Kumar, 2022). India constitutes 1.78% of the world's forest area (Figs. 11.1 and 11.2).

India's total forest and tree cover is 827,357 km^2, accounting for 25.17% of the country's geographical area. The forest cover is 7,15,343 km^2 (21.76%), and the Tree Cover is 1,12,014 km^2 (3.41%) (Forest Survey of India, 2023).

Forests play a pivotal role in maintaining ecological balance and providing indispensable ecosystem services that support life on Earth (Fig. 11.3). They act as carbon sinks, mitigating climate change by absorbing approximately 7.6 billion metric tons of CO_2 annually (Pan et al., 2011). Additionally, forests regulate the hydrological cycle, enhance soil fertility, and prevent erosion, ensuring agricultural productivity and water quality (Bonan, 2008a, b). They harbor over 80% of terrestrial species,

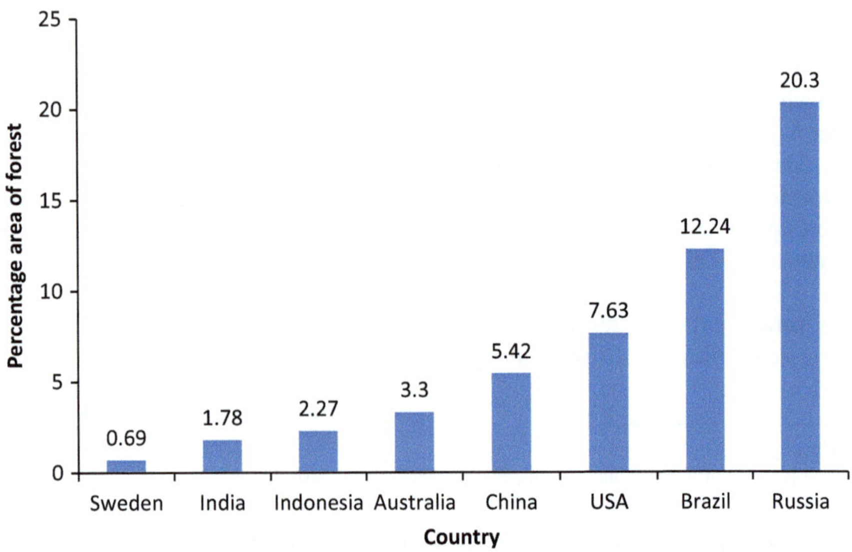

Fig. 11.1 Share of Global Forest area in 2020- Country-wise (*Source* Richie, 2021)

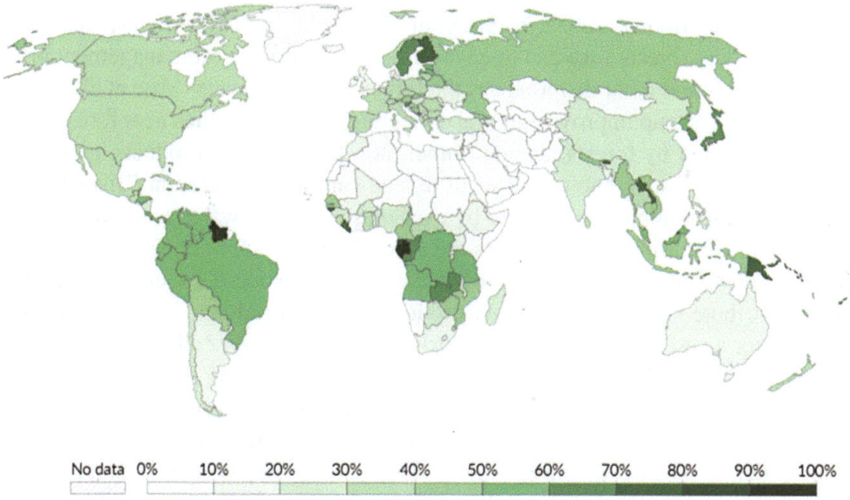

Fig. 11.2 Share of global forest area, 2020- World (Richie, 2021)

fostering biodiversity and resilience against ecological disruptions (FAO, 2020a, b, c). Moreover, forests contribute to microclimate stabilization and air purification by sequestering pollutants, thus enhancing human and ecosystem health. These ecological benefits underscore the necessity of sustainable forest management to ensure the provision of these services for future generations (Garg et al., 2016).

Forests provide numerous benefits and services, including ecological, social, economic, and aesthetic, such as a home for biodiversity, food, medicines, and other forest products, as well as regulation of the hydrological cycle, soil conservation,

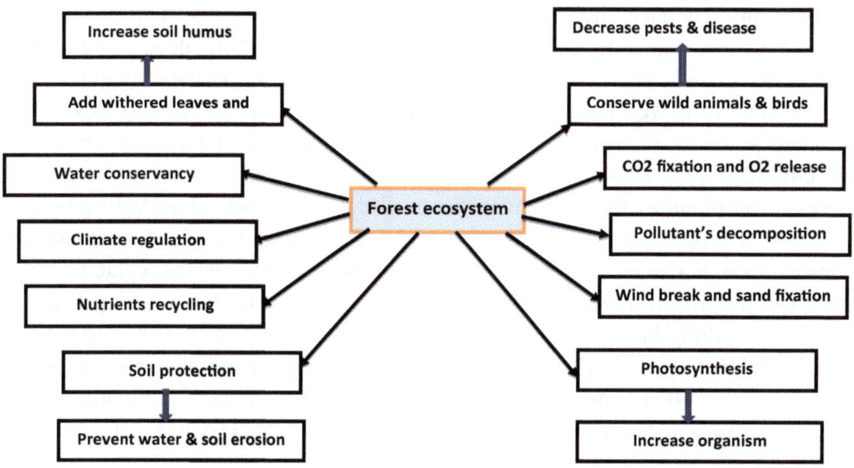

Fig. 11.3 Ecological benefits of Forest ecosystem (Xue & Tisdell, 2001)

and recreation (Bonan, 2008a, b; Singh et al., 2023a, b). The structure and functions offered by forests vary due to environmental factors, forest management practices, land use, and other disturbances the forest may experience (Vose et al., 2018). Forest resources help improve the rural economies of communities near forests. The resources offered by forests include timber, non-timber forest products, as well as services such as water, habitat for wildlife, and carbon sequestration (Bisui & Shit, 2024). They also impact the climate by exchanging carbon, water, and other chemicals with the atmosphere. The forests are a storehouse of ~45% of terrestrial carbon (Bonan, 2008a, b), which is one-third of the carbon dioxide (Le Gear et al., 2023). They contribute ~50% of net primary production (terrestrial) and have the potential to sequester a vast amount of carbon yearly. They have a lower albedo, which helps to mitigate the effect of snow's high albedo, in turn increasing the solar heating of land (Bonan, 2008a, b). Forests are home to nearly 90% of terrestrial biodiversity. This tropical forest is more important in terms of both species richness and is home to many endemic species (Barwa & Singh, 2024). The forest provides several benefits, such as protecting fisheries, watersheds, and soils. Forests are also a source of raw materials, mainly for rural poor who rely on forest products for their basic livelihood needs. However, today, with approximately 30% of the global land area forested, the deforestation rate of 13 million hectares per year poses a significant threat to the world's forests. This is particularly evident in tropical countries. Other forest types, such as temperate and boreal forests, are also threatened due to their industrialized management practices for timber production (Schmitt et al., 2009).

In 2005, the Millennium Ecosystem Assessment was implemented, integrating its findings into policies that helped to understand and acknowledge the roles of forest ecosystems in various functions, as well as the timber and non-timber goods and services they provide to people. In recent times, with increased urbanization and ecological degradation, the demand for recreational use of forests has been on the rise (Mann et al., 2024). The use of forests by humans affects various climatic changes and feedback, and the misuse of them can impact ecosystem services. At the same time, managing the forest effectively can help mitigate climate change and its associated effects (Bonan, 2008a, b). They also help promote, maintain, and restore health and well-being. Social forestry encompasses various activities related to forest products, the rural environment, and communities engaged in subsistence agriculture (Singh, 2024a). It often has external assistance for development, which benefits communities and their environment. Today, this concept has evolved beyond its original scope and now focuses on practices such as forest management and conservation. It also recognizes the impact of substitutions, such as replacing wood fuels with improved stoves (Singh et al., 2024). The straw from the field, previously used for cow dung-based fuel, now helps by adding structure and nutrients to the soil, thereby enhancing its quality. It also reduces fuel wood consumption, saving forests. Social forestry often benefits the most disadvantaged sections of communities that rely on forests, including women, the rural poor, and landless users (Hyde et al., 2000).

Various other factors, such as socio-economic conditions, technology, and organizational capacity, will determine the overall ability of society to adapt to climate change and its associated issues. However, certain general principles work across the forest despite their differences. A viable workforce in local communities that works in and for forests focusing on building forest resilience, can help take timely and effective actions to neutralize or minimize the effect of climate change (Ruhil et al., 2024). Conversion of forest to forest is a decision made by humans and impacts the ecosystem services provided by forests, as well as the health of the forest ecosystem. Keeping forest lands covered with forest is a significant challenge today (Vose et al., 2018).

Climate change poses a serious threat to the survival and persistence of species in forest ecosystems worldwide. Climate change will alter forests in numerous ways, impacting the services derived from forest ecosystems. Decreased forest ecosystems will also impact dependent communities, reducing their access to essential resources such as food and clean water, and ultimately leading to increased poverty and social conflict (Obeng et al., 2011). Climate change will cause irreversible damage to forest ecosystems, threatening the biodiversity and ecosystem services they provide. The time required for forests to adapt will also be longer, requiring appropriate adaptation strategies. Forest planning as well as development programs are to be formulated that address the impact of climate change on forests and associated ecosystem services (Ravindranath et al., 2006).

11.2 Types of Forests Globally

Generally, forest classification based on latitudes comprises northern boreal forests followed by temperate, subtropical, and tropical zones. These forests do not entirely cover these regions; they also include areas modified by human presence, such as roads, cities, and farms. The forest in these areas is not necessarily homogeneous. They are of different sizes and provide habitat for various species. These differences exist due to the history of the places. For example, the time since the area's last fire can imply significant differences in forest types within a homogeneous region. However, the broad forest category is most likely expected in that region (Joshi et al., 2024).

11.2.1 Boreal Forests

These forests are located in the Northern Hemisphere. They are spread from Scandinavia to the eastern part of Siberia, passing through Alaska and the northern part of Canada. They covered about a third of the total forested land globally and have faced a 20% reduction in area over the last hundred years due to anthropogenic interferences and actions. These regions have the harshest climate for tree survival, with temperatures falling as low as −50 °C. The growing season is limited to only 4–5 months, during which the maximum temperature reaches up to 15–20 °C. The low temperature, especially at night, did not hold moisture in the air, making it very dry. Boreal trees are limited in number. The conifers growing in the region have narrow crowns and flexible branches, which favor light absorption at low sun angles and help in shedding snow. Pines (Pinus) and spruces (Picea) are the most common species in these forests. Some deciduous species include larches (*Larix*), birch (*Betula*), aspen (*Populus*), and willows (*Salix*). Lichens reside in the forest's understory and serve as a food source for reindeer, caribou, etc. The region is also known to have various berries. Due to the extreme weather conditions, the native flora and fauna of the boreal forest remain largely intact. The harsh climate makes it difficult for any invasive species to adapt. The significant disturbances faced by these forests are wildfires and native insect outbreaks (Landsberg & Waring, 2014).

11.2.2 Temperate Forests

These forests are found in the temperate regions of the globe, characterized by a mix of coniferous and broad-leaved hardwoods. They are mixed to various degrees. So overall, temperate forests can be classified as.

11.2.2.1 Temperate Coniferous

These forests are home to one of the world's oldest trees, representing only a tiny portion of the world's forests. These trees grow straight and, therefore, have been exploited by the wood industries. Conifers dominate the mountainous areas of North America, Europe, China, and certain parts of the mountains in Korea, Japan, as well as in Nicaragua, Guatemala, and Mexico. Pinus species is predominantly found in the southern hemisphere. These forests have a greater number of species compared to boreal forests. The summers here are warm, and winters are cold and wet, with yearly rainfall ranging from 500 to 2500 mm and more. Common species found in these forests include pine (Pinus), fir (Abies), spruce (Pice*a*), hemlock (*Tsuga*), and cedar (*Thuja*).

11.2.2.2 Mixed Temperate

These forests contain a mix of deciduous and evergreen species, extending from the Southeastern United States and Europe through northern Iraq, Iran, and China. These forests are characterized by species such as oaks (*Quercus*), maples (*Acer*), beeches (*Fagus*), and hickories (*Carya*). However, today, due to threats such as land use change, which has introduced pathogens like chestnut blight, the area under mixed temperate forest has decreased by up to 40% compared to the original area.

11.2.2.3 Broad-Leaved Evergreen

They are scattered throughout Japan, Chile, New Zealand, and Australia. They are still in continuity in Australia, where the Eucalyptus species is spread worldwide. These forests have been heavily exploited for timber and have suffered significant damage. Furthermore, they have been cleared for agricultural purposes and face severe threats.

11.2.3 Tropical Forest

This forest covers around 50% of the world's total forest area and is also known as an evergreen forest. They are found in regions with minimal solar and seasonal fluctuations. The single location with the maximum evergreen forest is the Amazon Basin in South America, followed by the forests of central Africa, specifically the Congo region, Gabon, Cameroon, and the southeastern parts of Nigeria, as well as forests in Southeast Asia. These are the most diverse terrestrial ecosystems, with the highest number of species per unit area. The soil here is nutritionally poor due to high rainfall, but it can sustain rich forests because of a high turnover rate, as the organic matter falls from the trees and decays rapidly. The rainforest canopy is layered, with the upper layer comprising tall trees known as emergent. A middle layer follows, and the understory consists of small trees, shrubs, and other vegetation. The standing biomass in this forest varies from one location to another, depending on the soil type, the frequency of past disturbances, and other factors.

However, the forest is under much pressure today due to anthropogenic factors. It is believed that 14 million hectares of forest are destroyed yearly. These forests are an example of systems that thrive in a stable state if the disturbance does not affect the functioning canopies. However, gaps created in the forest result in its degradation, and recovery is often achieved through replanting. It is particularly challenging when native pollinators and seed dispersers are absent, making their conservation extremely important.

11.2.4 Plantation and Managed Forests

Plantations are vast areas where a single tree species is planted solely for wood production. Much of the timber comes from these plantations rather than natural forest areas. They comprise less than 3% of the world's forests but account for 25% of the wood product requirements. A significant advantage of these plantations is that they help fulfill wood requirements without harming the native forests. So, plantations are single-species stands managed for wood production and are easier to harvest, with a wood yield per unit land area that exceeds that of native forests (Singh et al., 2023a).

11.3 Benefits of Forests

11.3.1 Direct Benefits

The direct benefits of forests extend beyond their ecological roles to have a substantial impact on human livelihoods, healthcare, and economies. From provisioning essential resources to enabling economic development through employment and tourism, forests are indispensable to sustainable socio-economic development. Recognizing and investing in these benefits can enhance the well-being of communities while supporting broader environmental goals. Forests are among the most vital natural resources, offering numerous direct benefits that support human livelihoods, economies, and ecological stability. These direct benefits can be categorized into provisioning, economic, and cultural services (Jangra et al., 2024).

11.3.1.1 Provisioning Services

Forests provide essential resources, including timber, fuelwood, and non-timber forest products (NTFPs), which encompass medicinal plants, fruits, nuts, and resins. Timber is a cornerstone of the global construction and furniture industries, while fuelwood serves as a primary energy source for millions of rural households, particularly in developing countries. NTFPs significantly contribute to the livelihoods of local and indigenous communities, providing subsistence and income-generating opportunities (Joshi et al., 2020).

For example, in India, the forest-dependent population derives over 20% of their income from forest resources, highlighting the direct economic role of forests. Moreover, forest products account for significant export revenues in tropical and temperate regions, making them critical to national economies (Vedeld et al., 2007).

11.3.1.2 Employment Generation

The forestry sector provides direct employment opportunities in logging, sawmilling, and paper production. According to the Food and Agriculture Organization, forestry activities employ over 13 million people globally, with an additional 41 million indirectly linked to the sector through downstream industries. This employment fosters rural development, reduces migration to urban areas, and strengthens local economies (FAO, 2020a, b, c).

11.3.1.3 Food and Nutrition Security

Forests are crucial for ensuring food security, particularly in rural and forest-dependent communities. Forest fruits, nuts, and bush meat supply vital nutrients and serve as dietary supplements, especially during agricultural lean seasons. Sunderland et al. (2013) emphasize that forests provide 20–25% of household dietary needs in tropical regions, thereby significantly reducing the risk of malnutrition.

11.3.1.4 Medicinal Value

Approximately 80% of the global population relies on traditional medicine derived from forest plants for primary healthcare (WHO, 2019). Iconic examples include quinine from Cinchona trees, used to treat malaria, and paclitaxel from Pacific yew trees, used in cancer therapy. These medicinal resources underline the economic and healthcare value of forest ecosystems.

11.3.1.5 Cultural and Recreational Services

Forests serve as sites for tourism, recreation, and spiritual practices. Eco-tourism, particularly in tropical forests, generates billions in revenue annually. For instance, Costa Rica earned over USD 3.5 billion from eco-tourism in 2020, with a significant share attributed to forest-based activities (UNWTO, 2021). Such recreational benefits often coincide with conservation efforts, promoting sustainable forest management.

11.3.2 Indirect Benefit

The indirect benefits of forests highlight their multidimensional contribution to socio-economic development. Forests act as unseen yet indispensable catalysts for sustainable development by regulating climate, conserving soil and water, supporting biodiversity, and enriching human culture. These benefits underscore the need for robust conservation policies to ensure that forest ecosystems continue to thrive and help humanity. Forests are sources of direct economic gains, such as timber and non-timber forest products, and provide invaluable indirect benefits that contribute to socio-economic development. These indirect benefits often include ecological and social services that are critical for sustaining life and enhancing human well-being.

11.3.2.1 Climate Regulation

Forests act as significant carbon sinks, absorbing approximately 2.6 billion tons of CO_2 annually, significantly mitigating climate change (FAO, 2020a, b, c). Forests reduce the frequency and intensity of extreme weather events by regulating local and global temperatures (Kumar et al., 2020). Additionally, forests regulate hydrological cycles, influencing rainfall patterns that are essential for agriculture and water availability in downstream communities (Ellison et al., 2017).

11.3.2.2 Soil Conservation and Fertility

Forests play a pivotal role in preventing soil erosion by stabilizing the soil with their extensive root systems. They also contribute organic matter, enhancing soil fertility over time. This indirect benefit is vital for sustaining agriculture in adjacent lands and supporting food security and local economies (Chazdon, 2008).

11.3.2.3 Hydrological Cycle Regulation

Forests maintain groundwater recharge through transpiration and canopy interception and regulate surface water flow. This function helps reduce the risk of floods and droughts, particularly in tropical and subtropical regions (Calder, 2007). Moreover, forests improve water quality by filtering pollutants, thereby benefiting communities that depend on freshwater resources.

11.3.2.4 Biodiversity Conservation

Forests are biodiversity hotspots, supporting 80% of terrestrial species (WWF, 2021). This biodiversity underpins ecosystem services, such as pollination, pest control, and genetic resources, which are essential for agriculture, pharmaceuticals, and other industries. These services, though indirect, are crucial for economic activities and human well-being (Joshi et al., 2021).

11.3.2.5 Cultural and Aesthetic Values

Forests hold immense cultural significance, offering a range of benefits, including recreational, spiritual, and aesthetic value. These values promote mental health and offer opportunities for eco-tourism, making a substantial contribution to the global economy. For instance, forest-based tourism generates billions of dollars annually, creating jobs and supporting local livelihoods (UNEP, 2019).

11.3.2.6 Health Benefits

Forests play a crucial role in air purification by absorbing pollutants and releasing oxygen into the atmosphere. This function significantly reduces the incidence of respiratory diseases in both urban and rural populations. Additionally, the biodiversity of forests serves as a source of medicinal plants, supporting traditional and modern healthcare systems (WHO, 2013).

11.4 Forests as the Catalyst of Socio-Economic Development

Forests play a crucial role in both social and economic development. The resources available through forests play a pivotal role in uplifting the rural economy, particularly in the communities near forests. They offer various goods and services. Forest resources, particularly non-timber forest products (NTFPs), supplement household earnings mainly to cover income shortages. Multiple studies have demonstrated the role that forest-based income plays in sustaining rural livelihoods, especially in forest fringe areas (Jagger et al., 2014; Kamanga et al., 2009; Lax & Köthke, 2017; Vaughan, 2020). A study assessing forest-based income across 17 developing nations found that forest resources contributed 22% to the average household income (Mendako et al., 2022; Nerfa et al., 2020). Therefore, it can be said that despite ongoing deforestation and forest degradation for agricultural and other purposes, forests still play a critical role in ensuring food security. Therefore, it is crucial to develop such policies and management programs that meet the needs of locals, provide the benefits from forests, and promote sustainable practices. In India, around 68% of the rural community relies on forests for their day-to-day needs, including fuelwood and timber, as well as non-timber forest products (NTFPs), and forests contribute 57% of their income (Bisui & Shit, 2024). Similar work has been observed by Fardusi et al. (2011) in Bangladesh, where forest-adjacent communities rely on bamboo, wild vegetables, and cotton wool grass. Other studies have shown reliance on various edible non-timber forest products (NTFPs), including mushrooms, spices, honey, palm fruit, vegetables, and fruits (de Sousa et al., 2018; Mahonya et al., 2019; Matias et al., 2018; Ndumbe et al., 2019). This highlights the need for a proper strategy in forest management (Bisui & Shit, 2024), as it can influence the carbon sequestered by the trees, ultimately affecting how the forest responds to climate change and the services it provides to people (Bernier & Schoene, 2009).

Over the past decade, the therapeutic value of forests has garnered considerable attention, and researchers have been working to understand the impact forests have on an individual's physical and mental well-being. Forest-based therapeutic practices, also known as forest therapy or forest bathing, have been practiced in Asia for centuries and have gained popularity more recently in Europe and the United States. This aims to promote wellness through the use of nature-based practices. The trend of such practices is increasing, but there is a lack of understanding of the societal and institutional dimensions, as well as the economic effects, of forest therapy. Although the demand for forest therapy is rising, its socio-economic potential for forest owners, health benefits, and integration into policies and management strategies remain unexplored. Nature-based therapy is an ancient concept that has been part of the cultural landscape of Asia and Europe for a long time (Mann et al., 2024). For example, in Japan, concepts like forest bathing have been practiced due to the Buddhist connection to nature and associated practices, as well as the idea of nature's healing power in the Shinto religion. Japan also promoted the concept of Shinrin-yoku, which translates to "forest bathing," in the early 1980s to highlight the benefits of forests on society's well-being and happiness. (Mann et al., 2024; Miyazaki, 2018). The healing power of forests has inspired many legends and myths. Hippocrates referred to the benefits of salicylic tea extracted from willow around 400 BCE, whereas the compound was refined as aspirin in the late nineteenth century. The societal lockdowns due to the COVID-19 pandemic helped people realize the importance of outdoor exercising. This led to increased awareness about local green spaces and a broader appreciation for nature (Le Gear et al., 2023). Many cases have been mentioned below to understand the role of forests in socio-economic development-

11.4.1 Case Studies

Forests are a cornerstone of socio-economic development, providing essential resources, ecological services, and employment opportunities. Below are notable case studies from both global and Indian contexts that illustrate the multifaceted contributions of forests to socio-economic development.

11.4.1.1 Global Case Studies

The Great Green Wall Initiative, Africa

The Great Green Wall (GGW) project is an ambitious initiative aimed at combating desertification and promoting sustainable livelihoods across 11 countries in the Sahel region. Initiated by the African Union, this project aims to restore 100 million hectares of degraded land and sequester approximately 250 million tons of carbon by 2030. Communities along the wall benefit from improved agricultural productivity, sustainable forestry practices, and job creation opportunities. For instance, women-led cooperatives in Niger have developed agroforestry models that combine millet with drought-resistant trees, thereby boosting food security and economic independence (Singh, 2024b).

Community Forest Management in Nepal

Nepal's community forestry program is a globally recognized model for participatory forest management. Local communities manage approximately 22% of Nepal's forest area, making a significant contribution to rural development in the country. These forests generate revenue through the sale of timber, non-timber forest products (NTFPs), and ecotourism. For example, the collaborative management of the Churia Hills region has reduced illegal logging and increased income through sustainable harvesting (Singh, 2025).

REDD+ in Latin America

The REDD + (Reducing Emissions from Deforestation and Forest Degradation) initiative has been successfully implemented in Brazil and Costa Rica. The Costa Rica Payment for Ecosystem Services (PES) program compensates landowners for maintaining forest cover. This initiative has restored over 1 million hectares of degraded land and created a $300 million annual market for forest-based services.

Costa Rica's Payments for Ecosystem Services (PES) Program

It is one of the earliest, and most cited examples of forest conservation through economic incentives. The scheme compensates landowners for maintaining forest cover, promoting biodiversity conservation, regulating water, and sequestering carbon. This approach has successfully integrated environmental sustainability with economic development, providing financial benefits to local communities (Singh, 2024d).

The Oluwa Forest Reserve in Nigeria

The Oluwa Forest Reserve in Nigeria holds substantial economic importance for the region. It supports biodiversity and ecotourism, provides timber and non-timber forest products, and plays a crucial role in mitigating carbon sequestration and climate change. The forest also offers watershed protection, supports local agriculture and livelihoods, and creates job opportunities, thereby contributing to the socio-economic development of the local communities.

11.4.1.2 National Case Studies

Joint Forest Management in West Bengal: A Collaborative Model for Socio-Economic Growth

Joint Forest Management (JFM) initiatives in West Bengal have emerged as exemplary models of how community participation can enhance socio-economic development while promoting forest conservation. The Arabari Forest Range project, launched in the 1980s in the Paschim Medinipur district, is often regarded as the birthplace of the Joint Forest Management (JFM) initiative in India (Singh, 2024c). Under this initiative, local villagers were granted access to forest resources and financial incentives to sustain Sal forests. Over 600 families were protecting and regenerating forests, resulting in increased forest cover and enhanced biodiversity.

The socio-economic benefits were evident from these initiatives. The households earned income through employment in forest-related activities, including silviculture, bamboo harvesting, and nursery development. Women's self-help groups

(SHGs) significantly generated supplementary income by producing and selling Sal leaf plates and honey. The Arabari case demonstrates that decentralized forest governance improves livelihoods and ensures ecological sustainability (Singh et al., 2018).

Sacred Groves in Meghalaya: Traditional Knowledge and Biodiversity Conservation

Sacred groves in Meghalaya represent a unique intersection of cultural traditions and forest conservation. These community-protected forest patches, often associated with religious or spiritual significance, have long been maintained by indigenous Khasi tribes. The Mawphlang Sacred Grove, situated near Shillong, spans approximately 78 hectares and harbors a rich biodiversity, including endemic and rare plant species such as the *Schima wallichii* tree and various medicinal herbs.

From a socio-economic perspective, the sacred groves contribute to eco-tourism, attracting researchers, nature enthusiasts, and tourists. Local communities benefit through employment opportunities as guides, souvenir sellers, and homestay operators. Furthermore, preserving sacred groves has encouraged the Khasi youth to engage in traditional knowledge systems, fostering a sense of cultural pride and environmental stewardship.

Forest Rights Act (FRA) Implementation in Odisha: Empowering Tribals Through Land Rights

Implementing the Forest Rights Act of 2006 in Odisha has been instrumental in empowering tribal communities by granting them legal ownership of forest land. In Kandhamal district, over 12,000 households have received individual and community forest rights, enabling them to manage forest resources sustainably.

This has had a transformative socio-economic impact. Tribals now cultivate non-timber forest products (NTFPs), such as tamarind, mahua flowers, and hill brooms, which are sold in local and regional markets. Community-led initiatives in value addition, such as producing tamarind paste and mahua oil, have increased household incomes. The FRA has also reduced conflicts over land rights, fostering greater harmony between forest-dwelling communities and the Forest Department. This case highlights the crucial role of legal frameworks in fostering equitable socio-economic development.

Agroforestry Practices in Punjab: Integrating Forestry and Agriculture

Punjab's agroforestry initiatives demonstrate the potential of integrating forestry with agriculture to achieve socio-economic benefits. Farmers in the state have increasingly adopted poplar and eucalyptus plantations on their farmland, driven by schemes such as the Sub-Mission on Agroforestry under the National Agroforestry Policy (2014).

The benefits of these practices are multifaceted. Farmers earn additional income from selling timber and pulpwood, which is in high demand in the plywood and paper industries. Simultaneously, agroforestry enhances soil fertility and provides shade for crops such as wheat and mustard, thereby improving agricultural productivity. Success stories from districts like Hoshiarpur underscore how agroforestry contributes to rural livelihoods, promotes sustainable land use, and meets industrial timber demand.

Tiger Reserves and Ecotourism in Madhya Pradesh: A Conservation-Driven Economy

Tiger reserves such as Bandhavgarh and Kanha in Madhya Pradesh exemplify the role of forests in fostering a conservation-driven economy. These reserves attract thousands of domestic and international tourists each year, generating substantial revenue through park fees and associated hospitality services (Singh, 2021).

Local communities have directly benefited from ecotourism initiatives. Tribal members are employed as guides, drivers, and naturalists, while women's SHGs manage eco-lodges and handicraft enterprises. Revenue-sharing mechanisms under the Wildlife Protection Act (1972) ensure that a portion of tourism income is reinvested in community development projects, such as schools, healthcare, and infrastructure. These reserves demonstrate how wildlife conservation can be effectively integrated with socio-economic development to create a mutually beneficial scenario (Kumar & Singh, 2020).

The environmental and social conditions resulting from climate change and the adaptation of forests to it in the coming years will have a significant impact on forestry practices worldwide. According to the IPCC (2007), the warming of the climate and the increase in greenhouse gases due to anthropogenic factors will also lead to changes in day and night temperatures, seasonal temperatures, drought frequency, heatwaves, floods, storms, sea levels, and other related phenomena. This warming has already caused many changes in the forest; hence, the need for sustainable forest management is even greater (Bernier & Schoene, 2009). Suitable forest management steps are suggested to be taken to maintain the forest as well as increase the resource production for socio-economic development-

- Educating the people so that their livelihood options can diversify and they have alternative sources of income beyond forest-dependent activities (Zikargae et al., 2022). Their livelihood needs to be diversified into alternative income-generating activities to reduce their overdependence on forests (Bisui & Shit, 2024).
- Establishment of the market channel so that the communities get a fair price for the products which not only helps to boost their income but also promotes sustainable forest development (Belcher, 2005)

- Incorporating technological advancements, data-gathering technologies, digital networks, and automation will help increase the efficiency of forest managers and improve decision-making related to forest genetics, silvicultural practices, and harvesting. Techniques related to processing and value addition to forest products will also help enhance their economic durability and the first dependent livelihood (Fardusi et al., 2017; Singh et al., 2022).
- The initiative should focus on skill development, capacity building, and empowerment of marginalized communities. This will help reduce socio-economic disparities among people and promote inclusive growth (Thorbecke & Charumilind, 2002).
- Various conservation measures, such as reforestation and community-led initiatives, must be taken to ensure the ecological balance and long-term availability of forests and their resources (Bisui & Shit, 2024).
- Improving collaboration between local communities, government, non-governmental organizations, and other stakeholders. This will help better address the challenges faced by people and the forest and implement solutions for forest management and sustainable livelihood development (Edwards et al., 2019).
- Increasing awareness among local communities about the value of forest resources and the importance of sustainable practices. Empowering them to gain knowledge and develop decision-making capability to actively participate in conservation efforts and secure their livelihoods (Chou, 2019).
- Policy frameworks must be developed to focus on sustainable livelihood practices and effective forest management. The government must focus on the requirements of communities that depend on forests, facilitate access to resources, and support sustainable practices (Saxena, 2016).

11.5 Conclusion

Forests cover nearly 30% of the Earth's land area and play a crucial role in the planet's biosphere by regulating many ecological processes directly and indirectly. They are home to Earth's terrestrial biodiversity and play a key role in regulating the climate. Forests provide numerous direct and indirect benefits. Directly, they serve as critical sources of resources, including timber, fuelwood, and non-timber forest products, which support livelihoods, generate employment, and contribute to food security and healthcare. Forests also hold immense cultural and recreational value, fostering eco-tourism and spiritual practices that enrich economies and societies. Indirectly, forests play a pivotal role in regulating the global climate, conserving soil and water, and supporting biodiversity, all essential for maintaining ecological balance and human survival. Although less visible, these services are foundational to sustaining agriculture, ensuring water quality, and mitigating the impacts of climate change. Additionally, the cultural, aesthetic, and health-related benefits of forests enhance mental and physical well-being, contributing to a holistic socio-economic development framework. Forests are crucial for socio-economic development, especially for nearby communities. Communities rely on forests for various products, whether for consumption or as a source of income. Forests maintain the climate, sequester carbon, and support biodiversity, underscoring their vital role in sustaining both people and the Earth. Today, developing effective policies and strategies to improve and protect forests is essential. We need an action-oriented approach that enables people to rely on forests without causing further degradation. Furthermore, sustainable forest management practices can benefit communities by providing a continuous flow of goods and services from forests and contributing to the mitigation of climate change.

References

Barwa, S., & Singh, H. (2024). Climate-induced changes in forest ecosystems of indian himalayan region: A meta-analysis. In *urban forests, climate change and environmental pollution: Physio-biochemical and molecular perspectives to enhance urban resilience* (pp. 839–866). Springer.

Belcher, B. M. (2005). Forest product markets, forests, and poverty reduction. *International Forestry Review, 7*(2), 82–89.

Bernier, P. & Schoene, D. (2009). Adapting forests and their management to climate change: an overview, 5(11).

Bisui, S., & Shit, P. K. (2024). Assessing the role of forest resources in improving rural livelihoods in West Bengal of India. *Regional Sustainability, 5*(2), Article 100141. https://doi.org/10.1016/j.regsus.2024.100141

Bonan, G. B. (2008a). Forests and climate change: Forcings, feedbacks, and the climate benefits of forests. *Science, 320*(5882), 1444–1449.

Bonan, G. B. (2008b). Forests and climate change: Forcings, feedbacks and the climate benefits of forests. *Science, 320*, 1444.

Calder, I. R. (2007). *Forests and water—ensuring forest benefits offset forestation costs*. IWMI Research Report.

Chazdon, R. L. (2008). Beyond deforestation: Restoring forests and ecosystem services on degraded lands. *Science, 320*(5882), 1458–1460.

Chou, T. C. (2019). *Design for empowerment: Bolstering knowledge and skills to aid sustainable awareness and actions. PhD Dissertation*. Carnegie Mellon University.

Edwards, P., Sharma-Wallace, L., Barnard, T., et al. (2019). Sustainable livelihoods approaches to inform government-local partnerships and decision- making in vulnerable environments. *New Zealand Geographer, 75*(2), 63–73.

Ellison, D., Futter, M. N., & Bishop, K. (2017). Trees, forests, and water: Cool insights for a hot world. *Global Environmental Change, 43*, 51–61.

FAO. (2020b). *Global forest resources assessment 2020*. Food and Agriculture Organization of the United Nations.

FAO. (2024). *The State of the World's Forests 2024–Forest-sector innovations towards a more sustainable future*. FAO.

FAO. (2020). *Global forest resources assessment 2020*. Food and Agriculture Organization of the United Nations.

FAO. (2020). *Global forest resources assessment 2020: Main report*. Food and Agriculture Organization of the United Nations.

Fardusi, M. J., Rahman, M. H. & Roy, B. (2011). Improving livelihood status through collection and management of forest resources: An experience from Sylhet Forest Division, Bangladesh. *International Journal of Forest Usufructs Management, 12*(2), 59–76.

Fardusi, M. J., Chianucci, F., & Barbati, A. (2017). Concept to practice of geospatial-information tools to assist forest management and planning under precision forestry framework: A review. *Annals of Silvicultural Research, 41*(1), 3–14.

Forest Survey of India. (2023). *India state of forest report*. (Ministry of Environment Forest and Climate Change) Kaulagarh road, P.O. IPE Dehradun–248195, Uttarakhand. India. ISBN 978–81–950073–9–4.

Garg, A., Pal, D., Singh, H., & Pandey, D. C. (2016). A comparative study of NDBI, NDISI and NDII for extraction of urban impervious surface of Dehradun [Uttarakhand, India] using Landsat 8 imagery. *International Conference on Emerging Trends in Communication Technologies (ETCT), 2016*, 1–5.

Hyde, W. F., Kohlin, G., & Amacher, G. S. (2000). Social forestry reconsidered. *Silva Fennica, 34*(3), 285–314.

Intergovernmental Panel on Climate Change (IPCC). 2007. Climate change 2007: synthesis report. IPCC Fourth Assessment Report. Geneva, Switzerland.

Jagger, P., Luckert, M. M. K., Duchelle, A. E., et al. (2014). Tenure and forest income: Observations from a global study on forests and poverty. *World Development, 64*, 43–55.

Jangra, A., Kumar, K., Maikhuri, S., Bhandari, M. S., Pandey, S., Singh, H., & Barthwal, S. (2024). Unveiling stress-adapted endophytic bacteria: Characterizing plant growth-promoting traits and assessing cross-inoculation effects on *Populus deltoides* under abiotic stress. *Plant Physiology and Biochemistry, 210*, Article 108610.

Joshi, R., Singh, H., Chhetri, R., & Yadav, K. (2020). Assessment of carbon sequestration potential in degraded and non-degraded community forests in Terai region of Nepal. *Journal of Forest and Environmental Science, 36*(2), 113–121.

Joshi, R., Singh, H., Chhetri, R., Poudel, S. R., & Rijal, S. (2021). Carbon sequestration potential of community forests: A comparative analysis of soil organic carbon stock in community managed forests of Far-Western Nepal. *Eurasian Journal of Soil Science, 10*(2), 96–104.

Joshi, R., Sharma, B., Singh, H., Dhakal, N., Ayer, S., & Maraseni, T. (2024). Poplar plantation as an agroforestry approach: Economic benefits and its role in carbon sequestration in North India. *Journal of Resources and Ecology, 15*(4), 880–888.

Kamanga, P., Vedeld, P., & Sjaastad, E., (2009). Forest incomes and rural livelihoods in Chiradzulu District, Malawi. *Ecological Economics, 68*(3), 613–624.

Kumar, M., & Singh, H. (2020). Agroforestry as a nature-based solution for reducing community dependence on forests to safeguard forests in rainfed areas of India. In *Nature-based solutions for resilient ecosystems and societies* (pp. 289–306). Springer.

Kumar, A., Dwivedi, G. K., Tewari, S., Sah, V. K., Singh, H., Kumar, P., Kumar, N., & Kaushal, R. (2020). Soil organic carbon pools under *Terminalia chebula* Retz. based agroforestry system in Himalayan foothills, India. *Current Science, 118*(7), 1098–1103.

Landsberg, J., & Waring, R. (2014). Forest types around the world. *Forests in Our Changing World, 21–46.*https://doi.org/10.5822/978-1-61091-497-0_2

Lax, J., & Köthke, M. (2017a). Livelihood strategies and forest product utilisation of rural households in Nepal. *Small-Scale Forestry, 16*(4), 505–520.

Le Gear, K., et al. (2023). Deep roots: Realising the public health benefits of exposure to forest environments. *Advances in Integrative Medicine, 10*, 86–88.

Mahonya, S., Shackleton, C. M. & Schreckenberg, K. (2019). Non-timber forest product use and market chains along a deforestation gradient in southwest Malawi. *Frontiers in Forests and Global Change*https://doi.org/10.3389/ffgc.2019.00071

Mann, C., Hernandez-Morcillo, M., Ikei, H., & Miyazaki, Y. (2024). The socio-economic dimension of forest therapy: A contribution to human well-being and sustainable forest management. *Tree, Forest and People, 18*, Article 100731. https://doi.org/10.1016/j.tfp.2024.100731

Matias, D. M. S., Tambo, J. A., Stellmacher, T., et al., (2018). Commercializing traditional non-timber forest products: An integrated value chain analysis of honey from giant honey bees in Palawan, Philippines. *Forest Policy Eco*nomics, *97*, 223–231

Mendako, R. K., Tian, G., Ullah, S., et al. (2022). Assessing the economic contribution of forest use to rural livelihoods in the Rubi-Tele Hunting Domain, DR Congo. *Forests, 13*(1), 130. https://doi.org/10.3390/f13010130

Miyazaki, Y. (2018). *Shinrin-Yoku: The Japanese way of forest bathing for health and relaxation.* Octopus Publishing Group.

Ndumbe, L. N., Ingram, V., Tchamba, M., et al. (2019). From trees to money: The contribution of Njansang (Ricinodendron heudelotii) products to value chain stakeholders' financial assets in the South West Region of Cameroon. *Forests, Trees and Livelihoods., 28*(1), 52–67.

Nerfa, L., Rhemtulla, J. M., & Zerriffi, H. (2020). Forest dependence is more than forest income: Development of a new index of forest product collection and livelihood resources. *World Development, 125*, Article 104689. https://doi.org/10.1016/j.worlddev.2019.104689

Obeng, E., Owusu-Sekyere, E., Dumenu, W., Nutakor, E., Samar, S., & Opoku, E. (2011). *Climate change and forest dependent livelihoods; Vulnerabilities and impacts.* Technical Research Report. Forestry Research Institute of Ghana-Forest Livelihood and Sustainable Governance Division.

Pan, Y., Birdsey, R. A., Fang, J., Houghton, R., Kauppi, P. E., Kurz, W. A., & Hayes, D. (2011). A large and persistent carbon sink in the world's forests. *Science, 333*(6045), 988–993.

Ravindranath, N. H., Joshi, N. V., Sukumar, R., & Saxena, A. (2006). Impact of climate change on forests in India. *Current Science, 90*(3), 354–361.

Ritchie, H. (2021). "Forest area" Published online at OurWorldinData.org. 'https://ourworldindata.org/forest-area' [Online Resource].

Ruhil, T., Singh, H., & Barthwal, S. (2024). Long-term physiological monitoring for detecting climate change impacts on forest ecosystems. In *Forests and climate change: Biological perspectives on impact, adaptation, and mitigation strategies* (pp. 817–834). Springer.

Saxena, A., Guneralp, B., Bailis, R., et al. (2016). Evaluating the resilience of forest dependent communities in Central India by combining the sustainable livelihoods framework and the cross scale resilience analysis. *Current Science, 110*(7), 1195–1207.

Schmitt C. B., Belokurov A., Besançon C., Boisrobert L., Burgess N.D.,Campbell A., Coad L., Fish L., Gliddon D., Humphries K., Kapos V., Loucks C., Lysenko I., Miles L., Mills C., Minnemeyer S., Pistorius T., Ravilious C., Steininger M. & Winkel G. (2009). *Global ecological forest classification and forest protected area gap analysis.* Analyses and recommendations in view of the 10% target for forest protection under the Convention on Biological Diversity (CBD). 2nd revised edition. Freiburg University Press, Freiburg, Germany

Singh, H. (2024a). Climate change and forests: Biological perspectives. In *Forests and climate change: biological perspectives on impact, adaptation, and mitigation strategies* (pp. 1–17). Springer.

Singh, H. (2024b). Exploring adaptive modulation in plant functional traits and its impact on the productivity of *Acacia auriculiformis* under CO_2-enriched environment. *Industrial Crops and Products, 210*, Article 118186.

Singh, H. (2024c). *Forests and climate change: Biological perspectives on impact, adaptation, and mitigation strategies.* Springer Nature.

Singh, H. (2024d). *Urban forests, climate change and environmental pollution: Physio-biochemical and molecular perspectives to enhance urban resilience.* Springer Nature.

Singh, H. (2025). The role of plant functional traits in mitigating urban climate and environmental pollution. In *Plant Functional Traits* (pp. 297–308). Elsevier.

Singh, H., & Kumar, M. (2022). Climate change and its impact on Indian Himalayan forests: current status and research needs. In *Climate change: Impacts, responses and sustainability in the Indian Himalaya* (pp. 223–242). Springer.

Singh, H., Verma, A., & Kumar, M. (2018). *Can Elevated CO_2 alter the growth dynamics, morphology, physiology and biochemistry of the forestry tree species?* (pp. 288–305). Climate Change and Agriculture. Discovery Publishing House.

Singh, H., Kumar, N., Singh, R., & Kumar, M. (2023a). Assessing the climate change impact on the habitat suitability of the range-restricted bird species (*Catreus wallichii*) in the Indian Himalayan ecosystem. *Environmental Science and Pollution Research, 30*(57), 121224–121235.

Singh, M., Kumar, P., Singh, H., Kumar, A. K. A., Kumar, R., & Kumar, R. (2023b). *Neolamarckia cadamba*: A comprehensive review on its physiological, ecological, phytochemical and pharmacological perspectives. *Ecology Environment and Conservation, 29*, 241–250.

Singh, R., Gehlot, A., Akram, S.V., et al. (2022). Forest 4.0: Digitalization of forest using the Internet of Things (IoT). J. King Saud Univ. *Computer Information Science, 34*(8), 5587–5601.

Singh, H., Kumar, N., & Kumar, A. (2024). Enhancing resource use efficiency in crops through plant functional traits. In *Plant functional traits for improving productivity* (pp. 97–117). Springer.

de Sousa, F. F., Vieira-da-Silva, C.& Barros, F. B. (2018). The (in)visible market of miriti (Mauritia flexuosa L.f.) fruits, the "winter acai", in Amazonian riverine communities of Abaetetuba, Northern Brazil. *Global Ecology and Conservation, 14*, e00393. https://doi.org/10.1016/j.gecco.2018.e00393.

Sunderland, T. C., et al. (2013). Forests, food and nutrition: A review of the evidence. *Forests and Food Security Journal, 1*(1), 23–35.

Thorbecke, E., & Charumilind, C. (2002). Economic inequality and its socio-economic impact. *World Development, 30*(9), 1477–1495.

UNEP. (2019). *The economics of ecosystems and biodiversity for forests.*

UNWTO. (2021). Tourism and forest ecosystems. United Nations World Tourism Organization.

Vaughan, A. (2020). World braces for economic impact. *New Scientist (1971), 245*(3272), 10. https://doi.org/10.1016/S0262-4079(20)30477-2

Vedeld, P., Angelsen, A., Bojö, J., Sjaastad, E., & Berg, G. K. (2007). Forest environmental incomes and the rural poor. *Forest Policy and Economics, 9*(7), 869–879.

Vose, J. M., D. L. Peterson, G. M. Domke, C. J. Fettig, L. A. Joyce, R. E. Keane, C. H. Luce, J. P. Prestemon, L. E. Band, J. S. Clark, N. E. Cooley, A. D'Amato, & J. E. Halofsky. (2018): Forests. In D. R. Reidmiller, C. W. Avery, D. R. Easterling, K. E. Kunkel, K. L. M. Lewis, T. K. Maycock, & B. C. Stewart (Eds.), *Impacts, risks, and adaptation in the United States: Fourth national climate assessment*, Volume II. U.S. Global Change Research Program, Washington, DC, USA, pp. 232–267. https://doi.org/10.7930/NCA4.2018.CH6.

WHO. (2013). *Traditional medicine strategy 2014–2023.* World Health Organization.

WHO. (2019). *Traditional medicine strategy 2014–2023.* World Health Organization.

WWF. (2021). *Living Planet Report 2021.* World Wide Fund for Nature.

Xue, D., & Tisdell, C. (2001). Valuing ecological functions of biodiversity in Changbaishan Mountain Biosphere Reserve in Northeast China. *Biodiversity and Conservation.* 10.https://doi.org/10.1023/A:1016630825913

Zikargae, M. H., Woldearegay, A. G., & Skjerdal, T. (2022). Empowering rural society through non-formal environmental education: An empirical study of environment and forest development community projects in Ethiopia. *Heliyon, 8*(3), Article e09127. https://doi.org/10.1016/j.heliyon.2022.e09127

Chapter 12
Mitigating the Impacts of Climate Change Through Forestry

Garima Mishra, Kanishka Singh, Sachin Shah, Manish Singh, and Hukum Singh

Abstract Forestry plays a pivotal role in mitigating the adverse impacts of climate change by acting as a natural carbon sink, regulating the hydrological cycle, preserving biodiversity, and enhancing both ecosystem and landscape resilience. This chapter examines the multifaceted role of forests in climate stabilization through carbon sequestration, temperature regulation, biodiversity conservation, and the promotion of ecosystem resilience. It highlights key forestry-based approaches such as afforestation, reforestation, and agroforestry and assesses their effectiveness in contributing to climate change mitigation and adaptation strategies. Evidence indicates that sustainable forest management significantly enhances the long-term carbon storage potential of forests. The chapter also examines various national and international initiatives that utilize forestry-based interventions to achieve climate objectives. Despite their considerable potential, the implementation of such interventions is often hindered by land-use conflicts, limited financial resources, institutional challenges, and sociopolitical constraints. To address these barriers, we advocate for integrated, science-based, and landscape-level forestry approaches, supported by enabling policies and active community participation. Ultimately, forestry stands out as a cost-effective, scalable, and nature-based solution for advancing global climate action and achieving the Sustainable Development Goals.

G. Mishra
Genetics and Tree Improvement Division, ICFRE-Forest Research Institute, Dehradun, Uttarakhand, India

K. Singh
IT & GIS Division, ICFRE-Forest Research Institute, Dehradun, Uttarakhand, India

S. Shah
Silviculture and Forest Management Division, ICFRE-Eco Rehabilitation Centre, Prayagraj, Uttar Pradesh, India

M. Singh (✉)
Centre for Environmental Assessment and Climate Change, G.B. Pant National Institute of Himalayan Environment, Kosi-Katarmal, Almora, Uttarakhand, India
e-mail: manish.s92@gmail.com

H. Singh
Plant Physiology Discipline, Forest Research Institute, Dehradun, Uttarakhand, India

H. Singh (ed.), *Forestry for a Carbon-Neutral and Sustainable Future*, Sustainability Solutions, https://doi.org/10.1007/978-3-032-06427-1_12

Keywords Urban forests · Forest-based approach · Urban heat island · Climate-smart forests · Sustainable forest management · Sustainable development goals

12.1 Introduction

Climate change is a major environmental and socio-economic challenge of the modern era, with extensive and far-reaching consequences for ecosystems, economies, and human well-being. The Intergovernmental Panel on Climate Change (IPCC) has warned that global temperatures could rise by more than 1.5 °C above pre-industrial levels by the end of this century. The rising temperature will have significant consequences, including alterations in rainfall patterns, increased frequency of extreme weather events, loss of biodiversity, desertification, rising sea levels, and more frequent wildfires (IPCC, 2023). All of these changes threaten the stability of both natural ecosystems and human societies (Steffen et al., 2018).

Forests, comprising almost 31% of the Earth's terrestrial surface, are one of the most important ecosystems on the planet, playing a crucial role in climate regulation, particularly in the context of climate change (FAO, 2020). Forests are known to sequester carbon effectively. They absorb up to 7.6 million metric tons of CO_2 every year (Pan et al., 2011). Trees absorb and store atmospheric CO_2 through photosynthesis, within their biomass and the soil for many years. However, deforestation, forest degradation, and land-use change have diminished the sequestration capacity of the forests, resulting in the release of even greater emissions, in addition to the critical loss of ecosystem services (Curtis et al., 2018). It is suggested that global deforestation accounts for approximately 10% of annual anthropogenic CO_2 emissions, with tropical deforestation likely contributing even more (Baccini et al., 2017).

Despite these challenges, forests remain a scalable natural key to climate change mitigation. Sustainable forest management, reforestation, afforestation, and conservation efforts have the potential to enhance carbon sequestration, aid in biodiversity conservation, regulate the water cycle, and promote climate resilience (Griscom et al., 2017). Afforestation and reforestation programs, in particular, have gained traction as essential strategies in global efforts to achieve net-zero emissions. Chazdon and Guariguata (2016) suggested that improved forest management practices, such as selective logging, reduced-impact logging, and agroforestry, may significantly enhance the carbon sequestration capacity of existing forests while supporting sustainable livelihoods for local communities (Singh et al., 2023b). The global climate crisis similarly demands nature-based solutions, with forestry playing a central role in carbon sequestration, biodiversity conservation, and climate resilience (Barwa & Singh, 2024).

Moreover, forests influence regional and global climates beyond their role in carbon sequestration. They play a crucial role in regulating temperatures through evapotranspiration, which helps cool the atmosphere, and they contribute to the formation of precipitation patterns that sustain agriculture and freshwater supplies (Bonan, 2008).

Forest ecosystems also act as buffers against climate-induced extreme events, such as hurricanes, landslides, and heatwaves, by stabilizing soils and moderating local climates (Locatelli et al., 2020). Given these multiple benefits, forests have increasingly been integrated into international climate policies, including the Paris Agreement and the United Nations' REDD + (Reducing Emissions from Deforestation and Forest Degradation) framework, which seeks to incentivize forest conservation as a means to combat climate change (Angelsen et al., 2018).

However, while forestry-based climate solutions are promising, their effectiveness depends on addressing significant challenges such as illegal logging, land tenure conflicts, and the socio-economic drivers of deforestation. Many tropical and subtropical regions, which hold the highest potential for forest-based carbon sequestration, face governance issues, making it challenging to enforce conservation policies and sustainable management practices (Nepstad et al., 2014). Forests may not be able to function as reliable carbon sinks if they lack adaptive management practices (Allen et al., 2010).

Figure 12.1 emphasises the global effects of climate change and highlights the specific role of forestry as a natural climate solution. This chapter examines the role of forests in mitigating climate change, identifies the barriers and limitations to forest-based mitigation, and, importantly, outlines the policies and practices required to enhance forest-based mitigation. It posits that forests can effectively reduce carbon emissions, increase ecosystem resilience, and improve sustainability for nature and humanity when the conservation, restoration, and sustainable management of forests are a core component of global climate policies (Singh, 2024a).

12.2 Urban Forestry: Enhancing Climate Resilience and Urban Sustainability

Urban forestry serves as a vital link between city planning and environmental protection. It offers several benefits for cities that are experiencing increased problems due to climate change and urbanization. Urban forests, comprising trees, shrubs, and green spaces within cities, play a crucial role in the ecosystem, contributing to climate resilience, public health, and biodiversity conservation (Noss, 1990; Van den Bosch, 2017).

Fig. 12.1 An integrated approach to the Mitigation of climate change through forestry

12.2.1 *Urban Forests and the Urban Heat Island (UHI) Effect*

One of the most apparent effects of urbanisation is the urban heat island (UHI) effect. The UHI effect occurs when the built-up environment is substantially warmer than the surrounding rural areas (Singh et al., 2024). This situation arises from the absorptive thermal properties of anthropogenic materials, such as asphalt and concrete, and typically lacks vegetation in the urban environment. Urban forests significantly reduce UHI effects by creating shade and allowing for average evapotranspiration, which cools the surrounding air. Urban forests have the potential to decrease temperatures by 2–5 °C, depending on factors such as tree density, canopy cover, and species composition of urban green spaces (Singh et al., 2024).

12.2.1.1 Case Study: New York City's Urban Heat Island Mitigation Strategy

A prime example of urban forestry's role in reducing the UHI effect is New York City's Urban Heat Island Mitigation Strategy, initiated in the early 2000s. New York, like many major cities, has experienced elevated temperatures due to its dense urban development, which is exacerbated by climate change. To counter this, the city launched a comprehensive plan that included increasing tree canopy cover and implementing green roofs. According to Schroeder et al. (2020), the city's "MillionTreesNYC" initiative, which aimed to plant one million trees by 2030, has been central in cooling the urban environment, reducing the UHI effect, and improving public health by lowering heat-related illnesses. The initiative has also been linked to reduced energy consumption, as tree canopy coverage lowers the demand for air conditioning in both residential and commercial buildings (Singh & Kumar, 2022).

12.2.2 *Air Quality and Carbon Sequestration*

Urban forests enhance local air quality by filtering airborne pollutants, including particulate matter (PM2.5), nitrogen dioxide (NO_2), and sulfur dioxide (SO_2). Trees perform natural air filtration, capturing pollutants from the air and transporting them to their leaves and bark. Additionally, urban forests play a crucial role in carbon storage. While it will not be equal to the carbon stored in large forests, urban trees sequester significant quantities of carbon dioxide (CO_2) and support climate mitigation strategies. For example, it is estimated that trees in urban environments in the U.S. store approximately 708 million tons of carbon (Livesley et al., 2016).

12.2.2.1 Case Study: Toronto's Green Streets Program

In Toronto, Canada, the Green Streets program focuses on planting trees along major streets and highways to improve air quality and reduce the UHI effect. This program is part of a broader effort to enhance Toronto's environmental sustainability and resilience to climate change. Trees planted under this program help absorb pollutants from vehicle emissions, thereby improving the quality of life. McPherson et al. (2013) note that these urban trees are particularly effective in capturing carbon and reducing the city's carbon footprint. In addition to improving air quality, these trees provide valuable ecosystem services, such as stormwater management, which reduces the burden on the city's drainage system during heavy rainfall.

12.2.3 Water Management and Flood Mitigation

Urban trees play a crucial role in water management by intercepting rainfall, reducing stormwater runoff, and enhancing water infiltration into the soil. This function is vital in cities prone to flooding due to the presence of impervious surfaces. Green roofs, urban wetlands, and permeable pavements, when integrated with urban forestry practices, enhance water retention and filtration. Singapore's "City in a Garden" initiative exemplifies this approach by combining green infrastructure with water-sensitive urban design (Borelli et al., 2023).

12.2.4 Green Corridors and Rooftop Afforestation for Urban Sustainability

Green corridors, networks of interconnected green spaces, and rooftop afforestation (also known as green roofs) are innovative strategies for enhancing urban sustainability. Green corridors enhance biodiversity and provide ecological connectivity by linking fragmented habitats. They also help mitigate climate change by providing cooling effects and promoting carbon sequestration. Rooftop afforestation, also known as green roofs, offers additional green space in densely built urban environments, thereby reducing energy consumption and enhancing environmental quality (Gill et al., 2007; Panagakos, 2016; Ruhil et al., 2024).

12.2.4.1 Case Study: The High Line in New York City

The High Line in New York City is a notable example of an urban green corridor that repurposes a derelict elevated railway line into a public green space, enhancing biodiversity and promoting urban sustainability. The High Line was developed as part of a plan to connect the many parks and green spaces in the Chelsea neighbourhood to link them into a contiguous green corridor. Tzoulas et al. (2007) have documented the High Line's role as an important habitat for urban wildlife, including birds, insects, and small mammals, while also mitigating some of the urban heat island effect. The High Line has offered educational opportunities for the public to learn about the importance of urban greening in mitigating climate change (Joshi et al., 2024).

12.2.4.2 Case Study: Singapore's Rooftop Gardens

Singapore is renowned for its extensive use of rooftop gardens, which are integral to the city-state's strategy to increase green cover and promote sustainability. The government has incentivized the construction of green roofs on residential and commercial buildings, leading to a significant increase in green space per capita. According to Getter and Rowe (2006), these rooftop gardens provide numerous benefits, including cooling effects, improved air quality, and stormwater management. They also contribute to carbon sequestration, as the city's green roofs absorb and store significant amounts of CO_2. The rooftop garden project in Singapore demonstrates how green roofs can enhance city resilience to climate change and improve living conditions in densely populated areas (Singh et al., 2023a).

12.2.5 Innovative Approaches to Urban Forestry

Rooftop afforestation, vertical gardens, and the establishment of green corridors are transformative strategies for urban forestry. Vertical gardens not only increase green cover but also improve building insulation and aesthetics. Green corridors connect fragmented urban green spaces, enabling wildlife movement and promoting ecological balance within cities. Community-driven urban greening projects, such as the "MillionTreesNYC" campaign, highlight the importance of public participation in enhancing urban forestry (Garg et al., 2016; Kumar & Singh, 2020).

12.2.6 Challenges and Opportunities

Despite its benefits, urban forestry faces challenges such as limited space, high maintenance costs, and vulnerability to urban stressors like pollution and soil compaction. However, technological advancements such as remote sensing for tree health monitoring and predictive analytics for urban forestry planning offer new opportunities to overcome these barriers. Collaborative efforts among municipal authorities, private stakeholders, and local communities are crucial to maximizing the potential of urban forestry (Singh et al., 2018).

12.3 Forests and Biodiversity: An Interconnected Approach

Forests represent some of the most biologically diverse ecosystems globally, comprising over 80% of terrestrial biodiversity, which includes plants, animals, fungi, and microorganisms (Thompson et al., 2009). Forest biodiversity is crucial for sustaining ecosystem stability and productivity, as well as for the ability of terrestrial ecosystems to respond and adapt to environmental changes.

12.3.1 The Importance of Biodiversity in Forest Ecosystems

Biodiversity ensures ecosystem resilience by fostering complex interactions among species. Diverse forests are more resilient to pests, diseases, and climatic fluctuations because they support a wide range of functional traits. For instance, mixed-species forests often exhibit higher productivity and carbon sequestration rates compared to monocultures (Pan et al., 2011).

Biodiversity supports critical ecosystem services, including pollination, seed dispersal, and nutrient cycling, are essential for maintaining healthy ecosystems. The disappearance of essential species can disrupt these processes, leading to cascading ecological effects. For example, the decline of pollinators in forested regions threatens the reproduction of many tree species, ultimately reducing forest regeneration capacity (Laurance et al., 2014).

12.3.1.1 Case Study: Madagascar's Rainforests and Endemic Species Protection

Madagascar, an island known for its unique biodiversity, is home to numerous endemic species that are highly vulnerable to habitat loss, deforestation, and climate change. In response to these threats, conservation efforts have focused on protecting Madagascar's rainforests, which support a diverse array of threatened species. Gardner et al. (2013) highlight the importance of preserving biodiversity hotspots in Madagascar, such as the rainforests of the eastern slopes, to maintain resilient ecosystems. The island's distinct plant and animal life are especially vulnerable to variations in temperature and rainfall, which makes the protection of these forests critical for maintaining global biodiversity (Jangra et al., 2024).

12.3.2 Climate Adaptation Through Forest Biodiversity

Diverse forests are better equipped to adapt to changing climatic conditions. Genetic diversity within tree species enhances their ability to survive under shifting temperature and precipitation patterns. For example, forests in the boreal region are undergoing significant changes in species composition due to rising temperatures, underscoring the importance of maintaining genetic variability (Singh, 2024c).

12.3.3 Protection of Endemic and Threatened Species

Forests are critical habitats for endemic and threatened species. Biodiversity hotspots such as the Amazon rainforest, the Congo Basin, and the Western Ghats are reservoirs of unique genetic resources. In these areas, conservation initiatives prioritize community involvement, habitat preservation, and anti-poaching measures. Protected areas, including national parks and wildlife sanctuaries, play a crucial role in conserving biodiversity. However, integrating conservation goals with sustainable development is vital to ensure long-term success.

12.3.4 Community-Based Conservation Models

Community-led conservation initiatives have emerged as effective strategies for preserving forest biodiversity. Nepal's community forestry programs empower local communities to contribute to the sustainable management of forest resources. The community forestry programs have resulted in a net gain in forest cover and an enhancement of biodiversity, while providing livelihoods for local communities (Acharya, 2002).

12.4 Innovations and Technologies in Forestry

Technological advancements are transforming forestry practices, enabling more effective monitoring, conservation, and management of forest resources. Innovations in remote sensing, artificial intelligence (AI), and biotechnology are at the forefront of this transformation.

12.4.1 Remote Sensing and Geographic Information Systems (GIS)

Satellite imagery and GIS have revolutionized forest monitoring by providing high-resolution data on forest cover, biodiversity, deforestation, and land-use changes (Turner et al., 2015). Platforms like Global Forest Watch provide policymakers and conservationists with real-time data, enabling timely interventions against illegal logging and deforestation (Hansen et al., 2013).

Drones equipped with multispectral and thermal sensors are being increasingly used for forest assessments. They provide precise data on tree health, canopy density, and wildlife populations, thereby reducing the need for labor-intensive fieldwork. For instance, drones have been employed in the Amazon rainforest to monitor deforestation hotspots and map areas for reforestation (Chazdon et al., 2016).

12.4.1.1 Case Study: The Amazon Rainforest Monitoring Project

In the Amazon, the use of satellite imagery has played a crucial role in monitoring deforestation and forest degradation. The Amazon Rainforest Monitoring Project utilizes satellite imagery and GIS to track illegal logging activities and land-use changes in the region. This initiative has provided policymakers and conservation organizations with valuable data, enabling them to respond quickly to illicit activities and develop more effective forest management strategies. Liu et al. (2025) highlight the success of this approach in identifying and preventing deforestation hotspots in the Amazon.

12.4.2 Artificial Intelligence and Machine Learning

Artificial intelligence (AI) and machine learning techniques (MLT) examine extensive datasets to forecast forest expansion trends, wildfire hazards, and pest infestations. These tools are particularly valuable in managing climate-related risks. For example, AI-based systems have been used to detect early signs of forest degradation by analyzing satellite imagery and climatic data (Liu et al., 2025). AI-powered platforms also facilitate decision-making in sustainable forest management. Predictive models optimize timber harvesting schedules and identify areas suitable for conservation or afforestation. These technologies enhance efficiency while minimizing environmental impact.

12.4.2.1 Case Study: AI for Forest Monitoring in California

In California, AI is being utilized to track forest health and identify indications of pest infestations, diseases, and wildfires at the earliest stages. The California Department of Forestry and Fire Protection (CAL FIRE) has collaborated with leading technology companies to develop AI-based applications that utilize satellite images and field surveys to identify high-risk areas in forested regions. Over the last decade, AI-based applications have played a crucial role in helping forest managers identify areas to protect and address, as well as in reducing the likelihood of catastrophic wildfires. Keeley and Syphard (2021) state that CAL FIRE has improved its decision-making processes and wildfire prevention activities.

12.4.3 Biotechnology in Forestry

Biotechnology offers promising solutions for enhancing forest productivity and resilience. Genetically modified (GM) trees with traits such as faster growth, pest resistance, and enhanced wood quality are being developed to meet the increasing demand for timber and bioenergy. Additionally, biotechnology is being used to conserve endangered tree species through tissue culture and cloning techniques (Place et al., 2016).

Biotechnology also contributes to efforts aimed at carbon sequestration. Researchers are investigating methods to increase the carbon storage capacity of trees by modifying their genetic composition. However, the ecological risks and ethical implications of deploying GM trees necessitate stringent regulatory frameworks (Singh, 2024d).

12.4.3.1 Case Study: Genetically Modified Trees for Drought Resistance

In Australia, biotechnology has been used to develop genetically modified (GM) trees that are more resistant to drought and extreme heat. Zhao et al. (2019) report that GM eucalyptus trees have been shown to survive better in dry conditions, making them a potential solution for reforestation in arid regions. They demonstrated the potential for biotechnology to enhance forest resilience to climate change by improving the growth and survival of trees in challenging environments (Singh, 2025).

12.5 Agroforestry as a Climate Solution

Agroforestry, defined as the deliberate integration of trees and/or shrubs with cropland, offers a multifaceted approach to mitigating and adapting to climate change, providing an attractive alternative to traditional monoculture agriculture. Agroforestry systems, as an integrated land-use system, can not only increase carbon sequestration but also provide resilience to agricultural landscapes and communities (Araya et al., 2023). Agroforestry systems enhance sustainability and climate resilience in agriculture by integrating trees with crops and/or livestock, resulting in synergistic processes that optimize resource use, nutritional status, biological health, and biodiversity (Nair, 2007). Agroforestry systems sequester carbon because they have a higher carbon storage potential than pasturelands or monoculture croplands (Nair et al., 2010). The carbon storage potential of agroforestry systems are due to both above-ground biomass, such as the tree trunk and leaf area, and below-ground biomass (extensive root systems and higher soil organic matter) (Araújo et al., 2011).

Agroforestry serves as a mitigation tool for nutrient retention, soil health, air quality, insect management, and fire control, while also providing cultural services that enhance recreational, aesthetic, and cultural values (Udawatta et al., 2019). Moreover, by producing timber, fruits, nuts, and other tree-related items as additional sources of income diversification, farmers have minimized their sensitivity to market fluctuations and potential adverse impacts of climate change on crops (Lasco et al., 2013).

The adoption of agroforestry practices holds particular significance for Mediterranean drylands, regions grappling with escalating land degradation and water scarcity due to climate change (Pulighe et al., 2019). Agroforestry practices also provide a means to address land degradation and desertification by enhancing soil structure, increasing water infiltration, and reducing soil erosion (Udawatta et al., 2017). Incorporating drought-resistant tree species into agricultural landscapes in agroforestry systems can enhance water availability for crops, reduce soil erosion, and create microclimates that buffer against extreme temperatures. Furthermore, diversifying income sources through tree-based products provides economic stability for farmers, reducing their vulnerability to market fluctuations and climate-related crop failures. Improving water management in agriculture through agroforestry is crucial for achieving higher crop yields, crop diversification, and multiple cropping seasons, ultimately enhancing food availability and stability for communities (Sarwar, 2025). Water reservoirs enhance the efficiency of water use in agriculture, enabling smarter usage and achieving better and more efficient water utilization (Sarwar, 2025).

Agroecological practices, which integrate traditional knowledge with modern agricultural techniques, are essential for sustainable land management (Ranjithkumar, 2025). Agroforestry enhances soil organic carbon, improves nutrient availability, and boosts soil fertility due to the presence of trees, which also enhances soil microbial dynamics, positively influencing soil health (Pralhad et al., 2020). This method enhances resource efficiency, reduces dependence on external inputs, and promotes biodiversity, thereby supporting climate change mitigation and adaptation while improving livelihoods and food security (Singh, 2024b). Furthermore, the shift towards organic farming practices presents a key solution to the challenges associated with diminishing agricultural land, food insecurity, and the environmental impacts of conventional agriculture (Ranjithkumar, 2025). These methods enhance soil fertility, reduce erosion, and promote the sustainable use of natural resources, thereby protecting the environment and preserving long-term agricultural productivity (Vikas & Ranjan, 2024). Implementing strategies such as crop rotation, cover cropping, and integrated nutrient management improves the quality and health of the soil.

Integrating crop and animal production through agroforestry makes farms more efficient and profitable. Incorporating livestock into agroforestry systems enables farmers to enhance nutrient cycling, reduce fertilizer inputs, and diversify their income streams. Additionally, integrating trees and crops in space and time enhances food and nutrition security, reduces environmental damage, and provides a sustainable alternative to single-crop production (Nair, 2007). Agroforestry enhances food security, mitigates climate change, and reduces environmental impact by establishing diversified and resilient farming systems (Vikas & Ranjan, 2024). Moreover, agroforestry techniques that combine indigenous and scientific knowledge can improve ecosystem functioning and provide sustainable solutions.

To realize the full potential of agroforestry as a climate solution, enabling policies and institutional frameworks are essential. Governmental support, incentives, and regulations can encourage the implementation of agroforestry practices, thereby promoting their widespread adoption across various agricultural landscapes (Pantera et al., 2021). Additionally, favorable agricultural policies play a crucial role in promoting climate change resilience and enhancing food security by reducing production costs and increasing farmers' motivation (Lee et al., 2024). The integration of specialized funds, experts, and agricultural technologies, facilitated by well-designed policies, enhances agricultural production, particularly in regions prone to droughts (Lee et al., 2024). Eliminating obstacles to agroforestry implementation at both national and local levels, including ambiguous land and tree ownership, unfavourable regulations, and insufficient intersectoral coordination, is essential for promoting sustainable development.

12.6 Challenges in Using Forestry for Climate Change Mitigation

Climate consistently interacts with natural ecosystems and human enterprises, disrupting work, food, health, the economy, and overall well-being. Since the Industrial Revolution, the excessive use of fossil fuels and substantial land use changes have caused disrupted weather patterns, whose future implications frequently overwhelm any mitigation potential from the already compromised capacity of ecosystems. There has been a significant increase in discussions around creating strategies and actions to mitigate and adapt to climate change, with increased attention from politicians and scientists, particularly in techniques involving nature-based solutions (Psistsaki et al., 2024).

12.6.1 Land Use Conflicts and Competing Resources

While land is widely used, it is heavily contested on a global scale, often leading to conflicts over its utilization (Froese & Schilling, 2019). The connection among climate change, resource availability, and conflict is intricate and indirect, shaped by various elements including political and social dynamics (Scheffran et al., 2012). The needs and requirements vary among users, some view it as an ecosystem for agricultural production and other primary activities. In contrast, others see it as a space for construction, human settlements, or the exploitation of natural resources for economic purposes (Joshi et al., 2021). These differing goals, priorities, and interests among users often result in conflicts. Many of these conflicts arise from the improper and irrational use of land, which frequently contributes to various forms of land degradation (Angelkovic & Stojanovska, 2024).

Changes in ecosystems and their functions are primarily driven by land-use conversions and alterations caused by human activities. This prompts researchers to explore effective strategies for managing land resources, with a focus on region-specific policies. The exploitation of ecologically fragile land through human activities often leads to land-use conflicts, creating a clash between two competing objectives- Sustainable conservation and Land utilization (Pambudi, 2023). The significant impacts on rural communities in the Global South are primarily driven by large-scale land acquisitions and land-based climate change mitigation policies, which can either support or hinder their livelihoods and may lead to conflicts. Two key climate change strategies—namely, biofuel production and REDD +—can influence land conflicts (Hunsberger et al., 2017). Identifying land use conflicts serves as the foundation for preventing, mediating, and resolving them, which remains the central focus of land use conflict research (Izakovicova et al., 2018).

12.6.2 Climate Risks Like Wildfires, Pests, and Diseases Affecting Forests

Examining how climate affects different disturbances, their repercussions on forests, and the potential escalation due to climate change provides a basis for investigating approaches to mitigate the impacts of climate change, as outlined in Table 12.1.

Table 12.1 Table summarizing various climate-driven risks affecting forests

Climate risk	Key characteristics	Impact on forests	Additional notes	References
Fire	• Frequency, size, intensity, and seasonality • Shaped by weather, climate, forest structure, and composition	• Accelerates nutrient cycling • Causes tree mortality • Alter the successional pathway • Induces seed Germination • Depletes the soil seed bank • Increases landscape heterogeneity • Volatilizes soil nutrients	• Highly responsive to climate change • Impacts can overshadow the direct effects of climate change on species distribution and migration	Dale et al. (2001); Stocks et al. (1998)
Drought	• Influenced by soil properties, topographic exposure, species composition, life stage, frequency, duration, and severity	• Reduces Net Primary Production (NPP) and water use • Slows the decomposition process • Causes organic matter to build up on the forest floor • May increase fire frequency and intensity or disrupt nutrient cycling	• Effects depend on seasonal and annual climatic variations • Resilience depends on the adequacy of current drought adaptations	Dale et al. (2001); Hanson and Weltzin (2000)
Winds	• Increased frequency of extreme events, such as strong winds, winter storms, and droughts • Region-specific effects on commercial forestry	• Massive losses to commercial forestry • Reduced access to forestland • Increased costs for road and facility maintenance • Direct damage to trees (e.g., branch breaking, trunk breakage, crown loss, stand destruction) • Challenges in logging due to wetter winters and early thaws	– Warmer climates can lead to a faster buildup of growing stocks, increasing susceptibility to high wind damage	Kirilenko and Sedjo (2007)

(continued)

Table 12.1 (continued)

Climate risk	Key characteristics	Impact on forests	Additional notes	References
Insects	• Affected by changes in temperatures and precipitation • Influences the survival, reproduction, and distribution of herbivores and pathogens	• Shifts insect and pathogen boundaries • Modifies the physiology and defense mechanisms of trees • Poses risks to tree species in new habitats (e.g., mountain pine beetles threaten white-bark pines)	– Tree species currently intolerant to insect outbreaks may face increased threats under even moderate warming (eg, 20 °C rise)	Logan and Powell (2001); Dale et al. (2001)

12.6.3 Balancing Economic Development and Forest Preservation

Forests are widely recognized as essential to national economies, providing a diverse array of resources and services, including inputs for production, environmental benefits, food, fuel, medicines, household items, building materials, and raw materials for industrial use. Different stakeholders perceive these various values and services in distinct ways. There is a growing recognition of the economic value of forests in development strategies, which is becoming reflected in national policies and programs that affect the stocks, quality, and distribution of forests (Kumar et al., 2020). They are acknowledged not only as productive capital assets but also as key components of public infrastructure. Recent advancements in national accounting have enabled the inclusion of forest resources as productive capital stocks, facilitating the assessment of how variations in these resources can influence a nation's overall productive capacity (Cooper & Macfarlane, 2023).

Understanding the relationship between economic growth and environmental well-being is crucial for developing effective and sustainable strategies and policies. Insights from modern literature can inform policymakers and practitioners in developing approaches that strike a balance between economic development and forest conservation. Policies that encourage environmentally responsible practices not only enhance ecological well-being but also drive innovation and create new employment opportunities, thereby contributing to economic efficiency. Adopting a holistic view that acknowledges the interconnection between economic growth and environmental conservation is crucial for developing integrated and sustainable strategies. Collaboration between policymakers and stakeholders is vital for achieving economic progress that aligns with ecological stewardship, thereby fostering sustainable development (Raman et al., 2024).

12.7 The Future of Forestry in Climate Change Mitigation

Climate change has been acknowledged as a worldwide political concern for roughly fifty years, during which there have been some gradual mitigation effects. Key milestones in this process include the establishment of the UNFCCC at the 1992 United Nations Conference on Environment and Development (UNCED), also known as the Earth Summit; the signing of the Kyoto Protocol in 1997; and the establishment of the Paris Agreement in 2015 (Psistaki et al., 2024). Forest-based climate mitigation has emerged as a key component of the Paris Agreement, possibly accounting for as much as a quarter of the expected net emission cuts (Moreau et al., 2022).

12.7.1 The Potential of Reforestation to Meet Carbon Neutrality Goals

Forests act as vital carbon sinks by absorbing carbon dioxide via photosynthesis and retaining it within their biomass and soils. Expanding and restoring forested areas through afforestation and reforestation efforts can strengthen this function, helping to counterbalance global CO_2 emissions (Harris et al., 2021). Reforestation is a crucial approach for capturing atmospheric CO_2, playing a key role in achieving carbon neutrality by strengthening the capacity of forests as carbon sinks (Raman et al., 2024). Reforestation helps mitigate climate change by absorbing greenhouse gases and aligns with Sustainable Development Goal 13 (Climate Action). Research emphasizes the importance of developing region-specific models to optimize carbon sequestration in different climatic contexts (Raman et al., 2024).

Large-scale reforestation initiatives aim to rejuvenate degraded areas, enhance soil quality, and promote water conservation, thereby indirectly contributing to carbon sequestration and the provision of ecosystem services (Zhang et al., 2023). Sophisticated tools, such as Artificial Intelligence, remote sensing, and carbon capture technologies, including BECCS (Bioenergy with Carbon Capture and Storage), can enhance forest management and increase the effectiveness of carbon removal (Psistaki et al., 2024). Policies that support reforestation should align with the international framework, such as the Paris Agreement. It is recommended that multinational partnerships collaborate to combat deforestation and expand forest coverage (Raman et al., 2024).

12.7.2 Roles of Forestry in Achieving Sustainable Development Goals (SDGs)

Forests are highly effective at sequestering carbon, and products derived from forests typically offer more sustainable alternatives to other materials. Properly managed forests and their resources have immense potential to support countries in achieving the SDGs (Ma et al., 2022). Forestry has considerable appeal in sustainable development, as the concept of sustainable development, which originated in the eighteenth century, was originally closely associated with forestry. Currently, this sector remains crucial in driving the transition to a sustainable society (Sala et al., 2000).

The SDGs have attracted considerable academic attention across diverse fields, including sustainability science and development economics, as well as areas such as international relations and global governance. The SDGs can collectively be seen as integral to a broader conversation regarding a global socio-ecological framework (Safarli, 2024). In response to increasing challenges and the need for global advancement, the United Nations introduced the 2030 SDGs and 169 targets, serving as a blueprint for sustainable global progress. The SDGs address every aspect of society, environment, and economy with a focus on safeguarding, restoring, and enhancing ecosystems while promoting human well-being (De Jong et al., 2018).

Forest ecosystems, as a fundamental component of terrestrial ecosystems, play a mandatory role in achieving the Sustainable Development Goals (SDGs). The sustainable management of forests and the enhancement of their ecosystem services are closely linked to the objectives of the SDGs. For example, SDG 15 (Life on Land) is closely associated with forestry, emphasizing goals such as safeguarding, restoring, and ensuring the sustainable use of terrestrial ecosystems, promoting sustainable forest management, combating desertification, reversing land degradation, and conserving biodiversity (Ma et al., 2022). The essential function of forests in fulfilling the Sustainable Development Goals (SDGs) is well recognized. Consequently, it is crucial to delve deeper into these relationships and enhance our understanding of how forests and forest-related activities can be leveraged to promote the SDGs (De Jong et al., 2018).

12.7.3 Vision for Integrating Forestry with Other Climate Solutions

Climate change, also known as global warming, poses a significant challenge globally, affecting both our society and the forests on which we rely. Human activities, such as altering land use and burning fossil fuels, have contributed to climate change, influencing the contemporary perception of forests. Traditionally, forests offered resources and services to society; however, they are now increasingly recognized for their potential to help mitigate and adapt to climate change (Joshi et al., 2020). Hence, forests play a crucial role in "Natural Climate Solutions" (NCS) or "Nature-Based Solutions" (NBS), serving as strategies for tackling climate change by minimizing emissions from deforestation and/or capturing and storing carbon dioxide via photosynthesis and sequestration (Griscom et al., 2017). In an NCS scenario, we incentivize landowners to protect or restore forests by integrating forests into a carbon marketplace (e.g., credits). This allows forest-based carbon sequestration to reduce global emissions while mobilizing additional funding for further climate action (Cooper et al., 2023). Agroforestry is another NCS and NBS wherein trees, crops, and/or livestock systems integrate to produce food and reduce the impact of climate change. Agroforestry offers NCS and NBS benefits through preserving and enhancing food security with existing land, increasing carbon sequestration, improving biodiversity, and preventing soil erosion. Therefore, Agroforestry, as both an NCS and an NBS, is a solution to climate change challenges.

The integration of forests into urban planning, through the development of green spaces, parks, and green roofs, can help combat climate change by absorbing carbon and reducing the heat island effect. This approach also enhances resilience to extreme weather events such as floods and heatwaves (Akbari et al., 2001; Gill et al., 2007). Preserving forest biodiversity is essential for building climate resilience. Forests rich in biodiversity are better equipped to adapt to climate change, capture carbon more efficiently, and deliver vital ecosystem services. Incorporating biodiversity conservation into forestry practices promotes both climate stability and ecological well-being (Cooper et al., 2023).

12.8 Conclusion

Forests are indispensable natural assets in the global response to climate change, offering scalable, cost-effective, and ecosystem-based solutions for both mitigation and adaptation. Their capacity to sequester atmospheric carbon through biomass accumulation and soil retention, regulate microclimates and hydrological cycles, support biodiversity, and enhance ecosystem resilience positions them as essential components of climate stabilization strategies. The science-based implementation of approaches such as afforestation, reforestation, agroforestry, and sustainable forest management can substantially enhance long-term climate mitigation potential. These interventions not only increase carbon sinks but also deliver co-benefits, including improved soil quality, enhanced water retention, and diversified rural livelihoods. However, unlocking the full potential of forest-based solutions demands overcoming persistent challenges, including land-use conflicts, inadequate financial mechanisms, weak institutional capacity, and sociopolitical constraints that limit equitable participation. Addressing these issues requires a paradigm shift toward integrated, landscape-level forestry frameworks that are ecologically informed and socially inclusive. Policy coherence at both national and international levels is critical. Effective alignment with climate agreements such as the Paris Accord, complemented by robust governance structures and community engagement, will help ensure the durability and scale of forestry contributions to climate targets. As global efforts intensify to curb temperature rise and build resilience to climate extremes, forests must be elevated as a central pillar of climate policy. Strengthened forestry efforts not only drive climate action but also contribute directly to achieving several of the United Nations' Sustainable Development Goals (SDGs), including poverty alleviation (SDG 1), access to clean water (SDG 6), climate action (SDG 13), and conservation of terrestrial ecosystems (SDG 15). The way forward lies in recognizing forests not merely as ecological service providers but as strategic natural infrastructure essential for a climate-resilient future.

References

Acharya, K. P. (2002). Twenty-four years of community forestry in Nepal. *International Forestry Review, 4*(2), 149–156.

Adopted, I. P. C. C. (2014). Climate change 2014 synthesis report. *IPCC: Geneva, Szwitzerland*, 1059–1072.

Akbari, H., Pomerantz, M., & Taha, H. (2001). Cool surfaces and shade trees to reduce energy use and improve air quality in urban areas. *Solar Energy, 70*(3), 295–310.

Allen, C. D., Macalady, A. K., Chenchouni, H., Bachelet, D., McDowell, N., Vennetier, M., & Cobb, N. (2010). A global overview of drought and heat-induced tree mortality reveals emerging climate change risks for forests. *Forest Ecology and Management, 259*(4), 660–684.

Angelkovic, D. T., & Stojanovska, M. (2024). Conflicts in land use and management. *SilvaWorld, 3*(2), 52–56. https://doi.org/10.61326/silvaworld.v3i2.278

Angelsen, A., Martius, C., De Sy, V., Duchelle, A. E., Larson, A. M., & Pham, T. T. (2018). *Transforming REDD+: Lessons and new directions.* CIFOR.

Araújo, M. S., Bolnick, D. I., & Layman, C. A. (2011). The ecological causes of individual specialisation. *Ecology letters, 14*(9), 948–958.

Araujo, A. S. F., Leite, L. F. C., Iwata, B. D. F., Lira, M. D. A., Xavier, G. R., & Figueiredo, M. D. V. B. (2012). Microbiological process in agroforestry systems. A Review. *Agronomy for Sustainable Development, 32*, 215–226.

Araya, Y. N., Emmott, A., Rawes, W., & Zuza, E. J. (2023). Promoting climate-smart sustainable agroforestry to tackle social and environmental challenges: The case of macadamia agroforestry in Malawi. *Journal of Agriculture and Food Research, 14*, Article 100846.

Baccini, A., Walker, W., Carvalho, L., Farina, M., Houghton, R. A., & Schelhas, J. (2017). Tropical forests are a net carbon source based on aboveground measurements of gain and loss. *Science, 358*(6360), 230–234.

Barwa, S., & Singh, H. (2024). Climate-induced changes in forest ecosystems Of Indian Himalayan region: a meta-analysis. In: *Urban forests, climate change and environmental pollution: Physio-biochemical and molecular perspectives to enhance urban resilience* (pp. 839–866). Springer.

Bonan, G. B. (2008). Forests and climate change: Forcings, feedbacks, and the climate benefits of forests. *Science, 320*(5882), 1444–1449.

Borelli, S., Conigliaro, M., & Di Cagno, F. (2023). *Urban forests: a global perspective.* Food & Agriculture Org.

Chazdon, R. L., & Guariguata, M. R. (2016). Natural regeneration as a tool for large-scale forest restoration in the tropics: Prospects and challenges. *Biotropica, 48*(6), 716–730.

Cooper, L., & MacFarlane, D. (2023). Climate-smart forestry: Promise and risks for forests, society, and climate. *PLOS Climate, 2*(6), Article e0000212.

Curtis, P. G., Slay, C. M., Harris, N. L., Tyukavina, A., & Hansen, M. C. (2018). Classifying drivers of global forest loss. *Science, 361*(6407), 1108–1111.

Dale, V. H., Joyce, L. A., McNulty, S., Neilson, R. P., Ayres, M. P., & Flannigan, M. D. (2001). Hanson, PJ, Irland, LC, Lugo, AE, Peterson, CJ, Simberloff, D., Swanson. *Bioscience, 51*(9), 723–734.

De Jong, W., Pokorny, B., Katila, P., Galloway, G., & Pacheco, P. (2018). Community forestry and the sustainable development goals: A two way street. *Forests, 9*(6), 331.

Elmqvist, T., et al. (2003). The role of biodiversity in maintaining forest resilience. *Forest Ecology and Management, 179*(2–3), 201–216.

FAO. (2020). *The state of the world's forests 2020: Forests, biodiversity, and people.* Food and Agriculture Organization of the United Nations.

Freer-Smith, P. H., Muys, B., Bozzano, M., Drössler, L., Farrelly, N., Jactel, H., & Orazio, C. (2019). *Plantation forests in Europe: Challenges and opportunities* (Vol. 9, pp. 1–52). Joensuu, Finland: European Forest Institute.

Froese, R., & Schilling, J. (2019). The nexus of climate change, land use, and conflicts. *Current Climate Change Reports, 5*, 24–35.

Gardner, C. J., Nicoll, M. E., Mbohoahy, T., Oleson, K. L., Ratsifandrihamanana, A. N., Ratsirarson, J., & Davies, Z. G. (2013). Protected areas for conservation and poverty alleviation: Experiences from Madagascar. *Journal of Applied Ecology, 50*(6), 1289–1294.

Garg, A., Pal, D., Singh, H., & Pandey, D. C. (2016). A comparative study of NDBI, NDISI and NDII for extraction of urban impervious surface of Dehradun [Uttarakhand, India] using Landsat

8 imagery. *International Conference on Emerging Trends in Communication Technologies (ETCT), 2016*, 1–5.

Getter, K. L., & Rowe, D. B. (2006). The role of extensive green roofs in sustainable development. *HortScience, 41*(5), 1276–1285.

Gill, S. E., Handley, J. F., Ennos, A. R., & Pauleit, S. (2007). Adapting cities for climate change: The role of the green infrastructure. *Built Environment, 33*(1), 115–133.

Griscom, B. W., Adams, J., Ellis, P. W., Houghton, R. A., Lomax, G., Miteva, D. A., & Fargione, J. (2017). Natural climate solutions. *Proceedings of the National Academy of Sciences, 114*(44), 11645–11650.

Hansen, M. C., Potapov, P. V., Moore, R., Hancher, M., Turubanova, S. A., Tyukavina, A.,... & Townshend, J. R. (2013). High-resolution global maps of 21st-century forest cover change. *Science, 342*(6160), 850–853.

Hanson, P. J., & Weltzin, J. F. (2000). Drought disturbance from climate change: response of United States forests. In *The science of the total environment* (Vol. 262).

Harris, N. L., Gibbs, D. A., Baccini, A., Birdsey, R. A., De Bruin, S., Farina, M., & Tyukavina, A. (2021). Global maps of twenty-first century forest carbon fluxes. *Nature Climate Change, 11*(3), 234–240.

Hunsberger, C., Corbera, E., Borras, S. M., Jr., Franco, J. C., Woods, K., Work, C., & Vaddhanaphuti, C. (2017). Climate change mitigation, land grabbing and conflict: Towards a landscape-based and collaborative action research agenda. *Canadian Journal of Development Studies/ revue Canadienne D'études Du Développement, 38*(3), 305–324.

IPCC. (2023). Climate change 2023: synthesis report. *Intergovernmental panel on climate change*. https://www.ipcc.ch

Izakovičová, Z., Miklós, L., & Miklósová, V. (2018). Integrative assessment of land use conflicts. *Sustainability (Switzerland), 10*(9). https://doi.org/10.3390/su10093270

Jangra, A., Kumar, K., Maikhuri, S., Bhandari, M. S., Pandey, S., Singh, H., & Barthwal, S. (2024). Unveiling stress-adapted endophytic bacteria: Characterizing plant growth-promoting traits and assessing cross-inoculation effects on *Populus deltoides* under abiotic stress. *Plant Physiology and Biochemistry, 210*, Article 108610.

Joshi, R., Singh, H., Chhetri, R., & Yadav, K. (2020). Assessment of carbon sequestration potential in degraded and non-degraded community forests in Terai region of Nepal. *Journal of Forest and Environmental Science, 36*(2), 113–121.

Joshi, R., Singh, H., Chhetri, R., Poudel, S. R., & Rijal, S. (2021). Carbon sequestration potential of community forests: A comparative analysis of soil organic carbon stock in community managed forests of Far-Western Nepal. *Eurasian Journal of Soil Science, 10*(2), 96–104.

Joshi, R., Sharma, B., Singh, H., Dhakal, N., Ayer, S., & Maraseni, T. (2024). Poplar plantation as an agroforestry approach: Economic benefits and its role in carbon sequestration in North India. *Journal of Resources and Ecology, 15*(4), 880–888.

Keeley, J. E., & Syphard, A. D. (2021). Large California wildfires: 2020 fires in historical context. *Fire Ecology, 17*, 1–11.

Kirilenko, A. P., & Sedjo, R. A. (2007). Climate change impacts on forestry. *Proceedings of the National Academy of Sciences, 104*(50), 19697–19702.

Kumar, M., & Singh, H. (2020). Agroforestry as a nature-based solution for reducing community dependence on forests to safeguard forests in rainfed areas of India. In *Nature-based solutions for resilient ecosystems and societies* (pp. 289–306). Springer.

Kumar, A., Dwivedi, G. K., Tewari, S., Sah, V. K., Singh, H., Kumar, P., Kumar, N., & Kaushal, R. (2020). Soil organic carbon pools under *Terminalia chebula* Retz. based agroforestry system in Himalayan foothills, India. *Current Science, 118*(7), 1098–1103.

Lasco, R. D., Veridiano, R. K. A., Habito, M., & Pulhin, F. B. (2013). Reducing emissions from deforestation and forest degradation plus (REDD+) in the Philippines: will it make a difference in financing forest development?. *Mitigation and Adaptation Strategies for Global Change, 18*(8), 1109–1124.

Lasco, R. D., Delfino, R. J. P., Catacutan, D. C., Simelton, E. S., & Wilson, D. M. (2014). Climate risk adaptation by smallholder farmers: The roles of trees and agroforestry. *Current Opinion in Environmental Sustainability, 6*, 83–88.

Laurance, W. F., Andrade, A. S., Magrach, A., Camargo, J. L., Valsko, J. J., Campbell, M., & Laurance, S. G. (2014). Long-term changes in liana abundance and forest dynamics in undisturbed Amazonian forests. *Ecology, 95*(6), 1604–1611.

Lee, C. C., Zeng, M., & Luo, K. (2024). How does climate change affect food security? Evidence from China. *Environmental Impact Assessment Review, 104*, Article 107324.

Liu, H., Shu, L., Liu, X., Cheng, P., Wang, M., & Huang, Y. (2025). Advancements in artificial intelligence applications for forest fire prediction. *Forests, 16*(4), 704.

Livesley, S. J., McPherson, E. G., & Calfapietra, C. (2016). The urban forest and ecosystem services: Impacts on urban water, heat, and pollution cycles at the tree, street, and city scale. *Journal of Environmental Quality, 45*(1), 119–124.

Locatelli, B., Pramova, E., Di Gregorio, M., Brockhaus, M., Chávez, D. A., Tubbeh, R., & Perla, J. (2020). Climate change policy networks: Connecting adaptation and mitigation in multiplex networks in Peru. *Climate Policy, 20*(3), 354–372.

Logan, J. A., & Powell, J. A. (2001). Ghost forests, global warming, and the mountain pine beetle (Coleoptera: Scolytidae). *American Entomologist, 47*(3), 160–173.

Ma, Z., Hu, C., Huang, J., Li, T., & Lei, J. (2022). Forests and forestry in support of sustainable development goals (SDGs): A bibliometric analysis. *Forests, 13*(11), 1960.

McPherson, E. G., Xiao, Q., & Aguaron, E. (2013). A new approach to quantify and map carbon stored, sequestered and emissions avoided by urban forests. *Landscape and Urban Planning, 120*, 70–84.

Moreau, L., Thiffault, E., Cyr, D., Boulanger, Y., & Beauregard, R. (2022). How can the forest sector mitigate climate change in a changing climate? Case studies of boreal and northern temperate forests in eastern Canada. *Forest Ecosystems, 9*, Article 100026.

Nair, P. R. (2007). The coming of age of agroforestry. *Journal of the Science of Food and Agriculture, 87*(9), 1613–1619.

Nair, P. R., Nair, V. D., Kumar, B. M., & Showalter, J. M. (2010). Carbon sequestration in agroforestry systems. *Advances in Agronomy, 108*, 237–307.

Nepstad, D., McGrath, D., Stickler, C., Alencar, A., Azevedo, A., Swette, B., & Hess, L. (2014). Slowing Amazon deforestation through public policy and interventions in beef and soy supply chains. *Science, 344*(6188), 1118–1123.

Noss, R. F. (1990). Indicators for monitoring biodiversity: A hierarchical approach. *Conservation Biology, 4*(4), 355–364.

Nowak, D. J., & Crane, D. E. (2002). Carbon storage and sequestration by urban trees in the USA. *Environmental Pollution, 116*(3), 381–389.

Pambudi, A. S. (2023). Environmental land use conflicts and ecosystem services: a paper review. *Indonesian Journal of Applied Environmental Studies, 4*(2), 58–63. https://doi.org/10.33751/injast.v4i2.7851

Pan, Y., Birdsey, R. A., Fang, J., Houghton, R., Kauppi, P. E., Kurz, W. A., & Hayes, D. (2011). A large and persistent carbon sink in the world's forests. *Science, 333*(6045), 988–993.

Panagakos, G. (2016). Green corridors basics. *Green transportation logistics: The quest for win-win solutions* (pp. 81–121). Springer International Publishing.

Pantera, A, Mosquera-Losada, M. R., Herzog, F., & Den Herder, M. (2021). Agroforestry and the environment. *Agroforestry Systems, 95*(5), 767–774.

Place, F., Garrity, D., Mohan, S., & Agostini, P. (2016). *Tree-based production Systems for Africa's Drylands*. World Bank Publications.

Pralhad, B. S., Rajendran, P., Divya, M. P., Rajeswari, R., Thangamani, G., & Ramaha, C. (2020). Assessing the effect of different agroforestry practices on soil physico-chemical properties and microbial activity. *International Journal of Current Microbiology and Applied Sciences, 9*(12), 2802–2816.

Psistaki, K., Tsantopoulos, G., & Paschalidou, A. K. (2024). An overview of the role of forests in climate change mitigation. *Sustainability, 16*(14), 6089.

Pulighe, G., Bonati, G., Colangeli, M., Traverso, L., Lupia, F., Altobelli, F., & Napoli, M. (2019). Predicting streamflow and nutrient loadings in a semi-arid Mediterranean watershed with ephemeral streams using the SWAT model. *Agronomy, 10*(1), 2.

Raman, R., Manalil, S., Dénes, D. L., & Nedungadi, P. (2024). The role of forestry sciences in combating climate change and advancing sustainable development goals. *Frontiers in Forests and Global Change, 7*, 1409667.

Ranjithkumar, A. (2025). Assessing the impact of agricultural land reduction on future global food security: Challenges and sustainable solutions. *Indian Journal of Natural Sciences, 15*, 89066–89083.

Ruhil, T., Singh, H., & Barthwal, S. (2024). Long-term physiological monitoring for detecting climate change impacts on forest ecosystems. In *Forests and climate change: Biological perspectives on impact, adaptation, and mitigation strategies* (pp. 817–834). Springer.

Safarli, A. J. (2024). The economics of sustainable development: balancing growth and environmental conservation. *ELMİ ƏSƏRLƏR*, 63.

Sala, O. E., Stuart Chapin, F. I. I. I., Armesto, J. J., Berlow, E., Bloomfield, J., Dirzo, R., & Wall, D. H. (2000). Global biodiversity scenarios for the year 2100. *Science, 287*(5459), 1770–1774.

Sarwar, S. (2025). Consequences of land utilization, agriculture and water to handle the food security issues. *Land Degradation & Development.*

Scheffran, J., Brzoska, M., Kominek, J., Link, P. M., & Schilling, J. (2012). Disentangling the climate-conflict nexus: Empirical and theoretical assessment of vulnerabilities and pathways. *Rev. Eur. Stud., 4*, 1.

Schroeder, H. W., et al. (2020). Urban heat island mitigation strategies: Insights from the 2019 heatwave. *Urban Climate, 32*, Article 100602.

Singh, H., Verma, A., & Kumar, M. (2018). *Can elevated CO2 alter the growth dynamics, morphology, physiology and biochemistry of the forestry tree species?* (pp. 288–305). Climate Change and Agriculture. Discovery Publishing House.

Singh, H., Kumar, N., Singh, R., & Kumar, M. (2023a). Assessing the climate change impact on the habitat suitability of the range-restricted bird species (*Catreus wallichii*) in the Indian Himalayan ecosystem. *Environmental Science and Pollution Research, 30*(57), 121224–121235.

Singh, M., Kumar, P., Singh, H., Kumar, A. K. A., Kumar, R., & Kumar, R. (2023b). Neolamarckia cadamba: a comprehensive review on its physiological, ecological, phytochemical and pharmacological perspectives. Ecol Environ Conserv, 29, 241–250.

Singh, H., & Kumar, M. (2022). Climate change and its impact on Indian Himalayan forests: Current status and research needs. In *Climate change: Impacts, responses and sustainability in the Indian Himalaya* (pp. 223–242). Springer.

Singh, H., Kumar, N., & Kumar, A. (2024). Enhancing resource use efficiency in crops through plant functional traits. In *Plant functional traits for improving productivity* (pp. 97–117). Springer.

Singh, H. (2024a). Climate change and forests: biological perspectives. In *Forests and climate change: Biological perspectives on impact, adaptation, and mitigation strategies* (pp. 1–17). Springer.

Singh, H. (2024b). Exploring adaptive modulation in plant functional traits and its impact on the productivity of *Acacia auriculiformis* under CO_2-enriched environment. *Industrial Crops and Products, 210*, Article 118186.

Singh, H. (2024c). *Forests and climate change: Biological perspectives on impact, adaptation, and mitigation strategies.* Springer Nature.

Singh, H. (2024d). *Urban forests, climate change and environmental pollution: Physio-biochemical and molecular perspectives to enhance urban resilience.* Springer Nature.

Singh, H. (2025). The role of plant functional traits in mitigating urban climate and environmental pollution. In *Plant functional traits* (pp. 297–308). Elsevier.

Steffen, W., Rockström, J., Richardson, K., Lenton, T. M., Folke, C., Liverman, D., & Schellnhuber, H. J. (2018). Trajectories of the earth system in the anthropocene. *Proceedings of the National Academy of Sciences, 115*(33), 8252–8259.

Stocks, B. J., Fosberg, M. A., Lynham, T. J., Mearns, L., Wotton, B. M., Yang, Q., Jin, J. Z., Lawrence, K., Hartley, G. R., Mason, J. A., & McKenney, D. W. (1998). Climate change and forest fire potential in Russian and Canadian boreal forests. *Climatic Change, 38*(1), 1–13.

Thompson, I., Mackey, B., McNulty, S., & Mosseler, A. (2009). Forest resilience, biodiversity, and climate change. In *Secretariat of the convention on biological diversity, Montreal. Technical Series no. 43. 1–67.* (Vol. 43, pp. 1–67).

Tzoulas, K., Korpela, K., Venn, S., Yli-Pelkonen, V., Kaźmierczak, A., Niemela, J., & James, P. (2007). Promoting ecosystem and human health in urban areas using Green Infrastructure: A literature review. *Landscape and Urban Planning, 81*(3), 167–178.

Udawatta, R. P., Rankoth, L. M., & Jose, S. (2019). *Agroforestry and biodiversity. Sustainability, 11*(10), 2879. https://doi.org/10.3390/su11102879

Udawatta, R. P., Gantzer, C. J., & Jose, S. (2017). Agroforestry practices and soil ecosystem services. In *Soil health and intensification of agroecosystems* (pp. 305–333). Academic press.

Van den Bosch, M., & Sang, Å. O. (2017). Urban natural environments as nature-based solutions for improved public health–A systematic review of reviews. *Environmental Research, 158*, 373–384.

Vikas, & Ranjan, R. (2024). Agroecological approaches to sustainable development. *Frontiers in Sustainable Food Systems, 8*, 1405409.

Zhang, X., Busch, J., Huang, Y., Fleskens, L., Qin, H., & Qiao, Z. (2023). Cost of mitigating climate change through reforestation in China. *Frontiers in Forests and Global Change, 6*.https://doi.org/10.3389/ffgc.2023.1229216

Zhao, J., et al. (2019). Genetic engineering in tree resilience: Progress and prospects. *Tree Physiology, 39*(8), 1225–1238.

Chapter 13
Carbon Neutrality and Environmental Sustainability: Barriers and Future Directions

Navneet Shukla, Chandan Das, and Ragini Saini

Abstract The impact of climate change is receiving increasing global attention. Forests play a crucial role in achieving carbon neutrality by sequestering atmospheric carbon dioxide and storing it in biomass and soil. Consequently, afforestation, reforestation, and natural regeneration have become essential components of global climate change mitigation strategies. This chapter highlights the vital role of forests as global carbon sinks, contributing significantly to environmental sustainability. It also identifies key barriers that may hinder the achievement of carbon neutrality through forests, including ecological, technical, institutional, and socio-economic challenges. To address these obstacles, the chapter explores future directions that emphasize innovative forest management practices, technological advancements, and participatory approaches that foster both international and local collaboration. Overall, this chapter advocates for a multidisciplinary approach to achieving carbon neutrality through forestry, paving the way for a more resilient and sustainable future.

Keywords Carbon neutrality · Environmental sustainability · Carbon sequestration · Afforestation · Climate change mitigation

13.1 Introduction

Earth's carbon has been emitted from various sources, including volcanic eruptions in its early history, as well as recent contributions such as fossil fuel burning and land-use changes (Jacobson, 2002). As human beings evolved, they utilized more environmental resources to lead comfortable lives and protect themselves from various environmental stresses (Ellis et al., 2009). Among major greenhouse gas

N. Shukla (✉) · R. Saini
Institute of Environment and Sustainable Development, Banaras Hindu University, Uttar Pradesh, Varanasi, India
e-mail: nvneet4@gmail.com

C. Das
School of Environmental Sciences, Jawaharlal Nehru University, New Delhi, India

H. Singh (ed.), *Forestry for a Carbon-Neutral and Sustainable Future*, Sustainability Solutions, https://doi.org/10.1007/978-3-032-06427-1_13

(GHG) emission sectors, the Agriculture, Forestry, and Other Land Use (AFOLU) accounts for approximately 22% of global GHG emissions, with nearly half originating from the Land Use, Land-Use Change, and Forestry (LULUCF) sub-sector, particularly due to deforestation (Calvin et al., 2023). In the context of increasing carbon dioxide (CO_2) emissions, the ability of land and oceans to act as carbon sinks is becoming less effective due to the accelerating emission rates (Singh et al., 2018) A carbon budget imbalance occurs when emissions exceed the absorption capacity of natural sinks, jeopardizing efforts to achieve carbon neutrality. Every budget has some input and output attributes. The global carbon budget is compiled from data on carbon emissions and reductions from various sources and sinks. In the global carbon budget, sources include fossil fuel emissions resulting from combustion and carbonation processes, as well as CO_2 emissions from land-use change and forest degradation (Joshi et al., 2021). The sinks include removal through afforestation, biochar, harvested wood products (HWP), bioenergy with carbon capture and storage (BECCS), and other carbon dioxide removal techniques through geosphere capture. In 2024, atmospheric CO_2 concentration reached 422.5 ppm, a 52% increase from pre-industrial levels, with ocean CO_2 uptake stagnating since 2016 due to climate variability, while the land CO_2 sink, which contributes 30% of total emissions absorption, showed reduced capacity in 2023 due to El Niño impacts and Canadian wildfires (Friedlingstein et al., 2024).

Achieving the Paris Agreement's goal of limiting global temperature rise to 2 °C ideally, 1.5 °C depends heavily on strengthening carbon sink strategies to meet carbon neutrality. Figure 13.1 illustrates the world's CO_2 emissions, along with their historical trends and future projections. It highlights the importance of emissions reduction control to ensure that global warming is constrained within three temperature scenarios: 1.5, 1.7, and 2.0 °C (Kumar & Singh, 2020).

Between 2000 and 2020, global CO_2 emissions experienced a steady increase, reaching their highest point in the 2020s, followed by a slight reduction as the years progressed. In the most dramatic of the pathways, a 1.5 °C scenario, where net-zero is expected by 2040, the carbon budget stands at 380 $GtCO_2$, supporting the drastic measures that would be necessary. The second measure, at 1.7°C, which is on the higher end of the budget, supports a significant alteration to near zero but has an implementation period of approximately 20 years, from the end of 2040 to the beginning of 2060 (Ruhil et al., 2024) Following the peak ambition of a 2.0 °C carbon budget, which remains around 1230 $GtCO_2$, emissions levels were last targeted to reach zero by 2080 while building towards more realigned budgets through adjustments. Figure 13.1 illustrates the relationship between the carbon budget and the allowable temperature increase. Indeed, higher warming thresholds mean greater carbon budgets and net-zero emission targets set up in the distant future. However, the timing increases the risk of dire climatic consequences and thus compounds the need for urgent and sustained action to reduce CO_2 emissions across the board. Many governments and corporations are now implementing sustainability measures in business models, demonstrating that economic growth and environmental responsibility can co-exist (Chen et al., 2022a).

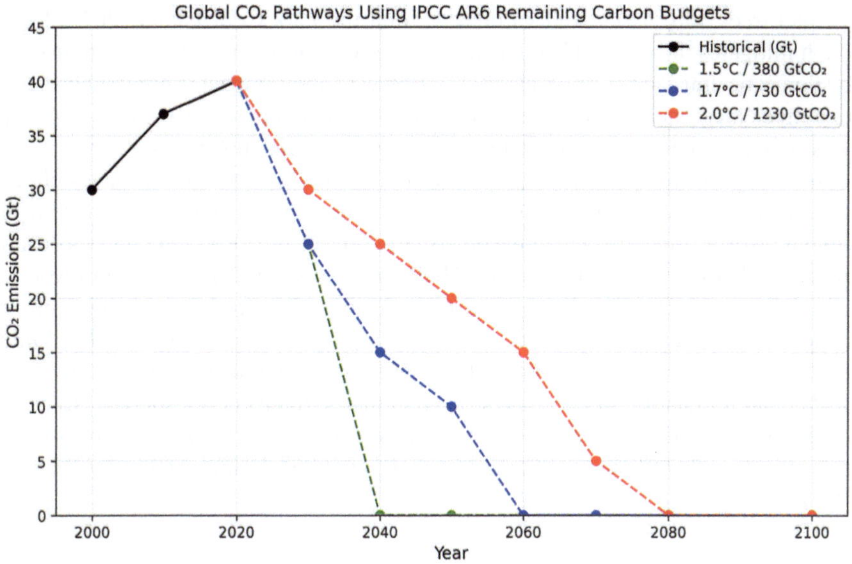

Fig. 13.1 Global CO_2 pathways using the intergovernmental panel on climate change sixth assessment Report IIPCC AR6 remaining carbon budgets. *Source* Friedlingstein et al., 2024

The IPCC Glossary defines carbon neutrality as the state where the removal of carbon dioxide (CO_2) equals the emissions of CO_2 related to a specific activity. According to this definition, CO_2 emissions are considered anthropogenic, meaning that human activities are the primary cause of them. "Greenhouse Gas neutrality" is a more precise definition of "carbon neutrality" that incorporates the requirement of balancing removal measures with anthropogenic GHG emissions, measured in metric-weighted CO_2 equivalents (Singh, 2021).

To promote a healthy planet, environmental sustainability seeks to preserve biodiversity, reduce pollution, and conserve resources. Ensuring that current requirements are met without compromising the needs of future generations requires the prudent management of ecosystems and resources. According to Robert (1995), environmental sustainability is more precisely defined as a set of limitations on the four main activities that control the size of the human economic subsystem: pollution and waste assimilation on the sink side and the use of renewable and nonrenewable resources on the source side. Hence, carbon neutrality strategies directly promote environmental sustainability and help economies grow responsibly (Singh, 2025).

Forests and soil carbon sequestration play a crucial role in mitigating climate change. Practices such as afforestation, reforestation, and revegetation (ARR) enhance carbon capture by facilitating additional aboveground biomass (AGB) and belowground biomass contributions (BGB), which increase soil organic carbon (SOC) stocks through the decomposition of organic material. Soil carbon sequestration offers benefits not only in reducing CO_2 emissions but also in improving soil quality, water quality, and biodiversity (Nair, 2011). Additionally, urban areas, such as street trees, city parks, and gardens, exhibit the Park Cool Island (PCI) effect, which mitigates the Urban Heat Island (UHI) effect caused by vehicular, industrial, and domestic emissions (Teo et al., 2022). However, their impact is minimal because street trees are generally scattered and have irregular canopy densities. Sustainable forest management has been the main principle in national forest policies, which holds ecological, economic, and social sustainability. The main challenge is defined as balancing the social, ecological, and economic aspects of forest management without favoring one aspect over the others (Vierikko et al., 2008). A country's development and Gross Domestic Product (GDP) are also significant factors in favoring one of these aspects. Developing and underdeveloped countries tend to prioritize economic and social factors over ecological aspects (Zhang et al., 2024). While some nations are proactively coming together to promote green energy alternatives, the demand for energy far exceeds the availability of green energy sources. To overcome this, some AFOLU and ARR projects are emerging as models for prioritizing economic growth without sacrificing social and ecological aspects (Sen & Ganguly, 2017).

13.2 The Role of Forests in Carbon Neutrality

Forests are significantly impacted by global warming and climate change, which both promote and threaten their growth. Forests can aid in both carbon sequestration and climate change mitigation due to their capacity to transfer energy between the land and the atmosphere, as well as the enormous volumes of CO_2 they absorb during photosynthesis. Recently, nature-based solutions have gained increasing attention from the scientific community and politicians as efforts to develop policies and plans for climate change mitigation and carbon sequestration (Joshi et al., 2020).

Forests are the primary continental carbon sink on a global scale. As a forest carbon sink, carbon dioxide is taken up by plants and fixed in biomass, soil, dead wood, and litter, which lowers the atmospheric CO_2 concentration. On a local scale, however, certain forests act as sources of carbon, while others act as sinks. Numerous man-made and natural variables influence the net carbon flux. Specifically, human actions can both generate and stimulate the release of carbon dioxide through deforestation and unsustainable forest management, and store it through afforestation, improved forest management, and avoided deforestation activities. The forest functions as a "carbon sink" when its stock of carbon grows and as a "carbon source" when its stock decreases as a result of carbon being released into the atmosphere. Thus, to understand the role of forests in carbon sequestration, it is essential to comprehend how trees contribute to the carbon cycle.

For forestry to be considered sustainable, the carbon stock in forests must remain constant or increase over time. Pan et al. (2011) stated that the world's forest stocks are around 860 GtC, of which 380 GtC (44%) is found in the soil, 360 GtC (42%) in live biomass (both above and below ground), 75 GtC (9%) in dead wood, and 45 GtC (5%) in litter. According to the Global Forest Resources Assessment, 2020, the worldwide forest carbon stock in 2020 was estimated to be 662 Gt, of which 44% was attributed to tree mass (including wood, bark, leaves, and the root system), 45% to soil organic matter, and the remaining portion to deadwood and litter. Nevertheless, during specific periods between 1990 and 2020, several studies have found that forests served as weak net carbon sources globally. This was most likely caused by widespread deforestation and accelerated forest degradation, primarily due to human activities. Geographically, the majority of the stock (470 GtC, 55%) is located in tropical forests, with the remainder distributed among temperate (120 GtC, 14%) and boreal (270 GtC, 32%) forests. In boreal forests, stocks are primarily found in the soil (60%), compared to 20% in the biomass, while in tropical forests, stocks are primarily found in the biomass (56%), compared to 32% in the soil (Malhi et al., 2006).

The study has focused on measuring the capacity of trees to sequester CO_2 and assessing how effectively afforestation and reforestation strategies mitigate climate change. The potential carbon sequestration capacity of aboveground woody biomass—that is, carbon stored in trees, branches, leaves, and other aboveground vegetation was determined by Teo et al., (2021). The report indicates that approximately 17.6% of the metropolitan areas under study are eligible for afforestation. At this pace, which is equivalent to approximately 10.9 Million Hectares, 82.4 ± 25.7 million tons of CO_2 equivalent could be sequestered yearly.

Li et al. (2015) observed that as tree age increased, the ecosystem's overall carbon stock also rose. They found that as trees grew older, the rate of CO_2 sequestration increased progressively until it peaked in the middle age of trees and then significantly decreased, indicating the influence of tree aging on carbon sequestration rates. Additionally, Bernal et al. (2018) noted that trees exhibited an increasing trend in CO_2 removal rates from colder to warmer areas. (Wang & Huang, 2020) reported the highest rate of carbon sequestration from aboveground tree biomass in tropical temperature zones, whereas the highest rate of carbon sequestration from forest soils in warm temperate climate regions. Moreover, it was found that several factors, including latitude, mean annual rainfall, and the number of frost-free days, significantly influenced the rate of carbon sequestration in the soil.

In addition to their influence on the carbon cycle, forests also impact the climate through biophysical processes, including canopy roughness, evapotranspiration, and albedo (the reflection of solar radiation), primarily at the local and regional levels. Forests have a lower albedo than low-vegetation areas, grazing fields, cultivated lands, soil, and snow due to their dark color (Bonan, 2008). As a result, a larger proportion of incident solar radiation is absorbed by forests, contributing to local and/or regional warming.

Carbon sequestration is one of the many ecological, economic, and social roles that forests play. In addition to carbon sequestration, preserving and restoring forests protect and increase biodiversity, thereby enhancing ecosystem resilience. Furthermore, forests provide a renewable and sustainable supply of wood and biomass products that can be utilized as substitutes for materials with larger carbon footprints and fossil fuels (*IPCC Glossary*).

Globally, forest ecosystems act as net carbon sinks, helping to slow the rise in the average global temperature. However, carbon stocks in various forest ecosystems worldwide exhibit significant variations due to several factors, including biodiversity, tree characteristics, climatic conditions, and forest management practices. It has been noted that under some circumstances, high temperatures and increased vapor pressure deficits might disrupt the equilibrium of trees by affecting photosynthesis and respiration. This may make forests less effective in absorbing carbon.

13.3 Barriers to Achieving Carbon Neutrality Through Forestry

13.3.1 Ecological Barriers

Forests store large amounts of carbon that is released into the atmosphere as CO2 when they are cut down. Deforestation results in an additional loss of above-ground biomass, thereby disrupting carbon neutrality and compromising biodiversity (Li et al., 2022). Frequent conversion of natural forests to agricultural land for the benefit of crops is widely increasing (Clement Akais, 2012). The introduction of invasive species, along with disturbance, leads to the destruction of native species and the deterioration of ecosystem stability, thus hindering carbon neutrality. Moreover, attacks by pests lead to the destruction of forests and a reduction in ecosystem services, ultimately resulting in a decrease in carbon sequestration by the forests (Freer-Smith & Webber, 2017).

13.3.2 Economic and Financial Barriers

Financial and economic barriers need to be considered, as they pose constraints to reforestation programs (Zeng et al., 2020). Private companies do not focus on investing in natural regeneration projects (Löfqvist et al., 2023). Methodological, socio-economic, and implementation challenges are responsible for regulating carbon offsets (Pan et al., 2022). Carbon offset deals with the reduction of carbon emissions in one location by compensating for the emissions in another location by the same enterprise or organization (Lovell & Liverman, 2010). Costs related to carbon offsetting for agro-companies are the most significant barrier to carbon neutrality (Acampora et al., 2023). The economic barrier can be overcome by adopting accurate technological innovations and simultaneously reducing carbon emissions.

13.3.3 Technological Barriers

The technological barriers are also a major hindrance to achieving carbon neutrality. Measurement of soil organic carbon (SOC) using satellite data has limitations, including inaccurate soil depth estimates and statistical methods and procedures employed in the laboratory (Nieto et al., 2024). Digital barriers to carbon neutrality are rarely discussed. Yet, they are one of the most critical challenges for achieving net-zero emissions, which can be addressed with innovative solutions and robust efficiency (Bhatia et al., 2024). Currently, Virtual Reality (VR) technology is gaining attention for forest management; however, it has challenges such as resolution and image processing features (Uusitalo & Orland, 2001). Nevertheless, innovation in forest management technology is underway to break this barrier and aid in achieving carbon neutrality (Brown et al., 2020).

13.3.4 Policy and Governance Barriers

The need for carbon sequestration is alarming, and the current policies and their implementation for carbon sequestration are insufficient. In the continental United States and Alaska, carbon emissions from agricultural activities are 43 Tg/year, while only 17 Tg/year of carbon is sequestered by the soil. Consequently, policies must be strong enough to address the actual reality of carbon sequestration (Lal et al., 2003). Coordination among different policy stakeholders is lacking both vertically and horizontally in forest management governance (Rayner et al., 2013). High government effectiveness and strong policy implementation, combined with low corruption, led to a reduction in deforestation and, consequently, more systematic forest management (Moreira-Dantas & Söder, 2022). Policies for forest management have been introduced in Peru; however, mismanagement in logging practices remains prevalent (Smith et al., 2006). Illegal logging is commonplace in Malaysia due to inadequate forest law enforcement, characterized by insufficient staff and equipment (Mohd Noor et al., 2021). Corruption plays a critical role in facilitating illegal activities in the forests, which need to be tackled with a focus on the communities that are directly dependent on forest products (Richards et al., 2003). Government policies and regulations are often loose for rural communities that rely heavily on forest products, particularly in areas with significant forest cover in Ethiopia, which is crucial for food security (Yami & Mekuria, 2022).

13.3.5 Social and Cultural Barriers

Socio-cultural barriers significantly affect carbon neutrality, which is often overlooked in policy and decision-making. Private forest landowners face challenges in adopting models and tools that enhance the effectiveness of carbon sequestration (Khanal et al., 2019). Co-management strategies between indigenous peoples and the government often lead to more authoritarian rule, thereby marginalizing the indigenous groups' forest rights and resulting in poor forest management (Armesto et al., 2001). Forest management programs and education are lacking in the Baka region, which can help determine and understand forest management for local indigenous communities (Carson et al., 2018). Awareness and education are highly lacking for sustainable practices and forest management. Nearly 70% of local communities in Iran showed little interest in participating in sustainable forest management practices (Savari et al., 2020).

13.3.6 Interplay of Barriers to Carbon Neutrality

Diverse barriers, including ecological, economic, technological, policy, governance, social, and cultural barriers, contribute to increased carbon emissions and hinder carbon neutrality, as discussed above. These barriers are directly or indirectly interrelated with each other and act as a cause and effect for the different obstacles (Fig. 13.2). The geo-sequestration technology of carbon sequestration faces barriers not only in technology but also in public acceptance and the economic implications of the technological setup, which play a vital role in the adoption of Carbon Capture and Storage (CCS) technology (Davies et al., 2013). The real barrier also lies in selling CCS technology to developing countries and not utilizing the CCS for carbon sequestration in the country of origin of the CCS technology (Bowen, 2010). Due to problems in forest policy, science, and technology for forest management, new models have been developed popularly known as the Sustainable Hyrcanian Forest Management Model (SHFMM) in Iran and Share Holding Integrated Forestry Tenure (SHIFT) in China that focused on community participation and role of diverse stakeholders in forest management leading to carbon neutrality (Goushehgir et al., 2022).

13.4 Future Directions for Achieving Carbon Neutrality Through Forestry

IPCC recommends attaining net-zero CO_2 emissions by 2050, sometimes known as "reaching carbon neutrality." The 2015 Paris Agreement called on signatories to achieve global carbon neutrality by 2050–2100 and limit the global temperature rise to below 2 °C compared to pre-industrial levels (Zhao et al., 2024). Achieving carbon neutrality is a global challenge that requires innovative and multidisciplinary approaches (Johnson, 1992).

Fig. 13.2 Barriers to carbon neutrality that are interconnected with each other

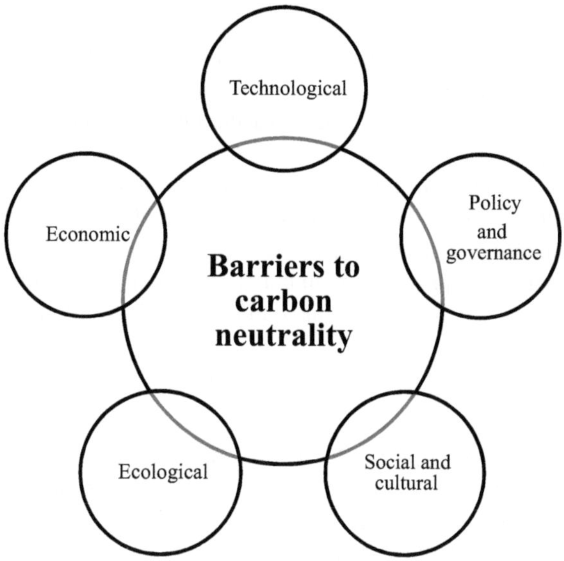

13.4.1 Advancements in Afforestation and Reforestation

13.4.1.1 Selection of Suitable Species

The choice of tree species is crucial for storing carbon in the soil (Kreutzer et al., 1986). The quantity and stability of stored carbon are influenced by variations in biomass production, rooting depth, and litter quality (Johnson, 1992). For instance, compared to conifers with shallow roots, deciduous species are more effective at storing carbon in the soil. Mixed-species forests confer additional advantages over monocultures, as ecosystem resilience increases in response to large events such as storms and pest infestations (Pretzsch, 2005).

The creation of mixed-species forests decreases the likelihood of large-scale events. Some site improvement techniques involve nitrogen fertilization, liming, and peatland drainage, which vary in their effect on the soil carbon. They may also increase productivity and carbon input, but may accelerate soil organic matter (SOM) decay, thereby risking carbon loss. For instance, the addition of nitrogen fertilizers will spur tree growth but decrease soil carbon by increasing microbial activity (Paul & Clark, 1989).

13.4.2 Innovations in Forest Management

Accurate quantification of forest carbon stock is crucial for monitoring and evaluating forest ecosystems, biodiversity, resources, and the effectiveness of mitigation efforts. Traditionally, forest carbon stock has been measured by ground-based forest inventories, which are labor-intensive and limited by spatial differences. Advancements in remote sensing technologies and machine learning offer hopeful alternatives for large-scale, efficient, and accurate assessments of forest carbon stock (Illarionova et al., 2024).

To conduct research in a forested area, a management-level inventory should be available. This data can serve as a reference measurement for training machine learning models. Algorithms such as the Extreme Gradient Boosting (XGBoost) can be employed in both classification and regression settings for predicting forest characteristics, including stand age, tree height, and basal area (Chen & Guestrin, 2016).

There are two approaches for estimating the growing stock volume (Illarionova et al., 2024).

1. **Direct Approach**: In this method, remote sensing data are used to directly generate maps of forest carbon properties, including density, diversity, volume, and carbon stock.
2. **Hierarchical Approach**: This method utilizes available and predicted inventory classifications, along with conversion factors and equations, to examine forest characteristics, including above- and below-ground biomass.

To estimate carbon stocks, the hierarchical technique incorporates intermediate forecasts of forest attributes, which can improve the accuracy and resilience of the assessments.

For growing stock timber volume (1), the hierarchical method incorporates height and basal area; for the biomass carbon mass (2), it includes age, species, height, and basal area.

$$\text{Timber stock (GSV)} = \text{Basal area} \times \text{Height}^{(1)}$$

$$\text{Carbon mass} = \text{GSV} \times \text{BCEF(fr)} \times \text{Carbon content coefficient}^{(2)}$$

where GSV stands for Growing Stock Volume and BCEF stands for Biomass Conversion and Expansion Factor. The integration of both direct and hierarchical approaches using earth observation imagery, provides a more accurate and holistic solution for carbon stock estimation, even with limited data for large-scale assessments (Illarionova et al., 2024b).

13.4.3 Long-Term Vision and Policies

There have been numerous strategies and stages of development in the global effort to achieve carbon neutrality. Some countries are at the forefront, while others are only in the planning or debate stage. The push to cut carbon emissions is accelerating, from those that have already accomplished carbon neutrality to those that have made bold promises for the future. Higher education institutions are becoming involved, tax systems are being implemented, and carbon–neutral legislation is being passed in various countries. The main advancements in the worldwide movement, as per the review analysis by Chen et al., (2022), towards carbon neutrality are summarized in Table 13.1.

There is now a need to integrate adaptation and mitigation strategies. In the case of forestry, the plantation and its maintenance exemplify the synergy between adaptation and mitigation, which mitigates climate change through carbon capture and adapts to the adverse impacts of climate change (Moomaw et al., 2019). Moreover, efficient agroforestry practices also act as adaptation and mitigation strategies for climate resilience by improving soil and water quality. In urban areas, rooftop green walls are being utilized as mitigation and adaptive methods in buildings to counter the urban heat island effect (Grafakos et al., 2019). Along with the green transport system, which features tree cover and solar panels along the road, also serves as a mitigation and adaptation strategy (Zhao et al., 2018).

Table 13.1 Global carbon neutrality progress: important advancements and approaches

Category	Details
Nations achieved carbon neutrality	Bhutan, Gabon, Guinea-Bissau, Suriname (self-reported, accomplished before 2021). Smaller nations are leading the way in reducing carbon emissions
Various pledging dates	Target years range from 2070 (India) to 2030 (Côte d'Ivoire). Varying degrees of ambition and viability due to differences in economies and infrastructures
Rules and regulations	Seventeen countries, including New Zealand, Denmark, and Germany, have enacted carbon–neutral laws that integrate carbon neutrality into their legal frameworks
Nations discussing carbon neutrality	Ninety-three countries are still in the process of drafting or negotiating carbon neutrality plans, indicating that global discussions are still ongoing
Innovative funding strategies	A Pigouvian tax system, which uses tax legislation to promote carbon abatements, has been adopted by a few Nordic nations
Participation in postsecondary education	Encouraging the development and dissemination of information about carbon neutrality, Australian universities are highlighting the role that education plays in fostering sustainable development

13.5 Case Studies and Best Practices

Voluntary carbon markets (VCMs) assist carbon-emitting firms by providing a cost-effective means of mitigating carbon emissions through the promotion of forest management and conservation (Ranjan, 2024). The VCM is projected to grow at a rate of 27% by the next decade (Ranjan, 2024). Additionally, green bonds are expected to reach USD 5.6 trillion by 2035, up from USD 1 trillion in 2024. Moreover, the fear of greenwashing is also deterring investors from considering participation in VCM.

In India, NGOs and project development companies serve as a link between communities and the government for the initiation and implementation of carbon mitigation projects through Afforestation, Reforestation, and Revegetation (ARR). In Uttar Pradesh, six districts- Bareilly, Saharanpur, Meerut, Lucknow, Gorakhpur, and Moradabad–signed a Memorandum of Understanding (MOU) with The Energy and Resources Institute and Value Network Ventures Advisory Services Private Limited for the implementation of the project named as Strengthening Rural Livelihood through Carbon Finance to Agroforestry Practices. This project enhances the efficiency of the agroforestry model by introducing a new income source, carbon credit transfer, as compensation for crop area loss resulting from the disadvantages of trees in agroforestry practices.

The Surui Forest Carbon Project, along with its case study by Zwick (2019), represents both the commitment and the drawbacks of a carbon–neutral and environmentally sustainable society. Led by the Paiter-Surui community of the Amazon, the project, which began in 2009, aims to halt forest degradation through the innovative utilization of carbon credits that support the sustainable development of livelihood and income-generating activities while respecting indigenous ways of life. The project funded six community-led initiatives, generating nearly 3 million VCS-certified offsets, equivalent to 64,000 cars off the road for a year.

These involved agroforestry practices, the cultivation and harvesting of medicinal plants, and explained how economic viability can be correlated with ecological sustainability. The project's internal challenges were the conflict between governance and policy. External pressure came from illegal mining, which reversed the situation and undermined the achievements made in 2018 when it was suspended. A sound institutional framework is vital. The Surui Forest Carbon Project offers valuable insights for future directions in achieving the interlinked goals of carbon neutrality and environmental sustainability, including the integration of traditional knowledge, fostering community engagement, and securing robust support from national and international frameworks and standards.

Forests play a crucial role in mitigating the impacts of climate change. The forest fixes CO_2 during photosynthesis and stores it in biomass and soil, which act as natural carbon sinks that reduce greenhouse gas emissions. In particular, China's extensive reforestation initiatives have increased biomass levels in the forests, enhancing their capacity to store carbon and improving their resilience. These initiatives align with global commitments, including the Paris Agreement, which emphasizes the importance of forest restoration and conservation as key strategies for mitigating climate change. Forests can absorb only a finite amount of carbon, and trees take time to grow, thus delaying the climatic benefits of afforestation and reforestation activities. Faubert et al. (2020) investigated whether a proposed liquefied natural gas (LNG) facility could offset its greenhouse gas (GHG) emissions. Using a lifecycle methodology, they measure emissions during the LNG production, transportation, and use stages to determine the project's overall environmental effect. The analysis emphasizes the need to coordinate these projects with Canada's and Quebec's greenhouse gas (GHG) reduction pledges, highlighting the importance of cooperation between communities, businesses, and governments. With proactive planning and creative solutions, this study demonstrates that economic development and environmental sustainability can coexist in large-scale industrial enterprises, providing a reproducible framework for emissions management (Faubert et al., 2020).

Sustainable forestry can be considered a crucial factor in mitigating climate change, serving as an essential tool to reduce carbon emissions through improved land-use practices. In Uttar Pradesh, India, for example, cooperation between local governments, NGOs, and private organizations pushed the agenda of agroforestry by incorporating carbon credit schemes. Such initiatives not only contribute to the development of sustainable land use but also provide economic motivation to farmers through compensation for crop losses caused by trees. Thus, these efforts demonstrate the close relationship between rural livelihoods and climate action. Hence, it is clear that community-centered solutions, combined with resilient institutional frameworks, offer a solution to the barriers to carbon neutrality and a path toward environmental sustainability.

13.6 Conclusion

Environmental sustainability and carbon neutrality are critical components of global efforts to mitigate climate change. Forests, as vital carbon sinks, are increasingly threatened by deforestation, biodiversity loss, and inadequate policy implementation. This chapter underscores the need to address a range of ecological, financial, technological, governance-related, and socio-cultural challenges to strengthen the capacity of forests for carbon sequestration. Achieving these goals requires a combination of afforestation, reforestation, natural regeneration, and the integration of modern innovations. Furthermore, establishing robust policy frameworks, promoting active community participation, and providing financial incentives are crucial for achieving long-term carbon neutrality targets. The importance of coordinated international efforts and interdisciplinary approaches lies in striking a balance between environmental protection and economic development to secure a sustainable legacy for future generations.

References

Acampora, A., Ruini, L., Mattia, G., Pratesi, C. A., & Lucchetti, M. C. (2023). Towards carbon neutrality in the agri-food sector: Drivers and barriers. *Resources, Conservation and Recycling, 189*, Article 106755. https://doi.org/10.1016/j.resconrec.2022.106755

Armesto, J. J., Smith-Ramirez, C., & Rozzi, R. (2001). Conservation strategies for biodiversity and indigenous people in Chilean forest ecosystems. *Journal of the Royal Society of New Zealand, 31*(4), 865–877. https://doi.org/10.1080/03014223.2001.9517681

Bernal, B., Murray, L. T., & Pearson, T. R. H. (2018). Global carbon dioxide removal rates from forest landscape restoration activities. *Carbon Balance and Management, 13*(1), 22. https://doi.org/10.1186/s13021-018-0110-8

Bhatia, M., Meenakshi, N., Kaur, P., & Dhir, A. (2024). Digital technologies and carbon neutrality goals: An in-depth investigation of drivers, barriers, and risk mitigation strategies. *Journal of Cleaner Production, 451*, Article 141946. https://doi.org/10.1016/j.jclepro.2024.141946

Bonan, G. B. (2008). Forests and climate change: forcings, feedbacks, and the climate benefits of forests. *Science, 320*(5882), 1444–1449. https://doi.org/10.1126/science.1155121

Bowen, F. (2010). Barriers to carbon capture and storage may not be obvious. *Nature, 464*(7286), 160–160. https://doi.org/10.1038/464160c

Brown, M., Ghaffariyan, M. R., Berry, M., Acuna, M., Strandgard, M., & Mitchell, R. (2020). The progression of forest operations technology and innovation. *Australian Forestry, 83*(1), 1–3. https://doi.org/10.1080/00049158.2020.1723044

Calvin, K., Dasgupta, D., Krinner, G., Mukherji, A., Thorne, P. W., Trisos, C., Romero, J., Aldunce, P., Barrett, K., Blanco, G., Cheung, W. W. L., Connors, S., Denton, F., Diongue-Niang, A., Dodman, D., Garschagen, M., Geden, O., Hayward, B., Jones, C., & Péan, C. (2023). *IPCC, 2023: Climate Change 2023: Synthesis Report. Contribution of Working Groups I, II and III to the Sixth Assessment Report of the Intergovernmental Panel on Climate Change, Core Writing Team, H. Lee and J. Romero (eds.). IPCC, Geneva, Switzerland.* (First). Intergovernmental Panel on Climate Change (IPCC). https://doi.org/10.59327/IPCC/AR6-9789291691647

Carson, S., Kentatchime, F., Nana, E., Njabo, K., Cole, B., & Godwin, H. (2018). Indigenous peoples' concerns about loss of forest knowledge: Implications for forest management. *Conservation and Society, 16*(4), 431. https://doi.org/10.4103/cs.cs_17_105

Chen, L., Msigwa, G., Yang, M., Osman, A. I., Fawzy, S., Rooney, D. W., & Yap, P.-S. (2022). Strategies to achieve a carbon neutral society: A review. *Environmental Chemistry Letters, 20*(4), 2277–2310. https://doi.org/10.1007/s10311-022-01435-8

Chen, T., & Guestrin, C. (2016). XGBoost: A scalable tree boosting system. In *Proceedings of the 22nd ACM SIGKDD International Conference on Knowledge Discovery and Data Mining*, pp. 785–794. https://doi.org/10.1145/2939672.2939785

Clement Akais, O. (Ed.). (2012). *Global Perspectives on Sustainable Forest Management.* IntechOpen.

Davies, L. L., Uchitel, K., & Ruple, J. (2013). Understanding barriers to commercial-scale carbon capture and sequestration in the United States: An empirical assessment. *Energy Policy, 59*, 745–761. https://doi.org/10.1016/j.enpol.2013.04.033

Ellis, B. J., Figueredo, A. J., Brumbach, B. H., & Schlomer, G. L. (2009). Fundamental dimensions of environmental risk: The impact of harsh versus unpredictable environments on the evolution and development of life history strategies. *Human Nature, 20*(2), 204–268. https://doi.org/10.1007/s12110-009-9063-7

Faubert, P., Bouchard, S., Morin Chassé, R., Côté, H., Dessureault, P.-L., & Villeneuve, C. (2020). Achieving carbon neutrality for a future large greenhouse gas emitter in Quebec, Canada: A case study. *Atmosphere, 11*(8), 810. https://doi.org/10.3390/atmos11080810

Freer-Smith, P. H., & Webber, J. F. (2017). Tree pests and diseases: The threat to biodiversity and the delivery of ecosystem services. *Biodiversity and Conservation, 26*(13), 3167–3181. https://doi.org/10.1007/s10531-015-1019-0

Friedlingstein, P., O'Sullivan, M., Jones, M. W., Andrew, R. M., Gregor, L., Hauck, J., Le Quéré, C., Luijkx, I. T., Olsen, A., Peters, G. P., Peters, W., Pongratz, J., Schwingshackl, C., Sitch, S., Canadell, J. G., Ciais, P., Jackson, R. B., Alin, S. R., Alkama, R., & Zheng, B. (2022). Global carbon budget 2022. *Earth System Science Data, 14*(11), 4811–4900. https://doi.org/10.5194/essd-14-4811-2022

Friedlingstein, P., O'Sullivan, M., Jones, M. W., Andrew, R. M., Hauck, J., Landschützer, P., Le Quéré, C., Li, H., Luijkx, I. T., Olsen, A., Peters, G. P., Peters, W., Pongratz, J., Schwingshackl, C., Sitch, S., Canadell, J. G., Ciais, P., Jackson, R. B., Alin, S. R., & Zeng, J. (2024). *Global Carbon Budget 2024*. ESSD–Global/Energy and Emissions. https://doi.org/10.5194/essd-2024-519

Global Forest Resources Assessment. (2020). Food and agriculture organization of the united nations (FAO). Accessed 31 Oct 2023. https://www.fao.org/3/I8661EN/i8661en.pdf

Goushehgir, Z., Feghhi, J., & Innes, J. L. (2022). Challenges facing the improvement of forest management in the hyrcanian forests of Iran. *Forests, 13*(12), 2180. https://doi.org/10.3390/f13122180

Grafakos, S., Trigg, K., Landauer, M., Chelleri, L., & Dhakal, S. (2019). Analytical framework to evaluate the level of integration of climate adaptation and mitigation in cities. *Climatic Change, 154*(1–2), 87–106. https://doi.org/10.1007/s10584-019-02394-w

Illarionova, S., Tregubova, P., Shukhratov, I., Shadrin, D., Efimov, A., & Burnaev, E. (2024). Advancing forest carbon stocks' mapping using a hierarchical approach with machine learning and satellite imagery. *Scientific Reports, 14*(1), 21032. https://doi.org/10.1038/s41598-024-71133-8

IPCC Glossary. (n.d.). https://apps.ipcc.ch/glossary/

Jacobson, M. Z. (2002). *Atmospheric Pollution: History, Science, and Regulation* (1st Ed.). Cambridge University Press. https://doi.org/10.1017/CBO9780511802287

Johnson, D. W. (1992). Effects of forest management on soil carbon storage. *Water, Air, & Soil Pollution, 64*(1–2), 83–120. https://doi.org/10.1007/BF00477097

Joshi, R., Singh, H., Chhetri, R., & Yadav, K. (2020). Assessment of carbon sequestration potential in degraded and non-degraded community forests in Terai region of Nepal. *Journal of Forest and Environmental Science, 36*(2), 13–121. https://doi.org/10.7747/JFES.2020.36.2.113

Joshi, R., Singh, H., Chhetri, R., Poudel, S. R., & Rijal, S. (2021). Carbon sequestration potential of community forests: A comparative analysis of soil organic carbon stock in community managed forests of Far-Western Nepal. *Eurasian Journal of Soil Science, 10*(2), 96–104. https://doi.org/10.18393/ejss.825066

Khanal, P. N., Grebner, D. L., Straka, T. J., & Adams, D. C. (2019). Obstacles to participation in carbon sequestration for nonindustrial private forest landowners in the southern United States: A diffusion of innovations perspective. *Forest Policy and Economics, 100*, 95–101. https://doi.org/10.1016/j.forpol.2018.11.007

Kreutzer, K., Deschu, E., & Hösl, G. (1986). Vergleichende Untersuchungen über den Einfluß von Fichte (Picea abies L. Karst.) und Buche (Fagus sylvatica L.) auf die Sickerwasserqualität. *Forstwissenschaftliches Centralblatt, 105*(1), 364–371. https://doi.org/10.1007/BF02741744

Kumar, M., & Singh, H. (2020). Agroforestry as a nature-based solution for reducing community dependence on forests to safeguard forests in rainfed areas of India. *In* S. Dhyani, A. Gupta & Karki M. (Eds.) *Nature-based Solutions for Resilient Ecosystems and Societies. Disaster Resilience and Green Growth.* Springer, Singapore. https://doi.org/10.1007/978-981-15-4712-6_17

Lal, R., Follett, R. F., & Kimble, J. M. (2003). Achieving soil carbon sequestration in the United States: A challenge to the policy makers. *Soil Science, 168*(12), 827–845. https://doi.org/10.1097/01.ss.0000106407.84926.6b

Li, Y., Zhao, M., Motesharrei, S., Mu, Q., Kalnay, E., & Li, S. (2015). Local cooling and warming effects of forests based on satellite observations. *Nature Communications, 6*(1), 6603. https://doi.org/10.1038/ncomms7603

Li, Y., Brando, P. M., Morton, D. C., Lawrence, D. M., Yang, H., & Randerson, J. T. (2022). Deforestation-induced climate change reduces carbon storage in remaining tropical forests. *Nature Communications, 13*(1), 1964. https://doi.org/10.1038/s41467-022-29601-0

Löfqvist, S., Garrett, R. D., & Ghazoul, J. (2023). Incentives and barriers to private finance for forest and landscape restoration. *Nature Ecology & Evolution, 7*(5), 707–715. https://doi.org/10.1038/s41559-023-02037-5

Lovell, H., & Liverman, D. (2010). Understanding carbon offset technologies. *New Political Economy, 15*(2), 255–273. https://doi.org/10.1080/13563460903548699

Malhi, Y., Wood, D., Baker, T. R., Wright, J., Phillips, O. L., Cochrane, T., Meir, P., Chave, J., Almeida, S., Arroyo, L., Higuchi, N., Killeen, T. J., Laurance, S. G., Laurance, W. F., Lewis, S. L., Monteagudo, A., Neill, D. A., Vargas, P. N., Pitman, N. C. A., & Vinceti, B. (2006). The regional variation of aboveground live biomass in old-growth Amazonian forests. *Global Change Biology, 12*(7), 1107–1138. https://doi.org/10.1111/j.1365-2486.2006.01120.x

Mohd Noor, M. N. H., Kadir, R., Universiti Malaysia Terengganu, & Muhamad, S. (2021). Issues of forest enforcement against illegal logging and forest offences in peninsular malaysia. *Journal of sustainability science and management*, *16*(7), 260–272. https://doi.org/10.46754/jssm.2021.10.019

Moomaw, W. R., Masino, S. A., & Faison, E. K. (2019). Intact forests in the United States: Proforestation mitigates climate change and serves the greatest good. *Frontiers in Forests and Global Change, 2*, 27. https://doi.org/10.3389/ffgc.2019.00027

Moreira-Dantas, I. R., & Söder, M. (2022). Global deforestation revisited: The role of weak institutions. *Land Use Policy, 122*, Article 106383. https://doi.org/10.1016/j.landusepol.2022.106383

Nair, P. K. R. (2011). Methodological challenges in estimating carbon sequestration potential of agroforestry systems. In B. M. Kumar & P. K. R. Nair (Eds.), *Carbon Sequestration Potential of Agroforestry Systems* (Vol. 8, pp. 3–16). Springer Netherlands. https://doi.org/10.1007/978-94-007-1630-8_1

Nieto, L., Houborg, R., Tivet, F., Olson, B. J. S. C., Prasad, P. V. V., & Ciampitti, I. A. (2024). Limitations and future perspectives for satellite-based soil carbon monitoring. *Environmental Challenges, 14*, Article 100839. https://doi.org/10.1016/j.envc.2024.100839

Pan, Y., Birdsey, R. A., Fang, J., Houghton, R., Kauppi, P. E., Kurz, W. A., Phillips, O. L., Shvidenko, A., Lewis, S. L., Canadell, J. G., Ciais, P., Jackson, R. B., Pacala, S. W., McGuire, A. D., Piao, S., Rautiainen, A., Sitch, S., & Hayes, D. (2011). A large and persistent carbon sink in the world's forests. *Science, 333*(6045), 988–993. https://doi.org/10.1126/science.1201609

Pan, C., Shrestha, A., Innes, J. L., Zhou, G., Li, N., Li, J., He, Y., Sheng, C., Niles, J.-O., & Wang, G. (2022). Key challenges and approaches to addressing barriers in forest carbon offset projects. *Journal of Forestry Research, 33*(4), 1109–1122. https://doi.org/10.1007/s11676-022-01488-z

Paul, E. A., & Clark, F. E. (1989). *Soil microbiology and biochemistry*. Academic press.

Pretzsch, H. (2005). Diversity and productivity in forests: evidence from long-term experimental plots. In M. Scherer-Lorenzen, C. Körner, & E.-D. Schulze (Eds.), *Forest Diversity and Function* (Vol. 176, pp. 41–64). Springer-Verlag. https://doi.org/10.1007/3-540-26599-6_3

Ranjan, R. (2024). Balancing greenwashing risks and forest carbon sequestration benefits: A simulation model linking formal and voluntary carbon markets. *Forest Policy and Economics, 168*, Article 103317. https://doi.org/10.1016/j.forpol.2024.103317

Rayner, J., McNutt, K., & Wellstead, A. (2013). Dispersed capacity and weak coordination: The challenge of climate change adaptation in Canada's forest policy sector. *Review of Policy Research, 30*(1), 66–90. https://doi.org/10.1111/ropr.12003

Resh, S. C., Binkley, D., & Parrotta, J. A. (2002). Greater soil carbon sequestration under nitrogen-fixing trees compared with eucalyptus species. *Ecosystems, 5*(3), 217–231. https://doi.org/10.1007/s10021-001-0067-3

Richards, M., Wells, A., Del Gatto, F., Contreras-Hermosilla, A., & Pommier, D. (2003). Impacts of illegality and barriers to legality: A diagnostic analysis of illegal logging in Honduras and

Nicaragua. *International Forestry Review, 5*(3), 282–292. https://doi.org/10.1505/IFOR.5.3.282.19153

Robert, G. (1995). *The Concept of Environmental Sustainability*, 1–24.

Ruhil, T., Singh, H., & Barthwal, S. (2024). Long-term physiological monitoring for detecting climate change impacts on forest ecosystems. In *forests and climate change: biological perspectives on impact, adaptation, and mitigation strategies* (pp. 817–834). Singapore: Springer Nature Singapore.

Savari, M., Eskandari Damaneh, H., & Eskandari Damaneh, H. (2020). Factors influencing local people's participation in sustainable forest management. *Arabian Journal of Geosciences, 13*(13), 513. https://doi.org/10.1007/s12517-020-05519-z

Sen, S., & Ganguly, S. (2017). Opportunities, barriers and issues with renewable energy development–A discussion. *Renewable and Sustainable Energy Reviews, 69*, 1170–1181. https://doi.org/10.1016/j.rser.2016.09.137

Singh, H., Verma, A., & Kumar, M. (2018). Can Elevated CO_2 alter the growth dynamics, morphology, physiology and biochemistry of the forestry tree species? *Climate change and agriculture.* Discovery Publishing House, New Delhi, pp. 288–305.

Singh, H. (2021). Process-based carbon sequestration study with reference to the energy-water-carbon flux in a forest ecosystem. *Climate impacts on sustainable natural resource management*, 336–351.

Singh, H. (2025). The role of plant functional traits in mitigating urban climate and environmental pollution. In *Plant Functional Traits* (pp. 297–308). Elsevier.

Smith, J., Colan, V., Sabogal, C., & Snook, L. (2006). Why policy reforms fail to improve logging practices: The role of governance and norms in Peru. *Forest Policy and Economics, 8*(4), 458–469. https://doi.org/10.1016/j.forpol.2005.08.001

Teo, H. C., Zeng, Y., Sarira, T. V., Fung, T. K., Zheng, Q., Song, X. P., Chong, K. Y., & Koh, L. P. (2021). Global urban reforestation can be an important natural climate solution. *Environmental Research Letters, 16*(3), Article 034059. https://doi.org/10.1088/1748-9326/abe783

Teo, Y. H., Makani, M. A. B. H., Wang, W., Liu, L., Yap, J. H., & Cheong, K. H. (2022). Urban heat island mitigation: GIS-based analysis for a Tropical City Singapore. *International Journal of Environmental Research and Public Health, 19*(19), 11917. https://doi.org/10.3390/ijerph191911917

Uusitalo, J., & Orland, B. (2001). Virtual forest management: possibilities and challenges. *International Journal of Forest Engineering, 12*(2), 57–66. https://doi.org/10.1080/14942119.2001.10702447

Vierikko, K., Vehkamäki, S., Niemelä, J., Pellikka, J., & Lindén, H. (2008). Meeting the ecological, social and economic needs of sustainable forest management at a regional scale. *Scandinavian Journal of Forest Research, 23*(5), 431–444. https://doi.org/10.1080/02827580802284693

Wang, S., & Huang, Y. (2020). Determinants of soil organic carbon sequestration and its contribution to ecosystem carbon sinks of planted forests. *Global Change Biology, 26*(5), 3163–3173. https://doi.org/10.1111/gcb.15036

Yami, M., & Mekuria, W. (2022). Challenges in the Governance of community-managed forests in Ethiopia: Review. *Sustainability, 14*(3), 1478. https://doi.org/10.3390/su14031478

Zeng, Y., Sarira, T. V., Carrasco, L. R., Chong, K. Y., Friess, D. A., Lee, J. S. H., Taillardat, P., Worthington, T. A., Zhang, Y., & Koh, L. P. (2020). Economic and social constraints on reforestation for climate mitigation in Southeast Asia. *Nature Climate Change, 10*(9), 842–844. https://doi.org/10.1038/s41558-020-0856-3

Zhang, S., Xu, G., Shu, Y., Zhu, J., & Cheng, W. (2024). Comparing developed and emerging nations' economic development with environmental footprint for low-carbon competitiveness. *Heliyon, 10*(14), Article e34039. https://doi.org/10.1016/j.heliyon.2024.e34039

Zhao, C., Yan, Y., Wang, C., Tang, M., Wu, G., Ding, D., & Song, Y. (2018). Adaptation and mitigation for combating climate change–from single to joint. *Ecosystem Health and Sustainability, 4*(4), 85–94. https://doi.org/10.1080/20964129.2018.1466632

Zhao, X., Mignone, B. K., Wise, M. A., & McJeon, H. C. (2024). Trade-offs in land-based carbon removal measures under 1.5 and 2 °C futures. *Nature Communications, 15*(1), 2297. https://doi.org/10.1038/s41467-024-46575-3
Zwick, S. (2019). *The surui forest carbon project- A case study*. Forest Trends.

Chapter 14
Green Belt Development: A Key Pathway to a Sustainable and Carbon Neutral Future

Bipin Kumar Singh, K. S. Shashidhar, Ayushi Srivastva, Ashish Kumar, Sejal Sahu, Siya Yadav, Pooja, and Hukum Singh

Abstract Green belt development represents a strategic approach to sustainable urban planning and climate resilience. As multifunctional landscapes comprising agricultural lands, forests, parks, and open spaces that encircle urban areas, green belts act as ecological buffers, mitigating urban sprawl, enhancing biodiversity, and safeguarding critical ecosystem services. They significantly contribute to carbon neutrality by sequestering atmospheric CO_2, regulating microclimates, and mitigating the urban heat island effect. Moreover, green belts improve air and water quality, support groundwater recharge, and offer natural defences against environmental hazards such as floods and wildfires. Socioeconomically, green belts preserve the cultural and historical identity of peri-urban regions, promote recreational opportunities, and stimulate sustainable land use through urban regeneration and brownfield redevelopment. Despite policy and implementation challenges, green belt development remains a vital land-use strategy to balance urban expansion with environmental conservation. This chapter underscores the role of green belts in fostering low-carbon, livable cities and highlights their potential as a cornerstone in achieving global sustainability and climate mitigation targets.

Keywords Carbon sequestration · Greenbelt policy · Housing in brownfield · Safeguarding · And Air quality management

B. K. Singh (✉)
Department of Agroforestry, Acharya Narendra Deva University of Agriculture and Technology, Kumarganj, Ayodhya, Uttar Pradesh, India
e-mail: singhbipin9832@gmail.com

B. K. Singh · K. S. Shashidhar · A. Srivastva · A. Kumar · S. Sahu · S. Yadav
Rajiv Gandhi South Campus, Banaras Hindu University, Mirzapur, Uttar Pradesh, India

Pooja
Department of Silviculture and Agroforestry, Dr. Y.S. Parmar University of Horticulture and Forestry, Solan, Himachal Pradesh, India

H. Singh
ICFRE-Forest Research Institute, PO New Forest Dehradun, Dehradun, Uttarakhand, India

H. Singh (ed.), *Forestry for a Carbon-Neutral and Sustainable Future*, Sustainability Solutions, https://doi.org/10.1007/978-3-032-06427-1_14

14.1 Introduction

As the world grapples with the challenges posed by climate change, rapid urbanization, and ecological degradation, sustainable development has emerged as a pressing global priority. Urban areas are responsible for approximately 70% of global carbon emissions and consume nearly two-thirds of the world's energy (UN-Habitat, 2020). The uncontrolled growth of urban areas, commonly known as urban sprawl, has given rise to numerous environmental and social issues. Between 1990 and 2018, the global urban population increased by nearly 78%, resulting in a corresponding surge in demand for housing, transportation, and infrastructure (World Bank, 2019).

Green belt development provides a strategic solution to mitigate the negative impacts of urban sprawl while enhancing ecological resilience and promoting sustainability. A green belt is typically a designated zone surrounding urban areas, where development is restricted to protect natural landscapes and encourage sustainable land use. These areas act as carbon sinks, absorbing atmospheric carbon dioxide and helping cities reduce their overall emissions. Forests in green belts can sequester an average of 2.6 gigatons (Gt) of CO_2 per year, accounting for approximately 30% of the global anthropogenic emissions (Bhattacharya et al., 2020). In addition to capturing carbon, green belts play a crucial role in conserving biodiversity by providing essential habitats for a wide range of plant and animal species. For example, the London Green Belt, established in 1938, spans over 516,000 hectares, supports a diverse range of wildlife, and serves as a natural barrier to urban expansion (Amati & Yokohari, 2007). Similarly, the National Capital Region (NCR) Green Belt in India helps mitigate air pollution in Delhi, one of the world's most polluted cities, by filtering particulate matter and lowering urban temperatures. The benefits of green belt development go beyond environmental impacts, significantly enhancing urban quality of life. These areas provide recreational spaces, boost public health, and strengthen social connections. Access to green spaces can reduce stress levels by 30% and increase physical activity by up to 50%, significantly improving overall well-being (WHO, 2017). Green belts also contribute to local economic growth by attracting tourism, increasing property values, and generating green jobs in industries like forestry, eco-tourism, and sustainable agriculture. In economic terms, preserving green belts can lead to long-term cost savings by reducing the financial burden of managing urban flooding, heat waves, and pollution-related health issues. For example, the implementation of green infrastructure in Copenhagen's green belt has saved the city an estimated $15 million annually in flood management costs (EEA, 2021).

14.2 History of Greenbelt Development

The Green Belt concept has a rich and evolving history, deeply rooted in various traditions and policies across different historical and cultural contexts. It was first mentioned in biblical times when the Old Testament proposed the development of green belts around Levite towns in ancient Israel, which was later interpreted by Moses Maimonides as applicable to all cities in Israel (Wenham, 1999). Following the Islamic Tradition era, in the seventh century, the Prophet Muhammad established a green belt around Medina by prohibiting the removal of trees within a 12-mile radius, thereby setting the framework for environmentally sensitive urban planning (Bsoul et al., 2022). In 1580, Elizabeth I of England banned new construction within a 3-mile radius of London to curb the spread of the plague, a measure aimed at maintaining green spaces around urban areas. The progression into the modern period saw the systematization of Green Belt Policies in which the UK pioneered modern green belt planning. Proposals emerged in the late nineteenth century, with the London Society's 1919 plan advocating for a continuous belt to limit urban sprawl, marking the initiation of regulated urban planning for environmental protection. Currently, the Concept of Global Sustainable Development for green belts is integrated into sustainable development, encompassing green spaces and structures. Developed and developing nations, including European countries, are implementing "Green Structure Planning" to harmonize development with environmental protection (Forshaw 1943) (Table 14.1).

14.2.1 Greenbelts: Objectives and Principle

Greenbelts function as essential ecological buffers, often described as the "lungs" of urban areas due to their role in enhancing air quality, regulating local climates, mitigating urban heat island effects, and supporting biodiversity. A prominent example is the Aravalli Green Wall Project, which spans Haryana, Gujarat, and Delhi, aligning with the Union Environment Ministry's vision of establishing green corridors across India to combat land degradation and desertification. This ambitious project spans an area of approximately 6 million hectares, with plans to develop a 1,400-km-long and 5-km-wide green belt along the Aravalli Hills. Inspired by Africa's Great Green Wall, initiated in 2007 and extending from Senegal to Djibouti, the Aravalli Project reflects a global effort to restore degraded landscapes and strengthen ecological resilience.

Historically, green belt concepts have been integral to urban planning, as seen in the London Green Belt, proposed by the Greater London Regional Planning Committee in 1935. Under this policy, local authorities were empowered to acquire land for preservation as open spaces, subject to stringent restrictions on its development or sale without state approval. This pioneering move has influenced similar initiatives worldwide (Fig. 14.1).

Table 14.1 History of green belt development in chronological order

Year	Event
1898	Garden City movement—Ebenezer Howard proposes Garden Cities surrounded by Green Belts
1926	Formation of CPRE, one of whose earliest campaigns was against urban sprawl
1935	First Green Belt was proposed in an official planning policy by the Greater London Regional Planning Committee to "provide a reserve supply of public open space and recreational areas and to establish a Green Belt or girdle of open space"
1938	The local government designates the Sheffield Green Belt
1938	Green Belt (London and Home Counties) Act
1947	Town and Country Planning Act allowed local authorities to control changes in the use of land from undeveloped to developed uses
1955	The Green Belt policy for England was set out in the Ministry of Housing and Local Government Circular 42/55, which invited local planning authorities to consider establishing Green Belts in their areas
1959	Metropolitan Green Belt fully designated in local plans
1977	Professor Wangari Maathai founded the Green Belt Movement (GBM) in Kenya to address the needs of rural women
1986	Completion of the M25 motorway, running largely through the Metropolitan Green Belt
1988	Circular 42/55 was replaced with Planning Policy Guidance Note 2
1995	PPG2 was amended to add positive objectives for Green Belt land
2001	The current version of PPG2 was issued

Fig. 14.1 Objectives of greenbelt development

In addressing climate change, greenbelts play a crucial role in carbon sequestration and regulating microclimates (Singh et al., 2018; Singh et al., 2025a). The Miyawaki forest initiative in Chennai has facilitated the creation of dense green zones in urban spaces, effectively offsetting carbon emissions. Globally, Japan has utilized green buffer zones to mitigate the adverse impacts of rising temperatures and typhoons, demonstrating how greenbelt strategies can be innovatively applied to combat global warming while enhancing local resilience (Verma et al., 2023). Greenbelts act as critical tools for curbing urban sprawl by promoting compact and sustainable urban planning (Singh et al., 2024a, 2024b). In Bengaluru, the Bangalore Development Authority introduced a greenbelt policy in 1984 to regulate unplanned urban expansion and restrict land use for non-agricultural purposes. Although this policy faces enforcement challenges, significant efforts have been made to strike a balance between urban growth and environmental preservation (Xie et al., 2020).

Similarly, Toronto's Greenbelt in Canada, spanning over 2 million acres, demonstrates how greenbelts can simultaneously limit urban sprawl while supporting agricultural practices and eco-tourism. These initiatives demonstrate the crucial role that greenbelts play in maintaining a balance between urbanization and environmental conservation on both local and global scales (Joshi et al., 2020). Additionally, greenbelts enhance the urban quality of life by providing recreational spaces, promoting mental well-being, and enhancing aesthetic value. For example, the Lalbagh Botanical Garden in Bengaluru serves as a serene retreat for city residents, providing relief from urban chaos (Wang et al., 2019). Internationally, the Central Park in New York City exemplifies the importance of urban green spaces. As a public greenspace, it significantly contributes to a city's livability, particularly in addressing the challenges posed by climate change and environmental inequities (Lee et al., 2011). Its significance extends beyond recreation, positioning it as a critical infrastructure for urban sustainability. These examples illustrate the multifaceted roles that greenbelts play in enhancing urban living conditions and promoting environmental harmony, underscoring their global relevance and impact.

14.2.2 Approaches for the Development of Greenbelts

The development of greenbelts follows standardized approaches aimed at reducing pollution, providing environmental benefits, and enhancing the aesthetic value of land. These approaches integrate the principles of environmental suitability, strategic location, accessibility, and the selection of appropriate plant species to ensure the functionality and sustainability of greenbelt initiatives (Joshi et al., 2024).

14.2.2.1 Climate and Environmental Suitability

The selection of plant species for greenbelts is crucial, with a focus on local and indigenous, fast-growing tree species that are well-adapted to the region's climatic conditions. These species are preferred due to their ability to establish rapidly and their ecological compatibility (Anake et al., 2022). Vegetation should be planted in areas devoid of prior greenery to ensure adequate row spacing for optimal growth and effectiveness. Additionally, unpalatable shrub species are recommended to minimize interference from herbivores during the growth phase. Greenbelt development must adhere to standards set by regulatory bodies, such as the Central Pollution Control Board (CPCB), which emphasizes compliance with industry criteria and consideration of prevailing climatic conditions (Barwa & Singh, 2024). Species with economic, medicinal, or fruit-yielding potential are particularly encouraged as they provide additional benefits beyond environmental improvement (Patel et al., 2022; Karada et al., 2025). For instance, tree species with continuous foliage and a height of 10 m or greater are ideal for creating effective buffers around industrial facilities (Minz et al., 2021; Singh et al., 2025b). Among the modern methodologies, the Miyawaki method is an effective technique for greenbelt development. Introduced by the Japanese botanist Dr. Akira Miyawaki in the 1970s, this method involves high-density planting of native species in multi-layered arrangements, thereby simulating natural forests. First implemented in India in 2009 through collaboration between Japanese Ambassador Yoshiji Nogami and the National Environmental Engineering Research Institute (NEERI), this approach has proven effective in creating dense, fast-growing greenbelts with significant ecological benefits.

14.2.2.2 Location and Accessibility

The effectiveness of greenbelts significantly depends on their location and accessibility. Strategically located greenbelts near emission sources, such as highways, power plants, or construction sites, help intercept pollutants before they disperse into adjacent areas or sensitive ecosystems (Obi et al., 2021). They are particularly effective when situated around pollution-sensitive regions, such as agricultural lands, and serve as protective barriers. Integrating greenbelts with existing green spaces, such as parks, forests, and wildlife corridors, enhances ecological connectivity, enables wildlife movement, and supports biodiversity (Shannigrahi et al., 2000). Accessibility is equally important for ensuring that greenbelts fulfill their social, ecological, and environmental functions. Public access is crucial for promoting recreational and fitness activities, whereas operational accessibility is necessary for maintenance and management. Facilities such as hiking trails, cycling paths, and picnic areas within the green belt can enhance community engagement in ecological conservation. Furthermore, greenbelts should be easily accessible from major transportation networks, such as highways or public transit hubs, to facilitate public use and regular maintenance (Zhao, 2021).

14.2.2.3 Plant Species Suitable for Establishing Greenbelt Development

Selecting climate-resilient plantation species is crucial for adapting to and mitigating urban climate change and environmental pollution through forestry initiatives. Creating a green belt involves selecting plant species that can thrive in urban settings, enhance biodiversity, and provide essential ecological services such as carbon sequestration, air filtration, and soil stabilization (Mondal et al., 2011). The effectiveness of a green belt in intercepting and retaining atmospheric pollutants depends on several factors, *including the* shape, size, moisture level, surface texture, and nature (soluble or insoluble) of both particulate matter and gases, as well as the intercepting plant parts. Key characteristics to consider when selecting plants for greenbelts include resilience during the dry season and in warm climates, resistance to pests and diseases, the ability to withstand flooding or waterlogging, and versatility in phenological timing. Plantation species chosen based on these useful characteristics exhibit adaptive qualities that help them withstand or alter the natural stresses caused by climate change (Kumar et al., 2022). Plants are the primary components of green buffers, serving as sinks and living filters to reduce air pollution through processes such as absorption, adsorption, detoxification, accumulation, and metabolization. They achieve this without significant foliar damage or growth decline, thereby enhancing the air quality and contributing oxygen to the atmosphere (Warren, 1973). In urban and industrial zones, air toxins can be adsorbed, ingested, or accumulated in plant tissues, potentially causing varying degrees of damage if they are toxic. Delicate species are more susceptible to damage, whereas tolerant species are less susceptible. Sensitive plants serve as early warning markers of contamination, although tolerant species contribute to reducing the overall contamination level, making a difference in keeping the air cleaner. The performance of the tree species under the greenbelt was evaluated by computing their air pollution tolerance index (APTI) and anticipated performance index (API) (Mondal et al., 2011).

The air pollution tolerance index (APTI) was proposed by Singh and Rao (1983) to assess the tolerance or resistance of plants to air pollution. The air pollution tolerance index was calculated using the following formula:

$$\text{APTI} = \frac{A(T + P) + R}{10}$$

where:

A =Ascorbic Acid (mg/g)

T =Total Chlorophyll (mg/g–fresh weight)

P = pH of the leaf extract

R = Relative water content of leaf (%)

14.3 Evolving Greenbelts: A Contemporary Debate on Preservation and Progress

Greenbelts, defined as areas of vegetation or uncultivated land designated around urban landscapes, have traditionally been regarded as pivotal for preserving the ecological balance, mitigating urban sprawl, and fostering sustainable development (Xie et al., 2020). However, in the twenty-first century, the relevance and efficacy of greenbelt policies have become a subject of intense debate. This controversy arises from the interplay of rapid urbanization, climate change, and competing demands for land allocation, which challenges the traditional role of greenbelts in contemporary urban planning and environmental management (Singh, 2025). The debate surrounding greenbelt policies can be broadly categorized into two perspectives: their role as planning mechanisms and their function in environmental management.

14.3.1 Greenbelts as a Planning Mechanism

A key area of contention is the effectiveness of greenbelts as a tool in urban planning. Proponents argue that greenbelts promote resource conservation by encouraging urban renewal, efficient land use, and co-location of homes, services, and employment opportunities within existing urban areas (Obi et al., 2021). Limiting urban expansion, greenbelts are believed to prevent unchecked suburban sprawl and preserve rural landscapes (Xie et al., 2020). However, critics highlight the unintended consequences of such policies, including restrictions on affordable housing availability and the displacement of development pressures to areas beyond the green belt, resulting in leapfrog development patterns that undermine sustainability goals (Kumar et al., 2022).

Furthermore, the debate extends to whether alternative mechanisms such as purposefully designed urban green spaces, ecological corridors, and compact urban planning might offer more effective solutions for integrating green infrastructure into cities. These approaches, which emphasize multifunctionality and adaptability, can potentially achieve greater ecological, social, and economic benefits (Mondal et al., 2011). Additionally, questions have been raised regarding the rigidity of current greenbelt policies. Critics argue that introducing more flexibility in terms of permitted development, particularly for sustainable projects, could better balance environmental conservation with urban growth demands (Zhao et al., 2021).

14.3.2 Greenbelts as Environmental Management Tools

The role of greenbelts in environmental management is another focal point for debate. Advocates argue that Greenbelts serve as crucial regulatory tools for managing land use in urban peripheries, protecting biodiversity, and maintaining ecological services, including air and water purification, carbon sequestration, and microclimate regulation (Kumar & Singh, 2020). They also emphasized the potential of greenbelts to mitigate land degradation and desertification, particularly in regions facing the dual pressures of urbanization and climate change (Chahal et al. 2022a). However, critics question whether effective planning and management strategies have adequately supported the land use objectives of greenbelt policies. Concerns include the degradation of greenbelt areas owing to neglect, encroachment, and inadequate enforcement of regulations (Kumar et al., 2022). To address these challenges, innovative approaches are required to enhance the efficiency and multifunctionality of greenbelts. These could include the integration of agroforestry practices, restoration of degraded ecosystems, and development of green infrastructure that connects urban areas with natural habitats.

Moreover, the debate underscores the need for a holistic and adaptive approach to green belt planning and management. This involves reconciling competing land-use demands, fostering community engagement, and leveraging technological advancements, such as GIS and remote sensing, to improve the monitoring and evaluation of greenbelt performance (Shannigrahi et al., 2000).

14.4 Contemporary Challenges in Greenbelt Dynamics

Greenbelts face numerous challenges in the twenty-first century, driven by rapid urbanization, economic pressure, and governance shortcomings. These challenges have increasingly called into question the efficacy of greenbelt policies in balancing conservation and development objectives, particularly in both developing and developed countries (Singh et al., 2024a, 2024b). Rapid population growth and urbanization have exerted immense pressure on green belt areas. For instance, in India, the Mumbai Metropolitan Region has witnessed widespread encroachment and illegal construction of designated greenbelts, undermining their ecological and social functions. Globally, the UK's Green Belt policy has faced criticism for limiting residential development, which has contributed to skyrocketing property prices in urban areas. These pressures highlight the tension between accommodating growing urban populations and preserving green belt spaces. Poorly planned and fragmented greenbelts often lead to unintended consequences such as leapfrog development and unregulated urban expansion.

In India, the green belt surrounding the Delhi-NCR has failed to fulfill its intended purpose, as unregulated construction activities have proliferated beyond its boundaries, creating fragmented urban landscapes (Jain et al., 2011). Similarly, in South Africa, inconsistent enforcement of greenbelt regulations has resulted in disorganized urban planning and weakened ecological benefits (Chihambakwe et al., 2024). Effective land management strategies and stricter enforcement mechanisms are essential for addressing these issues and enhancing the functionality of greenbelts. Greenbelts often face conflicts with economic priorities such as infrastructure development and housing demand. For instance, Bangalore's Peripheral Ring Road Project threatens significant portions of the city's greenbelt as authorities prioritize alleviating traffic congestion while preserving green spaces. Globally, China has adjusted its greenbelt policies in urban centers to accommodate economic zones, illustrating the complex challenge of balancing urban growth with environmental conservation (Sun et al., 2021). These examples underscore the need for integrated planning approaches that consider both economic and environmental objectives.

Maintaining greenbelts requires significant financial investments and robust governance, both of which are often inadequate. In India, the Yamuna floodplains in Delhi, designated as green zones, have been heavily encroached upon for farming and urban development, resulting in significant degradation of these critical ecological areas. Globally, Brazil's efforts to preserve green buffer zones around cities, such as São Paulo, face challenges from illegal land use and deforestation, which continue to erode these green spaces. Strengthening governance frameworks and ensuring community participation are critical to mitigating these issues.

14.5 Advancements and Future Directions in Greenbelt Development

In recent years, the integration of greenbelts into urban planning has become a prominent focus in addressing environmental challenges while supporting urban growth. Modern greenbelt strategies are increasingly emphasizing the importance of connecting these green spaces with urban infrastructure to create more sustainable, ecologically rich urban environments. A notable example of this integration is found in Pune, where the city has successfully developed biodiverse corridors linking urban parks and greenbelts. This connection fosters effective land-use planning and contributes to the city's overall ecological resilience. On a global scale, Singapore's "City in a Garden" vision exemplifies another effective approach to integrating greenbelts into the urban fabric. This strategy has enabled the city-state to strike a balance between urbanization and nature, promoting environmental sustainability while maintaining a high standard of living. Incorporating community involvement into greenbelt development has yielded significant positive outcomes. For example, the Delhi Ridge restoration project actively engages local NGOs and residents in afforestation efforts, which not only helps restore vital green spaces but also fosters a sense of ownership and responsibility among citizens. Similarly, in Rwanda, the Green Belt Initiative, which forms part of the country's climate resilience strategy, engages local communities in restoring degraded lands, contributing to both environmental and social benefits (MOE, 2011). These projects highlight the value of community-driven approaches to greenbelt management, demonstrating that local participation can enhance the effectiveness and sustainability of green space initiatives.

The efficacy of greenbelts is also contingent on the adaptability of policies that guide their management. Dynamic policy frameworks that can adapt to changing urban needs have proven essential for the continued success of greenbelts. For example, Kerala's buffer zone policy, which governs protected areas, strikes a balance between ecological preservation and livelihood concerns, illustrating how policy flexibility can accommodate both conservation and development needs. Similarly, in the Netherlands, the use of multi-functional greenbelts supports a variety of activities, including agriculture, tourism, and conservation, allowing greenbelts to serve diverse purposes while contributing to environmental sustainability. These examples underscore the importance of adaptive policies in maintaining the relevance and effectiveness of greenbelt systems in the face of evolving urban dynamics. Technological advancements have also played a significant role in monitoring and maintaining greenbelts, ensuring their long-term sustainability. For example, Hyderabad's Haritha Haram Project utilizes GIS mapping to track tree plantations within designated green zones, offering an efficient and data-driven approach to greenbelt management. Globally, countries like Germany have adopted satellite imagery to monitor urban greenbelts, helping to prevent encroachments and maintain the integrity of green spaces (Kumar and Singh, 2020). The integration of technology into greenbelt monitoring is becoming increasingly crucial in protecting these spaces from urban pressures, such as encroachment and degradation.

The debate over greenbelts often centers around the tension between preservation and development. Proponents of conservation argue that greenbelts are invaluable ecological assets that provide critical services such as air purification, habitat for wildlife, and carbon sequestration (Obi et al., 2021). They assert that destroying these green spaces would result in irreversible environmental damage and exacerbate the impacts of climate change, which already demands increased investment in green infrastructure. On the other hand, advocates for development emphasize the need to expand cities to accommodate growing populations, arguing that greenbelt policies can create inefficiencies in urban planning, such as limiting available land for housing and infrastructure (Lisina et al., 2024). They also suggest that innovative alternatives, such as vertical gardens and urban forests, could replace traditional greenbelts, offering sustainable solutions that meet urban needs while still providing ecological benefits.

To resolve the tension between preservation and development, a balanced approach is necessary. Greenbelt policies must be flexible, accommodating both urban growth and environmental conservation (Pongsiri, 2012). Public–private partnerships can play a key role in the development and maintenance of greenbelts, ensuring that both businesses and NGOs contribute to their sustainability and effectiveness. Furthermore, cities must adopt mixed-use urban planning that integrates green spaces at every level of development, promoting both ecological sustainability and high-quality urban living (Marx, 2019). Additionally, raising community awareness about the value of greenbelts can inspire grassroots efforts to protect these spaces and encourage local involvement in their conservation.

Looking ahead, the future of greenbelts in the twenty-first century hinges on their ability to adapt to modern challenges, including rapid urbanization, economic pressures, and climate change. Recent examples from India and around the world demonstrate the potential of greenbelt policies to contribute to both environmental and social resilience. Adopting innovative strategies, such as integrating greenbelts with urban infrastructure, engaging communities in conservation, and utilizing technology for monitoring, can help greenbelts continue to serve as crucial ecological buffers while supporting sustainable urban development.

14.6 Greenbelt: Driving Towards Sustainable Future

Greenbelts are more than mere aesthetic enhancements to urban landscapes. They are vital components of sustainable cities, providing a range of environmental, social, and economic benefits. These natural buffers not only enhance the beauty of neighborhoods but also play a crucial role in addressing urban challenges, including air quality, water management, climate change, and public health. Greenbelts act as powerful natural air filters, significantly improving urban air quality. Trees with dense canopies reduce particulate matter (PM10 and PM2.5) concentrations, while mixed forests of coniferous and broad-leaved species maximize the retention of pollution (Liu et al., 2023). The structure and diversity of plant communities, including trees, shrubs, and herbs, enhance their effectiveness in capturing pollutants and providing cooling benefits. Greenbelts also serve as effective sinks for pollutants. For example, a 500-m-wide greenbelt near a petroleum refinery demonstrated removal efficiencies of 36–40% for SO_2, NOx, and SPM, as well as an impressive 84–94% for THC, VOC and CO (Gupta et al., 2008).

Additionally, soil carbon pools and respiration dynamics in urban green spaces significantly contribute to urban carbon sinks, making greenbelts essential for achieving the Sustainable Development Goals. Urban forests intercept rainfall, allowing gradual infiltration into the soil and reducing surface runoff (Rahman et al., 2023). Their root systems act as natural sponges, retaining excess stormwater and filtering out toxins before they reach waterways. Mitigating stormwater runoff and improving water quality through greenbelts plays a crucial role in maintaining urban hydrological balance (Stan et al., 2018). Healthy root systems in urban forests also enhance soil conservation by increasing soil biodiversity, which is crucial for nutrient cycling and overall soil health (Hossack et al., 2004). These functions collectively support ecosystem stability, making greenbelts indispensable for sustainable urban development. Access to urban green spaces significantly enhances mental and physical well-being. These sanctuaries promote psychological restoration, alleviate depression and anxiety, and improve sleep quality, especially for vulnerable populations like cancer survivors (Jeon et al., 2024). Proximity to nature has been shown to reduce stress levels and enhance mental clarity, fostering healthier lifestyles.

Greenbelts also promote physical activity, providing spaces for jogging, yoga, and other recreational activities that contribute to overall health (Jeon et al., 2024). Beyond individual benefits, these spaces strengthen community bonds by providing venues for social interactions, fostering a sense of unity in increasingly fragmented urban environments. Greenbelts play a key role in combating climate change. They sequester significant amounts of carbon dioxide (CO_2), with studies in areas like Luohe, China, showing sequestration rates of 1.30 t C ha^{-1} yr^{-1}. Landscape connectivity and greening rates further enhance these benefits. Additionally, greenbelts help mitigate the urban heat island effect, a phenomenon in which cities experience elevated temperatures due to their infrastructure and human activities (Gaur & Singh, 2014). The creation of shade and cooling urban environments by greenbelts may reduce energy demands for cooling and improve overall livability (Singh (2024a).

14.6.1 Urban Forest in New Urban Agenda (NUA) and Sustainable Development Goals (SDGs)

In 2015, the global community sanctioned two key development agreements: the Paris Climate Change Agreement and the 2030 Agenda for Sustainable Development, both of which focus on urban sustainability (Seeliger et al., 2013). The 2030 Agenda, which builds on the Millennium Development Goals, calls on nations to mobilize efforts to remove poverty, reduce inequality, and address climate change, ensuring that no one is left behind. It includes 17 Sustainable Development Goals (Das et al., 2023). The sustainability goals of the Paris Agreement are heavily reliant on the role of cities, a fact highlighted during the 22nd session of the United Nations Framework Convention on Climate Change (UNFCCC), held in Marrakech, Morocco, in 2016. A key outcome of this conference was the adoption of the New Urban Agenda (NUA), which outlines a global strategy to address the challenges of urbanization in the coming decades ("The Role of Cities in Low(er)-Carbon Transition", 2023). The NUA advocates for the development of urban plans centered on enhancing the quality of life for residents, ensuring they can thrive rather than merely survive. It is built on three interconnected pillars: environmental sustainability, inclusive and sustainable urban economies, and the principle of leaving no one behind. The agenda is grounded in the belief that well-planned and managed urbanization can serve as a powerful driver of sustainable development in both developed and developing nations. Additionally, it underscores its alignment with the 2030 Agenda for Sustainable Development, emphasizing its role in advancing the global sustainability framework (Aleha et al., 2024).

According to the NUA and the SDGs, particularly SDG 11, urban green spaces, such as urban forests and greenbelts, are essential for enhancing urban living conditions, fostering community cohesion, and promoting environmental sustainability. As a result, countries are committed to developing safe, welcoming, accessible, and environmentally sustainable public spaces (Fig. 14.2).

Fig. 14.2 The role of urban forests in achieving sustainable development goals (Aleha et al., 2024)

14.6.2 Greenbelts for Carbon Neutral Future

A carbon–neutral future refers to achieving a balance between the carbon dioxide (CO_2) emissions released into the atmosphere and the amount removed or offset. This concept is pivotal in addressing climate change, requiring significant reductions in greenhouse gas (GHG) emissions to as close to zero as possible, with any remaining emissions offset through measures such as reforestation, carbon capture, or renewable energy credits (Gan & Zhang, 2022). Transitioning to carbon neutrality requires widespread societal shifts, including a shift away from fossil fuels and the adoption of renewable energy sources, such as solar, wind, and hydropower. Additionally, energy efficiency technologies, the electrification of transportation and heating systems, and

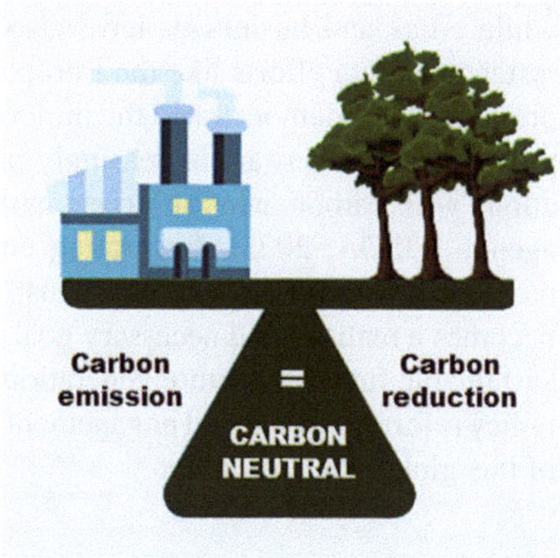

Fig. 14.3 The concept of carbon neutral

the implementation of sustainable agricultural and forestry practices are fundamental components of this transition. Advanced innovations, such as carbon capture and storage (CCS) and direct air capture (DAC), also play critical roles in extracting CO_2 directly from the atmosphere (Adun et al., 2024) (Fig. 14.3).

Achieving a carbon–neutral future aligns with international frameworks, such as the Paris Agreement, which aims to limit the global temperature rise to well below 2 °C above pre-industrial levels, with an aspirational goal of 1.5 °C. This global commitment has motivated many nations, cities, and organizations to set ambitious carbon neutrality targets, often by the mid-twenty-first century. For instance, the European Union has pledged to achieve net-zero emissions by 2050 under the European Green Deal, while the United States has committed to similar timelines through various climate initiatives. Additionally, private sector actors like Microsoft and Amazon have adopted carbon–neutral goals by leveraging renewable energy, optimizing operational sustainability, and investing in carbon offset programs (USDA, 2023).

The transition toward carbon neutrality requires a comprehensive strategy involving multiple components. A critical element is the significant reduction of emissions across key sectors, including energy, transportation, agriculture, and industry. This consists of transitioning to renewable energy systems, such as solar, wind, and hydropower, alongside the adoption of energy-efficient technologies to minimize resource consumption. Carbon sequestration also plays a vital role, relying on natural systems, such as forests, soils, and oceans, as well as engineered solutions like Carbon Capture and Storage (CCS), to absorb and store atmospheric CO_2 (Kale et al., 2024). Remaining emissions, which are challenging to eliminate, can be offset through projects such as afforestation, renewable energy development, or investments in innovative technologies that reduce emissions in other areas. Effective policy and governance mechanisms are essential to achieving carbon neutrality. International agreements, such as the Paris Agreement, provide a framework for nations to establish carbon–neutral objectives, ensuring alignment with global efforts to mitigate climate change. Many governments have set national-level commitments, while cities and businesses have also stepped forward with their initiatives. For instance, global efforts like the European Green Deal and various U.S.-led climate programs have demonstrated the importance of coordinated policy action. Similarly, private corporations are increasingly prioritizing sustainability, aligning their operations with carbon–neutral targets, and contributing to the global decarbonization agenda (USDA, 2023). Integrating emissions reductions, advanced sequestration methods, offset strategies, and robust policy frameworks, a carbon–neutral future becomes a realistic and necessary goal for mitigating climate change and ensuring a sustainable future for future generations. The interplay of technological innovation, policy reform, and societal engagement will be crucial in navigating the complexities of this global transformation.

14.6.3 Green Belts in Carbon Sequestration

Green belts play a fundamental role in carbon sequestration, offering significant contributions to climate change mitigation by absorbing and storing atmospheric CO_2. These areas, particularly those rich in vegetation and healthy soils, act as vital carbon reservoirs that stabilize the global carbon cycle. The trees of green belts are exceptionally efficient in sequestering carbon through the process of photosynthesis, storing it within their trunks, branches, leaves, and roots. Forested regions, often integral to green belts, collectively absorb approximately 2.6 gigatons of CO_2 annually, accounting for nearly one-third of anthropogenic emissions. Urban green belts also make a notable contribution to carbon sequestration. For instance, a study conducted in Kalaburagi, India, reported carbon stocks of 89.436 tons and CO_2 stocks of 327.871 tons from over 500 urban trees, highlighting their significant potential to capture and store atmospheric carbon (Shreyas & DJ, 2024). In addition to vegetation, soils within green belts play a crucial role in carbon storage. Healthy soils are capable of sequestering organic carbon, playing a critical role in stabilizing the global carbon cycle (Figs. 14.4 and 14.5).

It has been reported that soils globally sequester up to 3.8 gigatons of carbon annually, making them indispensable in reducing atmospheric CO_2 levels. This underscores the importance of maintaining soil health within green belts to optimize their carbon storage potential. Urban green belts, bolstered by afforestation and reforestation initiatives, further enhance carbon capture capabilities and help offset local emissions. For example, reforestation projects in urban and peri-urban areas have been documented to absorb substantial amounts of CO_2. Initiatives like the USDA's Climate Solutions Act promote urban afforestation and leverage carbon markets to maximize the sequestration benefits of green belts (USDA, 2023). These efforts not only enhance local air quality but also align with global objectives for reducing carbon emissions.

Recent data emphasizes the critical role of terrestrial ecosystems, including green belts, in global carbon management. According to the Global Carbon Budget (2023), land ecosystems sequestered 3.8 gigatons of CO_2 in 2022, offsetting a portion of the

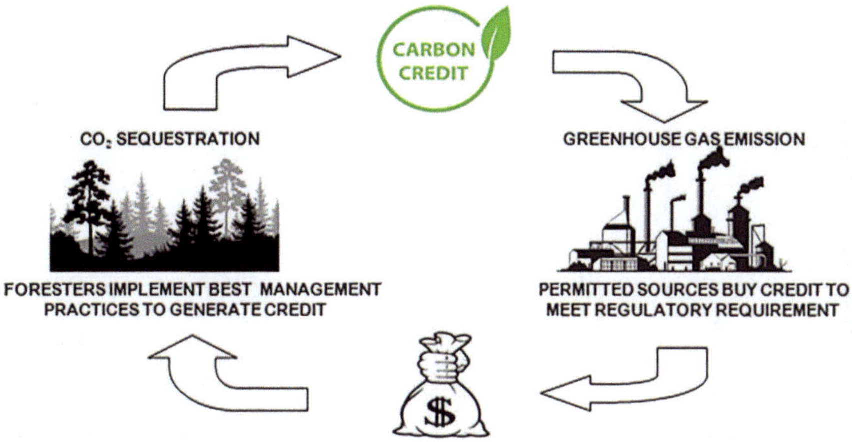

Fig. 14.4 Generation of carbon credit from urban vegetation

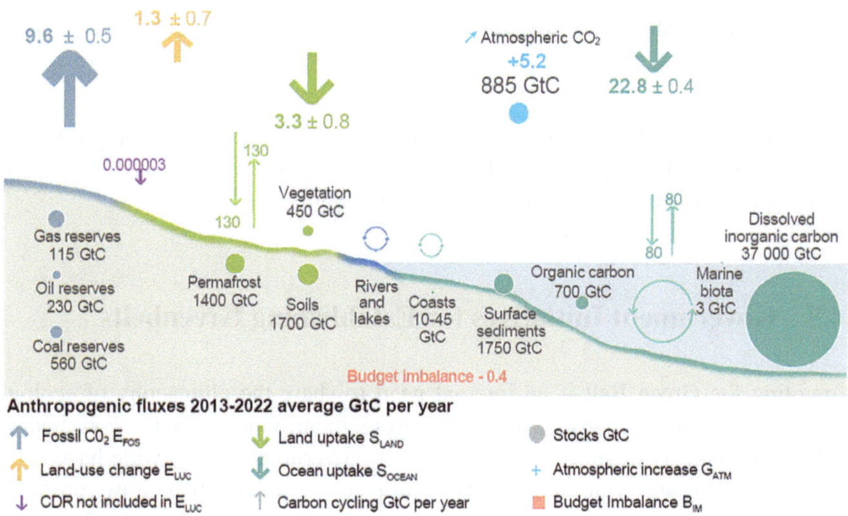

Fig. 14.5 Global carbon cycle (Global carbon budget 2023)

40.7 gigatons of CO_2 emitted by human activities in the same year. Urban initiatives have also gained traction in integrating green belts into climate action plans, as seen in China, where urban reforestation projects are projected to sequester approximately 0.5 gigatons of CO_2 by 2030 (USDA, 2023). Additionally, forest-based carbon markets are emerging as a dual ecological and economic strategy, with the U.S. Department of Agriculture highlighting the growing adoption of forestry-based carbon credits by landowners to support climate mitigation efforts (USDA, 2023) (Fig. 14.6).

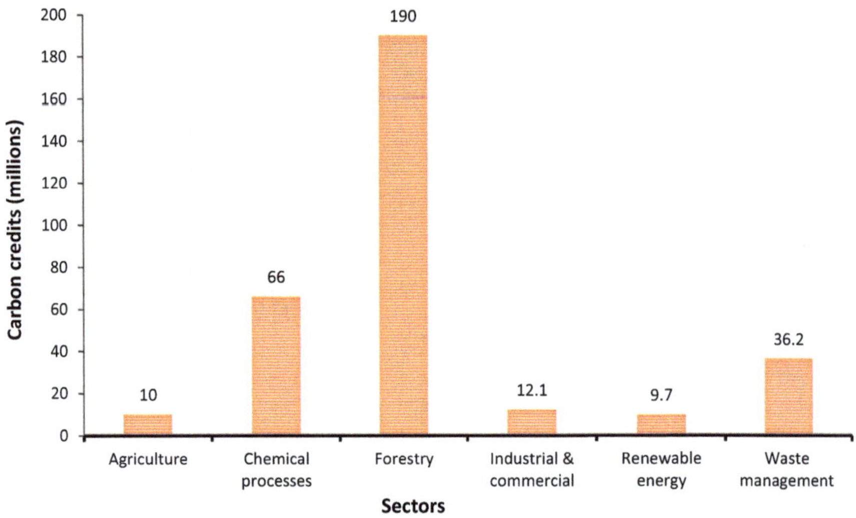

Fig. 14.6 Carbon credits issued from forestry in the United States from 2013 to 2022 (USDA, 2023)

Through their ability to absorb and store carbon, green belts not only mitigate climate change but also provide additional benefits, including improved air quality, biodiversity conservation, and enhanced urban livability.

14.7 Government Initiatives for Establishing Greenbelts

Promoting the Green Belt is an internal need to share the philosophy of ecological civilization and achieve sustainable development. Guided by the concepts of ecological civilization and green development, governments worldwide have been actively initiating and promoting green belt projects to combat urban sprawl, preserve biodiversity, and promote sustainable development (Brauteseth et al., 2023). Green development has become a common goal and a key component of global governance. Promoting the Belt and Road Initiative aligns with the international trend of pursuing green, low-carbon, and circular development. It is an effective means to reinforce the stimuli for continuous and healthy economic development. To promote the Green Belt, efforts will be made to incorporate the principles of resource efficiency and environmental friendliness into the entire process of international cooperation on production capacity and equipment manufacturing. It is a key move to serve and forge communities of shared interests, common responsibility, and common destiny (Li et al., 2024). Challenged by more severe global and regional environmental issues, countries also need a favourable ecological environment to achieve economic and social development. It is a shared responsibility of countries worldwide to prevent and mitigate environmental pollution and ecological damage. Promoting the Green Belt is conducive to fostering pragmatic cooperation, encouraging green investment, trade, and financial systems, thereby achieving a win–win situation for economic growth and environmental protection and building communities of shared interests, common responsibility, and common destiny. Below are some typical government initiatives towards green belts.

14.7.1 Policy Frameworks and Legislation

Green belts are integral to urban planning, incorporated into city master plans by governments to regulate urban expansion and limit sprawl, as evidenced by their role in cities like Beijing, which demonstrates how green belts can significantly constrain urban growth. Zoning laws play a critical role in this process, designating areas as green belts with legal protections to restrict industrial and commercial development. These regulations vary across countries, with notable examples in the UK, Canada, and Russia, which focus on securing green belt zones from urban encroachment (Lisina et al., 2024). In India, cities like Bangalore utilize zoning laws to earmark green belt areas, controlling urban sprawl and maintaining ecological balance (Eswar, 2021). Environmental Protection Acts further reinforce green belt policies by preserving green spaces and promoting environmental sustainability. These laws encourage industries to adopt sustainable practices and comply with environmental standards (Kumar & Dwivedi, 2023). Countries worldwide have adopted policies to establish green belts around cities, restricting urban development to prevent encroachment into rural or natural areas. The UK's Green Belt policy, implemented since the mid-twentieth century, is a prominent example. It ensures cities like London and Manchester preserve surrounding green spaces, safeguarding agricultural land, promoting biodiversity, and reducing pollution. Integrating urban planning policies, zoning laws, and environmental protection acts enables governments globally to aim to create sustainable urban environments that balance development with ecological preservation.

14.7.2 Reforestation and Afforestation Programs

Governments worldwide are implementing large-scale tree-planting campaigns and restoration projects to enhance green belt areas, addressing environmental challenges and promoting ecological balance. These initiatives aim to improve air quality, sequester carbon, and restore degraded lands. China's reforestation programs have transformed vast expanses into thriving green belts, mitigating air pollution and fostering biodiversity in urban regions (Yao et al., 2024). Restoration projects focus on rehabilitating degraded green belt lands by converting abandoned areas into parks or forests, thereby significantly boosting local biodiversity. Governments often encourage these efforts within green belts to ensure long-term environmental sustainability. In India, the Green India Mission under the National Action Plan on Climate Change exemplifies such initiatives. It aims to increase forest cover and enhance ecosystem services through afforestation in urban areas and green belts. These combined efforts demonstrate the transformative potential of tree planting and restoration initiatives in creating sustainable and resilient green belt ecosystems.

14.7.3 Community Participation and Awareness

The success and sustainability of green belts rely heavily on active community participation and the support of private landowners. Governments and organizations have implemented various strategies to engage the public and incentivize landowners to contribute to the preservation and maintenance of these vital ecological spaces. Involving local communities in green belt conservation is a key approach to fostering environmental stewardship. Initiatives often include awareness drives, workshops, and citizen science programs that empower individuals to take an active role in maintaining green spaces (Kumar et al., 2020). A notable example is the program in Cikelat Village, Java, where workshops and skill-sharing sessions have raised environmental awareness and encouraged eco-friendly practices among residents. These community-driven efforts demonstrate the potential of public engagement to drive sustainable outcomes (Sidik et al., 2024). Governments have also adopted measures to encourage private landowners to maintain green belts on their property. Incentives such as subsidies or tax benefits make conservation financially viable for landowners. For instance, Australia offers concessional tax treatments for conservation covenants, aligning with international sustainability obligations. This model has proven effective in promoting private land conservation while supporting broader environmental goals (Stoianoff & Kelly, 2007).

14.7.4 Economic Measures

Green belts are essential for sustainable urban and environmental planning, and their development relies on strategic funding, infrastructure integration, and innovative use cases such as eco-tourism. Governments and organizations worldwide are employing diverse approaches to ensure the long-term sustainability and multifaceted utility of green belts. Financial support is a cornerstone of green belt development, with governments allocating funds and providing incentives for the creation and maintenance of these areas. Many nations employ a combination of public funds, private investments, and green financing mechanisms, such as carbon credits, to fund green belt initiatives (Sternik, 2024). These funding models not only ensure the financial sustainability of green belts but also align with global environmental goals. Green belts are increasingly integrated into urban development strategies through programs like the Green Infrastructure Program in the United States. This initiative promotes the creation of green spaces, including green belts, to manage stormwater and mitigate the environmental impact of urbanization. Reducing stormwater runoff through green infrastructure minimizes flooding risks and enhances water quality by filtering pollutants (Kriebel et al., 2024). These efforts underscore the role of green belts as functional components of urban ecosystems. Green belts also offer economic opportunities through eco-tourism. Developing these areas into sustainable tourism destinations generates revenue while preserving natural habitats. For instance, the implementation of green practices in eco-tourism in Belgrade has enhanced the tourist experience while promoting environmental conservation (Šimičević et al., 2024). This approach highlights the dual benefits of green belts as both ecological assets and economic drivers.

14.7.5 Integrated Urban Development Programs

Green belts are increasingly recognized as critical elements of sustainable urban development, particularly within the frameworks of smart cities and green infrastructure. Smart city initiatives leverage technology and innovative planning to create sustainable urban environments (Singh (Ed.) (2024). Green belts are integral to these plans, as they effectively limit urban expansion and mitigate sprawl. For example, urban green belts (UGBs) in cities like Beijing have significantly reduced sprawl rates, showcasing their role in controlling unchecked urban growth while maintaining ecological balance. Green belts are often designed to include parks, urban forests, and waterways, serving both ecological and recreational purposes. The European Union's Green Infrastructure Strategy emphasizes creating interconnected green spaces, including green belts, to address critical urban challenges such as heat islands, air pollution, and biodiversity loss. These interlinked spaces not only enhance urban resilience but also provide accessible recreational areas for residents, aligning with the goals of the EU Biodiversity Strategy for 2030. Globally, countries have integrated green belts into national urban development strategies. Japan's Comprehensive National Land Development Plan exemplifies this approach, incorporating green belts within cities to promote ecological balance and recreational spaces. These green belts enhance urban livability by providing shade, improving air quality, and supporting biodiversity (Yokohari et al., 2000).

14.7.6 Biodiversity Conservation Programs

Green belts serve as vital areas for the protection of wildlife and biodiversity, playing an essential role in preserving and enhancing natural habitats. Governments and organizations globally recognize the ecological significance of green belts and implement targeted policies to safeguard and restore biodiversity within these zones. Green belts are designated as critical areas for the conservation of wildlife and the preservation of biodiversity. These spaces provide refuge for various species and serve as ecological corridors, connecting fragmented habitats and supporting the migration and survival of species (Neugarten et al., 2024). Many countries have adopted Biodiversity Action Plans (BAPs) to integrate green belts into their conservation strategies. For instance, the UK Biodiversity Action Plan (1992–2012) focused on preserving natural habitats within green belts, encouraging the restoration of native species, and improving the ecological health of these areas. The biodiversity in green belt planning has helped maintain ecological balance and enhance the resilience of natural systems.

14.7.7 International Collaboration

The development of green belts is increasingly aligned with global sustainability frameworks and regional cooperation to address ecological challenges on a broader scale. The adoption of global goals and the creation of cross-border green belts may help governments and organizations contribute to environmental conservation and sustainable development worldwide. Green belt initiatives often align with international frameworks such as the United Nations' Sustainable Development Goals (SDGs). Specifically, SDG 11: Sustainable Cities and Communities emphasizes the need for inclusive, safe, resilient, and sustainable urban spaces. Incorporating green belts into urban planning supports these objectives by fostering environmental sustainability, reducing urban sprawl, and enhancing urban resilience. Transnational green belts address ecological concerns at regional or global scales, promoting environmental quality and fostering international collaboration. For example, China's cross-border e-commerce pilot zones have demonstrated how integrated policies can significantly reduce industrial pollution, indirectly supporting ecological balance across borders (Li et al., 2024). Such initiatives demonstrate the potential of cross-border green belts to address shared environmental challenges while fostering sustainable economic activities.

14.7.8 Public–Private Partnerships (PPP)

Governments often collaborate with private companies, NGOs, and local communities to maintain and expand green belts. For example, Singapore's Parks and Recreation Master Plan involves partnerships with private stakeholders to create green spaces within and around urban areas. Parks are viewed as vital assets that contribute to national identity and social cohesion, reflecting Singapore's commitment to sustainability and economic growth (Henderson, 2013).

14.7.9 Indian Government Initiatives

India has implemented several significant initiatives to address environmental challenges, enhance forest cover, and promote ecological sustainability through the development and expansion of green belts. These programs, at both national and state levels, demonstrate a comprehensive approach to combating desertification, preserving biodiversity, and encouraging sustainable practices. The National Mission for a Green India (GIM) is one of the eight missions under the National Action Plan on Climate Change (NAPCC). It focuses on protecting, restoring, and enhancing India's diminishing forest cover while emphasizing biodiversity conservation, water resource management, and ecosystem services. GIM aims to integrate sustainable practices into forest and landscape management, contributing to climate resilience and ecological balance (Singh and Kumar, 2022). The Great Green Wall of India is a visionary initiative by the government to create a 1,400 km long and 5 km wide green belt from Gujarat to the Delhi-Haryana border. This project aims to combat desertification and land degradation, particularly in the vulnerable Aravalli region. Establishing a continuous green corridor, the initiative seeks to restore ecological health and mitigate the adverse effects of climate change. Launched in October 2023, the Green Credit Programme is a market-based mechanism designed to incentivize voluntary environmental actions across various sectors. The program encourages individuals, societies, industries, and private companies to engage in sustainable practices, such as tree planting on degraded wastelands. The offering of green credits fosters a culture of environmental responsibility and active participation in ecological restoration. At the state level, initiatives like Haryana's Green Wall project complement national efforts. This project involves creating a five km-wide green belt along the Aravalli Mountain range, aiming to combat desertification, restore degraded land, and promote ecological sustainability. State-level programs exemplify localized approaches to addressing broader environmental challenges. These initiatives reflect India's commitment to environmental conservation and sustainable development through the creation and expansion of green belts across the nation.

14.8 Conclusion

It is concluded that the greenbelts are foundational to sustainable urban planning and climate resilience, offering a wide range of ecological, social, and economic benefits. They serve as critical buffers against urban sprawl, contribute to carbon sequestration, enhance biodiversity, and regulate urban microclimates. Integrating nature into cityscapes, greenbelts support public health, foster environmental stewardship, and strengthen cities' adaptive capacity to climate change. Despite challenges such as encroachment, policy rigidity, and land-use pressures, greenbelts remain a vital strategy for achieving carbon–neutral, livable, and resilient urban environments. Future success will depend on adaptive governance, community involvement, and innovative ecological approaches that align greenbelt development with global sustainability goals. As cities continue to grow, there is a need to prioritize greenbelts as indispensable tools for harmonizing urban expansion with environmental conservation, which may direct us to achieve ecological sustainability.

References

Adun, H., Ampah, J. D., Bamisile, O., Ozsahin, D. U., & Staffell, I. (2024). Sustainability implications of different carbon dioxide removal technologies in the context of Europe's climate neutrality goal. *Sustain Produc Consump, 47*, 598–616. https://doi.org/10.1016/j.spc.2024.04.003

Aleha, A., Zahra, S. M., Qureshi, S., Shah, S., Marri, S. A., & Khan, M. (2024). Urban forests and their contribution to sustainable urban development in a global context: A case study of Multan. *Pakistan. Front Clim, 6*, 1275102. https://doi.org/10.3389/fclim.2024.1275102

Amati, M., & Yokohari, M. (2007). The establishment of the London greenbelt: Reaching consensus over purchasing land. *Journal of Planning History, 6*(4), 311–337. https://doi.org/10.1177/1538513207302695

Anake, W. U., Bayode, F. O., Jonathan, H. O., Omonhinmin, C. A., Odetunmibi, O. A., & Anake, T. A. (2022). Screening of plant species response and performance for green belt development: Implications for semi-urban ecosystem restoration. *Sustain, 14*(7), 3968. https://doi.org/10.3390/su14073968

Barwa, S., & Singh, H. (2024). Climate-induced changes in forest ecosystems of Indian Himalayan region: A meta-analysis. In *urban forests, climate change and environmental pollution: physio-biochemical and molecular perspectives to enhance urban resilience* (pp. 839–866). Cham: Springer Nature Switzerland.

Bhattacharya, A., Saikia, K., Takhelmayum, M., & Sarkar, P. (2020). Carbon sequestration in the bio-edaphic ecosystem of National Highway-27 in Guwahati, Assam. *India. Heliyon, 6*(9), Article e04969.

Brauteseth, B. L., & Bani, D. (2023). Transplanting China's ecological civilization to the belt and road initiative: What will it take? *J Int Develop, 35*(4), 639–654. https://doi.org/10.1002/jid.3698

Bsoul, L., Omer, A., Kucukalic, L., & Archbold, R. H. (2022). Islam's perspective on environmental sustainability: A conceptual analysis. *Social Sciences, 11*(6), 228. https://doi.org/10.3390/socsci11060228

Chahal, R, Verma, S. K., & Singh, A. K. (2022a). Effect of biofertilizers nutrient uptake by green gram. In Vigna radiata (L.) Wilczek, under agri-horti system in Vindhyan region of Eastern Uttar Pradesh. *International Journal Plant Soil Science 34*(23), 686–691. https://doi.org/10.9734/IJPSS/2022/v34i232476

Chihambakwe, F., & Moyo, I. (2024). An Assessment of urban greening challenges and opportunities associated with urban densification in the EThekwini Municipality. *South African Geogr J, 106*(3), 268–289. https://doi.org/10.1080/03736245.2023.2272905

Das, N., & Mondal, D. (2023). Sustainable development goal 10: reducing inequality-evidence from India and its States. *Scholars Journal of Arts Humanities and Social Science, 12*, 328–335. https://doi.org/10.36347/sjahss.2023.v11i12.001

EEA. (2021). Urban adaptation to climate change in Europe. European Union.

Eswar, M. (2021). The green belt of Bangalore: planning and the socio-economic context. *Theoretical and Empirical Researches in Urban Management, 16*(2), 21–38.

Forshaw, J. H., & Abercrombie, P. (1943). County of London plan.

Gan, S., & Zhang, W. (2022). Building carbon neutrality goals break down strategies for sustainable energy development. *Int J Emerg Electric Power Syst, 23*(6), 899–911. https://doi.org/10.1515/ijeeps-2022-0152

Gaur, R., & Singh, H. (2024). Urban tree responses to climate change and environmental pollution. In *Urban forests, climate change and environmental pollution: Physio-biochemical and molecular perspectives to enhance urban resilience* (pp. 279–305). Cham: Springer Nature Switzerland.

Gupta, R. B., Chaudhari, P. R., & Wate, S. R. (2008). Overview on attenuation of industrial air pollution by greenbelt. *Journal of Industrial Pollution Control, 24*(1), 1–8.

Henderson, J. C. (2013). Urban parks and green spaces in Singapore. *Managing Leisure, 18*(3), 213–225. https://doi.org/10.1080/13606719.2013.796181

Hossack, I., Robertson, D., Tucker, P., Hursthouse, A., & Fyfe, C. (2004). A GIS and web-based decision support tool for the management of urban soils. *Cybernet Syst an Int J, 35*(5–6), 499–509. https://doi.org/10.1080/01969720490451878

Jain, M., Taubenböck, H., & Namperumal, S. (2011). Seamless urbanisation and knotted city growth: Delhi Metropolitan Region, pp. 853–862.

Jeon, Y., Kim, Y. S., Kim, G. Y. & Yeon, P. S. (2024). Effects of urban forest therapy on depression, anxiety, and sleep quality in people experiencing cancer. *Journal People Plants Environmental 27*(4), 303–317. https://doi.org/10.11628/ksppe.2024.27.4.303

Joshi, R., Singh, H., Chhetri, R., & Yadav, K. (2020). Assessment of carbon sequestration potential in degraded and non-degraded community forests in Terai region of Nepal. *Journal of Forest and Environmental Science, 36*(2), 13–121.

Joshi, R., Sharma, B., Singh, H., Dhakal, N., Ayer, S., & Maraseni, T. (2024). Poplar plantation as an agroforestry approach: Economic benefits and its role in carbon sequestration in north India. *Journal of Resources and Ecology, 15*(4), 880–888. https://doi.org/10.5814/j.issn.1674-764x.2024.04.009

Kale, I., Kumar, N. M., Singh, H., Ali, S., Singh, A. K., & Hareesh, D. (2024). Urban ecosystem services, futuristic trends in agriculture engineering & food sciences. Vol. 3 Book 2, IIP Series, *3*, 117–128. https://doi.org/10.58532/V3BCAG2P1CH13

Karada, M. S., Agnihotri, D., Singh, A. K., Singh, B. K., Mishra, R., Prasad, S., & Singh, Y. (2025). Optimizing chickpea yield and profitability through phosphorus application and PSB inoculation in a custard apple (Annona squamosa) based agri-horti system. *Plant Science Today, 12*(4): 01–07. https://doi.org/10.14719/pst.8912

Kriebel, M., Liu, T., Ramirez-Marquez, J. E., & Prigiobbe, V. (2024). Modelling green infrastructure as a flood mitigation strategy in an urban coastal area. *Urban Ecosyst, 28*(2), 9. https://doi.org/10.1007/s11252-024-01660-9

Kumar, A., Dwivedi, G. K., Tewari, S., Jaipaul, S. V. K., Singh, H., Kumar, P., Kumar, N., & Kausal, R. (2020). Soil organic carbon pools under *Terminalia chebula* Retz. based agroforestry system in Himalayan foothills, India. *Current Science, 118*, 1098–1103. https://doi.org/10.18520/cs/v118/i7/1098-1103

Kumar, R., Kumar, R., Karmakar, S., Kumar, A., Singh, A. K., Kumar, A., & Singh, J. (2023). Impact of amide fertilizer on carbon sequestration under the agroforestry system in the eastern plateau region of India. *Sustain, 15*(12), 9775. https://doi.org/10.3390/su15129775

Kumar, A., & Dwidvedi, A. K. (2023). The development and effects of environmental law on green governance. *Journal Law Sustainability Development, 11*(6), e1182-e1182. https://doi.org/10.55908/sdgs.v11i6.1182

Kumar, M., & Singh, H. (2020). Agroforestry as a nature-based solution for reducing community dependence on forests to safeguard forests in rainfed areas of India. *In:* S. Dhyani, A. Gupta, M. Karki (Eds.) *Nature-based Solutions for Resilient Ecosystems and Societies. Disaster Resilience and Green Growth.* Springer, Singapore. https://doi.org/10.1007/978-981-15-4712-6_17.

Kumar, A., Singh, H., Kumari, G., Bisht, S., Malik, A., Kumar, A., Singh, M., Raturi, B. S., Thakur, A., & Kaushal, R. (2022). Adaptive resilience of roadside trees to vehicular emissions via leaf enzymatic, physiological, and anatomical trait modulations. *Environmental Pollution, 313*, 120191. https://doi.org/10.1016/j.envpol.2022.120191

Lee, A. C., & Maheswaran, R. (2011). The health benefits of urban green spaces: A review of the evidence. *J Public Heal, 33*(2), 212–222. https://doi.org/10.1093/pubmed/fdq068

Li, B., Zhang, Z., & Zhang, Z. (2024). The impact of the Belt and Road Initiative on green innovation and innovation modes: Empirical evidence from Chinese listed enterprises. *Frontiers in Environmental Science, 11*, 1323888. https://doi.org/10.3389/fenvs.2023.1323888

Lisina, N., Ushakova, A., & Ivanova, S. (2024). Green belt legislation regulation: Comparative legal research. *Laws, 13*(5), 58. https://doi.org/10.3390/laws13050058

Liu, C., Dai, A., Ji, Y., Sheng, Q., & Zhu, Z. (2023). Effect of different plant communities on fine particle removal in an urban road greenbelt and its key factors in Nanjing. *China. Sustain, 15*, 156. https://doi.org/10.3390/su15010156

Marx, A. (2019). Public-private partnerships for sustainable development: Exploring their design and its impact on effectiveness. *Sustain, 11*(4), 1087. https://doi.org/10.3390/su11041087

Minz, S. D., Singh, A. K., Kumar, N. M., & Singh, B. K. (2021). Effect of crop geometry and nitrogen management on growth attributes of pearl millet (Pennisetum glaucum L.) under guava based Agri-Horti system. *The Pharma Inno Journal, 10*(9), 2191–2195.

MOE, Ministry of Environment, Republic of Rwanda. (2011). *Green Growth and Climate Resilience Strategy: National Strategy on Climate Change and Low-Carbon Development.*

Mondal, D., Gupta, S., & Datta, J. K. (2011). Anticipated performance index of some tree species considered for green belt development in an urban area. *Int Res J Plant Sci, 2*(4), 99–106.

Neugarten, R. A., Chaplin-Kramer, R., Sharp, R. P., Schuster, R., Strimas-Mackey, M., Roehrdanz, P. R., & Rodewald, A. D. (2024). Mapping the planet's critical areas for biodiversity and nature's contributions to people. *Nature Commun, 15*(1), 261. https://doi.org/10.1038/s41467-023-43832-9

Obi, N. I., Nwalusi, D. M., Ibem, E. O., & Okeke, O. F. (2021). Assessment of the role of greenbelts in environmental and socio-economic development of urban areas in southeast Nigeria. *Civ Engi Archit, 9*(2), 545–557.

Patel, A. K., Bohra, J. S., Dewangan, S., & Singh, A. K. (2022). Effect of fertility levels on sesame (Sesamum indicum) varieties under agri-horti system. *The Indian Journal Agriculture Science, 92*(8), 1015–1017. https://doi.org/10.56093/ijas.v92i8.98465

Pongsiri, N. (2012). Public-private partnerships and urban infrastructure development in southeast Asia. *Urbanization in Southeast Asia Issues & Impacts*, 139–153.

Rahman, M. A., Pawijit, Y., Xu, C., Moser-Reischl, A., Pretzsch, H., Rötzer, T., & Pauleit, S. (2023). A comparative analysis of urban forests for storm-water management. *Science and Reports, 13*(1), 1451. https://doi.org/10.1038/s41598-023-28629-6

Seeliger, L., & Turok, I. (2013). Towards sustainable cities: Extending resilience with insights from vulnerability and transition theory. *Sustain, 5*(5), 2108–2128. https://doi.org/10.3390/su5052108

Shannigrahi, A. S., & Sharma, R. C. (2000). Environmental factors in green belt development-an overview. *Indian Journal of Environmental Protection, 20*(8), 602–607.

Shreyas, S. K., & Dwarka, D. J. (2024). Trees as reliable carbon capture and storage in urban spaces: A case study in Kalaburagi. *India.* https://doi.org/10.1101/2024.07.22.604147

Sidik, A. D. W. M., Edwinanto, E., Padilah, A., Hak, A., Gulo, M., Aghniya, G. I., & Agustin, Y. P. (2024). Empowering communities through enhancing environmental awareness: A case study of Cikelat Village, Sukabumi, West Java in community service activities. *Journal Innovation Computer Engagement, 5*(1), 18–34. https://doi.org/10.28932/ice.v5i1.7413

Simicevic, D., Lekovic, J., & Ilic, M. (2024). Green practices in management of sustainable tourist destinations: the case of Belgrade. In *Tourism International Scientific Conference Vrnjacka Banja-TISC, 8*(1), 87–97. https://doi.org/10.52370/TISC2487DS

Singh, H., & Kumar, M. (2022). Climate change and its impact on Indian Himalayan forests: current status and research needs. In S. Rani, R. Kumar, (Eds.) *Climate Change* (pp. 223–242) Springer Climate. Springer, Cham. https://doi.org/10.1007/978-3-030-92782-0_11

Singh, R., Prakash, P., Singh, A. K., Shukla, S., Sharma, P., Chauhan, V., & Singh, B. K. (2025a). Altitudinal variation in carbon stocks of *Quercus leucotrichophora* forests in the Western Himalayas:

Implications for climate change mitigation. *Trees, Forests and People, 22*, 101059. https://doi.org/10.1016/j.tfp.2025.101059

Singh, R., Chavan, S. B., Tomar, A., Singh, H., Chauhan, V., Paul, N., & Singh, A. K. (2025b). Species variation in air pollution tolerance, performance, and dust retention of urban roadside trees: implications for urban greening and green corridor planning. *Air Quality, Atmosphere & Health*, 1–17. https://doi.org/10.1007/s11869-025-01841-1

Singh, H., Verma, A., & Kumar, M. (2018). Can elevated CO_2 alter the growth dynamics, morphology, physiology and biochemistry of the forestry tree species? *Climate change and agriculture*. Discovery Publishing House, (pp. 288-305). New Delhi.

Singh, A. K,, Shukla, S., Singh, M. K., Singh, A., Singh, R., & Singh, B. K. (2024a). Urban forests: importance, challenges and opportunities. *Urban Forests, Climate Change and Environmental Pollution,* 23–45. https://doi.org/10.1007/978-3-031-67837-0_2

Singh, R., Chavan, S. B., Kakade, V. D., Morade, A. S., Singh, A. K., Rawale, G. B., Uthappa, A. R., Keerthika, A., Chichaghare, A. R., Gurav, S., & Reddy, K. S. (2024b). Ecosystem services provided by Urban and Peri-Urban Forests. In *Urban Forests, Climate Change and Environmental Pollution* (pp. 417–445). Springer, Cham. https://doi.org/10.1007/978-3-031-67837-0_20

Singh, H. (2024a). Urban forests, climate change and environmental pollution: physio-biochemical and molecular perspectives to enhance urban resilience. Springer Nature. https://doi.org/10.1007/978-3-031-67837-0

Singh, H., (Ed.) (2024). Forests and climate change: Biological perspectives on impact, adaptation, and mitigation strategies. Springer, Singapore. https://doi.org/10.1007/978-981-97-3905-9. ISBN:978-981-97-3904-2

Singh, H. (2025). The role of plant functional traits in mitigating urban climate and environmental pollution. *Plant Functional Traits* (pp. 297–308). Elsevier.

Stan, V. J. T., Underwood, S. J., Friesen, J. (2018). Urban forestry: An underutilized tool in water management. In *Advances in chemical pollution, environmental management and protection* (Vol. 3, pp. 35–61). Elsevier. https://doi.org/10.1016/bs.apmp.2018.04.003

Sternik, S. G. (2024). Global practices in financing environmental investment-construction projects. *E3S Web Conferences, EDP Sciences,* (535), 01018. https://doi.org/10.1051/e3sconf/202453501018

Stoianoff, N. P., & Kelly, A. H. (2007). Conserving native vegetation on private land: Subsidizing sustainable use of biodiversity. *Critical Iss Environ Taxat, 4*, 299–315. https://ssrn.com/abstract=1399192

Sun, L., Fertner, C., & Jørgensen, G. (2021). Beijing's first green Belt-A 50-year long Chinese planning story. *Land, 10*(9), 969. https://doi.org/10.3390/land10090969

UN-Habitat. (2020). World Cities Report 2020: The value of sustainable urbanization.

USDA. (2023). USDA releases assessment on agriculture and forestry in Carbon Markets. USDA Press. Release No. 0214.23.

Verma, K., Sharma, P., Bhardwaj, D. R., Kumar, R., Kumar, N. M., & Singh, A. K. (2023). Land and environmental management through agriculture, forestry and other land use (AFOLU) system. *Land Environ Manag through Forest, 4*, 247–271. https://doi.org/10.1002/9781119910527.ch10

Wang, R., Zhao, J., Meitner, M. J., Hu, Y., & Xu, X. (2019). Characteristics of urban green spaces in relation to aesthetic preference and stress recovery. *Urban Forestry & Urban Greening, 41*, 6–13. https://doi.org/10.1016/j.ufug.2019.03.005

Warren, J. L. (1973). Green space for air pollution control. School of Forest Resources, Technical Report No. 50, North Carolina State University, Raleigh, North Carolina *50*(4), 118.

Wenham, G. (1999). The old testament and the environment: A response to Chris Wright. *Transfor J, 16*(3), 86–92. https://doi.org/10.1177/026537889901600305

WHO. (2017). Urban green spaces and health: A review of evidence. World Health Organization, pp. 1–22.

World Bank. (2019). Urban development indicators. Accessed worldbank.org. https://www.worldbank.org

Xie, X., Kang, H., Behnisch, M., Baildon, M., & Krüger, T. (2020). To what extent can the green belts prevent urban sprawl-a comparative study of Frankfurt am Main. *London and Seoul. Sustain, 12*(2), 679. https://doi.org/10.3390/su12020679

Yao, L., Liu, T., Qin, J., Jiang, H., Yang, L., Smith, P., & Piao, S. (2024). Carbon sequestration potential of tree planting in China. *Nature Communications, 15*(1), 8398. https://doi.org/10.1038/s41467-024-52785-6

Yokohari, M., Takeuchi, K., Watanabe, T., & Yokota, S. (2000). Beyond greenbelts and zoning: A new planning concept for the environment of Asian mega-cities. *Landscape Urban Plan, 47*(3–4), 159–171. https://doi.org/10.1016/S0169-2046(99)00084-5

Zhao, W., Wang, Y., Chen, D., Wang, L., & Tang, X. (2021). Exploring the influencing factors of the recreational utilization and evaluation of urban ecological protection green belts for urban renewal: A case study in Shanghai. *International Journal of Environmental Research and Public Health, 18*(19), 10244. https://doi.org/10.3390/ijerph181910244

Chapter 15
Forest Degradation and Deforestation: Drivers and Solution

Pooja, Rupali Singh, Vivek Chauhan, and Alok Kumar Singh

Abstract Forest degradation and deforestation are considered a significant threat to the environment, with a profound impact on global biodiversity, various ecosystem services, and climate stability. Globally, it has become one of the major causes of greenhouse gas emissions (GHG) besides the burning of fossil fuels. Numerous drivers contribute to forest degradation and deforestation, which can be categorized into two groups: proximate causes or direct drivers and underlying causes or indirect drivers. The proximate causes of these issues include agricultural expansion, illegal logging, infrastructure development, urbanization, mining, overgrazing, and uncontrolled forest fires, often exacerbated by demographic, economic, and governance factors, which are the underlying causes of forest degradation and deforestation. Knowledge and understanding of the drivers of forest degradation and deforestation are crucial in the development of effective policies and management practices. These drivers lead to the loss of habitat, loss of carbon sinks, reduction in biological diversity, and soil erosion, which in turn accelerate climate change. To address the challenges of forest degradation and deforestation, a combination of strategies is employed, including forest land restoration, sustainable agriculture, reforestation, and afforestation practices, as well as various conservation incentives and the implementation of participatory management programs. Additionally, effective policies, strict implementation of environmental regulations, and punishment rules, combined with technological innovation, offer an excellent pathway for reducing forest loss and restoring degraded ecosystems.

Keywords Forest degradation · Deforestation · Global biodiversity drivers · Strategies

---

Pooja (✉)
Department of Silviculture and Agroforestry, College of Forestry, Dr. Y S Parmar University of Horticulture and Forestry, Nauni-Solan, Himachal Pradesh, India
e-mail: vermapooja.pv97@gmail.com

R. Singh · V. Chauhan
ICFRE-Forest Research Institute Dehradun, Dehradun, India

A. K. Singh
Rajiv Gandhi South Campus, Banaras Hindu University, Mirzapur, Uttar Pradesh, India

H. Singh (ed.), *Forestry for a Carbon-Neutral and Sustainable Future*, Sustainability Solutions, https://doi.org/10.1007/978-3-032-06427-1_15

15.1 Introduction

Forests are recognized as a resourcc of global significance, covering approximately one-third of the Earth's land surface and harboring most of the Earth's terrestrial biodiversity. For example, they provide habitats for about 80% of amphibian species, 75% of bird species, and 68% of mammal species (Pooja et al., 2025; Vie et al., 2009). The forest contains 662 billion tonnes of carbon, which is more than half of the global carbon stock in soil and other vegetation. Additionally, it provides multiple ecosystem goods and services, including livelihood security, socio-economic development, ecosystem functioning, biodiversity maintenance, nutrient cycling, and climate regulation (Joshi et al., 2020; Singh et al., 2025). However, forests worldwide are under threat, which jeopardizes these benefits. Among the various threats, deforestation and forest degradation are two important dimensions of environmental change that are directly related to global warming and pose significant threats to the biodiversity and livelihoods of forest-dependent local communities.

Forest degradation is a global environmental concern, but its definition is problematic. The term 'forest degradation' lacks a universally accepted definition despite being a widely used term for over three decades. Definitions of forest degradation are generally considered to be more varied in their formulation than those of deforestation. Typically, it is a gradual process that takes place over a long period, where a forest loses its ability to provide goods and services to people and nature. According to Grandon et al. (2018), a degraded forest is one whose species composition, structure, and function have been significantly impaired or permanently lost due to human activities. These definitions primarily rely on a reduction in the capacity of forests to produce various ecosystem goods and services but fail to address how, when, and to what degree this reduction occurs, which ultimately leads to the degradation of forests (Kumar & Singh., 2020). The variation in existing definitions and its continuous use as a term indicate that declines in the full spectrum of forests essential functions, products, and services are significant concerns to a wide variety of actors, even though they may frame the term degradation (and indeed other forest concepts and definitions) in different perspective according to their cultural, educational and social background. For example, foresters may focus on the outlook for degradation in terms of reduced timber productivity.

In contrast, ecologists primarily focus on the decline in forest ecosystem processes or the resilience of forests (Thompson et al., 2009). Climate scientists may be most concerned with the declining capacity of forests to sequester and store carbon. At the same time, indigenous peoples may view degradation in the context of a decrease in socio-economic status and subsistence economies (Chazdon et al., 2016). Countries employ various definitions of degraded forest, making it infeasible to aggregate and compare data on forest areas at regional and global levels. Most definitions are based primarily on the presence of forest disturbances. However, some countries specify the type of disturbance, whether it is natural or anthropogenic, such as logging and fire, while others are more generic. The rate of forest degradation associated with fuelwood extraction, logging, and charcoal production has a significant impact on biodiversity and the livelihoods of people, resulting in a reduction of carbon storage, degradation of species habitats, and a decline in ecosystem services, such as water quality (Kumar and Singh., 2020). For instance, in dry tropical areas (e.g., Africa, Central America, and South Asia), the extraction of fuelwood and fodder has lowered the quality of the forest. The pressures of the triple planetary crises of climate change, biodiversity loss, and pollution, together with direct drivers of unsustainable forest management and use, are curtailing the role of forests in mitigating climate change, restoring ecosystems, and promoting sustainable livelihoods.

According to the definition provided in the FAO's Global Forest Resources Assessment (FRA), deforestation refers to "the conversion of forested land to other land uses, regardless of whether it is caused by anthropogenic activities or not." It essentially refers to a change in land use rather than a change in tree cover. The UNFCCC (United Nations Framework Convention on Climate Change) defines deforestation as human-induced activities that transform forested areas into non-forested lands. From this perspective, deforestation is understood as the loss of forest cover resulting from anthropogenic actions and activities. According to the FAO, tree plantations established primarily for timber production are classified as forests. Consequently, the conversion of natural forests into plantations is not considered deforestation, although it is still recorded as a loss of natural forest. However, the FAO does not recognize tree plantations that provide non-timber products as forests, except for rubber plantations, which are classified as forests. Deforestation is mainly a concern for developing tropical countries (Myers, 1994), contributing to the reduction of tropical forest areas (Barraclough & Ghimire, 2000), which results in biodiversity loss and exacerbates the greenhouse effect. (Angelsen et al., 1999). Since 2000, the world has lost approximately 10% of its tropical tree cover, with nearly 47,000 square miles (121,000 square kilometers) destroyed in 2019 alone, according to a 2020 report by The New York Times.

Often, deforestation occurs when people cut or clear the forest area to make way for agriculture or grazing. The Union of Concerned Scientists (UCS) estimates that an area the size of Switzerland (approximately 14,800 square miles, or 38,300 square kilometers) is lost to deforestation every year. Deforestation is a significant environmental issue with far-reaching consequences for both local communities and the planet. This process contributes to climate change by releasing stored carbon dioxide, disrupting ecosystems, causing desertification and soil erosion, and increasing the area's vulnerability to landslides and flooding. However, deforestation undoubtedly provides some private benefits, particularly to those directly involved in driving it. Benefits have included the creation of agricultural land and the resulting supply of many outputs (Joshi et al., 2024) Particular actors involved in the deforestation process may also profit from the sale of timber, as well as from the sale of cleared land itself, given that land may appreciate or maintain its value in inflationary economies.

15.2 Scenario of Deforestation and Forest Degradation

Every year, the ecosystem absorbs millions of tonnes of carbon dioxide emissions. When those trees are chopped down or burned, they not only release the carbon they are storing, but they also lose their ability to absorb carbon emissions. Since 1990, the globe has lost over 178 million hectares of forest, an area almost the size of Libya. However, the rate of net forest loss fell significantly throughout the period from 1990 to 2020 due to a reduction in deforestation in certain countries, as well as gains in forest area in others through afforestation and natural forest growth. According to the Global Forest Resource Assessment report (2020), an estimated 420 million hectares of forest were lost through deforestation between 1990 and 2020, although the rate slowed over the period. From 1990 to 2000, deforestation occurred at an average rate of 15.8 million hectares per year. This rate declined slightly to 15.1 million hectares per year from 2000 to 2010, then further dropped to 11.8 million hectares annually between 2010 and 2015, and subsequently to 10.2 million hectares per year from 2015 to 2020. Figure 15.1 illustrates the trends in the average annual rates of deforestation and forest expansion, which, combined, equal the net change in forest area.

The tropical domain contributed to nearly 90% of deforestation between 1990 and 2020. Between 2015 and 2020, the average annual rate of deforestation in the tropics was 9.28 million hectares, which is significantly lower than the average of 13.8 million hectares annually observed between 1990 and 2000. The annual rate of deforestation showed a substantial decline in the subtropical domain between the periods of 1990–2000 and 2015–2020, from 1.44 million ha to 0.50 million ha. The deforestation rate decreased slightly in the temperate domain between these two periods and was relatively low in the boreal domain (Table 15.1).

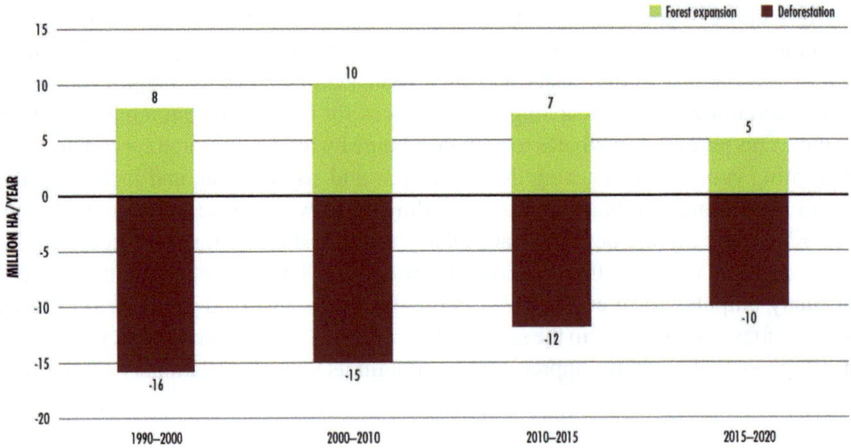

Fig. 15.1 Trends in the average annual rates of deforestation and forest expansion

Table 15.1 Deforestation rate, by climatic domain, for four periods spanning 1990–2020

Climatic domain	Deforestation (million ha/yr)			
	1990–2000	2000–2010	2010–2015	2015–2020
Boreal	0.10	0.09	0.13	0.06
Temperate	0.49	0.54	0.53	0.31
Sub-Tropical	1.44	1.35	0.88	0.50
Tropical	13.8	13.2	10.3	9.28
Total	15.8	15.1	11.8	10.2

Source-Global Forest Resource Assessment (2020)

According to World Bank estimates, approximately 3.9 million square miles (10 million square kilometers) of forest have been lost since the beginning of the twentieth century. Forests have lost 502,000 square miles (1.3 million square kilometers) in the last 25 years, which is more than the entire area of South Africa. According to Global Forest Watch, the globe lost 488 million hectares of forest cover between 2001 and 2023, representing a 12% decline since 2000. Of this loss, 23% occurred in regions where deforestation was the leading cause.

A process of deterioration that adversely impacts the forest's structural and functional qualities results in a degraded forest. Human activities, severe climatic events, fires, pests, diseases, and other environmental disturbances can cause the degradation of forests, thereby reducing the provision of goods and services, biodiversity values, productivity, and health. The Global Forest Resources Assessment (FRA) 2020, which incorporated data on degraded forests, revealed that 58 countries—accounting for an average 38% of the world's forest area—reported monitoring these degraded areas. Nearly one-third of these countries, located in Africa, encompassed 28% of the region's forest area. Reporting was most extensive in South America, where participating countries represented 79% of the region's forested area (Table 15.2). Understanding the patterns and rates of deforestation and forest degradation is a critical step in developing effective conservation and management strategies for any forested area.

15.3 Divers of Forest Degradation and Deforestation

Drivers of deforestation and forest degradation are believed to be diverse, complex, and interconnected. Drivers of deforestation and forest degradation occur at all scales (from global to local), and thus, strategies to address these drivers will also happen at all scales. Identification and estimation of drivers of deforestation and forest degradation are not only crucial for targeting national strategy to reduce the emissions from deforestation and degradation but have wide applications to sustainable development initiatives supporting local economies as well as conservation efforts seeking

Table 15.2 Number and forest area of countries reporting on forest degradation by region

Region	No. of reporting countries	Forest area of reporting countries (1000 ha)	% of total forest area
Africa	18	178 226	28
Asia	16	271 887	44
Europe	7	18 063	2
North and Central America	6	383 297	51
Oceania	3	39 519	21
South America	8	663 849	79
World	**58**	**1 554 841**	**38**

Source Global Forest Resource Assessment Report (2020)

to reverse or slow the significant downward trends in forest cover and quality (Bernhard et al., 2020). Drivers of deforestation and forest degradation can be categorized into two groups: direct or proximate drivers and indirect or underlying drivers.

15.3.1 Direct or Proximate Drivers

Direct drivers of forest degradation and deforestation are human activities and actions that originate from human choices of land use, which directly impact forest cover and result in the loss of carbon stock (Ojima et al., 1994; Turner et al., 1994). The primary drivers of deforestation are mainly large-scale processes, whereas the drivers of degradation are small-scale processes.

15.3.2 Underlying or Indirect Drivers

The underlying causes of deforestation and forest degradation are structural (or systemic) in nature. These can be viewed as a complex of social, political, economic, technological, and cultural variables that constitute the initial conditions in human-environmental relations (Geist & Lambin, 2002). These drivers may operate at various spatial levels and have an impact on the proximate drivers causing deforestation and forest degradation (Kissinger et al., 2012).

Therefore, a systematic and careful analysis of all direct and indirect drivers of forest degradation and deforestation is a prerequisite for effectively reducing emissions and thereby enhancing carbon stocks. The dynamics and causes, or drivers, of forest degradation and deforestation are very complex, multifaceted, and vary from place to place.

15.4 Different Direct or Proximate Drivers of Deforestation and Forest Degradation

15.4.1 Overharvesting and Unsustainable Extraction of Forest Products (Mostly Illegally)

Excessive extraction of forest products to meet persistently high demands for timber, fuelwood, fodder, and other forest products is a major cause of deforestation and forest degradation in most parts of the landscape. According to the study by Hosonuma et al. (2012), in Latin America and Asia, timber extraction and logging account for more than 70% of total degradation. In 2021, according to the World Wildlife Fund, 94% of deforestation reported in the Amazon (and surrounding) rainforests had links to illegal activity. Unregulated logging has contributed to the accumulation of timber and has catalyzed the global illicit logging trade (ILT). Driven by profits, ILT has intensified and is considered one of the major causes of global deforestation (Fayiah, 2021). According to Gregory et al. (2006), logged forests in the Amazon that are penetrated by logging roads are roughly 400% more likely to be deforested than unlogged forests. A direct consequence of the illegal logging trade (ILT) is the encouragement of individuals to fell forests, particularly long-standing, ecologically stable natural forests. Therefore, deforestation and degradation of forests driven by illegal logging and timber trafficking (ILT) harm the world's livelihoods. Unemployment and high demand for firewood have motivated many people to engage in illegal collection for sale. The use of fuelwood significantly contributes to forest degradation, as it remains a primary energy source for rural households and certain industrial sectors in developing countries. Approximately 2 billion people in the developing world still primarily rely on fuelwood and charcoal for cooking and heating in both rural and urban areas (FAO, 2017). Unsustainable extraction practices, including the intensive harvesting of reproductive parts such as flowers, seeds, and fruits, along with repeated visits, have led to the degradation of non-timber forest products, particularly high-value medicinal and aromatic plants, in many regions. These unsustainable methods of forest resource extraction have led to the loss of virgin forests, degradation of habitats for the most endangered wildlife, biodiversity loss, forest fires, soil erosion, and environmental problems, including water pollution, flooding, and landslides (Lawson & Macfaul, 2010).

15.4.2 Agricultural Expansion

Globally, agrarian expansion is considered one of the key drivers of deforestation. In recent years, agricultural expansion in tropical regions has led to the conversion of natural forests into croplands, accounting for approximately 80% of newly cultivated areas (Zhang et al., 2018). This phenomenon is especially evident in tropical developing countries, where extensive forest areas suitable for agriculture, favorable climates for high-value crops, and rapid population growth are key factors driving this transformation. To address the growing demand for agricultural land, the expansion of cultivation into forested areas has been reported across nearly all parts of the landscape. However, the significance of commercial versus subsistence agriculture varies depending on the geographical region. According to the secretariat of the United Nations Framework Convention on Climate Change, subsistence farming is responsible for 48% of deforestation, commercial agriculture for 32%, logging for 14%, and the removal of wood for fuel for 5%. According to Hosonuma et al. (2012), among the three different continents, commercial agriculture is the leading driver of deforestation in Latin America, accounting for 68%.

In contrast, in Africa and Asia, it accounts for around 35%. Local and subsistence agriculture is relatively equally distributed among these continents (27–42%), which makes sense since this type of land use remains widespread in all areas of the tropics and sub-tropics. Overall, agriculture accounts for approximately 80% of deforestation worldwide, which aligns with the study by Geist and Lambin (2002) for the 1980s and 1990s. Lim et al. (2017) in Myanmar also identified agricultural expansion as a direct and major driver of forest loss. It is anticipated that the agricultural industry will develop significantly over time, primarily due to improved access to global markets and increasing private sector investment, which encourage the farmers to prioritize high-value export crops such as palm oil, coffee, cocoa, soybeans, etc., establish monoculture plantations and also increase the pressure on forest or marginal land. The large area of forest tracts is cleared frequently, resulting in significant deforestation.

15.5 Forest Encroachment for Infrastructure Development, Resettlement, and Urban Expansion

According to Rimal et al. (2014), deforestation was driven by a combination of infrastructure development, resettlement, and urban expansion through various mechanisms. These drivers are more pronounced in the lower-altitude hilly areas and progressively less critical as the altitude increases. The unregulated construction of schools, hospitals, electric transmission lines, water storage tanks, army barracks, police camps, temples, roads, and recreational facilities (e.g., picnic spots) within forests is a common practice. Worldwide, road infrastructure is recognized as a significant driver of the spatial pattern of forest loss and subsistence agriculture (Lambin et al., 2003; Laurance et al., 2009). It often encourages illegal logging, mining, and further land conversion by making remote forests more accessible. Displacement from various disasters and development initiatives frequently leads to the clearing of forests for the establishment of new communities, agriculture, and other land uses. Urbanization and infrastructural improvements create forest "edges" that expose ecosystems to external pressures, leading to the degradation of interior forest areas.

15.6 Uncontrolled Forest Fires

One of the primary causes of forest degradation with global repercussions has been identified as uncontrolled forest fires, resulting in the loss of human life, built assets, biodiversity, habitats, and ecosystem productivity, as well as landscape degradation and the disruption of livelihoods. Frequent forest fires create a "window of disturbance" that is conducive to surface runoff, landslides, and soil erosion and also alter the soil's characteristics, seriously impairing or preventing the regeneration and development of seedlings. Apart from modifying the composition and structure of forests, these fires have the potential to introduce invasive species, threaten biological diversity, change soil fertility and water cycles, and ruin the livelihoods of those who live in and around forest areas. Each year millions of acres of forests are destroyed or degraded by fire. Worldwide, wildfires are responsible for approximately 4.8 million hectares of forest degradation, accounting for nearly 23% of forest degradation caused by all factors (Curtis et al., 2018). Data compiled by the Joint Research Centre of the European Commission indicates that approximately 7.20 billion hectares of land (including areas beyond forests) were burned between 2001 and 2018, averaging over 400 million hectares per year (Artes et al., 2019). However, the annual burned area from 2013 to 2018 was less than the long-term average (FAO 2020). Africa accounted for more than two-thirds of all wildfires between 2001 and 2018. Globally, the most significant areas burned were in the subregions of Eastern and Southern Africa, Western and Central Africa, Oceania (mainly Australia), Northern Africa, and South America, all surpassing 500 million hectares over the period (Table 15.3).

Table 15.3 Total land area burned in 2001–2018 by region or subregion

Region	Total land area burned
Eastern and Southern Africa	2.18 billion ha
Western and Central Africa	1.91 billion ha
Oceania	878.44 million ha
Northern Africa	737.11 million ha
South America	587.94 million ha
Europe (inc Russian Federation)	281.36 million ha
South and Southeast Asia	262.06 million ha
Western and Central Asia	184.95 million ha
North America	124.87 million ha
East Asia	77.64 million ha
Central America	15.55 million ha
Caribbean	4.53 million ha

Source Global Forest Resource Assessment (2020)

15.7 Overgrazing

According to Teague et al. (2011), overgrazing refers to the repeated, intense grazing over several years that results in the degradation of plant communities, a reduction in productivity and biodiversity, and a decline in the area's vigor due to an insufficient recovery period. Overgrazing and unregulated grazing in forests, commonly practiced by sedentary small farmers in the Siwaliks and nomadic herders in the high mountains, significantly hinder the regeneration and growth of seedlings. Over time, this repeated pressure disrupts the ecosystem's balance and reduces the forest's ability to regenerate, ultimately leading to forest degradation. The degradation of forests by such activities results in a loss of biodiversity, reduced soil fertility, declining ecosystem health, and the spread of invasive non-native plant species and weeds. The practice of grazing in forests was also widespread in the mid-hills in the past but has subsequently decreased significantly due to the introduction of the community-based forestry program, which aims to manage local forests sustainably and generally prohibits grazing.

15.8 Invasion by Alien Plant Species

Another major threat to global and local biodiversity is invasion by alien or non-native plant species (Pysek et al., 2004). According to the IUCN, the influence of alien species is immense, insidious, and usually irreversible, resulting in fragmentation, destruction, alteration, or complete replacement of habitats, which in turn have cascading effects on even more species and ecosystem processes. For example, in Eastern North America, invasive diseases and pests have led to significant changes in forest composition over the past century, including the decline of species such as hemlock, chestnut, and elm (McNeely et al., 2001). According to Mooney and Hofgaard (1999) invasive grasses also led to permanent losses of forests due to their higher susceptibility to fire. Deforestation and forest degradation are often associated with various human activities, including logging, infrastructure development, and agricultural expansion. Still, invasion by alien species adds a layer of complexity by contributing to the ongoing process of forest loss in ways that can exacerbate the effects of human-driven environmental pressures and also increase the vulnerability of forests to other pressures.

15.9 Mining

It is considered the fourth-largest driver of deforestation, after agriculture, infrastructure development, and urban expansion (Hansen et al., 2013). Its significance is rapidly increasing as mining activities grow in size and increasingly encroach upon

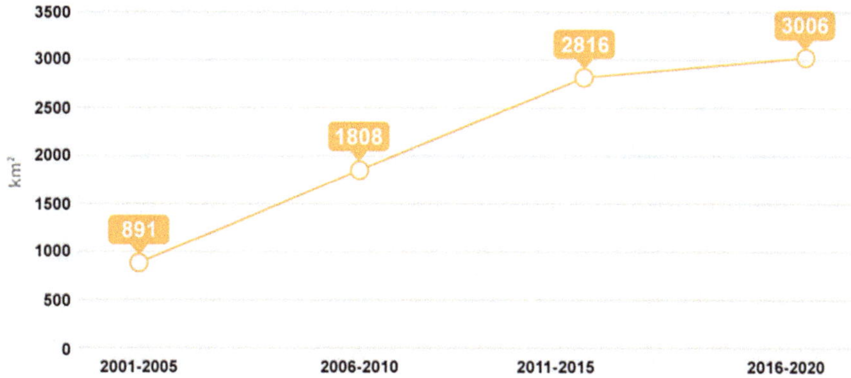

Fig. 15.2 Direct MRD in tropical rainforests from 2001 to 2020, by 5-year period *Source* WU (2022)

environmentally sensitive areas. Approximately 63% of the deforestation, approximately 8,600 km^2, caused by the expansion of global mining activities over the past 20 years occurred after 2010, while only 37% took place between 2001 and 2010 (WU, 2022). Tropical and subtropical moist broadleaf forests (also known as rainforests) encompass just 29% (nearly 30,000 km^2) of the world's mining areas. Yet, they account for 62%, or approximately 8,533 km^2, of total direct deforestation occurring within mining areas between 2000 and 2020 worldwide. Other vital ecosystems are also significantly affected, with boreal forests contributing 13% and temperate forests 9% of the total direct mining-related deforestation (MRD) (WU, 2022). Yet, tropical rainforests stand out as a hotspot of mining-related deforestation. The deforestation patterns presented in Fig. 15.2 indicate that the situation is deteriorating: more than 35% of all MRD evaluated for tropical rainforests over the previous 20 years occurred in the final five-year period (2016–2020).

Mining not only directly contributes to deforestation by expanding extractive operations, such as clearing forests for mining pits, but it also indirectly fosters deforestation by mandating the construction of transportation and energy infrastructure. These infrastructure developments divided enormous, continuous forest ecosystems into smaller fragments, increasing their vulnerability to degradation and biodiversity loss. Furthermore, it opens up previously inaccessible areas, which facilitates illicit logging and resource extraction, resulting in increased deforestation.

Hosonuma et al. (2012) considered separate proximate drivers for deforestation and forest degradation. The national driver estimations are presented for the three continents of Africa, Latin America, and (sub)tropical Asia (including Oceania) for deforestation and degradation (Fig. 15.3). Commercial agriculture (including livestock) is the most important driver of deforestation in Latin America, leading to around 2/3 of the total deforested area. In both Africa and (sub) tropical Asia, commercial agriculture accounts for around 1/3 of deforestation and is of similar importance as subsistence agriculture.

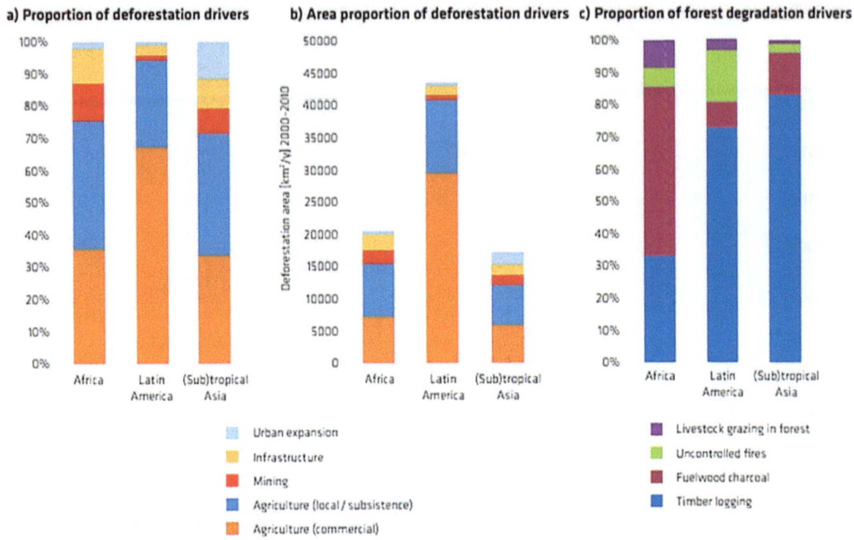

Fig. 15.3 Continental level estimation of deforestation and forest degradation proximate drivers *Source* Hosonuma et al., 2012

Mining plays a larger role in Africa and (sub) tropical Asia than in Latin America. Urban expansion is most significant in (sub)tropical Asia, perhaps due to the large population growth (Defries et al., 2010). Timber and logging activities account for more than 70% of total degradation in Latin America and (sub) tropical Asia. Fuelwood collection and charcoal production are the primary drivers of degradation on the African continent, whereas they are less prominent in (sub)tropical Asia and Latin America.

15.10 Underlying Drivers of Forest Degradation and Deforestation

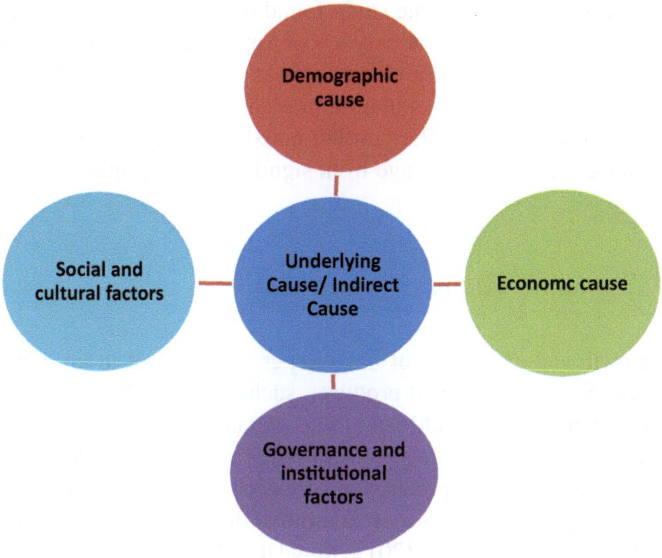

*Source*Geist & Lambin, (2002)

15.10.1 Demographic Factors

Population pressure is one of the most frequently cited underlying causes of forest degradation and deforestation. Growth and density of population are closely interrelated with increased demand for agricultural land, pressures on fuelwood, and easier access to remote forests due to infrastructure development, land tenure arrangements, agro-technological changes, and increased demand for forest products (Rademaekers et al., 2010). The growth of population, along with changing density and distribution patterns, is believed to have driven deforestation and degradation of the forest landscape. The exact role that population expansion plays in deforestation is a matter of debate, although many analysts agree that it does contribute in some way. Some believe that population dynamics, which include growth, density, and migration, are the primary drivers of deforestation. Others, however, hypothesize that these dynamics only serve to exacerbate or alleviate fundamental causes (poverty, land-use policies, and resource consumption). Population changes are thought to have a variety of effects on deforestation, including shifts in the number of rural families looking for land to cultivate, fuel wood, or timber, as well as changes in the demand for agricultural and forest products, even though it is challenging to correlate demographic change with patterns of forest use. The high reliance of households on forests for energy and timber and on agriculture for livelihoods, combined with an ever-increasing demand for food due to population growth, has led to agricultural expansions on forestland. One study found that population increase driven by high birth rates was responsible for only 8% of tropical deforestation cases. In 2000, the United Nations Food and Agriculture Organization stated that deforestation may result from "a combination of population pressure and stagnant economic, social, and technological conditions." They further noted that "the influence of population dynamics in a local context can range from significant to negligible".

15.10.2 Economic Factors

Economic growth and expansion of income per capita result in growing demands for agricultural and forest-derived products, such as paper. As a result, expanding levels of income per capita would tend to accelerate forest decline. As income per capita increases, there is a significant shift in the composition of demand for goods and services, with a growing preference for forest-related environmental services and non-destructive uses, such as recreation, that do not necessarily result in forest cover loss. Damette and Delacorte (2012) found that deforestation is positively correlated with economic development, timber harvesting, and demographic factors. The U-shaped Kuznets Curve provides the theoretical basis for understanding the relationship between per capita income and deforestation. This framework suggests that in less developed countries, economic growth initially has a detrimental impact on the environment, as seen in deforestation. However, once GDP per capita surpasses a certain threshold, environmental degradation, including deforestation, reaches a peak and no longer increases with further economic development. This phenomenon has the potential to enforce afforestation practices even in more developed countries (Adila et al., 2021; Caravaggio, 2022; Culas, 2007). Nguyen et al. (2023) conducted a study evaluating the effects of tourism on forested areas in 140 economies worldwide. Their results suggest that tourism has a positive short-term impact, but its long-term effects become harmful. The most significant negative effects of tourism were noticeable in the Caribbean, Latin America, and Africa. The push for economic development, combined with weak enforcement of forest laws, has resulted in the unplanned and unregulated construction of infrastructure within forest areas across the landscape. While the unregulated construction of rural roads is common in the mid-hills, the illegal building of schools, hospitals, electric transmission lines, and water storage tanks within forest boundaries is also widespread.

15.10.3 Governance-Related and Institutional Factors

One major underlying factor behind deforestation and forest degradation is the weak enforcement of laws and regulatory mechanisms, poor central planning, a lack of political will, inadequate coordination, conflicting policies, and overall poor governance of the forestry sector. Laws may exist, but they are not implemented due to a lack of political will and insufficient financial resources dedicated to their implementation. The government is either unwilling or unable to enforce forest laws and other environmental regulations, especially when large-scale projects are involved. The lack of a transparent reward and punishment system for government staff and Forest User Groups contributes to weak law enforcement, corruption, and a culture of impunity. Additionally, inter-departmental rivalries where trade and economic ministries overshadow those focused on social and environmental welfare represent a significant governance failure. This issue is not limited to under-resourced countries. The prolonged political instability and conflicts have also contributed to the acceleration of deforestation and forest encroachment processes. In-efficient distribution and marketing systems for forest products are another contributing factor, alongside the unsustainable harvesting of forest resources, which promotes overexploitation and reduces incentives for conservation and sustainable resource management.

15.10.4 Social and Cultural Factors

Cultural factors such as lack of concern by individuals regarding the welfare of others and future generations, non-profit orientation, continuation of conventional modes of resource utilization, farming practices (e.g., shifting cultivation), growing levels of demand and aspiration, and general disregard for the sacredness of nature are various factors that underpinning loss and degradation of forest in the landscape. Cultural, social, and economic patterns, as well as behavioral tendencies, change relatively slowly, even when economic and technological developments are introduced and adopted. Many contemporary characteristics in African cultures have their roots deep in pre-colonial cultural, social, and economic practices. These include production techniques and strategies, as well as land tenure, mentally and culturally determined attitudes towards land and other natural resources, human reproduction, social organization of labor, and cooperation.

15.11 Different Solutions for Forest Degradation and Deforestation

Deforestation and forest degradation are critical global challenges, so it requires a range of strategies and solutions to address these issues effectively.

15.11.1 Forest Landscape Restoration (FLR)

The livelihoods, well-being, food, water, energy security, and resilience of millions of people worldwide are threatened by deforestation and land degradation. Forest landscape restoration serves as a nature-based solution that helps countries reverse the impacts of deforestation and degradation while restoring the ecological, social, climatic, and economic benefits of forests. FLR addresses the underlying drivers of forest loss and involves not only the planting of trees but also aims to restore a healthy ecosystem and enhance human well-being across deforested or degraded landscapes (Kumar et al., 2020) It involves restoring entire landscapes to address current and future needs while providing a variety of benefits and land uses over time. For instance, it equips farming communities living in and around forests with knowledge of sustainable agricultural practices that do not depend on forest destruction. FLR engages with local communities, governments, the private sector, civil society, and academia as key stakeholders in the landscape. By encouraging the involvement of diverse stakeholders, FLR strikes a balance between social, economic, and environmental objectives, ensuring that costs and benefits are fairly distributed. Together, these groups identify, evaluate, and negotiate restoration options and land-use trade-offs transparently. In this way, FLR can deliver a wide range of benefits to stakeholders, including income, carbon storage, wood, energy, food, biodiversity conservation, and many more. There are several ongoing, crucial efforts aimed at restoring degraded land, combating deforestation, enhancing biodiversity, mitigating climate change, and supporting local communities. The Bonn Challenge is recognized as one of the most significant global initiatives for restoring degraded forest land, to bring 150 million hectares of deforested and degraded land into restoration by 2020 and 350 million hectares by 2030. Other efforts include REDD + (reducing emissions from deforestation and forest degradation), the Great Green Wall, the UN Decade on Ecosystem Restoration, and the global forest goal, among others.

Forest Landscape Restoration Will

- **Restore environmental functions**

  - Improving the water quality and preserving watersheds
  - Expanding or improving the habitat for endangered species
  - Expanding the range of species
  - Nutrient cycling
  - Stabilization of soils

- **Restore the benefits for communities, including**

  - Products for food and raw materials (eg:- Fruit, nuts, rubber, rattan, etc.)
  - Reduction in landslide risk
  - Providing recreation areas
  - Providing clean water

- Climate change

- **Reduction in biodiversity and vulnerability of forests such as**

- Pests/disease
- Fires
- Climate change

- **Tackle the root causes of forest loss and degradation to reverse the current trends**

- Consulting and involving local people
- Changing policies related to land use
- Substitute the harmful incentives with more positive ones

- **Forest Landscape Restoration involves**

- Forest landscape restoration includes a package of solutions, such as natural regeneration, enrichment planting, selective logging, reintroduction of native species, fencing, and agroforestry.
- Connecting forest fragments and establishing a link between protected areas and well-managed forests.
- Valuing forest goods and services to evaluate and quantify how stakeholders can directly benefit from these
- Planning with key stakeholders to identify solutions that are acceptable to all. It will entail collaborating with local communities, informing and training people, building capacity, and acknowledging traditional values and knowledge.

15.11.2 Establishment of Protected Areas

Protected areas are a globally recognized approach to conserving biodiversity, playing a pivotal role in safeguarding specific regions of land that hold ecological, natural, or cultural significance (Xu et al., 2019). A recent estimate indicates that the number of protected areas worldwide is skyrocketing, now covering over 23% of tropical forests (Spracklen et al., 2015) and more than 15.2% (20,455,273 km^2) of the global land surface (United Nations Environment Programme–World Conservation Monitoring Centre (UNEP-WCMC, 2018). Since the formation of protected areas is targeted at conserving forests, it is anticipated that this measure will influence an area's land-use conversion pattern, thereby avoiding deforestation (Cuenca et al., 2016). Several activities aimed at forest conservation and enhancing livelihoods have been initiated as part of protected forest management. The effectiveness of protected areas in preventing forest loss has, therefore, been the subject of considerable research (Gaveau et al., 2009). The studies suggest that protective areas are at least partially effective in reducing deforestation (Bruner et al., 2001; Joppa & Pfaff, 2010), whereas others have discovered that they offer little protection (Clark et al., 2013; Curran et al., 2004). However, One global study found that deforestation in the areas immediately surrounding protected area boundaries tends to level off or reach saturation at a radius of approximately five kilometers (Spracklen et al., 2015).

15.11.3 Sustainable Agriculture

As agriculture drives about 80% of tropical deforestation, preventing deforestation is a critical step toward achieving sustainable agriculture. Agricultural sustainability rests on the principle that we must use resources in a way that meets the needs of the present generation without compromising the ability of future generations to meet their own needs. Sustainable agriculture prevents deforestation by adopting various farming practices, such as organic farming, biodynamic agriculture, hydroponics, agroforestry, crop diversification, and integrated pest management, which results in the efficient use of existing land, thereby minimizing the need for forest clearance and expansion. Among these different farming practices, agroforestry, crop rotation, and diversification are the most promising approaches, which play an important role in reducing pressure on the forest. It is also noted that agroforestry holds significant promise for reducing deforestation and forest degradation, supporting rural livelihoods, providing habitats for species outside formally protected areas, and alleviating pressure on conservation areas from resource users (Abebe, 2005; Souza et al., 2012). Trees and shrubs grown in agroforestry practices have the potential to reduce the deforestation rate, improve income for smallholder farmers, and directly or indirectly contribute to soil fertility improvement. Such practices prevent the over-exploitation of the soil and help mitigate the demand for forest clearance, particularly in tropical regions where land conversion for agriculture is rampant.

15.12 Implementation of Participatory Forest Management Programs

Another important point is the greater participation of the community in natural resource conservation, which has resulted in better forest protection outcomes worldwide (Rasmussen et al., 2017). Deforestation and forest degradation continue to occur at an alarming rate, which significantly affects the livelihoods of communities that rely on forests. Approximately 1.6 billion people, or 25% of the global population, depend on forests for their income, subsistence needs, employment, and other livelihoods (The Global Forest Goals Report, 2021). Furthermore, the expropriation of forests by governments as state-owned forest reserves led to the gradual degradation of the remaining forests (Arnold & Perez, 2001). Therefore, participatory forest management programs emerged in the 1900s as an essential tool or choice to increase local people's involvement in decision-making. Community-based forestry initiatives have made a substantial contribution to reducing or controlling deforestation and forest degradation in the area, particularly in the mid-hills. To address the challenges of forest degradation and deforestation, the Ethiopian Government launched community-based forest management initiatives in the late 1990s. Nonetheless, the impact of community-based resource management is not fully understood, underscoring the need for a deeper understanding of the design, implementation, and governance of community-based forest management (CBFM) initiatives. Most CBFM programs are implemented by government or non-governmental organizations, often with the participation of local communities in their design and implementation.

15.12.1 Afforestation and Reforestation

Reforestation of deforested sites and enrichment plantations in degraded forest patches are vital in managing deforestation (Williams, 2002). This could be achieved by utilizing vacant or unused lands and waste lands for forest plantations, thereby avoiding the cutting down of trees in already established forests. This approach will help combat deforestation and its associated devastating impacts. Planting trees on roadside areas, along railway tracks, on contours, avenues, boundaries, and other non-forest areas will reduce pressure on primary and established forests, ultimately reducing deforestation. Reforestation and afforestation help restore damaged ecosystems by improving air and water quality, sequestering excess carbon in the atmosphere, preventing soil erosion, enhancing biodiversity, and improving groundwater recharge, among other benefits. A notable example of reforestation is the Amazon forest restoration initiative announced in 2017, which aims to rehabilitate approximately 30,000 hectares of forest, equivalent to over 70 million trees.

15.12.2 Piloting rEDD + and Payments for Environmental Services (PES)

Payments for Environmental Services (PES) are one of the approaches that countries can utilize to decrease deforestation and, therefore, earn payments for Reduced Emissions from Deforestation and Forest Degradation (REDD). Programs of PES have attracted considerable attention as a conservation tool in recent years and are likely to play an important role in many REDD strategies. Payments for ecosystem services (PES) often compensate landowners for managing their properties and providing ecosystem benefits that would not be available otherwise. According to

Werf et al. (2009) and Harris et al. (2012), globally, deforestation and forest degradation contribute approximately 10–12% of total CO_2 emissions. The Clean Development Mechanism (CDM), which emphasizes afforestation, has long included forests in debates about climate change. In contrast, REDD is a relatively recent concept that was originally introduced at the 2005 Montreal COP (Conference of the Parties) and focuses on preventing forest loss. At the 16th Conference of the Parties (COP-16) in 2010, the REDD framework was expanded to REDD + to encompass sustainable forest management, conservation, and the enhancement of forest carbon stocks. REDD + (Reducing Emissions from Deforestation and Forest Degradation) represents a Payment for Ecosystem Services (PES) mechanism designed to reduce carbon emissions associated with forest conversion and resource extraction in low-income countries. Payment for Ecosystem Services (PES) is a market-based mechanism for conservation financing, founded on the dual principles that beneficiaries of environmental services (e.g., clean water users) should bear the cost of these services, while those responsible for their provision (e.g., upstream land users) should receive compensation for their contributions (Wunder, 2005; Pagiola & Platais, 2007; Engel et al., 2008; Wunder et al., 2008). REDD can be viewed as a type of international Payment for Ecosystem Services (PES), through which countries are rewarded for the carbon sequestration services they offer to the global community. Countries seeking to participate in a REDD program encounter numerous challenges, with one of the most critical being the development of effective strategies to reduce deforestation. Although many details of how REDD will be implemented are still to be determined, virtually all proposals stipulate that REDD payments be conditional. Countries would only receive payments for verifiable reductions in deforestation and forest degradation, as well as for increases or stabilization in forest carbon stocks. Therefore, countries must adopt effective strategies to reduce emissions or enhance and stabilize carbon stocks to secure REDD funding.

15.13 Formulation and Implementation of Policy, Strategies, and Law Enforcement

Despite various limitations, forestry administration has made efforts to control deforestation and forest degradation and conserve forest resources under its jurisdiction through the implementation of specific policies and laws. Many national and international efforts have been made to reduce deforestation and increase transparency in the forestry sector. There should be enforcement of laws that regulate logging practices and land usage while protecting local endangered species. Kaimowitz et al. (1998) undertook a thorough analysis of policies to abate deforestation and concluded that the following types of policies can influence deforestation rates-

- Regulation of price and demand for tropical agriculture and forestry products.
- Increasing the expense and danger of production is linked with deforestation.
- Curbing speculation of land

- Enhancing the profitability of forest maintenance.
- Increasing the opportunity cost associated with labour and capital used in the clearing of forests.

Kaimowitz et al. (1998) and Wunder (2004) reported that the elimination of subsidies that drive deforestation and forest degradation is also a crucial step in combating these issues. They identified the following list of subsidies that can be eliminated:

- Subsidies for agriculture input that encourage the expansion of agriculture,
- Financial support for settlement and colonization scheme.
- Technical and advisory support, tax credits, and subsidies for taking activities on cleared land.
- Subsidies on road and transport.
- Subsidies to logging and other industries engaged in destructive exploitation.

15.14 Conclusion

Deforestation and forest degradation are complex global environmental issues, where deforestation involves the conversion of natural forests into non-forested land and forest degradation is characterized by changes in forest structure and function without altering land cover, resulting in reduced biomass and ecosystem services. Many factors influence the rate of deforestation and degradation; however, the relative contributions of these factors vary over regions and time. In high-profile international discussions, the concerning rates of deforestation and forest degradation in emerging nations have assumed a central role. These discussions have led to several interventions, technologies, protocols, and policies aimed at addressing these issues and achieving sustainable forest management. Tackling deforestation and degradation is not only an ecological necessity but also a crucial factor in preserving biodiversity, stabilizing the climate, and protecting cultural heritage. Therefore, effective policies, robust legal framework, innovative technologies, sustainable agriculture, and forestry practices are crucial to combat deforestation and forest degradation. Additionally, the engagement of local communities in forest conservation efforts is vital. Empowering local communities through education and training programs, as well as involving them in decision-making processes, can lead to more sustainable forest management practices.

References

Abebe, T. (2005). Diversity in homegarden agroforestry systems of Southern Ethiopia. Wageningen University and Research.

Adila, D., Nuryartono, N., & Oak, M. (2021). The environmental Kuznets curve for deforestation in Indonesia. *Economics and Finance in Indonesia, 67*, 3.

Angelsen, A. (1999). Agricultural expansion and deforestation: Modeling the impact of population, market forces and property rights. *Journal of Development Economics, 58*, 185–218.

Arnold, J. M., & Perez, M. R. (2001). Can non-timber forest products match tropical forest conservation and development objectives? *Ecological Economics, 39*, 437–447.

Artes, T., Oom, D., de Rigo, D., Houston Durrant, T., Maianti, P., Libertà, G., & San-Miguel-Ayanz, J. (2019). A global wildfire dataset for the analysis of fire regimes and fire behaviour. *Scientific Data, 6*, 296.

Barraclough, S. L., & Ghimire, K. B. (2000). Agricultural expansion and tropical deforestation: poverty, international trade and land use. *Earthscan.*

Bernhard, K. P., Smith, T. E., Sabuhoro, E., Nyandwi, E., & Munanura, I. E. (2020). Effects of integrated conservation–development projects on unauthorized resource use in Volcanoes National Park, Rwanda: A mixed-methods spatio-temporal approach. *Oryx, 55*, 613–624.

Bruner, A. G., Gullison, R. E., Rice, R. E., & da Fonseca, G. A. (2001). Effectiveness of parks in protecting tropical biodiversity. *Science, 291*, 125–128.

Caravaggio, N. (2022). Economic growth and forest transition in Latin America. *Forest Policy and Economics, 135*, Article 102667.

Chazdon, R. L, Brancalion, P. H., Laestadius, L., Bennett-Curry, A., Buckingham, K., Kumar, C., Moll-Rocek, J., Vieira, I. C. G., & Wilson, S. J. (2016). When is a forest a forest? Forest concepts and definitions in the era of forest and landscape restoration. *Ambio, **45***, 538–550.

Clark, N. E., Boakes, E. H., McGowan, P. J. K., Mace, G. M., & Fuller, R. A. (2013). Protected Areas in South Asia have not prevented habitat loss: A study using historical models of land-use change. *PLoS One, 8*, pe65298.

Cuenca, P., Arriagada, R., & Echeverría, C. (2016). How much deforestation do protected areas avoid in tropical Andean landscapes? *Environmental Science & Policy,* 56–66.

Culas, R. J. (2007). Deforestation and the environmental Kuznets curve: An institutional perspective. *Ecological Economics, 61*, 429–437.

Curran, L. M., Trigg, S. N., McDonald, A. K., Astiani, D., Hardiono, Y. M., & Siregar, P. (2004). Lowland forest loss in protected areas in Indonesian Borneo. *Science, 303*, 1000–1003.

Curtis, P. G., Slay, C. M., Harris, N. L., Tyukavina, A., & Hansen, M. C. (2018). Classifying drivers of global forest loss. *Science, 361*, 1108–1111. https://doi.org/10.1126/science.aau3445

Damette, O., & Delacote, P. (2012). On the economic factors of deforestation: What can we learn from quantile analysis? *Economic Modelling, 29*, 2427–2434.

Defries, R. S., Rudel, T., Uriarte, M., & Hansen, M. (2010). Deforestation driven by urban population growth and agricultural trade in the twenty-first century. *Nature Geoscience, 3*, 178–181.

Engel, S., Pagiola, S., & Wunder, S. (2008). Designing payments for environmental services in theory and practice: An overview of the issues. *Ecological Economics, 65*, 663–674.

FAO. (2017). What Woodfuels Can Do to Mitigate Climate Change; Food and Agriculture Organisation: Rome, Italy. http://www.fao.org/3/i1756e/i1756e00

FAO. (2018). Terms and definition- FRA 2020. Forest Resource Assessment Working Paper 188. Rome.

FAO. (2020). Global Forest Resources Assessment: Main report. Rome.

Fayiah, M. (2021). Uncertainties and trends in the forest policy framework in Sierra Leone: An overview of forest sustainability challenges in the post-independence era. *International Forestry Review, 23*, 139–150.

Gandon, A. V., Donoso, P. J., & Gerding, V. (2018). Forest degradation: When is a forest degraded? *Forests, 9*, 726.

Gaveau, D. L. A., Epting, J., Lyne, O., Linkie, M., Kumara, I., & Kanninen, M. (2009). Evaluating whether protected areas reduce tropical deforestation in Sumatra. *Journal of Biogeography, 36*, 2165–2175.

Geist, H., & Lambin, E. (2002). Proximate causes and underlying driving forces of tropical deforestation. *Bio Science, 52*, 143–150.

Gregory, P. A., Eben, N. B., Paulo, J. C. O., Michael, K., David, E. K., & Jose, N. M. S. (2006). Condition and fate of logged forests in the Brazilian Amazon. *Proceedings of the National Academy of Sciences, 103*, 129–147.

Hansen, M. C., Potapov, P. V., Moore, R., Hancher, M., Turubanova, S. A., Tyukavina, A., Thau, D., Stehman, S. V., Goetz, S. J., Loveland, T. R., & Kommareddy, A. (2013) High resolution global maps of 21st-century forest cover change. *Science, 342*, 850–853.

Harris, N. L., Brown, S., Hagen, S. C., Saatchi, S. S., Petrova, S., Salas, W., Hansen, M. C., Potapov, P. V., & Lotsch, A. (2012). Baseline map of carbon emissions from deforestation in tropical regions. *Science, 336*, 1573–1576.

Hosonuma, N., Herold, M., De Sy, V., De Fries, R. S., Brockhaus, M., Verchot, L., Angelsen, A., & Romijn, E. (2012). An assessment of deforestation and forest degradation drivers in developing countries. *Environmental Research Letters, 7*(4), Article 044009.

Joppa, L. N., & Pfaff, A. (2010). Reassessing the forest impacts of protection: The challenge of nonrandom location and a corrective method. *Annals of the New York Academy of Sciences, 1185*, 135–149.

Joshi, R., Singh, H., Chhetri, R., & Yadav, K. (2020). Assessment of carbon sequestration potential in degraded and non-degraded community forests in Terai region of Nepal. *Journal of Forest and Environmental Science, 36*(2), 13–121.

Joshi, R., Sharma, B., Singh, H., Dhakal, N., Ayer, S., & Maraseni, T. (2024). Poplar plantation as an agroforestry approach: Economic benefits and its role in carbon sequestration in north India. *Journal of Resources and Ecology, 15*(4), 880–888. https://doi.org/10.5814/j.issn.1674-764x.2024.04.009

Kaimowitz, D., Byron, N., & Sunderlin, W. (1998). Public policies to reduce inappropriate tropical deforestation. Agriculture and the Environment: Perspectives on Sustainable Rural Development. *World Bank, Washington, DC*, pp.302–322.

Kissinger, G. M, Herold, & De Sy, V. (2012). Drivers of Deforestation and Forest Degradation. *A synthesis report for REDD+Policymakers*. 48.

Kumar, A., Dwivedi, G. K., Tewari, S., Jaipaul, S. V. K., Singh, H., Kumar, P., Kumar, N., & Kausal, R. (2020). Soil organic carbon pools under *Terminalia chebula* Retz. based agroforestry system in Himalayan foothills, India. *Current Science, 118*, 1098–1103. https://doi.org/10.18520/cs/v118/i7/1098-1103

Kumar, M., & Singh, H. (2020). Agroforestry as a nature-based solution for reducing community dependence on forests to safeguard forests in rainfed areas of India. In S. Dhyani, A. Gupta, & Karki M. (Eds.), *Nature-based Solutions for Resilient Ecosystems and Societies. Disaster Resilience and Green Growth*. Springer, Singapore. https://doi.org/10.1007/978-981-15-4712-6_17.

Lambin, E. F., Geist, H. J., & Lepers, E. (2003). Dynamics of Land-Use and Land-Cover Change in Tropical Regions. *Annual Review of Environment and Resources, 28*, 205–241.

Laurance, W. F., Goosem, M., & Laurance, S. G. (2009). Impacts of roads and linear clearings on tropical forests. *Trends in Ecology and Evolution, 24*, 659–669.

Lawson, S., & MacFaul, L. (2010). Illegal Logging and Related Trade: Measuring the Global Response. *Chatham House: London, UK, 2010*, 13–18.

Lawson, S., Blundell, A., Cabarle, B., Basik, N., Jenkins, M., & Canby, K. (2014). *Consumer Goods and Deforestation: An Analysis of the Extent and Nature of Illegality in Forest Conversion for Agriculture and Timber Plantations; Forest Trend Report Series: Washington.* DC.

Li, Y., Liu, H., Zhu, X., Yue, Y., Xue, J., & Shi, L. (2021). How permafrost degradation threatens boreal forest growth on its southern margin? *Science of the Total Environment, 762*, 143–154.

Lim, C. L., Prescott, G. W., Alban, J. D. T., Ziegler, A. D., & Webb, E. L. (2017). Untangling the proximate causes and underlying drivers of deforestation and forest degradation in Myanmar. *Conservion. Biology, 31*, 1362–1372.

McNeely, J. A., Mooney, H. A., Neville, L. E., Schei, P., & Waage, J. K. (2001). A global strategy on invasive alien species. Gland, Switzerland and Cambridge, UK, World Conservation Union (IUCN).

Mooney, H. A., & Hofgaard, A. (1999). Biological invasions and global change. In Invasive Species and Biodiversity Management. *Based on a Selection of Papers Presented at the Norway/UN Conference on Alien Species, Trondheim, Norway, 24*, 139–148.

Myers, N. (1994). Tropical deforestation: rates and patterns. In: The causes of tropical of tropical deforestation. In K. Brown, & D. Pearce (Eds.), *The economic and statistical analysis of factors giving rise to the loss of the tropical forest*, (pp. 27–40). UCL Press.

Nguyen, C. P., Nguyen, B. Q., & Tran, D. T. L. (2023). Over two decades of severe deforestation: An economic perspective of tourism development. *Journal of Environmental Studies and Sciences, 13*, 83–104.

Ojima, D. S., Galvin, K. A., & Turner, B. L. (1994). The global impact of land-use change. *Bio Science, 44*, 300–304.

Pagiola, S., & Platais, G. (2007). *Payments for Environmental Services: From Theory to Practice.* World Bank.

Prakash, P., Kumar, P., Bishist, R., Kumar, P & Singh, R. (2025). Impact of altitudinal gradients on composition, biomass and soil nutrients of Cedrus deodara forests in the Northwestern Himalaya. In *Trees, forests and people*, (p. 100937).

Pysek, P., Davis, M. A., Daehler, C. C., & Thompson, K. (2004). Plant invasions and vegetation succession: Closing the gap. *Bulletin of the Ecological Society of America, 85*, 105–109.

Rademaekers, K., Eichler, L., Berg, J., Obersteiner, M., & Havlik, P. (2010). Study on the evolution of some deforestation drivers and their potential impacts on the costs of an avoiding deforestation scheme. Prepared for the European Commission by ECORYS and IIASA. Rotterdam, Netherlands.

Rasmussen, L. V., Watkins, C., & Agrawal, A. (2017). Forest contributions to livelihoods in changing agriculture-forest landscapes. *Forest Policy and Economics, 84*, 1–8.

Rimal, B., van Goor, W., Trines, E., & Karki, G. (2014). Draft report: Assessment of drivers of deforestation in Nepal. FCPF/REDD/S/QCBS-4: Analytical study on assessing the value of forests, the political economy of land use and the carbon emissions from the drivers of DD. Face the Future (the Netherlands) for the REDD-Forestry and Climate Change Cell, Ministry of Forests and Soil Conservation, Government of Nepal.

Romero-Uribe, H. M., Lopez-Portillo, J., Reverchon, F., & Hernández, M. E. (2022). Effect of degradation of a black mangrove forest on seasonal greenhouse gas emissions. *Environmental Science and Pollution Research, 29*, 11951–11965.

Singh, R., Prakash, P., Singh, A.K., Shukla, S., Sharma, P., Chauhan, V., & Singh, B.K. (2025). Altitudinal variation in carbon stocks of *Quercus leucotrichophora* forests in the Western Himalayas: Implications for climate change mitigation. In *Trees, forests and people* (p. 101059).

Souza, H. D. N., De Goede, R. G. M., Brussaard, L., Car- doso, I. M., Duarte, E. M. G., & Fernandes, R. B. A. (2012). Pro-tective shade, tree diversity and soil properties in coffee agroforestry systems in the Atlantic Rain-forest biome. *Agriculture Ecosystem & Environment, 146*, 179–196.

Spracklen, B. D., Kalamandeen, M., Galbraith, D., Gloor, E., & Spracklen, D. V. (2015). A Global Analysis of Deforestation in Moist Tropical Forest Protected Areas. *PLoS ONE, 10*, Article e0143886.

Teague, W. R., Dowhower, S. L., Baker, S. A., Haile, N., DeLaune, P. B., & Conover, D. M. (2011). Grazing management impacts on vegetation, soil biota and soil chemical, physical and hydrological properties in tall grass prairie. *Agriculture Ecosystems and Environment, 141*, 310–322.

Thompson, I., Mackey, B., Macnulty, S., & Mosseler, A. (2009). Forest resilience, biodiversityand climate change. A synthesis of biodiversity/resilience/stability relationship in forest ecosystems., Montreal: Secretariat of the convention on convention on biological diversity.

Turner, B. L., & Meyer, W. B. (1994). Global land-use and land-cover change: an overview. *Changes in land use and land cover: a global perspective, 4*(3).

UN-DESA. (2021). United nations department of economic and social affairs, united nations forum on forests secretariat. The global forest goals report 2021.

UNEP-WCMC (2018) United Nations List of Protected Areas. Supplement on protected area management effectiveness. UNEP-WCMC: Cambridge, UK.

Vasquez-Grandón, A., Donoso, P. J., & Gerding, V. (2018). Forest degradation: when is a forest degraded? *Forests 9*, 726.

Vie, J. C., Hilton-Taylor, C., & Stuart, S. N Eds. (2009). Wildlife in a changing world: an analysis of the 2008 IUCN Red List of threatened species. IUCN.

Werf, G. R., Morton, D. C., De Fries, R. S., Olivier, J. G., Kasibhatla, P. S., Jackson, R. B., Collatz, G. J., & Randerson, J. T. (2009). CO_2 emissions from forest loss. *Nature Geoscience., 2*, 737–738.

Williams, B., Johns, D., Williams, H., & Ledbetter, W. (2002). Converting pine plantations to bottomland hardwood wetlands: Lessons learned from a unique restoration effort at the Nature Conservancy Roy E. Larsen Sandyland Sanctuary in Southeast Texas. *Restoration Ecology, 20*, 88–95.

World Bank Blogs. (2016). Five forest figures for the international day of forests. https://blogs.worldbank.org/opendata/five-forest-figures-international-day-forests

WU. (2022). Own calculations based on global forest change data (Hansen et al. 2013), global mining area data (Maus et al. 2022) and global mining production data (SNL, 2022). Vienna, Austria: Vienna University of Economics and Business.

Wunder, S. S., & Engel, P. S. (2008). Taking stock: A comparative analysis of payments for environmental services programs in developed and developing countries. *Ecological Economics, 65*, 834–852.

Wunder, S. (2004). Policy options for stabilising the forest frontier: a global perspective. In *Land use, nature conservation and the stability of rainforest margins in Southeast Asia* (pp. 3–25). Berlin, Heidelberg: Springer Berlin Heidelberg.

Wunder, S. (2005). Payments for environmental services: Some nuts and bolts." Occasional Paper No.42. Bogor: CIFOR.

Xu, P., Wang, Q., Jin, J., & Jin, P. (2019). An increase in nighttime light detected for protected areas in mainland China based on VIIRS DNB data. *Ecological Indicators, 107*, Article 105615.

Zhang, Y., Prescott, G. W., Tay, R. E., Dickens, B. L., Webb, E. L., Htun, S., Tizard, R. J., Rao, M., & Carrasco, L. R. (2018). Dramatic cropland expansion in Myanmar following political reforms threatens biodiversity. *Scientific Reports,* 8, 1–10.

Chapter 16
Ecorestoration: An Alternative Approach for Achieving Carbon Neutrality and a Sustainable Future

Abhishek Kumar, Deepali Upadhyay, Shilpa Kumari, Aakash Kumar, Vatsala, Komal Rani, Mukesh Kumar, Darshita, Prabal Kumar, Manish Singh, and Sanjay Singh

Abstract Ecorestoration is a cost-effective strategy for achieving carbon neutrality and sustainable development goals. Eco-restoration's multiple benefits in biodiversity and environment improvement are linked to enhanced ecosystem service delivery. Ultimately, the restoration of degraded landscapes through initiatives such as eco-restoration can foster resilient ecosystems that are responsive to all current environmental stressors and climate change impacts. This chapter explores the integration of advanced technologies, including remote sensing and machine learning, to enhance

A. Kumar (✉)
Department of Biotechnology, Graphic Era (Deemed to Be University), Dehradun, Uttarakhand, India
e-mail: abhishek259kumar@gmail.com; abhishekkumar.bt@geu.ac.in

D. Upadhyay
Division of Genomic Resources, ICAR-National Bureau of Plant and Genetic Resources, New Delhi, India

S. Kumari
Division of Genetics, Indian Council of Agricultural Research, Indian Agriculture Research Institute, New Delhi, India

A. Kumar
Himalayan Institute for Learning and Leadership (HILL), UPES, Dehradun, Uttarakhand, India

Vatsala · Darshita · P. Kumar · M. Singh
ICFRE-Forest Research Institute, Dehradun, India

K. Rani
Genetics and Tree Improvement Division, Tropical Forest Research Institute, Jabalpur, Madhya Pradesh, India

M. Kumar
Botany Department, Government College Bahadurgarh, Jhajjar, Haryana, India

S. Singh
Centre of Excellence On Sustainable Land Management, Indian Council of Forestry Research and Education, Dehradun, Uttarakhand, India

H. Singh (ed.), *Forestry for a Carbon-Neutral and Sustainable Future*, Sustainability Solutions, https://doi.org/10.1007/978-3-032-06427-1_16

the accuracy of monitoring and managing restoration projects. In addition, it highlights the need to align appropriately with global carbon market frameworks that generate carbon credits, providing financial incentives and investments. Furthermore, the chapter highlights the pressing issues with current approaches and strategies for eco-restoration by examining case studies that demonstrate how it can be a practical and integrated approach towards achieving carbon neutrality, maintaining ecological balance, and securing the future for both natural and human systems.

Keywords Climate change · Ecorestoration · Ecosystem services · Sustainability

16.1 Introduction

Ecological restoration, or eco-restoration, refers to the "process of assisting the recovery of an ecosystem that has been degraded, damaged, or destroyed" (Society for Ecological Restoration, 2004). Eco-restoration is the practice by which natural landscapes can be reclaimed and restored to their respective structural and functional integrity. These involve the reintroduction of native species, the restoration of natural vegetation, and the remediation of soil to promote biodiversity and resilience. The rapid destruction of the environment and ecosystems has led to significant alterations in eco-restoration efforts, which aim to balance human development with environmental sustainability (Singh et al., 2024). Restored ecosystems can provide essential ecosystem services and support local communities by generating jobs, securing natural resources, and enhancing resilience to climate impacts through initiatives such as afforestation and reforestation, which help capture atmospheric carbon dioxide and aid in climate change mitigation (Nautiyal et al., 2022).

Eco-restoration is an effective tool for mitigating climate change. With increasing global temperatures and the frequency of extreme weather events, the proactive implementation of sustainability concepts to mitigate greenhouse gas emissions and stabilize ecosystems is essential (Singh, 2021). The most crucial application of eco-restoration is its long-term potential for carbon storage in terrestrial ecosystems, including forests, wetlands, and grasslands.

Reforested tropical rainforests, such as those in the Amazon, Southeast Asia, and Central Africa, store significant amounts of carbon as biomass in their soils. Brazil aims to restore 15 million hectares of forest, possibly restoring millions of tons of carbon dioxide. Similarly, China has numerous afforestation initiatives aimed at increasing carbon sequestration in afforested landscapes. Mangrove forests of South America, Asia, and Africa can store up to four times more carbon than tropical rainforests. Grasslands and agricultural systems also enhance carbon storage, and techniques such as intercropping in agroforestry help sequester more carbon. According to the United Nations Environment Programme (UNEP), restoring ecosystems could contribute up to one-third of the total carbon sequestration required to meet the Paris Agreement targets (UNEP, 2021). For example, forests have been identified as one of the most prominent carbon sinks, capable of absorbing approximately

2.6 billion tonnes of carbon dioxide annually, accounting for one-third of the total carbon emissions from fossil fuels. Degraded landscapes are often more susceptible to climate-related disturbances, such as droughts, floods, and wildfires.

16.1.1 Why Ecosystem-Based Approaches Are Essential

Ecosystem-based approaches are emerging as a crucial strategy for achieving carbon neutrality. Ecosystem-based strategies capitalize on the natural capacity of forest wetlands, grasslands, and marine ecosystems to store substantial amounts of carbon over extended periods (Singh, 2024c). For example, mangrove forests can sequester four times as much carbon per hectare as terrestrial forests, representing their potential for carbon sequestration.

In protecting and restoring natural habitats, ecosystem-based approaches safeguard species diversity, soil health, water quality, and landscape stability against climate-driven perturbations such as floods and erosion. Such benefits extend beyond carbon capture, creating robust ecosystems that contribute to climate adaptation and sustainability. Restoration of natural ecosystems also aligns with the concept of Nature-Based Solutions (NbS), which the International Union for Conservation of Nature (IUCN) defines as actions that harness nature to address societal challenges, such as climate change (Cohen-Shacham et al., 2016).

Ecosystem-based approaches can complement existing carbon reduction strategies to fill the gaps not covered by traditional carbon offsetting measures. Furthermore, investment in ecological health supports more sustainable economies, increasing the livelihood support of local communities that rely on these resources. For instance, community-driven reforestation projects can generate jobs, provide access to better food security, and build resilience to climate impacts, thereby reinforcing the social dimensions of sustainability alongside ecological and economic benefits.

16.2 Ecorestoration: An Ecosystem-Based Alternative Pathway

Technological advancements and modernization, rapid industrialization, urbanization, and anthropogenic activities have led to ecological degradation. Therefore, ecorestoration is a nature-based solution proposed to restore degraded ecosystems, benefiting the flora and fauna that inhabit these vital ecosystems. Diverse ecological practices that include sustainability are promoted for ecorestoration. Plantation forestry is one such approach adopted at national and international levels, providing various ecosystem services. There are different ecorestoration approaches discussed below:

16.2.1 Afforestation and Reforestation

Afforestation has various ecological benefits and is crucial for environmental sustainability. Afforestation measures help remove CO_2 from the environment and contribute to multiple targets aimed at combating climate change. Afforestation measures contribute to climate resilience by mitigating extreme weather events, such as droughts and floods, through these efforts. On the other hand, reforestation is the process of replanting trees on recently deforested land, whereas afforestation is the planting of trees on ground that has not been afforested in the last 50 years or more (Veldkamp et al., 2020; Duan & Abduwali, 2021). These two strategies could open up "win–win" policy choices if seen as complementary. Afforestation and reforestation are internationally recognized as measures for carbon sequestration that contribute to mitigating climate change. However, they also support adaptation to climate change by reducing human pressures, such as the destruction or degradation of habitat, which enhances landscape connectivity and reduces fragmentation to promote and allow species migration under climate change conditions (Kumar et al., 2021). Afforestation and reforestation can also contribute to preserving biodiversity hotspots by maintaining the hydrological flow of water and running biogeochemical cycles.

In cases where crop failure arises from changing climate, forests can be an essential source of safety nets for the local communities in terms of many resources, including wood and non-wood products. The hydrological ecosystem services of forests preserve the underground water table, regulate storm flows, and control erosion while managing water flow (Carvalho-Santos et al., 2014). The planting of native trees would create new habitats for resilient species and enhance biodiversity, particularly when multispecies plantations are established with a focus on native species and a deliberate avoidance of invasive ones (Kumar et al., 2024a, 2024b, 2024c, 2024d). Additionally, afforestation and reforestation activities help prevent soil degradation and reduce the risks of landslides and erosion (Kumar and Singh 2020). EU afforestation policies spanning from 1994 to 2015 resulted in the planting of nearly 2 million hectares of trees on agricultural land. India has committed to restoring approximately 21 million hectares of forest by 2030 as part of the Bonn Challenge, aligning with India's national forest policy. This target is to achieve tree cover on 33% of the country's land area. The Three North Afforestation Program (TNAP) of China was launched in 1978 and is one of the most extensive afforestation programs in the world, covering over 40% of China. It is estimated that 34 billion dollars is still required to complete the project (Zhu et al., 2017).

Large-scale afforestation has never been able to provide for regulating ecosystem services and biodiversity conservation when the primary focus was on productivity. Pérez-Silos et al. (2021) analysed 75 global studies and highlighted that a more integrated and multidisciplinary approach is required for afforestation measures. With approximately $44 trillion, or half of the global GDP, at risk from natural disasters, it is crucial to prevent deforestation, conserve intact forests, and restore vulnerable ecosystems to mitigate climate change threats. Closing the annual nature financing gap, which averages $711 billion, is essential to fostering a nature-positive economy and reversing the decline of natural ecosystems by 2030. Several global initiatives, including the Bonn Challenge, the Great Green Wall, the Forest and Landscape Restoration Mechanism (FLRM), the Global Restoration Initiative, and other start-up initiatives such as Funga, Grow Trees, the Open Forest Protocol, the Sankalp Taru Foundation, and Terraces, are focused on afforestation and restoration efforts.

16.2.2 Wetland and Peatland Restoration

Wetlands serve as important stopovers for migratory birds and are experiencing more significant losses in biodiversity than other terrestrial and marine ecosystems. Over 20–90 percent of today's coastal wetlands sequester carbon at rates up to 55 times faster than tropical rainforests, and they are on the verge of extinction by the end of the century. Wetland loss rates have been accelerating since 2000, depending on the degree of sea-level rise resulting from the climate crisis.

According to the Kunming-Montreal Global Biodiversity Framework (KM-GBF), biodiversity is valued, conserved, and restored, with targets set until 2050 to wisely maintain ecosystem services, sustain a healthy planet, and deliver benefits essential for all people. This vision of a world living in harmony with nature depends on conserving and restoring wetlands. To prevent, halt, and reverse the degradation of our planet's wetlands, the United Nations Decade on Ecosystem Restoration (2021–2030) presents a unique and timely opportunity to coordinate efforts and make significant progress globally.

Wetlands generally include peatlands and non-peat wetlands, although distinction is required to confirm this definition. Peat develops under wet, anaerobic conditions, making peatlands with multiple layers of soil. It is essential for ecological services, acts as a natural water filter, and is a vital carbon reservoir holding more carbon than forests or other biomass types. According to the Ramsar Convention on Wetlands, the technical report, Convention on Wetlands (2021), focused on peatland restoration using proper planning restoration techniques. The Ramsar Convention is a framework that focuses on the wise use of wetlands and their resources. Governments worldwide have intensified their efforts to safeguard and rejuvenate these natural spaces, which experts assert are vital for biodiversity protection and combating the climate crisis. Throughout much of the world, wetland environments mitigate the negative impacts of flooding and purify polluted water, which has declined by over 35 percent since 1970 (Mohanty et al. 2024).

16.2.3 Mangrove and Coastal Ecosystem Restoration

It is crucial to preserve the remaining mangrove cover worldwide. There is also a growing emphasis on mangrove rehabilitation and restoration to achieve several national and international goals, particularly during the UN Decade of Ecosystem Restoration (2021–2030). A potentially game-changing natural approach to mitigating climate change and enhancing coastal resilience is to restore mangroves. However, because of their low success rates and perceived risk, ambitious mangrove restoration programs have not been quickly scaled up or invested in. Successful mangrove restoration strategies have offered equitable benefits for many years.

Systematic conservation planning is a widely used framework that empowers and involves stakeholders to inform successful restoration outcomes (Rodríguez-Rodríguez et al., 2024). Song et al. (2023) observed that 370 mangrove restoration sites worldwide had a more efficient carbon storage potential than afforestation projects. The selection of tree species or the best approach to accomplish restoration has been a primary hypothesized question in mangrove restoration pathways that maximize carbon sequestration (e.g., what planting density should be applied to carbon and sedimentation, and whether to choose monoculture versus mixed-species) (Lovelock et al., 2022).

16.2.4 Grassland and Soil Restoration

Grasslands cover nearly 40% of the terrestrial biosphere and are home to a diverse array of animals, plants, and more than 1 billion people worldwide. Over the last 100 years, the degradation of grasslands has progressed through three stages: biotic and abiotic indicators, ecosystem functions, and ecosystem services (Sun et al., 2022). Grassland ecosystems are the most significant terrestrial carbon sinks, contributing to the balance of atmospheric greenhouse gas concentrations through the additional sequestration of carbon. Extreme climate phenomena and human activities have exacerbated the ecological degradation of these ecosystems.

Forest restoration has garnered significant attention within the United Nations Decade on Ecosystem Restoration. This attention ironically poses one more risk to grasslands, as restoration efforts may pursue randomly or thoughtlessly conceived tree planting in grassland and savannah ecosystems under the guise of restoration. For instance, the goal is to plant trees over more than 1 million km2 of Africa's grassland biomes by 2030. This practice bypasses the idea of conserving and restoring grasslands. The approach would suppress other woody species suitable for grazing and restore fire-affected regimes. Various eco-restoration approaches help maintain sustainable ecosystems (Fig. 16.1). Conversely, soil conservation and restoration are equally important, as they are closely intertwined. The conservation programs should also focus on soil quality and related aspects to ensure its long-term conservation. The connection between plant communities and soil microorganisms is crucial,

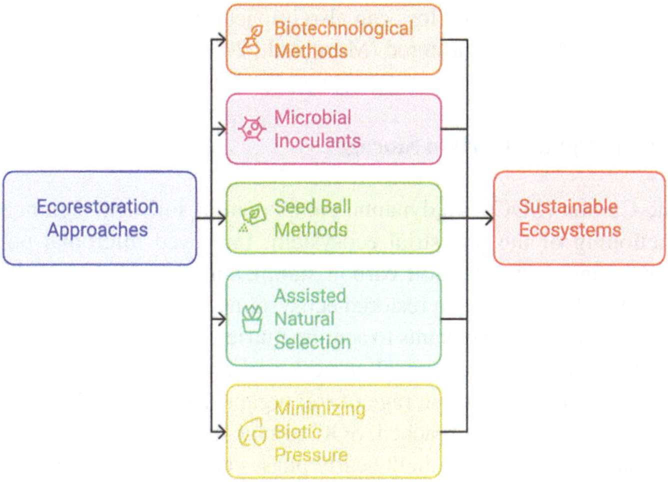

Fig. 16.1 Various ecorestoration approaches

and much research has now focused on understanding the synergies between soil microorganisms and plant communities (Farrell et al., 2020). According to Levin et al. (2022), eco-restoration activities involve a multidisciplinary team of professionals, including field laborers, ecologists, conservation scientists, planners, and other specialists, who are engaged in reforestation, soil restoration, invasive species management, and waterway rehabilitation. On a larger scale, afforestation and reforestation actions require massive labor inputs in planting trees, preparing the soil, and managing seedlings, thus generating much-needed employment in areas where employment opportunities are scarce. The World Resources Institute report suggests that reforestation projects alone may create millions of jobs worldwide, primarily in rural and poorer communities, where such employment opportunities can have a profound impact on local livelihoods (Mohan et al., 2021). In addition to man-labor, eco-restoration programs create technical and scientific employment opportunities. Professionals must assess the baselines of the environment, monitor changes in ecology, and determine adaptive management for restoration to be successful. As the trend of data-driven eco-restoration continues to increase, remote sensing, GIS, and environmental data analysts will play a crucial role in planning and monitoring restoration projects.

16.2.5 Biodiversity Importance in Carbon Sequestration

The biological diversity of an ecosystem is closely related to its capacity to sequester carbon. A rise in biodiversity can contribute to increased biomass output, improving carbon storage and biomass accumulation. Although biodiversity can be impacted by climate change, biodiversity loss can also impact climate change by altering the way carbon is stored and sequestered (Mori et al., 2021).

16.2.5.1 Soil Organic Carbon Storage

Soil organic Carbon (SOC) is a dynamic component of soil with significant effects on the functioning of the terrestrial ecosystem. Improved microbial populations, organic inputs, and particular soil carbon stabilization mechanisms contribute to enhancing soil carbon storage in restored ecosystems. Rhizosphere associations are also important, as they enable plants to acquire nutrients (Khan et al., 2023). Interactions between the dynamic ecological processes of photosynthesis, decomposition, and soil respiration result in the storage of soil organic carbon (SOC). After a certain level of ecological maturity is reached, SOC storage may stabilize or even decrease, although it usually rises quickly in the early phases of restoration (Liu et al., 2023). Over the past 150 years, human activity has altered these processes, resulting in the depletion of SOC storage and the escalation of global climate change (Ontl & Schulte, 2012).

16.3 Ecosystem Services Provided by Eco-Restoration

According to the UN Decade on Restoration, eco-restoration refers to assisting in the reclamation of ecosystems that have been degraded or destroyed, as well as conserving intact ecosystems. Eco-restoration is demonstrated through actions as diverse as new mangroves, grass or other plantings, assisted or natural regeneration, agroforestry, soil enrichment measures, or improved and sustainable management to accommodate a variety of land, aquatic, or marine uses (IUCN, 2022).

16.3.1 Habitat Restoration

Ecosystem restoration offers the opportunity to effectively halt and reverse degradation, improve ecosystem services, and restore biodiversity. It can enhance ecosystem health and create conditions that allow native species to thrive. It can also strengthen habitat complexity by introducing a vegetation layer that supports a diverse array of life forms. According to the UN Decade on Restoration, the restoration of 350 million hectares of degraded aquatic and terrestrial ecosystems could produce ecosystem services worth US$9 trillion by 2030. Approximately 60% of species extinctions could be avoided through the active restoration of 15% of converted lands (Strassburg et al., 2020).

16.3.2 Water Regulation and Quality Improvement

However, eco-restoration plays a vital role in restoring the connectivity of hydrological regimes and the interaction between biota and ecosystems (Polvi et al., 2020). However, by reducing soil moisture and water availability, extensive eco-restoration may harm the ecosystem (Feng et al., 2016; Zhao et al., 2021). For example, tropical and subtropical commercial fruit species thrive in areas with greater water conservation, but drought-tolerant savanna woody species should be planted in water-deficient areas. Healthy ecosystems with high biodiversity indirectly support higher levels of plant growth and carbon sequestration because they retain and increase water quality.

16.3.3 Climate Regulation and Resilience

To better combat the detrimental effects of climate change, restoration procedures should be adapted to increase ecosystem resilience (Huang et al., 2023). It has been reported that eco-restoration can improve biomass, air quality, water quality, and nutrient cycling while also regulating erosion and climate, and enhancing biodiversity and habitat for species (Liu et al., 2023). Approximately 13–26 gigatons of greenhouse gases might be removed from the environment through restoration. In addition to the adverse effects of climate change, restoration strategies must be maximized to enhance ecosystem resilience to counteract these effects more effectively (Huang et al., 2023).

16.3.4 Soil Health and Erosion Control

Eco-restoration enhances soil health by promoting the accumulation of organic matter, increasing nutrient availability, reducing surface runoff, and improving aeration and water infiltration. Restored ecosystems rich in organic matter tend to be more resilient to erosion. Research has shown that high-intensity restoration in water-scarce areas can cause acute drought and a decline in soil moisture, leading to reduced vegetation growth and a decline in ecosystem services (Huang et al., 2022; Schwärzel et al., 2020; Zhao et al., 2021).

16.4 Social and Economic Benefits of Eco-Restoration

16.4.1 Job Creation and Economic Incentives

Eco-restoration initiatives are significant job creators, providing indirect economic benefits across various sectors. Eco-restoration can increase biodiversity and trigger local economic growth, thereby generating opportunities for employment and financial benefits and encouraging societies to invest in the long-term health of their environment (Gann et al., 2019). Eco-restoration is a sustainable development initiative that enhances ecological regeneration and creates opportunities for job creation and socio-economic development.

16.4.2 Economic Stimulation Through Local Business Growth

Apart from providing direct employment, developing ecorestoration projects boosts local businesses by supplying the materials, tools, and services needed to implement the restoration work. Nurseries that sell native plants and seeds benefit significantly from the increased demand for indigenous flora, particularly in cases where they focus on habitat restoration and biodiversity enhancement (BenDor et al., 2015). Moreover, businesses that provide equipment and machinery for land processing, planting, and maintenance benefit economically from increased ecorestoration activities.

Economic benefits are derived from eco-restoration and propagate through other ancillary industries such as construction, transportation, and hospitality. Restoration sites typically require infrastructure development, including access roads, visitor facilities, educational centers, and other amenities, which also generate employment opportunities in construction and ancillary industries. In areas where eco-tourism is anticipated, increased local investments in facilities and services are expected to contribute to long-term economic growth.

16.4.3 Incentivizing Community Participation in Eco-Restoration

Eco-restoration initiatives motivate local communities to participate by offering them financial advantages and promoting community involvement in restoration projects. For example, participation in eco-restoration programs provides direct economic benefits for implementing sustainable practices, such as sustainable agriculture, by compensating community members for their efforts to preserve and restore local ecosystems (Datta et al., 2023). In Costa Rica, where landholders are compensated to restore degraded or preserve existing forest areas, the Payment for Ecosystem Services (PES) schemes have proven effective, yielding both ecological and economic benefits (Chand, 2024). Governments and non-governmental organizations also contribute to these programs by providing grants, tax breaks, or subsidies to people and communities who engage in eco-restoration. To maintain the sustainability of agricultural systems and improve agroforestry practices while preventing land degradation, these incentives are excellent motivators for residents to adopt restoration-friendly methods. For instance, the Compensatory Afforestation Fund Management and Planning Authority (CAMPA) in India provides financial incentives to state governments to prioritize eco-restoration initiatives by funding afforestation and restoration projects (Bharucha & Pretty, 2010). The main challenge is that eco-restoration is expensive. Planning and implementing restoration activities, as well as the continued effort of restoration, are very costly. Financial constraints may limit the scope of restoration activities and often decide the nature of the intervention to be made as well as the geography to be covered. Often costly, restorative methods include afforestation and reforestation; labor-intensive activities such as soil preparation, planting, and monitoring are expensive. Large restoration projects, for instance, can easily multiply in cost, as noted in sub-Saharan Africa, where slight budgetary constraints have hindered the ability to effectively restore degraded lands in their large states (Chazdon & Guariguata, 2016). Many eco-restoration projects are highly funded by the government or external sources, including international donors and non-governmental organizations (NGOs). International funding is a strong support base for such projects. However, it might be bound to a particular project and has time constraints, often posing a challenge to sustainability once the funding cycle is completed. A relevant example can be drawn from Southeast Asian forest restoration, where donor funds initially yielded some success. Still, as time passed and funds dwindled, the projects became unsustainable. Short-term finance disrupts the continuity necessary to enable ecosystem recovery, as successful eco-restoration requires sustained interventions spanning several years.

16.4.4 Ecorestoration Resilience

The ability of people in a community to recover and adapt to changes and shocks that include climate, scarce resources, and economic adversity (Uddin et al., 2021). Eco-restoration projects enhance resilience within the system by restoring ecosystem functions within the community, including water purification, soil fertility, and erosion control. For instance, restoring wetlands and riverbanks reduces flooding risks, which is a significant advantage for communities vulnerable to extreme weather. Restoration of degraded lands in the Tana River Basin, Kenya, has restored soil water-holding capacity, thereby reducing the risk of drought and enabling local crop farmers to maintain their crop yields during dry periods. Diverse ecosystems offer higher resilience to diseases, invasive species, and environmental changes, thereby making them a more reliable basis for community livelihoods (Joshi et al.,

2021). In varied and resilient ecosystems, there is better resistance to disturbances, and they can recover more quickly, allowing local communities to continue enjoying the ecosystem services they offer. Reestablishing mangrove forests in Southeast Asia has been a vital conservation effort in the region, as they protect coastal populations from storm surges, which are becoming increasingly frequent due to climate change.

16.4.4.1 Promoting Sustainable Livelihoods

Ecological restoration can directly enhance the livelihoods of those who depend on natural resources, providing a sustainable path for economic development. They also provide the requisite resources that drive local economies, and ecological restoration enables communities to regain access to fish, timber, non-timber forest products, and arable land (Ninan & Inoue, 2014). For instance, the reforestation of degraded forests in Brazil's Amazon Basin supports the livelihoods of local communities by increasing the volume of timber and NTFP harvests, which are much needed by indigenous and rural communities. Ecological restoration projects also support sustainable agriculture and forest management, reducing dependence on practices that cause deforestation and soil degradation. Agroforestry, for example, integrates trees into farm systems to offer shade, fertility, and supplementary income through fruits, nuts, and other forest products. Projects on eco-restoration through agroforestry have led to the diversification of farmers' income sources, thereby enhancing economic resilience and food security in India (Bharucha & Pretty, 2010). Sustainable harvesting can be encouraged through eco-restoration efforts to ensure long-term resource availability, allowing communities to utilize what nature has provided without depleting it completely. This will significantly benefit regions where overharvesting has led to resource depletion, threatening the wellness of those who live within and beyond that area.

16.4.4.2 Revitalization of Traditional Practices and Culture

Ecological restoration offers opportunities to revive and adopt traditional practices that are often more environmentally friendly than modern industrial approaches. Many generations have passed since Indigenous and rural communities acquired knowledge systems about the safe and resource-conserving use of the resource base and ecosystem conservation. Therefore, incorporating such practices in eco-restoration would reinforce and promote cultural heritage with a sense of community ownership and pride. For example, Aboriginal Australian traditional knowledge of forest management through fire has been integrated into forest restoration, where it continues to be effective in controlling the flora and preventing unwanted outbreaks of wildfires (Pascoe & Gammage, 2021). In West Africa, traditional soil and water conservation methods have played a significant role in rehabilitating degraded agricultural lands and enhancing soil fertility and crop productivity. Different types of ecorestoration techniques have been discussed in Table 16.1, along with their mechanisms and benefits individually.

16.5 Challenges and Limitations of Eco-Restoration

Besides financial resources, a successful eco-restoration project needs human resources, ecological know-how, and specialized equipment. Restoration efforts often call for multidisciplinary skills from ecologists, soil scientists, hydrologists, and local practitioners. However, most development areas lack trained staff, and crucial resources and technologies are inaccessible (Kumar and Singh, 2023). These factors may delay or hinder the efficiency of restoration processes, as most restoration plans developed by local authorities are often scientifically incorrect due to a lack of expertise.

Another resource allocation issue is the availability of seeds of native plant species and materials for use in the re-vegetation process. Restoring ecosystems with native species enhances biodiversity and generally makes an ecosystem more stable; however, obtaining native plant species can sometimes prove difficult due to the scarcity of nurseries and seed banks for native flora (Kumar et al., 2020). The lack of established seed banks for native flora is yet another limitation associated with large-scale restoration efforts, as has happened in Australia's drylands, where restoration projects have been unable to access nearly enough native plant materials to meet their goals.

Thus, logistical problems due to rugged terrain, adverse weather conditions, and transportation are significant limiting factors in distributing resources for large-scale restoration projects. Remotely located or inaccessible areas require additional resources to relocate materials and personnel, thereby increasing costs and straining budgets. For example, the eco-restoration project deployed in mountainous and coastal regions of mangrove areas typically faces logistical problems that require more time, money, and a larger workforce to overcome.

16.5.1 Scaling Issues in Large Eco-Restoration

Scaling issues for large-scale eco-restoration interventions often arise due to the financial, logistical, and policy constraints associated with small to landscape-level interventions. While small-scale projects are manageable with limited resources and specific, focused interventions, scaling up requires coordinated activities across larger geographic areas and the involvement of multiple stakeholders, including governments, communities, and the private sector. This scalability problem can even be illustrated in examples such as China's "Grain for Green" program, where it is challenging to transform arable farmland into forests across large areas without compromising the quality of the outcome (Liu et al., 2012).

Table 16.1 Summary of ecorestoration techniques, carbon sequestration mechanisms, and key sustainable benefits

Ecological focus	Primary techniques	Carbon sequestration mechanisms	Key quantifiable carbon potential	Other major sustainability benefits	References
Forest restoration	Afforestation, reforestation, Assisted Natural Regeneration (ANR), Agroforestry, native species planting	Photosynthesis, Biomass storage (trunks, branches), Soil carbon storage	Significant CO2 removal; Carbon stock increases after 10–15 years	Biodiversity enhancement, Soil stabilization, Water regulation, Economic benefits (timber, non-wood products, jobs)	Rajapakshe et al., (2024); Kewat et al., (2025a, 2025b); Kewat et al., (2025a, 2025b); Vatsala et al., (2025); Fayiah & Kumar, (2025)
Wetland restoration	Re-establishing hydrological regimes, Reintroducing native vegetation, Controlling invasive species, Hydrologic management, Nutrient assimilation systems	High productivity and slow decomposition in anaerobic soils, Peat formation, Permanent geologic storage	Up to 5 × more carbon than forests/unit area; 1.5 tons/ha/year; 2.55 million tons CO2 over century (Snohomish Estuary)	Water quality improvement (natural filters), Flood risk reduction, Shoreline stabilization, Habitat provision (fisheries)	Walker et al., (2021)
Soil regeneration	Conservation tillage, Cover cropping, Crop rotation, Organic amendments, Agroforestry, Vegetative cover, Terracing, Sediment control dams	Increased Soil Organic Carbon (SOC), Reduced erosion, Enhanced soil structure, Biomass contribution (agroforestry)	Increased SOC (e.g., + 20–30% with conservation tillage/ organic amendments)	Improved soil health & fertility, Enhanced agricultural productivity/yields, Reduced chemical inputs, Water retention & flood mitigation, Biodiversity promotion, Food security	Singh et al., (2025); Agrawal et al., (2025); Chand et al., (2024); Kumar et al., (2024a, 2024b, 2024c, 2024d); Kumar et al., (2024a, 2024b, 2024c, 2024d)

(continued)

Table 16.1 (continued)

Ecological focus	Primary techniques	Carbon sequestration mechanisms	Key quantifiable carbon potential	Other major sustainability benefits	References
Coastal restoration	Re-establishing hydrological regimes, Native species planting (mangroves, salt marshes, seagrasses), Silvofishery, Coastal greenbelt corridors	High sequestration rates, Primarily belowground storage in anaerobic soils/ sediments	Mangroves: 1.82–4.45t CO2-eq/ha/yr; Salt Marshes: 0.55–2.25t CO2-eq/ha/yr; Seagrasses: 0.33–1.43t CO2-eq/ha/yr; 10 × tropical forests	Coastal protection (storms, erosion), Fisheries support (nursery grounds), Biodiversity enhancement, Recreational opportunities, Income diversification	Richards et al., (2020)
Agroecology	Diversifying crops/ livestock, Conservation agriculture, Agroforestry, Cover cropping, Organic amendments, Natural pest management, Water harvesting	Enhanced soil carbon storage, Biomass storage (trees), Reduced GHG emissions from inputs	Can sequester up to 10.28 Gt CO2-eq/yr globally; Agroforestry: 3.3 t C/ha/yr	Increased agricultural production/yields, Reduced input expenses, Diversified income, Improved soil health, Climate resilience, Biodiversity promotion, Food security, Poverty reduction	Rani and Kumar, (2024); Rani et al., (2024)

Eco-restoration also needs to be sensitive to the issues of ecological variation, as what applies perfectly in one ecosystem or region may not be appropriate elsewhere. Applying a one-size-fits-all approach to large-scale restoration often leads to inefficiencies or unintended ecological consequences. For example, reforestation projects that deploy non-native tree species to meet their rapid targets tend to form monocultures, offering fewer environmental benefits and are thus more prone to disease outbreaks (Suding et al., 2015). Therefore, effective scaling demands localized adaptation and appreciable investment in finance to ensure site-specific intervention, thus increasing the cost and complexity of the project.

16.5.2 Policy and Regulatory Barriers

Successful ecorestoration programs are highly contingent upon policy frameworks and regulatory support that align with the country's and local environmental goals. Supportive policies might lead to successful ecorestoration through adequate funding, clear land-use rights, and incentives for sustainable practices. Generally, limiting or archaic regulations are one of the reasons restoration processes are not fully accomplished, as they restrict the area's scalability, community involvement, and long-term sustainability (Mansourian & Vallauri, 2014).

16.5.3 Role of Supportive Policies in Eco-Restoration

Supportive policies create momentum for the success of implementing an eco-restoration initiative by establishing an enabling environment for conservation, sustainable land management, and resource protection. Policies encouraging reforestation, biodiversity conservation, and emissions reduction can also support the bottom-up approach of an eco-restoration initiative by providing legal backing and incentives. The policies encourage PES by providing financial incentives to owners and communities to restore degraded landscapes and conserve biodiversity (Wunder, 2015).

Clear land tenure policies are also essential for eco-restoration. Stronger tenure security over the land will enhance the ability of local communities and private owners to carry out restoration activities without being evicted or restricted from accessing lands in the future. Insecure tenure often constrains communities from participating in restoration projects, as they fear being evicted or having their lands used differently after restoration efforts. In Brazil, clear legal structures that cede rights to local communities to the lands have been associated with success in various restored forests. This ensures a stake in restored ecosystems (Joshi et al., 2020).

Furthermore, policies that integrate the goals of eco-restoration into broader environmental and development plans facilitate coordination among parties and prevent land-use-related conflicts. NDCs, under the Paris Agreement, provide for eco-restoration as part of the methodologies to achieve reductions in greenhouse gas emissions. The Paris Agreement offers a framework under which Governments will prioritize and finance restoration (Seddon et al., 2020).

16.5.4 Limiting Rules

Restrictive regulations can limit eco-restoration by imposing rigid requirements, banning certain restoration techniques, or creating unnecessary bureaucratic steps that slow down the implementation of projects. Many environmental laws, for instance, are applied in various countries on the continent to protect existing natural resources, but they often lack provisions for encouraging active restoration. For example, in some nations, an EIA is required for projects involving eco-restoration; this delays and increases the planning process costs because restoration is an environmentally friendly approach.

Furthermore, stringent land use regulations may restrict access to degraded lands, particularly when these lands are designated as protected areas or have complex land tenure arrangements. Projects involving degraded lands within protected areas may require special permits or legislative amendments; such procedures often take years, and further delaying the restoration is no longer feasible. Reforestation activities in forest reserves are often delayed due to complex forestry regulations that restrict access to the reserves and prohibit restoration activities at certain sites.

In this regard, inflexible regulations may also fail to facilitate flexible approaches to the adaptability of the eco-restoration process. Eco-restoration is often site-specific, and specific methods must be applied to accommodate different approaches in local ecological conditions. For instance, laws regarding non-native species may seriously limit the introduction of some plant species even though they might fortify soil stabilization or provide habitat and restore degraded areas effectively. These restrictions sometimes limit the effectiveness of restoration, for instance, in arid regions, where native species may be insufficient to meet rapid restoration goals.

16.5.5 Policy Gaps in Addressing Climate Adaptation and Ecosystem Resilience

It is critical in the context of global climate change. Many countries lack comprehensive policies on climate adaptation and ecosystem resilience. Climate change adaptation can be a goal of restoration projects at the ecosystem level; however, restoration is seldom integrated with policy frameworks that set climate adaptation objectives. Without this integration, restoration efforts can be reversed by climate-induced events such as droughts, floods, and wildfires (Hobbs et al., 2021). Policies must fill this gap focused on climate-resilient eco-restoration, which prioritizes restoring natural buffers, such as mangroves, wetlands, and forests, to mitigate climate impacts. Policies aimed at coastal eco-restoration must support the rehabilitation of mangrove forests to safeguard against the effects of increased sea levels and storm surges, because mangroves, above all, offer critical ecosystem services and climate resilience (DasGupta & Shaw, 2015). However, most countries lack policies aimed at promoting such restoration. As a result, climate adaptation and building resilience is lost.

16.5.6 Barriers Related to the System of Administration and Bureaucratic Framework

Ecological restoration projects often face bureaucratic impediments that escalate costs, prolong durations, and lower productivity. This is a consequence of overlapping jurisdictions, conflicting policies, and voluminous paperwork that limit the flow of resources. Administrative impediments are often more challenging in developing countries where government agencies usually lack the capacity and capability to execute such projects. The situation worsens when restoration projects require approval from multiple agencies before commencing, as has been the case at times (Reid et al., 2017). For instance, in transboundary restoration projects, such as re-greening transboundary ecosystems, the involvement of multiple governmental structures with various priorities and needs complicates project management. (Mason et al., 2020). Significant deregulation and coordination among agencies will be necessary to eliminate bureaucratic delays in large-scale initiatives.

16.5.7 Shift Towards More Favorable Regulatory Frameworks

Existing law needs to be reformed, and adaptable policies should be implemented that encourage restoration without compromising environmental protection. Streamlining and simplifying procedures for acquiring permits may accelerate the speed of green restoration. Administrative barriers and vague instructions discourage stakeholders from participating in restoration activities. Therefore, removing bureaucratic barriers and providing clearly stated guidance in the established procedures attracts more people to participate in restoration exercises (McDonald et al., 2016). Furthermore, policies must adopt adaptive management techniques that allow for flexibility and adjustments in consideration of local ecological needs and fluctuating environmental conditions. Adaptive policy management can be achieved by embracing eco-restoration projects to explore novel restoration approaches and gather ecological feedback. For instance, Australia's dryland restoration policies have adapted to prioritize adaptive management, making it feasible to implement different strategies for specific site conditions and restoration aims (Lindenmayer et al., 2012).

Lastly, international cooperation is vital in harmonizing policies and approaches to regulation during cross-border efforts on eco-restoration activities. Transboundary ecosystems require cooperative policies that support restoration activities across national borders, ensuring the continuity and effectiveness of such eco-restoration efforts on a larger scale, specifically in river basins and migratory corridors.

16.5.8 Time and Scale Constraints

The ecological, socio-economic, and climate-resilience benefits of eco-restoration can take considerably longer and broader spatial scales to manifest, even while minor efforts at the local level may provide immediate benefits to the localized areas. Stakeholder commitment, financial continuity, and ecological variability—all crucial factors for achieving successful eco-restoration outcomes—are made more difficult by these lengthy and expansive procedures.

16.5.9 Challenges of Long-Term Project Sustainability

Thus, the biggest challenge to the long-term impact of eco-restoration is the sustainability of funding, political will, and community interest over a prolonged period. Most eco-restoration projects depend on budgets that run on short-term cycles of government grants, NGOs, or international donors. Upon depletion of the initial funds, projects may be left unfinished, thereby skimping on their ecological benefits and increasing the likelihood of the degradation cycle repeating itself. Restoration initiatives in the tropics frequently suffer from cash deficits at some point, even after taking action measures, as open-ended funds are often left to sparse follow-up and monitoring (Brancalion et al., 2014). Political priorities and economic interests can shift over time, affecting long-term project commitment. In many regions, competition from sectors such as agriculture, urban development, or resource extraction could threaten the lands set aside for restoration. For instance, restoration projects in the Amazon may be disrupted in Brazil by land conversion into cattle ranches and soy farms, among other commercial activities, primarily due to the erosion of political support. This critical aspect has been instrumental in achieving long-term success in maintaining political and community commitment—a feat that has not been possible for a long time, given the shifting economic and social circumstances. Another factor associated with stakeholder fatigue is the easy loss of interest from community members and local organizations, who may withdraw their support if they do not immediately reap some benefits from restoration. Engaging local communities in long-term restoration requires a clear understanding of both the short-term and long-term benefits to sustain their continued involvement. Some direct benefits include employment and income from sustainable harvesting, while the indirect benefits will focus on building resilience in the face of environmental challenges, as well as preserving biodiversity and ecosystem services. Community participation and motivation should be sustained over time to ensure the project's long-term success.

16.5.10 Constraints at Scale in Eco-Restoration

Eco-restoration projects at small scales are effective in rehabilitating certain areas but generally fail to contribute more meaningfully to broader environmental benefits, particularly in terms of resilience to ecosystem disturbance, biodiversity conservation, and climate mitigation. This requires large-scale restoration projects to produce meaningful landscape impacts, where fragmented habitats are connected, biodiversity corridors are created, and watersheds are restored to replenish resources for multiple communities. Examples include the Bonn Challenge, which has committed to restoring 350 million hectares of degraded land by 2030 (Chand et al., 2021; Kumar, 2021; Kumar, 2024; Singh et al., 2024). This suggests that landscape-scale restoration is necessary to achieve global targets while addressing the ecological, resource, and jurisdictional coordination challenges associated with scaling up eco-restoration. Large-scale projects necessitate a restorative strategy for each area, as ecological conditions can change significantly even within a single landscape. Restoration techniques applied to one ecosystem do not necessarily function effectively in another. For instance, in Brazil's Atlantic Forest, diverse ecological zones will require different reforestation techniques, species selections, and soil management practices to achieve a successful restoration outcome throughout the entire region (Rodrigues et al., 2009).

Additionally, large-scale eco-restoration is resource-intensive, requiring substantial financial, technical, and logistical resources to access and coordinate. Extensive seed banks, nurseries, and knowledgeable personnel are necessary for implementing and managing restoration work across large geographic areas. Furthermore, some remote and inaccessible areas may increase transportation and labor costs, making large-scale restoration even more challenging. Due to the lack of such resources, the demanding requirements of massive projects make large-scale eco-restoration difficult in low-income economies.

16.5.11 Ecological and Climate Uncertainties

Eco-restoration is inherently challenging because ecological systems are dynamic; however, with climate change, there is an added layer of uncertainty that has never been experienced before. While eco-restoration aims to restore ecological systems to a more natural and resilient state, it introduces climate variables that affect restoration in various ways. These include changes in weather patterns, regime shifts in temperature, increased frequency of extreme events, and long-term changes in species distributions, all of which make restoration projects challenging.

16.5.12 Climate Change and Ecosystem Functioning

Restored ecosystem design mimics the natural processes of nutrient cycling, water regulation, and carbon sequestration, thereby promoting a healthier environment. Climate change on long-time scales will affect these processes in unpredictable ways. For instance, warming can alter the timing of seasonal events, such as plant flowering and animal migration, in ways that can disrupt species interactions and disrupt the ecological balance of restored ecosystems. For instance, with increased temperatures, the species composition in restored forest ecosystems is likely to change, with heat-tolerant species gaining a competitive advantage over more sensitive ones (Kanta et al., 2024). Similarly, a higher temperature and altered precipitation patterns are likely to influence the growth rates of plants and the carbon sequestration capacity of restored ecosystems. Some restorations, such as reforestation aimed at increasing carbon sequestration, may not store the desired amounts of carbon if warming-related increases in mortality or decreased growth from restoration forests eventually overcome any immediate benefits.

Extreme events can undermine the years of effort invested in restorations, including soil erosion, damage to plant communities, and disturbance to ecological processes that have been re-established. For instance, restored forests can be sensitive to fire in drier and hotter climates. Even mature, restored forests can be easily overrun by wildfires, pushing restoration several steps backward. Restored wetlands or grasslands in dry climates may rapidly lose their hydrological equilibrium if rainfall remains erratic, degrading the habitat and thereby promoting the loss of species. This would seriously threaten the effectiveness of restoration methods, as such disturbances could not be predicted, and ecosystems might be unable to recover shortly after.

Flooding caused by heavy rainfall, as recently experienced in the Himalayan foothills (Nagamani et al., 2024), would inundate restored areas, undermine all restoration efforts, and alter landscape features. Ecosystems restored to a level of some ecological state may be susceptible to sudden perturbations that render them less resilient to such extreme events.

16.5.13 Species Distribution Shift and the Risk of Invasion

Climate change is altering the geographical distribution of species, including those targeted for focused restoration. Once well-adapted to a given geographic area, species become less suited to the new climatic conditions, causing a shift in vegetation and animal populations. Restored ecosystems can face challenges, as some species reintroduced or naturally recolonizing within the ecosystem may no longer thrive due to temperature shifts, changes in precipitation, or new pest pressures (Barton et al., 2019). For example, multiple species of trees may be planted to reestablish a forest. Still, those tree species may later not be adapted to new changes in growth conditions, such as increased temperatures or droughts, so the density of the forest is reduced or the forest structure changes. In other cases, invasive species may start to invade the restored area by taking advantage of changes in conditions caused by climate change. For instance, the overproduction of non-native grasses in restored prairies can alter soil composition, fire regimes, and nutrient cycling, thereby complicating restoration processes (Seabloom et al., 2023).

16.5.14 Uncertainty in Long-Term Restoration Objectives

Climate uncertainty does not facilitate the long-term outcome of eco-restoration projects. Restored ecosystems are unlikely to remain in a stable state for long periods if they have been exposed to new or unprecedented climatic stress. For instance, carbon sequestration over long periods that would otherwise have been possible through reforestation may not occur, depending on how much warming undermines the growth potential of key tree species. Similarly, wetlands restored for flood control might lose functionality if a new climate regime results in more frequent and intense flooding. Climate change could also cause ecological shifts, making it inappropriate for restoration practitioners to define "success" easily. In this regard, ecosystems may never revert to their pre-disturbance state, or, in many cases, they may not do so since species extinctions, altered climatic conditions, and the dominance of invasive species prevent ecosystem restoration.

16.6 Approaches to Mitigating Climate Uncertainties in Eco-Restoration

Since climate change is uncertain, the practices of ecologically restoring must be climate-sensitive and enhance the restored ecosystems' ability to adapt. Such practices emphasize the resilience-building of ecosystems, the integration of climate predictions, and adaptive management.

16.6.1 Focus on Ecosystem Resilience

It is possible to focus on increasing the resilience of the ecosystems to future climate conditions. Restoration efforts can utilize a diverse mix of species that are tolerant of a wide range of environmental conditions, thereby creating relatively more resilient ecosystems to the impacts of climate change. In areas that suffer from fires or drought, it could be selected for fire-resistant plant species or drought-tolerant vegetation, allowing restored ecosystems to withstand the impact of fires or drought.

16.6.2 Climate-Smart Restoration Approaches Climate

The climate-smart restoration approach incorporates future climate projections into restoration planning for species selection and strategies that are most likely to thrive under future climatic conditions. For example, selecting more heat-tolerant or drought-resistant species would enhance the long-term survival of restoration projects, even under changing climatic conditions. Such an approach should closely collaborate with climatologists to predict impacts and incorporate these predictions into the design of restoration work (Seddon et al., 2020).

16.6.3 Adaptive Management

Adaptive management is one of the most effective management techniques for handling climate uncertainties. This approach involves flexibility and continuous adjustments based on new data and changing situations. For instance, regular monitoring of restoration sites for changes in climate or ecosystem health conditions would enable better responses to the unpredictable scenario of climate change (Hobbs et al., 2021). Some changes may include species composition, recovery practices, or the introduction of new interventions to ensure that restoration targets remain viable.

16.6.4 Buffer Zones and Landscape Connectivity

Incorporating buffer zones or landscape connectivity can help protect ecosystems from extreme weather events and climate-related effects. For example, when restored forests are connected, they form larger networks of protected areas, allowing species to move and adapt to climate change. This connectivity would eliminate ecosystem fragmentation, as it could limit the ability of species to adapt to new environmental pressures (Primavera et al., 2019).

16.7 Global Carbon Stock Quantification Across Ecosystems

Measuring the amount of carbon stored in various ecosystems is crucial for addressing climate change and effectively managing our natural resources. The table presented here brings together case studies from various parts of the world, each illustrating how different landscapes, ranging from forests to croplands to wetlands, store carbon in their soil, vegetation, and biomass. These examples also highlight the different methods scientists use to estimate carbon stocks and the importance of these ecosystems in climate solutions. In Turkey's Kızıldağ region, researchers studied forests, croplands, and grasslands to estimate a total carbon stock of about 3.5 million tons (Kızıldağ et al., 2024). The researchers used forest inventories and GIS-based spatial modeling to assess carbon stored not only in trees, but also in soil, roots, and leaf litter. Meanwhile, in Bangladesh's Maheshkhali Island, scientists are focusing on homestead forests, which are small, community-managed patches of green belts surrounding homes. It was found that carbon storage varied depending on the diversity of tree species, emphasizing the importance of maintaining biodiversity.

Similarly, in Nepal's Karnali Province, oak forests under different management types (community-managed versus protected) were compared. This study highlighted the impact of human decisions and forest policies on carbon storage, utilizing field plots and soil samples to identify the differences. Similarly, a study in the Subarnarekha Estuary in India examined natural mangroves of different ages, which possess very high carbon levels, reaching up to 865 tons per hectare.

Urban ecosystems also play a vital role in supporting the ecosystem. In Sri Lanka, urban freshwater wetlands are found to store significant amounts of carbon, ranging between 14 and 66 tons of carbon per hectare, primarily in aboveground biomass. These studies highlights that even small green spaces in cities can contribute to climate mitigation. In Ecuador's northern coast, mangrove forests hold around 7.7 million tons of carbon. These results, derived from a combination of fieldwork and satellite mapping, demonstrate the vital role mangroves play in both climate and coastal protection (Hamilton et al., 2017). In the case of India, an example from the Thoria Watershed has been taken, showing how land use changes affect carbon storage over time (Roy and Dutta, 2018). Likewise, in China's Qiantang River headwaters, huge peatlands were studied, estimated to contain around 400 million tons of carbon, which would, upon damage, release nearly 3.7 million tons of CO_2 into the atmosphere. These case studies (Table 16.2) reveal that it is crucial to protect vital ecosystems alike.

Table 16.2 Global case studies on carbon quantification in different ecosystems

Country/Region	Ecosystem type	Carbon stock (t C ha^{-1} or total)	Key methods used	Reference
Turkey (Kızıldağ)	Mediterranean forest, grassland, cropland	Forest: ~ 120.8 t C ha¯1; Cropland: ~ 32.3 t C ha¯1; Total: ~ 3.5 million t C	Forest inventory,GIS spatial modeling, estimation of all carbon pools (above, below, soil, litter)	Kızıldağ et al., (2024)
Bangladesh (Maheshkhali Island)	Homestead forest ecosystems species diversity Biomass measurements, species-wise inventory, soil carbon sampling	Variable by plot and species diversity	Biomass measurements, species-wise inventory, and soil carbon sampling	Islam et al., (2021)
Nepal (Karnali Province)	Oak-dominated forests (community versus protection)	Biomass and soil carbon varied by forest type	Field plots, soil sampling (0–30 cm), statistical comparisons	Nepal et al., (2023)
India (Subarnarekha Estuary)	Natural mangroves with different stand ages	Ecosystem CO_2e: ~ 865 t/ha; Potential release upon loss:421–1160 t/ha	Biomass and soil sampling, age-wise carbon analysis	Roy and Dutta, (2018)
Sri Lanka (Colombo Wetlands)	Urban freshwater wetlands	AGB: 14–66 t C/ha; BGB: 2.5–10 t C/ha	Vegetation biomass estimation, root biomass sampling, and carbon content analysis	Perera et al., (2022) (Carbon Balance Manage.)
Ecuador (Northern Mangroves)	Mangrove estuarine forests	~ 7.7 million tons of standing carbon	Field-based mangrove biomass inventory, remote sensing, and GIS mapping	Hamilton et al., (2017)

(continued)

Table 16.2 (continued)

Country/Region	Ecosystem type	Carbon stock (t C ha^{-1} or total)	Key methods used	Reference
India (Thoria Watershed)	Cropland, woodland, settlements	2670 t (2000) → 2583 t(2020); 2602 t with + 20% woodland	InVEST model, LUCC analysis, field validation via 128 plots	Guo et al., (2024)
China (Qiantang River Headwaters)	Blanket bog (peatland)	~ 400 million t stored; Degraded peat emits ~ 3.7 Mt CO_2/year	Peat depth surveys, long-term carbon monitoring, estimation of annual carbon flux	Han et al., (2025); Yang et al., (2023)

16.8 Way Forward and Conclusion

Ecorestoration has emerged as a key strategy for attaining a carbon–neutral future and progressing towards a more sustainable one. When ecosystems are damaged, biodiversity varies significantly, and the planet's capacity to store carbon increases. Healthy forests, wetlands, and marine ecosystems serve as significant carbon sinks, absorbing vast amounts of CO_2. The additional benefit of ecological restoration is the development of robust ecosystems, which are more resistant to environmental stresses (Joshi et al., 2024). This ensures the development and creation of critical ecosystem services, such as clean water, fertile soils, and habitat for wildlife. The ecorestoration approach simultaneously brings ecological health into consonance with human well-being and aligns itself with global efforts toward sustainability. Technological innovations in eco-restoration are improving efficiency, accuracy, and scalability in restoration efforts. New drones and remote sensing technologies have made detection and decision-making easier. These advanced data analytics approaches and machine learning algorithms have led our ecologists to design and predict early ecosystem responses. These technologies identify the key priority restoration areas, provide real-time results, and optimize resource allocation. A strong collaboration between governments, NGOs, the private sector, and local communities is needed to ensure the scalability of restoration programs. The focus should be on policies and practices that support restoration activities.

Additionally, indigenous and local communities should be engaged in the restoration efforts. Education and public awareness campaigns are also necessary for mass sensitization and community awareness. A commitment to long-term monitoring and adaptive management will ensure that restored ecosystems continue to deliver the long-term benefits of restoration. A proper planning and strategic approach to eco-restoration is needed to develop climate resiliency and green infrastructure that imbibes nature-based solutions.

References

Barton, P. S., Evans, M. J., Foster, C. N., Pechal, J. L., Bump, J. K., Quaggiotto, M. M., & Benbow, M. E. (2019). Towards quantifying carrion biomass in ecosystems. *Trends in Ecology & Evolution, 34*(10), 950–961.

BenDor, T., Lester, T. W., Livengood, A., Davis, A., & Yonavjak, L. (2015). Estimating the size and impact of the ecological restoration economy. *PLoS ONE, 10*(6), Article e0128339.

Bharucha, Z., & Pretty, J. (2010). The roles and values of wild foods in agricultural systems. *Philosophical Transactions of the Royal Society b: Biological Sciences, 365*(1554), 2913–2926.

Brancalion, P. H. S., & Rodrigues, R. R. (2014). Seed size-number trade-off in Euterpe edulis in plant communities of the Atlantic Forest. *Scientia Agricola, 71*, 226–231.

Carvalho-Santos, C., Honrado, J. P., & Hein, L. (2014). Hydrological services and the role of forests: Conceptualization and indicator-based analysis with an illustration at a regional scale. *Ecological Complexity, 20*, 69–80.

Chand, H. B., Singh, S., Kumar, A., Kewat, A. K., Bhatt, R., & Bohara, R. (2021). Comprehensive overview of REDD+ in India: Status, opportunities and challenges. *Grassroots J Nat Res, 4*(3), 185–200.

Chand, H. B., Singh, S., Kumar, A., Kewat, A. K., Sharma, S., & Prabhakar, A. (2024). Unveiling nature's worth: Embracing social valuation in ecosystem services assessment. *Trends in Current Biology., 3*(1), 03–05.

Chand, H. B., Singh, S., Kumar, A., Kewat, A. K., Sharma, S., Prabhakar, A. (2024). Unveiling nature's worth: Embracing social valuation in ecosystem services assessment. *Trends in Current Biology 3*(1), 03–05.

Chazdon, R. L., & Guariguata, M. R. (2016). Natural regeneration as a tool for large-scale forest restoration in the tropics: Prospects and challenges. *Biotropica, 48*(6), 716–730.

Cohen-Shacham, E., Walters, G., Janzen, C., & Maginnis, S. (2016). *Nature-based solutions to address global societal challenges*. IUCN

Convention on Wetlands. (2021). *Global guidelines for peatland rewetting and restoration*. Ramsar Technical Report No. 11. Gland, Switzerland: Secretariat of the Convention on Wetlands.

DasGupta, R., & Shaw, R. (2015). An indicator based approach to assess coastal communities' resilience against climate related disasters in Indian Sundarbans. *Journal of Coastal Conservation, 19*, 85–101.

Datta, S., Hamim, I., Jaiswal, D. K., & Sungthong, R. (2023). Sustainable agriculture. *BMC Plant Biology, 23*(1), 588.

Duan, J., & Abduwali, D. (2021). Basic Theory and Methods of. *Silviculture*, 77.

Farrell, H. L., Léger, A., Breed, M. F., & Gornish, E. S. (2020). Restoration, soil organisms, and soil processes: Emerging approaches. *Restoration Ecology, 28*, S307–S310.

Fayiah, M., & Kumar, A. (2025). Role of Sustainable Land Management Practices in Biodiversity Conservation. In *Biodiversity Conservation and Land Degradation Neutrality: From Loss to Restoration* (pp. 155–174). Singapore: Springer Nature Singapore.

Feng, X., Fu, B., Piao, S., Wang, S., Ciais, P., Zeng, Z., & Wu, B. (2016). Revegetation in China's Loess Plateau is approaching sustainable water resource limits. *Nature Climate Change, 6*(11), 1019–1022.

Gann, G. D., McDonald, T., Walder, B., Aronson, J., Nelson, C. R., Jonson, J., & Dixon, K. (2019). International principles and standards for the practice of ecological restoration. *Restoration Ecology, 27*(S1), S1–S46.

Guo, Z., Sharma, H., Jadav, M., Hettiarachchi, U., Guha, C., Zhang, W., & Meinzen-Dick, R. S. (2024). Measuring above-ground carbon stock using spatial analysis and the InVEST model: Application in the Thoria Watershed, India. *Environmental Research Communications, 6*(11), Article 115036.

Hamilton, S. E., Lovette, J. P., Borbor-Cordova, M. J., & Millones, M. (2017). The carbon holdings of northern Ecuador's mangrove forests. *Annals of the American Association of Geographers, 107*(1), 54–71.

Han, D., Sun, Y., Cong, J., Wang, G., & Gao, C. (2025). Enhanced hydrological disturbance and atmospheric pollution on an ombrotrophic peatland (Greater Khingan Mountains, Northeast China) since 1950s inferred from diatom records. *Anthropocene, 50*, Article 100473.

Hobbs, A. (2021). *Do droughts drive deforestation*?

Huang, B., Yang, Y., Li, R., Zheng, H., Wang, X., Wang, X., & Zhang, Y. (2022). Integrating remotely sensed leaf area index with biome-BGC to quantify the impact of land use/land cover change on water retention in Beijing. *Remote Sensing, 14*(3), 743.

Huang, B., Lu, F., Wang, X., Wu, X., Zheng, H., Su, Y., & Ouyang, Z. (2023). The impact of ecological restoration on ecosystem services change modulated by drought and rising CO2. *Global Change Biology, 29*(18), 5304–5320.

Islam, K. N., Sultana, A., Wadley, D., Dargusch, P., Henry, M., & Naito, Y. (2021). Opportunities for inclusive and efficient low carbon food system development in Bangladesh. *Journal of Cleaner Production, 319*, Article 128586.

IUCN. (2022). Ecosystem restoration. Retrieved from: https://iucn.org/our-work/topic/ecosystem restoration#:~:text=About%20Ecosystem%20restoration,%2C%20aquatic%2C%20or%20marine%20uses

Joshi, R., Singh, H., Chhetri, R., & Yadav, K. (2020). Assessment of carbon sequestration potential in degraded and non-degraded community forests in Terai region of Nepal. *Journal of Forest and Environmental Science, 36*(2), 13–121.

Joshi, R., Sharma, B., Singh, H., Dhakal, N., Ayer, S., & Maraseni, T. (2024). Poplar plantation as an agroforestry approach: Economic benefits and its role in carbon sequestration in north India. *Journal of Resources and Ecology, 15*(4), 880–888. https://doi.org/10.5814/j.issn.1674-764x.2024.04.009

Joshi, R., Singh, H., Chhetri, R., Poudel, S. R., & Rijal, S. (2021). Carbon sequestration potential of community forests: A comparative analysis of soil organic carbon stock in community managed forests of Far-Western Nepal. *Eurasian Journal of Soil Science, 10*(2), 96–104. https://doi.org/10.18393/ejss.825066.

Kanta, C., Kumar, A., Chauhan, A., Singh, H., & Sharma, I. P. (2024). The interplay between plant functional traits and climate change. In *Plant Functional Traits for Improving Productivity* (pp. 41–58). Singapore: Springer Nature.

Kewat, A. K., Kumar, P., Singh, S., Kumar, A., Chand, H. B., & Kumar, R. (2025a). Temporal shifts in community dominance and regenerative capacity of tropical forests in central India. *Ecology, Environment & Conservation, 0971765X*, 31.

Kewat, A. K., Kumar, P., Singh, S., Kumar, A., Chand, H. B., Singh, K. P., & Kumar, R. (2025b). Revisiting the preservation plots of tropical forests: An assessment of 30 years temporal changes in vegetation structure, composition and distribution patterns. *Journal of Forest and Environmental Science, 41*(2), 145–168.

Khan, S., Ambika, Rani, K., Sharma, S., Kumar, A., Singh, S., & Singh, Y. (2023). Rhizobacterial mediated interactions in Curcuma longa for plant growth and enhanced crop productivity: A systematic review. *Frontiers in Plant Science, 14*, 1231676.

Kızıldağ, M., Abut, F., & Akay, M. F. (2024). Development of new electricity system marginal price forecasting models using statistical and artificial intelligence methods. *Applied Sciences, 14*(21), 10011.

Kumar, A., Dwivedi, G. K., Tewari, S., Jaipaul, S. V. K., Singh, H., Kumar, P., Kumar, N., & Kausal, R. (2020). Soil organic carbon pools under *Terminalia chebula* Retz. based agroforestry system in Himalayan foothills, India. *Current Science, 118*, 1098–1103. https://doi.org/10.18520/cs/v118/i7/1098-1103

Kumar, A., Singh, S., Chand, H. B., Kewat, A. K., & Singh, K. P. (2021). Road towards restoring land degradation in India: A UNCCD perspective. *Research Journal of Chemical and Environmental Sciences, 9*(4), 27–38.

Kumar, A., Singh, S., Kumar, D., Singh, R. K., Gupta, A. K., Premkumar, K., & Kewat, A. K. (2024a). Investigating the phenology and interactions of competitive plant species co-occurring with invasive *Lantana camara* in Indian Himalayan Region. *Scientific Reports, 14*(1), 400.

Kumar, A., Singh, S., Kumar, R., Kumar, M., Rani, K., & James, J. (2024b). Balancing conservation and development for achieving land degradation neutrality. *Trends in Current Biology., 2*(4), 17–24.

Kumar, M., Kumar, A., Kumar, S., Patle, G. T., Kumar, S., & Tanwar, A. (2024c). Role of cutting-edge technology in horticulture crops for shaping the future smart farming ecosystem: A comprehensive review. *Environment and Ecology, 42*(4B), 1933–1942.

Kumar, M., Patle, G. T., Lal, M., Kumar, A., Rani, K., Kumar, S., & Kumar, S. (2024d). An overview of crop residue management challenges, opportunities, and the way forward for sustainable food-energy security in India. *Ecology, Environment & Conservation, 0971765X*, 30.

Kumar, M., & Singh, H. (2020). Agroforestry as a nature-based solution for reducing community dependence on forests to safeguard forests in rained areas of India. In S.Dhyani, A. Gupta, M. Karki (Eds.), *Nature-based solutions for resilient ecosystems and societies*. Springer, Singapore: Disaster Resilience and Green Growth. https://doi.org/10.1007/978-981-15-4712-6_17.

Levin, S. (2022). Ecological complexity and the biosphere: the next 30 years. *Philosophical Transactions of the Royal Society B, 377*(1857), 20210376.

Lindenmayer, D. B., Likens, G. E., Andersen, A., Bowman, D., Bull, C. M., Burns, E., & Wardle, G. M. (2012). Value of long-term ecological studies. *Austral Ecology, 37*(7), 745–757.

Liu, Y., Wei, X., Guo, X., Niu, D., Zhang, J., Gong, X., & Jiang, Y. (2012). The long-term effects of reforestation on soil microbial biomass carbon in sub-tropic severe red soil degradation areas. *Forest Ecology and Management, 285*, 77–84.

Liu, J., Du, J., Zhang, C., Zhang, J., Yang, H., Donald, M. L., & Dong, T. (2023). Ecosystem service assessment under ecological restoration programs: A systematic review of studies from China. *Frontiers in Ecology and Evolution, 11*, 1152907.

Lovelock, C. E., Barbier, E., & Duarte, C. M. (2022). Tackling the mangrove restoration challenge. *PLoS Biology, 20*(10), Article e3001836.

Mansourian, S., & Vallauri, D. (2014). Restoring forest landscapes: Important lessons learnt. *Environmental Management, 53*(2), 241–251.

Mason, N., Ward, M., Watson, J. E., Venter, O., & Runting, R. K. (2020). Global opportunities and challenges for transboundary conservation. *Nature Ecology & Evolution, 4*(5), 694–701.

McDonald, T., Gann, G., Jonson, J., & Dixon, K. (2016). International standards for the practice of ecological restoration–including principles and key concepts. Soil-Tec, Inc., ©Marcel Huijser, Bethanie Walder. Washington, DC, USA: Society for Ecological Restoration.

Mohan, M., Rue, H. A., Bajaj, S., Galgamuwa, G. P., Adrah, E., Aghai, M. M., & Cardil, A. (2021). Afforestation, reforestation and new challenges from COVID-19: Thirty-three recommendations to support Civil Society Organizations (CSOs). *Journal of Environmental Management, 287*, Article 112277.

Mohanty, S., Pandey, P. C., Pandey, M., Srivastava, P. K., & Dwivedi, C. S. (2024). Wetlands contribution and linkage to support SDGs, its indicators and targets-A critical review. *Sustainable Development, 32*(5), 5348–5392.

Mori, A. S., Dee, L. E., Gonzalez, A., Ohashi, H., Cowles, J., Wright, A. J., & Isbell, F. (2021). Biodiversity–productivity relationships are key to nature-based climate solutions. *Nature Climate Change, 11*(6), 543–550.

Nagamani, K., Mishra, A. K., Meer, M. S., & Das, J. (2024). Understanding flash flooding in the Himalayan Region: A case study. *Scientific Reports, 14*(1), 7060.

Nautiyal, S., Singh, H., Kamboj, S., Kumar, A., Pal Panwar, V., & Bala, N. (2022). Carbon regulating services of a 30-year-old reclaimed limestone mine area in Foothills of Himalaya. *Indian Forester, 148*(3), 338–343. https://doi.org/10.36808/if/2022/v148i3/168187

Nepal, S., Kc, M., Pudasaini, N., & Adhikari, H. (2023). Divergent effects of topography on soil properties and above-ground biomass in Nepal's Mid-hill forests. *Resources, 12*(11), 136.

Ninan, K. N., & Inoue, M. (2014). Valuing forest ecosystem services: What we know and what we don't. *Valuing Ecosystem Services*, 189–226.

Ontl, T. A., & Schulte, L. A. (2012). Soil carbon storage. *Nature Education Knowledge, 3*(10).

Pascoe, B., & Gammage, B. (2021). *First knowledges country: Future fire, future farming*. Thames & Hudson Australia.

Perera, N., Lokupitiya, E., Halwatura, D., & Udagedara, S. (2022). Quantification of blue carbon in tropical salt marshes and their role in climate change mitigation. *Science of The Total Environment, 820*, 153313.

Pérez-Silos, I., Álvarez-Martínez, J. M., & Barquín, J. (2021). Large-scale afforestation for ecosystem service provisioning: learning from the past to improve the future. *Landscape Ecology, 36*(11), 3329–3343.

Polvi, L. E., Lind, L., Persson, H., Miranda-Melo, A., Pilotto, F., Su, X., & Nilsson, C. (2020). Facets and scales in river restoration: Nestedness and interdependence of hydrological, geomorphic, ecological, and biogeochemical processes. *Journal of Environmental Management, 265*, Article 110288.

Primavera, J. H., Friess, D. A., Van Lavieren, H., & Lee, S. Y. (2019). The mangrove ecosystem. *World seas: an environmental evaluation*, 1–34.

Rajapakshe, R., Karthigesu, J., Thavananthan, S., Sivachandiran, S., Navaneetham, V., & Sinnamani, S. (2024). Restoring a dry tropical forest through assisted natural regeneration: Enhancing tree diversity, structure, and carbon stock. *Trees, Forests and People, 17*, Article 100616.

Rani, K., & Kumar, A. (2024). Biotechnological innovations in urban forestry: adapting and mitigating climate change and environmental pollution. In *Urban forests, climate change and environmental pollution: Physio-biochemical and molecular perspectives to enhance urban resilience* (pp. 573–591). Cham: Springer Nature Switzerland.

Rani, K., Kumar, K., Navpute, G. V., Kumar, A., & Nishtha. (2024). Unveiling the microbiome marvels for its role in plant defence and resilience for sustainable agriculture. *Vegetos*, 1–12.

Reid, J. L., Wilson, S. J., Bloomfield, G. S., Cattau, M. E., Fagan, M. E., Holl, K. D., & Zahawi, R. A. (2017). How long do restored ecosystems persist? 1. *Annals of the Missouri Botanical Garden, 102*(2), 258–265.

Richards, D. R., Moggridge, H. L., Warren, P. H., & Maltby, L. (2020). Impacts of hydrological restoration on lowland river floodplain plant communities. *Wetlands Ecology and Management, 28*(3), 403–417.

Rodrigues, R. R., Lima, R. A., Gandolfi, S., & Nave, A. G. (2009). On the restoration of high diversity forests: 30 years of experience in the Brazilian Atlantic Forest. *Biological conservation, 142*(6), 1242–1251.

Rodríguez-Rodríguez, J. A., de Paula Costa, M. D., Wartman, M., Rasheed, A. R., Palacios, M., & Macreadie, P. (2024). Global trends in applying decision science in mangrove restoration: Are we missing some dimensions? *Ocean & Coastal Management, 254*, Article 107172.

Roy, A. K., & Datta, D. (2018). Analyzing the effects of afforestation on estuarine environment of river Subarnarekha, India using geospatial technologies and participatory appraisals. *Environmental Monitoring and Assessment, 190*(11), 645.

Schwärzel, K., Zhang, L., Montanarella, L., Wang, Y., & Sun, G. (2020). How afforestation affects the water cycle in drylands: A process-based comparative analysis. *Global Change Biology, 26*(2), 944–959.

Seabloom, E. W., Peace, A., Asik, L., Everett, R. A., Frenken, T., González, A. L., & Borer, E. T. (2023). Dead or alive: Carbon as a currency to integrate disease and ecosystem ecology theory. *Oikos, 2023*(7), Article e09880.

Seddon, N., Chausson, A., Berry, P., Girardin, C. A., Smith, A., & Turner, B. (2020). Understanding the value and limits of nature-based solutions to climate change and other global challenges. *Philosophical Transactions of the Royal Society B, 375*(1794), 20190120.

Singh, S., Kumar, A., & Kumar, A. (2024). Harmony in extraction: Navigating mining, land degradation and ecological restoration for sustainable futures. In *Forests and climate change: biological perspectives on impact, adaptation, and mitigation strategies* (pp. 513–534).

Singh, S., Kumar, A., & Gautam, M. K. (2025). Combating land degradation and desertification through forestry interventions. In *Biodiversity conservation and land degradation neutrality: From loss to restoration* (pp. 191–205). Singapore: Springer Nature Singapore.

Singh, H. (2021). Process-based carbon sequestration study with reference to the energy-water-carbon flux in a forest ecosystem. *Climate impacts on sustainable natural resource management*, 336–351.

Singh, H. (2024c). Climate change and forests: Biological perspectives. In *Forests and climate change: Biological perspectives on impact, adaptation, and mitigation strategies* (pp. 1–17). Singapore: Springer Nature Singapore.

Society for Ecological Restoration. (2004). *International science & policy working group.* Society for Ecological Restoration.

Song, S., Ding, Y., Li, W., Meng, Y., Zhou, J., Gou, R.,... & Lin, G. (2023). Mangrove reforestation provides greater blue carbon benefit than afforestation for mitigating global climate change. *Nature Communications, 14*(1), 756.

Strassburg, B. B., Iribarrem, A., Beyer, H. L., Cordeiro, C. L., Crouzeilles, R., Jakovac, C. C., & Visconti, P. (2020). Global priority areas for ecosystem restoration. *Nature, 586*(7831), 724–729.

Suding, K., Higgs, E., Palmer, M., Callicott, J. B., Anderson, C. B., Baker, M., & Schwartz, K. Z. (2015). Committing to ecological restoration. *Science, 348*(6235), 638–640.

Sun, J., Wang, Y., Piao, S., Liu, M., Han, G., Li, J., & Tsubo, M. (2022). Toward a sustainable grassland ecosystem worldwide. *The Innovation, 3*(4).

Uddin, M. S., Haque, C. E., Khan, M. N., Doberstein, B., & Cox, R. S. (2021). "Disasters threaten livelihoods, and people cope, adapt and make transformational changes": Community resilience and livelihoods reconstruction in coastal communities of Bangladesh. *International Journal of Disaster Risk Reduction, 63*, Article 102444.

UNEP. (2021). *Ecosystem restoration for people, nature, and climate.*

Vatsala, Kumar, A., Negi, A., Gautam, M. K., Singh, S., Kumar, A., & Chand, H. B. (2025). Bridging horizons: Global biodiversity framework and land degradation neutrality. In *Biodiversity conservation and land degradation neutrality: From loss to restoration* (pp. 35–51). Singapore: Springer Nature Singapore.

Veldkamp, E., Schmidt, M., Powers, J. S., & Corre, M. D. (2020). Deforestation and reforestation impacts on soils in the tropics. *Nature Reviews Earth & Environment, 1*(11), 590–605.

Walker, S., Bellingham, P. J., Kaine, G., Richardson, S., Greenhalgh, S., Simcock, R., & Lee, W. G. (2021). What effects must be avoided, remediated or mitigated to maintain indigenous biodiversity? *New Zealand Journal of Ecology, 45*(2), 1–12.

Wunder, S. (2015). Revisiting the concept of payments for environmental services. *Ecological Economics, 117*, 234–243.

Yang, Q., Liu, Z., & Bai, E. (2023). Comparison of carbon and nitrogen accumulation rate between bog and fen phases in a pristine peatland with the fen-bog transition. *Global Change Biology, 29*(22), 6350–6366.

Zhao, M., Zhang, J., Velicogna, I., Liang, C., & Li, Z. (2021). Ecological restoration impact on total terrestrial water storage. *Nature Sustainability, 4*(1), 56–62.

Zhu, J. J., Zheng, X., Wang, G. G., Wu, B. F., Liu, S. R., Yan, C. Z., & Ellison, A. M. (2017). Assessment of the world largest afforestation program: success, failure, and future directions. *BioRxiv*, 105619.

Chapter 17
From Forests to Carbon Neutrality: The Role of REDD+ In Climate Mitigation

Himanshu Joshi, Mahendra Singh Lodhi, Om Prakash Arya, and Harshit Pant

Abstract REDD+ plays a vital role in achieving carbon neutrality by addressing emissions from forest loss and enhancing carbon sinks. Forests absorb nearly one-third of global CO_2 emissions annually, yet deforestation and degradation remain significant sources of greenhouse gases. REDD+ provides a results-based framework for developing countries to reduce forest-related emissions while promoting biodiversity conservation, ecosystem services, and sustainable livelihoods. Under the UNFCCC, REDD+ has evolved into a structured mechanism implemented in three phases: readiness, implementation, and results-based payments. Its core principles include additionality, permanence, leakage prevention, and adherence to social and environmental safeguards. Global examples from Brazil, Indonesia, Nepal, and the Democratic Republic of Congo highlight the need for strong institutions, community participation, and effective monitoring, reporting, and verification systems. In the Indian context, the REDD+ finds alignment with India's global climate commitments, including the Net Zero target by the year 2070 and updated NDCs of 2022, specifically NDC-1, 3, and 5, respectively, under the Paris Agreement at UNFCCC. The National REDD+ Strategy (2018) outlines steps for phased implementation and linkage with international climate finance. Key institutions such as the MoEFCC, FSI, and state forest departments support monitoring, reporting & verification, forest reference emission levels, and safeguard compliance. National programs such as CAMPA, the Green India Mission, and the recently launched Green Credit Programme of MoEFCC provide institutional and financial platforms that can be leveraged to expand REDD+ initiatives in India. However, challenges remain around coordination, legal clarity on carbon rights, and equitable benefit-sharing. Ensuring social inclusion, land tenure security, and indigenous rights is critical for REDD+ to effectively support India's climate, biodiversity, and sustainable development objectives.

H. Joshi · M. S. Lodhi · O. P. Arya · H. Pant (✉)
Govind Ballabh Pant National Institute of Himalayan Environment, Kosi-Katarmal, Almora, Uttarakhand, India
e-mail: harshit.pantj@gbpihed.nic.in

H. Singh (ed.), *Forestry for a Carbon-Neutral and Sustainable Future*, Sustainability Solutions, https://doi.org/10.1007/978-3-032-06427-1_17

Keywords Deforestation · Land degradation · Climate change mitigation · Forest-based climate solutions · Carbon market · Access · Benefit-sharing mechanisms

17.1 Introduction

The climate emergency presents one of the most urgent and complex challenges of the twenty-first century. Anthropogenic greenhouse gas (GHG) emissions have already led to a global temperature rise of approximately 1.1 °C above pre-industrial levels, with escalating risks for ecosystems, livelihoods, and economies worldwide (IPCC, 2021). In response, the net-zero movement has emerged as a dominant global paradigm, compelling governments, corporations, and institutions to commit to carbon neutrality, aiming to balance anthropogenic emissions with atmospheric removals by mid-century. By 2024, over 140 countries, including major emitters like China, India, and the European Union (EU), have pledged net-zero targets, covering more than 90% of global Gross Domestic Product (GDP) and 88% of global emissions (UNEP, 2023). However, decarbonizing economies alone will be insufficient; achieving net-zero requires integrating natural climate solutions, particularly from land-use sectors that can provide substantial carbon sequestration at relatively low cost (Griscom et al., 2017).

Forests play a pivotal role in mitigating climate change by acting as vast carbon sinks. They absorb roughly one-third of anthropogenic CO_2 emissions annually, primarily through photosynthetic uptake and biomass accumulation (Pan et al., 2011). However, deforestation and forest degradation remain significant sources of emissions, contributing approximately 11% of global GHG emissions each year (Harris et al., 2021). Nature-based solutions (NbS), including forest conservation and restoration, have gained international traction as essential complements to technological mitigation strategies (Nautiyal et al., 2022). Among these, REDD+ (Reducing Emissions from Deforestation and Forest Degradation, plus conservation, sustainable management of forests, and enhancement of forest carbon stocks) stands out as the most institutionalized and widely adopted mechanism under the United Nations Framework Convention on Climate Change (UNFCCC) (Angelsen et al., 2018). It enables developing countries to receive results-based payments for verified reductions in forest emissions and carbon stock enhancement, thereby linking climate finance to verifiable ecosystem outcomes.

REDD+ is uniquely positioned to contribute to global carbon neutrality goals for several reasons. First, REDD+ targets deforestation and forest degradation, which are among the major contributors to greenhouse gas emissions from the land-use sector. By promoting scalable, science-based interventions such as afforestation, improved forest governance, and sustainable land management REDD+ tackles the root causes of forest carbon loss. Second, it bridges the mitigation-finance gap by enabling developing countries to access international funding for climate action. Third, REDD+ not only focuses on carbon mitigation but also delivers a wide range of co-benefits, including biodiversity conservation, improvement of forest-dependent livelihoods, and the strengthening of ecosystem-based resilience to climate change impacts (Duchelle et al., 2018, 2019; Phelps et al., 2012). By preserving forest habitats and promoting sustainable land-use practices, REDD+ contributes to ecological integrity while supporting social development goals.

Furthermore, recent studies on NbS highlight that approaches like REDD+ could supply up to 30% of the cost-effective emissions reductions needed by 2030 to keep global warming below the 2 °C threshold as outlined in the Paris Agreement (Griscom et al., 2017). REDD+ integrates safeguards for indigenous rights, gender equity, and ecological integrity, making it not only a mitigation tool but also a framework for just and inclusive climate governance (Pham et al., 2021; UNFCCC, 2023). In this context, REDD+ emerges as both a strategic mitigation instrument and a multi-functional platform aligned with sustainable development and equity goals. As countries formulate net-zero strategies, mainstreaming REDD+ into national carbon accounting systems and carbon markets becomes an imperative pathway for synergizing climate and development agendas.

17.2 REDD+ : Conceptual Evolution and Global Framework

REDD+ has evolved from a carbon-focused mitigation idea into a broader framework recognized by the UNFCCC for supporting sustainable development, biodiversity, and social co-benefits. The initial concept of REDD emerged from efforts to compensate tropical forest countries for emission reductions from deforestation. Over time, the concept of REDD+ evolved significantly beyond its initial focus on reducing emissions from deforestation. This transformation was shaped through successive rounds of formal negotiations under the UNFCCC, particularly following the adoption of the Bali Action Plan (2007), which recognized the role of forests in climate mitigation. These discussions laid the foundation for expanding REDD+ to include activities such as forest conservation, sustainable management, and carbon stock enhancement. The culmination of these efforts came with the establishment of the Warsaw Framework for REDD+ (2013), which formalized REDD+ as a structured, phased mechanism with defined rules on measurement, safeguards, and result-based finance. This institutional shift embedded REDD+ more firmly within the global climate governance architecture (Bekele & Zewdie, 2018; McDermott et al., 2022; Skutsch, 2017; Voigt et al., 2025). REDD+ formally includes five activities: (i) reducing emissions from deforestation; (ii) reducing emissions from forest degradation; (iii) conservation of forest carbon stocks; (iv) sustainable forest management (SFM); and (v) enhancement of forest carbon stocks. These broaden the scope of intervention from merely reducing forest loss to promoting resilient landscapes and equitable governance. Figure 17.1 illustrates key activities under REDD+ .

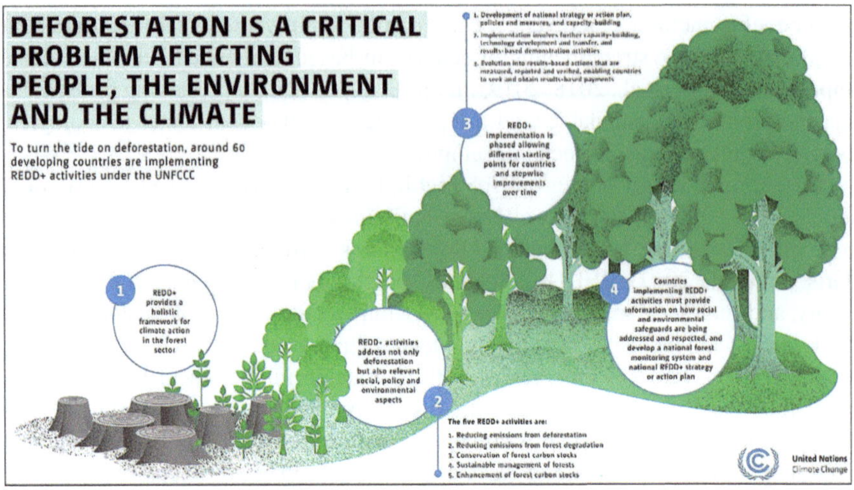

Fig. 17.1 Five activities under REDD+ . + Adapted from UNFCCC (2022): REDDinfographic

17.2.1 Key Components of REDD+

REDD+ implementation follows a structured, phased approach to guide countries from capacity building to measurable climate outcomes. The first phase, readiness, focuses on developing national strategies, legal frameworks, stakeholder engagement, and institutional capacities. In the second phase, implementation, countries roll out policies, pilot programs, and demonstration activities to reduce deforestation and forest degradation. Finally, the third phase, results-based payments (RBP), enables countries to access climate finance based on verified and reported reductions in forest-sector emissions (Angelsen et al., 2018; Duchelle et al., 2019). Supporting institutions include the UN-REDD Programme, the Forest Carbon Partnership Facility (FCPF), and national agencies. Since 2008, over 63 countries have engaged with REDD+ mechanisms.

A key institutional innovation within the REDD+ framework is the requirement for participating countries to submit Forest Reference Emission Levels (FRELs) or Forest Reference Levels (FRLs) to the UNFCCC. These reference levels serve as benchmarks against which future emissions reductions from the forest sector are assessed. FRELs/FRLs establish a transparent and standardized baseline for performance measurement (Johnson & Scheyvens, 2016; Sandker et al., 2022). As of 2024, over 1.7 billion hectares of tropical forests are included under FREL/FRL submissions (Kim, 2018). The Warsaw Framework mandated robust systems for Measurement, Reporting, and Verification (MRV) to ensure transparency and environmental integrity (McDermott et al., 2022; Parrotta et al., 2022). National Forest Monitoring Systems (NFMS), Safeguard Information Systems (SIS), and MRV processes are now central to REDD+ accountability. Countries receive payments only for verified and reported results. For example, 19 countries qualified for result-based finance by meeting these rigorous criteria (Vijitharan et al., 2024).

17.2.2 Principles of Additionality, Permanence, Leakage, and Co-Benefits

REDD+ is guided by core principles ensuring environmental and social soundness:

(i) **Additionality**: REDD+ interventions must achieve emission reductions that wouldn't occur otherwise.
(ii) **Permanence**: Reductions must be lasting and not subject to future reversal.
(iii) **Leakage**: Refers to the unintended displacement of deforestation or forest degradation activities from one area to another due to REDD+ interventions (while emissions may decrease within a targeted project site, they could increase in surrounding or unprotected areas as deforestation pressures shift geographically).
(iv) **Safeguards**: The Cancun Agreements (2010) and Warsaw Framework require countries to submit summaries explaining how seven social and environmental safeguards are addressed, including Free, Prior, and Informed Consent (FPIC), biodiversity, and indigenous rights (Kishwan, 2023).

These criteria position REDD+ as a multidimensional climate tool—not only targeting emissions, but fostering equity, biodiversity, and governance reform.

17.3 Carbon Neutrality and Forest Sector Contributions

The concept of carbon neutrality has rapidly become a central pillar in global climate governance. A growing number of countries, institutions, and cities have committed to net-zero targets, integrating carbon neutrality into policy frameworks, development plans, and international climate negotiations (Tan et al., 2022). While often used interchangeably, the concepts of carbon neutrality, net-zero, and climate neutrality differ in technical scope and implications. Carbon neutrality refers narrowly to balancing carbon dioxide (CO_2) emissions with equivalent removals, either through biological sequestration or engineered sinks. In contrast, net-zero emissions encompass all GHGs, mandating that emissions of methane (CH_4), nitrous oxide (N_2O), and other potent gases are also fully offset through equivalent removal strategies. This concept is embedded in the Paris Agreement, which calls for global GHG emissions to peak as soon as possible and then decline thereafter to net-zero in the second half of the twenty-first century. The overarching aim is to hold the increase in global average temperature to well below 2 °C above pre-industrial levels, while also pursuing efforts to limit the temperature rise to 1.5 °C, recognizing the vastly reduced risks at the lower threshold (Meinshausen et al., 2022). A more comprehensive term, climate neutrality, goes beyond GHGs and includes all anthropogenic climate forcers, such as aerosols and land albedo changes, seeking a state where human activities have no net effect on the climate system (IPCC, 2021; UNEP, 2023). While governments and intergovernmental bodies increasingly adopt the language of net-zero and climate neutrality in long-term emissions strategies, the term carbon neutrality continues to dominate discourse among corporate actors, private initiatives, and project-level interventions, including many REDD+ programs. This distinction is more than semantic: net-zero and climate neutrality often encompass all greenhouse gases and typically require deep decarbonization across sectors, whereas carbon neutrality can involve balancing only CO_2 emissions, often through offsetting mechanisms. Understanding these nuances is essential for assessing the credibility, comparability, and ambition of climate pledges(Joshi et al., 2021). It also helps contextualize the true mitigation value of REDD+ within broader climate frameworks, ensuring forest-sector efforts align with scientifically robust pathways.

17.3.1 Forests and Land Use in the Global Carbon Budget

Forests and land use play a dual role in the global carbon budget as both significant carbon sinks and prominent sources of emissions. Forest ecosystems globally absorb an estimated 7.6 gigatonnes of CO_2 annually, helping mitigate approximately 30% of anthropogenic emissions, primarily through natural processes such as photosynthesis and carbon storage in biomass and soils (Pan et al., 2011). Despite growing global awareness and mitigation efforts, land-use change especially deforestation and forest degradation remains a major driver of GHG emissions, contributing approximately 11% of annual global emissions. These emissions primarily originate from tropical regions, where forests are rapidly cleared for agriculture, logging, and infrastructure expansion (Harris et al., 2021). The dual nature of forests underscores the critical importance of protecting existing carbon stocks while enhancing degraded ecosystems through restoration. Emerging science on NbS places forest conservation and restoration at the forefront of cost-effective mitigation strategies. For instance, Griscom et al. (2017) estimated that forest-based NbS could contribute up to 30% of the global mitigation required by 2030 to meet the Paris Agreement's temperature targets. In this context, reducing emissions from the forest sector not only enhances carbon sequestration but also supports biodiversity, water regulation, and local livelihoods, reinforcing the multifunctional value of forests in achieving climate and development goals.

17.3.2 Role of rEDD+ in NDCs and Global Net-Zero Pathways

REDD+ emerges as a central instrument in operationalizing forest-sector contributions to national and global climate mitigation objectives. Recognized under Article 5 of the Paris Agreement, REDD+ supports developing countries in reducing emissions from deforestation and degradation, while also promoting conservation, sustainable forest management (SFM), and carbon stock enhancement. As of 2023, more than 100 countries have included the land use, land-use change, and forestry (LULUCF) sector in their Nationally Determined Contributions (NDCs), with over 60 explicitly referencing REDD+ activities or principles (UNFCCC, 2023). REDD+ thus plays a crucial role in helping countries meet NDC targets by offering structured, science-based pathways to MRV emissions reductions. Moreover, REDD+ provides access to results-based finance (RBF) from multilateral funds, such as the Green Climate Fund, and bilateral donors, helping to bridge the gap between national ambition and local implementation capacity. Countries such as Brazil, Indonesia, Ghana, and the Democratic Republic of the Congo have incorporated REDD+ into their national climate strategies, with some receiving significant international support for demonstrated emission reductions. The multidimensional design of REDD+ , which links carbon mitigation with co-benefits such as biodiversity conservation and community livelihoods, further reinforces its relevance as an integrated approach to achieving global net-zero trajectories (Dugasseh & Anderson, 2024; Kengoum et al., 2024).

17.3.3 Integration of rEDD+ into National Carbon Accounting Systems

The integration of REDD+ into national carbon accounting systems is fundamental to ensuring consistency, transparency, and credibility in climate reporting. To participate in REDD+ under the UNFCCC, countries must submit FRELs or FRLs, which serve as baselines against which progress is measured. These baselines must be developed using historical forest data and submitted alongside robust MRV frameworks that meet the methodological standards of the Enhanced Transparency Framework under the Paris Agreement (Kishwan, 2023). Integration efforts are ongoing in countries such as Mexico, Colombia, Indonesia, and Vietnam, where REDD+ MRV systems are being aligned with national greenhouse gas inventories. These efforts are supported by technical assistance and policy support from multilateral initiatives, such as the UN-REDD Programme and the FCPF. Moreover, the Lima REDD+ Information Hub, established by the UNFCCC, offers a transparent platform where countries publish their FRELs, MRV reports, and safeguard summaries, thereby promoting accountability and global comparability (UNFCCC, 2023). As carbon markets under Article 6 of the Paris Agreement mature, the accurate integration of REDD+ into national registries becomes even more crucial. This would allow countries to trade verified emission reductions under transparent accounting frameworks, creating opportunities for climate finance and international cooperation. By ensuring that each ton of reduced emissions is counted only once, such mechanisms safeguard against double-counting, a critical risk in global carbon markets. This, in turn, reinforces the environmental integrity and credibility of REDD+ and other mitigation outcomes under international agreements (Ahonen et al., 2022; Botero Tabares, 2025). The successful integration of REDD+ thus enhances not only national compliance but also the global credibility of forest-based mitigation in a carbon-constrained future.

17.4 Global Experiences and Case Studies

Global implementation of REDD+ reveals a diversity of approaches shaped by institutional capacity, ecological conditions, and political will. Countries such as Brazil, Indonesia, Nepal, and the Democratic Republic of the Congo (DRC) demonstrate how REDD+ has been tailored to diverse regional contexts to tackle local drivers of deforestation, involve forest-dependent communities, and secure results-based payments (RBP). Their varied experiences offer practical insights into the enabling conditions such as institutional readiness, policy alignment, and stakeholder participation as well as the challenges, including governance gaps and technical barriers, that influence the real-world effectiveness of REDD+ implementation.

In Brazil, REDD+ implementation has been anchored by the Amazon Fund, one of the largest results-based forest finance mechanisms globally. Operational since 2008 and managed by the Brazilian Development Bank (BNDES), the Fund has received over USD 1.3 billion primarily from Norway and Germany based on independently verified reductions in deforestation across the Legal Amazon. Brazil submitted forest emissions data consistent with its national MRV systems, allowing it to access RBPs through the UNFCCC framework. The São Félix do Xingu initiative, among others, demonstrated quantifiable reductions in deforestation, attributed to a combination of enforcement, land tenure clarification, and financial incentives (Duchelle et al., 2018). However, recent political shifts and institutional weakening of Brazil's environmental agencies have posed setbacks, demonstrating the fragility of REDD+ achievements in the absence of sustained governance and political commitment (Pham et al., 2021).

Indonesia has implemented REDD+ through a multi-layered strategy that includes peatland protection, subnational demonstration activities, and regulatory reforms. The country signed a USD 1 billion agreement with Norway in 2010, which eventually led to its first RBP of USD 103 million in 2020, following the achievement of verified reductions in forest emissions (UN-REDD Programme, 2020). A significant part of Indonesia's REDD+ strategy has focused on peatland ecosystems, which are carbon-dense but ecologically vulnerable. However, MRV challenges persist. These include inconsistent data at subnational levels, overlapping land claims, and limited integration of community-based monitoring (Pasgaard et al., 2016; Pham et al., 2021). The tension between centralized MRV protocols and localized land use realities underscores the need for flexible, multi-scale monitoring frameworks. Despite these impediments, Indonesia's REDD+ journey highlights how regulatory coherence, such as the Peatland Restoration Agency and presidential decrees, can strengthen national efforts (Caldecott et al., 2018).

Nepal's REDD+ experience is grounded in its globally recognized community forestry model, which covers over one-third of the country's forest area. With support from the FCPF and other partners, Nepal completed its readiness phase by 2018, submitting a national REDD+ strategy that aligns forest governance with climate mitigation objectives. Community forest user groups (CFUGs), often legally empowered and socially cohesive, play a critical role in forest monitoring and decision-making. While pilot projects have showcased local participation and early institutional preparedness, challenges remain in generating consistent carbon data, clarifying benefit-sharing mechanisms, and aligning REDD+ efforts with federal restructuring (Vijge et al., 2016). Nevertheless, Nepal exemplifies how decentralized governance can be an asset in REDD+ implementation, provided that local institutions are adequately supported and linked to national carbon accounting systems.

In the Democratic Republic of Congo (DRC), REDD+ has attracted significant international attention due to the country's vast tropical forests, which form part of the Congo Basin. DRC's REDD+ strategy has been supported by the Central African Forest Initiative (CAFI), the World Bank, and bilateral donors. However, forest governance challenges remain entrenched. Moonen et al. (2016) challenge the widely held assumption that subsistence agriculture is the primary driver of deforestation in the DRC. The study displayed that much of the forest loss is instead linked to economically motivated actors including small-scale entrepreneurs and commercial operators who possess greater market access, financial capital, and informal or formal land tenure claims. This shifts the deforestation narrative away from poverty-driven degradation toward structural economic drivers, which complicates the design and targeting of REDD+ interventions. Additionally, institutional fragmentation, tenure insecurity, and weak enforcement limit the ability to track, report, and verify carbon outcomes. While some pilot sites have demonstrated success, including community forest zoning and participatory land-use planning, DRC's experience underscores the critical need for sustained donor engagement, tailored land governance reforms, and capacity-building at subnational levels.

Across these country cases, several cross-cutting success factors and bottlenecks emerge. Effective REDD+ implementation depends heavily on political stability, institutional coherence, secure land tenure, and meaningful stakeholder engagement. Countries that have aligned national development strategies with REDD + goals, such as Brazil and Indonesia, have accessed significant RBPs. Conversely, where forest governance is fragmented or tenure systems are insecure, as in parts of the DRC, REDD+ outcomes remain fragile. Technical challenges, including MRV complexity, inconsistent baselines, and low carbon prices, continue to limit scalability. Equity concerns also persist, with studies highlighting the risk of elite capture, tokenistic participation, and marginalization of indigenous and forest-dependent communities (Pasgaard et al., 2016; Phelps et al., 2012). Overall, global experiences with REDD+ reveal that while the mechanism holds promise as a scalable nature-based climate solution, its success hinges on more than technical readiness. Long-term effectiveness requires adaptive governance, robust safeguards, and institutional commitment across sectors and scales. Lessons from these countries should inform the design of REDD+ strategies elsewhere, particularly in emerging economies with high forest cover and ambitious carbon neutrality goals.

17.5 REDD+ in India: Policy, Strategy, and Institutional Architecture

India holds a unique position in the global REDD+ landscape, both as a megadiverse country and as a party with substantial forest cover relative to its land area and population. According to the India State of Forest Report 2021, forests cover approximately 24.6% of the country's geographical area, with the total carbon stock estimated at 7204 million tonnes, showing an increase of 79.4 million tonnes compared to 2019 (FSI, 2021). This reflects not only the significance of India's forest ecosystems as carbon sinks but also the dynamic nature of land use and forest governance in the country. However, India's forests face increasing pressure from infrastructure development, land-use changes, and population growth. Expanding transportation networks, energy corridors, and urban expansion are contributing to forest fragmentation and habitat loss (Nayak et al., 2020). Meanwhile, overlapping land claims and intensified agriculture are shifting land away from forest cover, especially in regions with high demographic stress (Hazarika & Bhattacharjee, 2022). In this context, integrating REDD+ into national and subnational forest policies offers both mitigation potential and an opportunity to enhance forest-based livelihoods.

India officially adopted its National REDD+ Strategy in 2018, marking a significant milestone in aligning forest conservation with international climate mitigation frameworks. Developed by the Ministry of Environment, Forest and Climate Change (MoEFCC), the strategy is built on three key pillars: (i) readiness, (ii) implementation, and (iii) result-based actions. Its objectives include enhancing forest carbon stocks, improving biodiversity, supporting forest-dependent communities, and accessing international finance through results-based mechanisms. The strategy aligns with India's broader commitments under the Paris Agreement and its NDCs, particularly the NDC-3-to reduce Emissions Intensity of its GDP by 45 percent by 2030, from 2005 level, NDC-5-pledge to create an additional 2.5 to 3 billion tonnes of carbon dioxide equivalent sequestration through forest and tree cover by 2030, and NDC-7-To mobilize domestic and new & additional funds from developed countries to implement the above mitigation and adaptation actions given the resource required and the resource gap (GoI, 2022). While the strategy is comprehensive in design, its on-ground implementation has varied significantly across states, revealing systemic challenges in translating national ambitions into local action. Key bottlenecks include insufficient technical and institutional capacity, particularly at the state forest department level; complex and often contested land tenure systems that complicate benefit-sharing and forest governance; and inadequate participation of local and indigenous communities, which undermines both social legitimacy and effectiveness. (Chand et al., 2021).

Institutionally, REDD+ in India is coordinated by the MoEFCC, GoI, which serves as the nodal agency for national climate and forest policy. Key technical and implementing partners include the Forest Survey of India (FSI), responsible for assessing and monitoring forest carbon stocks; the Indian Council of Forestry Research and Education (ICFRE), which provides research and training; and State Forest Departments, which are responsible for implementation at the local level. These actors are expected to collaborate in establishing MRV systems, developing forest reference levels, and facilitating stakeholder consultations. While these institutions bring considerable technical capacity, challenges persist in coordination, data harmonization, and enabling effective participation from community institutions and civil society. The centralization of REDD+ responsibilities within MoEFCC has raised concerns about potential marginalization of decentralized forest governance frameworks, such as Joint Forest Management (JFM), which have traditionally empowered local forest user groups (Vijge & Gupta, 2014).

India has attempted to synergize REDD+ with existing forestry and climate programs. The Green India Mission (GIM), one of the eight missions under the National Action Plan on Climate Change (NAPCC), aims to enhance ecosystem services and increase forest-based carbon sequestration. Complementing this, the Compensatory Afforestation Fund Management and Planning Authority (CAMPA) provides financial resources for forest regeneration as a substitute for diverted forest land. Other programs, such as the National Afforestation Programme and the recently launched Green Credit Program (under tree plantation-based green credit) of the MoEFCC, also support afforestation and forest restoration. While these programs share similar objectives, they often operate in silos, limiting their collective potential to contribute to REDD+ outcomes (Bhattacharya & Mehra, 2021; Duraisami et al., 2022). A more integrated and harmonized implementation model, supported by clear monitoring protocols and shared benefit mechanisms, is necessary to realize synergistic benefits across these platforms.

Pilot REDD+ projects in India have been initiated in states like Himachal Pradesh, Madhya Pradesh, and Sikkim, each experimenting with readiness activities, community engagement, and forest monitoring systems. Himachal Pradesh, for example, developed a state-level REDD+ strategy focusing on forest-dependent communities, while Madhya Pradesh has piloted community-based MRV models. These subnational efforts have provided valuable lessons in participatory forest management, but their replication and scaling remain limited by funding constraints and institutional fragmentation. Additionally, varying levels of readiness across states have impeded the establishment of a standardized framework for benefit sharing, safeguards implementation, and data reporting. To operationalize REDD+ effectively at scale, India must address these disparities and foster inter-sectoral coordination between national and state-level actors.

India's REDD+ journey thus reflects both institutional readiness and structural challenges. The country possesses robust forest monitoring systems, technical expertise, and a dense institutional architecture. Yet, the effectiveness of REDD+ will depend on sustained capacity building, inclusive governance reforms, and stronger integration of REDD+ with India's broader environmental and development agenda. As India progresses toward its 2070 net-zero target, unlocking the full potential of REDD+ will require a convergence of policy innovation, financial mobilization, and participatory implementation.

17.6 Role of Community-Managed Forests in REDD+ implementation

Community-managed forests (CMF) play a pivotal role in REDD+ by integrating local knowledge, sustainable practices, and equitable benefit-sharing into forest conservation efforts. Indigenous and local communities, as custodians of the CMF, employ traditional management techniques that enhance carbon sequestration, water regulation, protect biodiversity, etc. (Bhatt et al., 2024). Securing land tenure rights and involving local communities in REDD+ initiatives are essential for achieving both long-term forest conservation and sustainable livelihood outcomes. Secure land tenure empowers communities to manage and protect forest resources, incentivizing sustainable use through mechanisms such as carbon credit payments and benefit-sharing schemes. Active participation in decision-making processes further enhances accountability, legitimacy, and local ownership of REDD+ actions. Studies show that without clear tenure and participatory frameworks, REDD+ projects often struggle with legitimacy and effectiveness (Awono et al., 2014; Lawlor et al., 2013). Moreover, locally adapted REDD+ strategies that respect customary land rights and incorporate community-based monitoring have shown promise in balancing carbon goals with equity and rural development (Atela et al., 2015; Corbera et al., 2020).

Additionally, community-led monitoring helps prevent illegal logging and leakage, where deforestation shifts to other areas. Hence, to maximize the effectiveness, especially in the context of CMF addressing unclear land tenure, inadequate funding, and unequal benefit-sharing mechanisms, is warranted (Bhatt & Jugran, 2024). Overall, empowering local forest governance is key to achieving climate mitigation, ecological resilience, and social equity under REDD+.

Empowering local communities to conserve and sustainably manage forests, CMF contributes significantly to reducing carbon emissions, enhancing forest carbon stocks, and promoting biodiversity conservation. Simultaneously, CMF supports numerous ecosystem services such as soil health, water regulation, etc., by maintaining healthy forests that buffer against climate-related impacts (Table 17.1). Through traditional ecological knowledge, participatory monitoring, and inclusive governance structures (e.g., Joint Forest Management (JFM) Committees in

Himachal Pradesh and Van Panchayats in Uttarakhand), CMF ensures community-led decision-making, equitable benefit-sharing, and the integration of indigenous practices into climate action. Thus, CMF serves as a nexus between climate mitigation and adaptation, offering a cost-effective, socially inclusive, and ecologically sustainable pathway for implementing REDD+ at the local and landscape level.

Table 17.1 Envisaged role of CMF in REDD+ and NbS implementation

Key functions	Role in REDD+ implementation	Role in NbS implementation
Forest conservation and management	Reduces deforestation and degradation, preserving forest cover, and carbon sinks	Maintains ecosystem structure and function, enhancing resilience to climate stressors
Ecosystem services	Enhances carbon sequestration, biodiversity, and livelihoods	Supports climate adaptation, water regulation, and habitat conservation
Community participation and MRV	Engages local communities in MRV of forest carbon and land use changes	Promotes participatory ecosystem monitoring, vulnerability mapping, and local adaptation planning
Traditional ecological knowledge	Informs sustainable forest management and safeguards	Guides ecosystem restoration and community resilience
Sustainable livelihoods	Provides income through NTFPs, ecotourism, and potential carbon financing under benefit-sharing schemes	Diversifies livelihoods via climate-resilient practices such as agroforestry and medicinal plant cultivation
Biodiversity conservation	Maintains habitat diversity and supports long-term ecosystem productivity and carbon storage	Enhances species and genetic diversity, supporting ecological resilience and adaptation capacity
Institutional frameworks	Leverages grassroots governance structures (e.g., Van Panchayats, JFMCs) for REDD+ readiness and implementation	Empowers the same institutions to implement decentralized, context-specific NbS initiatives
Social safeguards and equity	Protects customary rights, ensures inclusive decision-making, and promotes equitable benefit-sharing	Ensures climate adaptation strategies are socially just, gender-inclusive, and locally accepted
Disaster risk reduction	Mitigates climate-induced forest degradation	Strengthens natural buffers that reduce community vulnerability to extreme weather and slow-onset events
Policy integration and scaling	Supports national and sub-national REDD+ strategies and contributes to NDCs under the Paris agreement	Aligns with National Adaptation Plans (NAPs), SAPCCs, and community-level climate resilience goals

17.7 Contribution of REDD+ to India's Carbon Neutrality Commitments

REDD+ significantly contributes to India's long-term goal of achieving carbon neutrality by 2070 by enhancing forest-based carbon sinks and reducing emissions from deforestation and forest degradation (Gorain et al., 2025). As outlined in India's updated NDC (2022), the country aims to create an additional carbon sink of 2.5–3 billion tons of CO_2-equivalent through forest and tree cover. REDD+ supports this by promoting afforestation, sustainable forest management, and the conservation of ecosystems. The National REDD+ Strategy (2018), along with programs such as CAMPA and the Green India Mission, provides a framework for implementation (Gunawardena et al., 2024; Barman et al., 2025). With robust MRV, safeguards, and community involvement, REDD+ aligns closely with India's climate mitigation and adaptation objectives.

17.7.1 India's NDC and Net Zero 2070 Target

India's commitment to climate action has evolved significantly over the past decade, culminating in its pledge at COP26 to achieve net-zero emissions by 2070. This long-term target is embedded within India's broader climate architecture, particularly its Updated NDCs submitted in 2022 under the Paris Agreement. The revised NDC emphasizes two key quantified goals for 2030: (i) reducing emissions intensity of GDP by 45% compared to 2005 levels (NDC-3), and (ii) achieving about 50% cumulative electric power capacity from non-fossil fuel-based sources (NDC-4) (MoEFCC, 2022). A third, less quantified but equally critical component relates to enhancing forest and tree cover to serve as carbon sinks. In its fifth NDC (2022), India pledged to increase forest-based sequestration by 2.5 to 3 billion tonnes of CO_2 equivalent (CO_2e) by 2030 (GoI, 2022). This target remains central to India's net-zero transition, particularly in light of the limited short- and medium-term decarbonization options in sectors such as transport and heavy industry. REDD+ is positioned as a key strategy to help operationalize this forestry-based mitigation target by promoting sustainable forest management, enhancing carbon stocks, and integrating local livelihoods with global climate objectives. Importantly, India's 2018 National REDD+ Strategy explicitly links REDD+ to its NDC commitments, reinforcing the alignment between domestic forest governance and international climate obligations.

17.7.2 Readiness for Inclusion of Forest Offsets in India's Emerging Carbon Market

A critical pathway for scaling REDD+ contributions lies in integrating forest offsets into India's emerging carbon market architecture. The Indian government has initiated a domestic compliance carbon market under the Energy Conservation (Amendment) Act, 2022, with a plan to operationalize the Indian Carbon Market (ICM) by 2025. While the initial focus is on energy-intensive sectors, the ICM is expected to evolve into a multi-sectoral mechanism that includes nature-based offsets, including forest carbon credits. REDD+ stands to play a pivotal role in this transition, given its potential to generate verifiable emission reductions at scale, particularly in forest-rich states. India's institutional readiness for REDD+ implementation is underpinned by robust technical capacities, notably the FSI satellite-based forest monitoring systems and the MoEFCC's Green Credit Program, which is designed to incentivize ecological stewardship through tradable credits. These initiatives offer a promising infrastructure for integrating REDD+ into emerging market-based mechanisms, including voluntary and compliance carbon markets. However, significant challenges persist. Establishing credible MRV protocols that meet international standards remains complex, especially in a country with diverse forest types and governance structures. Ensuring additionality that emissions reductions would not have occurred without the REDD+ intervention is also difficult to validate. Furthermore, there is a risk of double counting emissions reductions that are simultaneously reported in India's national GHG inventory and sold as offsets, which could undermine environmental integrity. Compounding these challenges is the historically low price of forest carbon in voluntary markets, often insufficient to cover the costs of project implementation and safeguard compliance. This raises concerns about long-term economic viability, particularly in the absence of targeted policy support or sustained international climate finance (Rakatama et al., 2017). Institutional convergence will be essential, particularly between the MoEFCC, the Bureau of Energy Efficiency, and market regulators, to create a harmonized and high-integrity carbon market that can incorporate REDD+ activities transparently (Datta, 2009; Sud et al., 2012). If designed with environmental and social safeguards, REDD+ credits could serve as a credible offset mechanism that not only contributes to India's net-zero goal but also incentivizes conservation and rural development in forested landscapes.

17.8 MRV Systems, Safeguards, and Transparency

MRV systems, safeguards, and transparency are essential pillars of effective REDD+ implementation. MRV ensures accurate measurement of emission reductions through remote sensing, field data, and independent verification. Safeguards protect Indigenous rights, biodiversity, and community well-being, aligning with the UNFCCC

Cancun Agreements and national laws, such as India's Forest Rights Act. Transparency fosters trust by promoting open access to data, inclusive participation, and equitable sharing of benefits (Grassi et al., 2017; Lofts et al., 2021; Lyster, 2011; Silori et al., 2013; Trivedi, 2025). Together, these elements support credible, socially just, and environmentally sound REDD+ actions. In India, institutions such as the FSI and MoEFCC are working to integrate these frameworks into national and sub-national REDD+ strategies.

17.8.1 Global MRV Requirements and Best Practices

The effectiveness of REDD+ as a climate mitigation mechanism depends critically on the robustness of its MRV systems. Globally, MRV protocols are governed by decisions under the Warsaw Framework for REDD+ , which mandates that participating countries establish transparent, consistent, and accurate MRV systems that are aligned with the Intergovernmental Panel on Climate Change (IPCC) guidelines (Voigt et al., 2025). Best practices emphasize the development of NFMS that integrate remote sensing, ground-based inventories, and geospatial analysis. These systems must support the submission of FRELs and associated emissions reduction reports, which UNFCCC experts technically assess (Neeff, 2017). Countries such as Brazil, Mexico, and Vietnam have demonstrated advanced MRV systems capable of generating verifiable data for result-based payments. Best-in-class MRV systems extend beyond carbon accounting to include social and environmental safeguards, which help prevent unintended negative consequences such as community displacement or biodiversity loss. These systems feature dedicated reporting mechanisms for tracking compliance with safeguards, as required under the Cancun Agreements, and they actively promote community participation in data collection and monitoring processes. By embedding these dimensions, MRV frameworks ensure both technical accuracy and procedural legitimacy, fostering greater trust, transparency, and equity (Angelsen et al., 2018; Jagger et al., 2014). A strong MRV framework thus underpins both environmental integrity and financial credibility in REDD+ implementation. Figure 17.2 depicts the working principle and safeguards of REDD+ MRV.

17.8.2 India's MRV Status: FSI, Remote Sensing, and Field-Based Carbon Estimation

India has made considerable progress in establishing its own MRV infrastructure, led primarily by the FSI. The FSI conducts biennial remote sensing-based forest assessments, producing the India State of Forest Report, which includes estimates of growing stock, carbon stock, and land-use change (Kaul et al., 2009). These

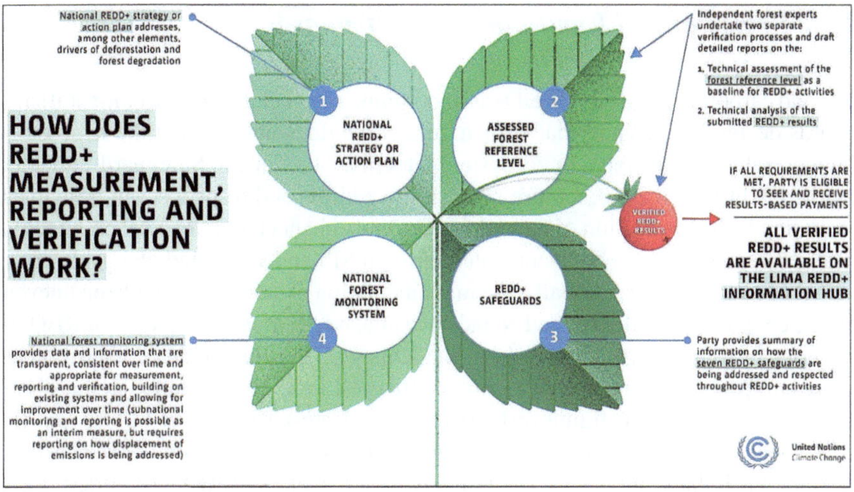

Fig. 17.2 Working principle of REDD+ MRV. + Adapted from: UNFCCC (2022): REDDinfographic

satellite-based surveys are complemented by field-based carbon estimation protocols, particularly through sample plot measurements and allometric models. FSI has also piloted methodologies for community-level carbon monitoring in collaboration with state forest departments. India's REDD+ readiness phase, supported by the UN-REDD Programme, enabled initial capacity-building for MRV design; however, integration with subnational reporting systems remains inconsistent. Despite India's significant progress in forest monitoring, challenges persist in harmonizing data collection methodologies across states, which often employ varied techniques and standards for forest assessments. These inconsistencies, combined with limited technical training at the local level, impede the development of a cohesive national monitoring framework. Moreover, India has yet to fully operationalize NFMS that is seamlessly integrated with MRV and safeguards information systems, stalling the creation of a unified REDD+ registry a critical requirement for accessing large-scale results-based finance (RBF) under the UNFCCC (Pandey, 2021; Bhattarai, 2024). Nevertheless, India possesses strong technical capabilities in remote sensing and carbon stock modeling, which provide a solid foundation for advancing MRV systems (Mitchell et al., 2017; Yang et al., 2025). Leveraging these capacities while addressing governance and data integration challenges will be essential for India to align fully with global REDD+ standards.

17.8.3 Social and Environmental Safeguards

Beyond carbon accounting, REDD+ is underpinned by social and environmental safeguards designed to ensure that climate mitigation does not come at the expense of human rights, biodiversity, or ecosystem integrity. The Cancun Safeguards, adopted by the UNFCCC in 2010, require countries to address and respect seven principles, including FPIC, protection of natural forests, and enhancement of social co-benefits. These safeguards are to be documented through SIS and submitted alongside emissions reduction reports. Globally, the implementation of safeguards has been uneven. While countries like Mexico and Nepal have integrated safeguards into their REDD+ governance frameworks, others face challenges in translating these legal obligations into practical action (McDermott & Ituarte-Lima, 2016; Paudel et al., 2015). In India, existing legal frameworks such as the Forest Rights Act (2006) and the Biological Diversity Act provide a normative basis for safeguards. However, their operational linkage with REDD+ remains underdeveloped. REDD+ policy documents have acknowledged the importance of FPIC and biodiversity, but practical guidelines for implementation, monitoring, and redressal are still lacking. Safeguard performance in India is further complicated by the intersection of state-led conservation programs and unresolved tenure disputes in forest-dependent regions (Vijge & Gupta, 2014).

17.8.4 Role of Community Participation and Traditional Knowledge

A critical enabler of safeguards and monitoring is the participation of forest communities and the incorporation of traditional ecological knowledge. Globally, community involvement in REDD+ MRV has been shown to increase cost-effectiveness, local ownership, and data reliability. In countries such as Tanzania and Nepal, community-based forest monitoring has been institutionalized through the use of participatory tools and formalized benefit-sharing mechanisms (Gilani et al., 2017; Khatun et al., 2015). In India, the JFM system and Community Forest Resource Rights (CFRR) under the Forest Rights Act provide legal and institutional pathways for involving local communities in REDD+ . Despite this, participation has often remained tokenistic rather than transformative, with decisions concentrated at higher administrative levels and limited feedback loops from the ground. Field studies have found that forest user groups possess deep knowledge of forest dynamics and degradation drivers, yet are rarely empowered to contribute meaningfully to MRV or safeguard design (Chand et al., 2021; Vijge & Gupta, 2014). Strengthening community participation will require capacity-building, financial incentives, and the creation of inclusive institutional spaces where local voices inform REDD+ strategy and implementation. In sum, robust MRV systems and enforceable safeguards are indispensable for the credibility and equity of REDD+ in India. While technical capacities exist, the integration of social, ecological, and procedural dimensions remains a work in progress. Moving forward, India must invest in strengthening decentralized monitoring systems, clarifying safeguards frameworks, and institutionalizing community engagement to fully realize the potential of REDD+ as a climate and development solution.

17.9 Finance and Carbon Market Linkages

The success and scalability of REDD+ as a climate mitigation instrument are intrinsically linked to the availability, predictability, and accessibility of climate finance. Globally, the REDD+ finance architecture has evolved through a combination of multilateral, bilateral, and market-based sources. Prominent among these are the Green Climate Fund (GCF), the FCPF, and various voluntary carbon markets (VCMs). The GCF provides performance-based payments to countries that demonstrate verified emissions reductions as dictated by UNFCCC guidelines, with Brazil and Indonesia among the key recipients. Bilateral agreements, such as Norway's partnerships with Guyana, Indonesia, and Liberia, have also played a key role, offering upfront financing for readiness activities and ex-post payments for verified outcomes (Helland & Mæstad, 2015; Angelsen, 2017). While these arrangements underscore donor commitment, they also expose REDD+ to vulnerabilities such as political shifts, donor fatigue, and inconsistent disbursement timelines (Pham et al., 2021; Schroeder et al., 2020). Meanwhile, VCMs have emerged as a complementary pathway for REDD+ financing, though concerns about credit integrity, pricing volatility, and permanence risks continue to challenge their scalability and legitimacy. Overall, REDD+ finance remains fragmented, with significant gaps between available resources and the estimated USD 20–30 billion per year needed for large-scale forest carbon mitigation globally (Angelsen et al., 2018).

17.9.1 India's Domestic Funding Sources: CAMPA, GIM, and Potential for Carbon Trading

In India, the REDD+ finance landscape is shaped by a combination of domestic public funding and emerging market mechanisms. The most prominent sources include the CAMPA, which disburses funds from developers required to compensate for forest diversion, and the GIM under the NAPCC. Together, these programs channel public finance into afforestation, forest restoration, and ecosystem enhancement activities, some of which align with REDD+ objectives. However, neither CAMPA nor GIM currently operates under carbon accounting frameworks, nor do they generate tradable carbon credits, limiting their contribution to results-based REDD+ finance. More recently, India has signaled its intention to develop a domestic compliance carbon market under the ICM framework, with potential for inclusion of nature-based offsets, including REDD+ . The Green Credit Programme, launched in 2023 by the MoEFCC, aims to incentivize environmental actions such as tree plantation, sustainable agriculture, mangrove conservation & restoration, air pollution reduction, etc., by issuing tradable credits, potentially bridging the gap between conservation and carbon finance (MoEFCC, 2022). These evolving instruments underscore a growing policy interest in integrating REDD+ with market-based mechanisms, although mechanism for the assessment and rollout of the green credits is still under development.

17.9.2 Payment-For-Performance Models and Private Sector Engagement

Payment-for-performance (PFP) models are central to the REDD+ financing logic, wherein countries or projects receive funds only upon verified emission reductions. Globally, PFP has encouraged institutional reforms, investment in MRV systems, and results-oriented governance. In India, however, readiness for PFP remains nascent. Although the country has submitted FRELs to the UNFCCC and possesses considerable MRV capacity through FSI, it has not yet received international REDD+ payments due to pending integration of safeguard reporting and full MRV validation. Moreover, the absence of a dedicated REDD+ financial mechanism or registry within India limits its ability to aggregate verified results and channel them into structured payments. Engagement from the private sector remains crucial for scaling finance, but it remains minimal, constrained by regulatory uncertainty, low carbon prices, and a lack of clear benefit-sharing frameworks (Dunlop & Corbera, 2016; Morita & Matsumoto, 2023; Wong et al., 2019). Lessons from countries like Peru and Ghana demonstrate that private sector actors are more likely to participate when legal clarity, baseline credibility, and market access are ensured. For India, unlocking private investment will depend on the standardization of forest-based carbon methodologies, assurance of credit integrity, and alignment with emerging carbon market frameworks under the ICM.

17.9.3 Challenges in Accessing and Utilizing Carbon Finance in India

Despite institutional progress, India faces several challenges in accessing and effectively utilizing REDD+ finance (Bhattarai et al., 2023; Chand et al., 2021). First, financial disbursement mechanisms, especially those involving CAMPA and GIM, are often slow, bureaucratic, and disconnected from REDD+ outcome metrics. Second, benefit-sharing arrangements are ill-defined, particularly regarding the role of forest communities and local institutions. Without robust safeguards and community consent, REDD+ finance risks reinforcing existing governance asymmetries and exacerbating tenure-related grievances. Third, the lack of coordination between national ministries, state forest departments, and carbon market regulators impedes the establishment of a unified REDD+ finance ecosystem. Moreover, India's plural forest tenure systems and ongoing land conflicts complicate efforts to link finance to forest performance in a socially equitable manner. These structural bottlenecks highlight the need for institutional convergence, capacity-building, and the formal integration of REDD+ into India's climate finance and carbon market policy architecture. If addressed strategically, REDD+ could serve as a cornerstone in India's nature-based net-zero strategy while delivering co-benefits for biodiversity and livelihoods.

17.10 Co-Benefits, Critiques, and Equity Considerations

Beyond its core objective of mitigating emissions from land-use change, REDD+ has been widely recognized for its potential co-benefits, including biodiversity conservation, livelihood enhancement, watershed protection, etc. These co-benefits elevate REDD+ from a narrow carbon-focused mechanism to the NbS framework that can support broader ecological and social outcomes. Empirical studies from Brazil, the DRC, and Southeast Asia confirm that REDD+ interventions, particularly those emphasizing sustainable forest management and conservation, can generate synergistic gains for endangered species habitats, freshwater provisioning, and climate adaptation (Smith et al., 2019; Kengoum et al., 2024). In regions where forests are integral to cultural and subsistence livelihoods, REDD+ also offers opportunities to enhance community incomes through forest stewardship, sustainable Non-Timber Forest Produce (NTFP) harvesting, and performance-based incentives. International REDD+ frameworks explicitly recognize such co-benefits and encourage their integration into safeguards, MRV, and benefit-sharing systems. However, the realization of these outcomes is contingent on strong institutional design, clarity in land tenure, and participatory implementation processes.

17.10.1 Critiques: Commodification of Forests, Tenure Insecurity, Elite Capture

Despite its promise, REDD+ has drawn significant critiques, particularly around the commodification of forests, tenure insecurity, and elite capture. Critics argue that by reducing forests to carbon storage units, REDD+ risks ignoring their multifunctional ecological roles and cultural significance, especially in Indigenous and forest-dwelling communities. Commodification may incentivize state-centric forest control and exclusionary conservation practices, leading to the dispossession of traditional users under the guise of global climate objectives. Tenure insecurity remains a central concern; in many REDD+ countries, land rights are ambiguously defined or weakly enforced, making it difficult to determine who holds the authority to trade carbon credits or receive payments. Moreover, case studies from Tanzania, Indonesia, and Vietnam highlight how REDD+ implementation has often reinforced existing power asymmetries, with benefits disproportionately accruing to government agencies, project developers, and elites, while communities bear the costs of restricted access and increased surveillance (Ickowitz et al., 2017; Pasgaard et al., 2016). These governance failures not only compromise equity but also undermine the legitimacy and long-term sustainability of REDD+ as a mitigation instrument.

In India, the concerns about tenure, equity, and rights are particularly salient. The Forest Rights Act (FRA, 2006) recognizes the customary rights of Scheduled Tribes and other traditional forest dwellers to forestland and resources, including rights of habitation, cultivation, and community forest governance. However, REDD+ planning and MRV systems have fallen significantly short in institutionalizing FRA provisions, thereby marginalizing the voices of tribal and gendered communities in forest carbon governance (Vijge & Gupta, 2014). For instance, policy framing around REDD+ in India has emphasized centralized monitoring and carbon accounting, often sidelining community-managed forests and ignoring the diversity of forest tenure arrangements. Gender dynamics further complicate the landscape. Women, who play critical roles in forest-based livelihoods and ecological knowledge systems, are frequently excluded from REDD+ negotiations and benefit-sharing frameworks. Additionally, most REDD+ documents in India insufficiently address the intersectionality of tribal identity, caste hierarchies, and land marginalization, thus risking injustices under the umbrella of climate mitigation.

17.10.2 Strengthening Equity and Justice in REDD+ implementation

To strengthen equity and justice in REDD+ implementation, both globally and in India, several interlinked strategies must be pursued. First, REDD+ design must embed tenure clarity and legal recognition of community rights as a foundational precondition, not a downstream safeguard. Countries with formalized community forest rights, like Nepal, offer compelling models for how tenure security enhances both equity and carbon outcomes. Second, inclusive governance mechanisms, such as multi-stakeholder platforms, gender-sensitive representation, and participatory MRV, are essential for reapportioning the voice and agency to marginalized actors (Mbano-Mweso, 2023; Riamit & Tauli-Corpuz, 2012). In India, this means aligning REDD+ implementation explicitly with JFM committees, CFR management bodies, and Gram Sabhas, thereby operationalizing legal pluralism in forest governance. Third, REDD+ benefit-sharing systems must be transparent, progressive, and context-sensitive, recognizing differentiated opportunity costs and contributions across communities. Flat-rate payments risk favoring wealthier actors and overlooking the needs of vulnerable households (Ickowitz et al., 2017). Ultimately, capacity-building efforts must extend beyond technical training to encompass rights literacy, institutional negotiation, and gender equity, ensuring that REDD+ is not only environmentally effective but also socially transformative (Dhungana et al., 2018). Without proactive integration of justice considerations, REDD+ risks perpetuating the very inequalities it aims to redress.

17.11 Conclusion and Way Forward

The urgency of the global climate crisis and the momentum toward carbon neutrality have amplified the importance of REDD+ as both a mitigation strategy and a tool for sustainable development. Moving forward, REDD+ must transcend its original function as a pilot-based forest intervention and evolve into a core pillar of climate action integrated within net-zero roadmaps, national development strategies, and global carbon markets. At the global level, REDD+ must be structurally embedded into national carbon neutrality plans and Article 6 mechanisms under the Paris Agreement. Greater methodological harmonization, transparency in reporting, and alignment with market integrity standards are critical to ensure credibility, comparability, and scalability. Countries should adopt jurisdictional approaches to consolidate fragmented projects and ensure landscape-level emission reductions, while building central registries and robust safeguard systems to support verifiable climate outcomes and equitable benefits.

For India, the path ahead requires a multi-pronged approach. First, inter-sectoral coordination must be institutionalized. REDD+ operates at the intersection of forest management, climate policy, rural livelihoods, and land governance. Effective implementation will require alignment among ministries and agencies, including the MoEFCC, FSI, State Forest Departments, and emerging carbon market regulators. Institutional convergence is key to reducing duplication and enhancing policy coherence. Second, India must invest in capacity building, especially in integrating advanced MRV technologies with participatory forest monitoring. Strengthening human and technical capacities at the state and local levels will enhance the credibility of carbon accounting and enable more inclusive implementation. Community engagement should move beyond consultation to co-management, valuing the contributions of local and Indigenous knowledge. Third, REDD+ must be positioned as a foundational component of India's emerging carbon market ecosystem. As the Indian Carbon Market develops, mechanisms must be established to generate high-quality forest carbon credits that adhere to integrity standards and safeguard principles. Legal clarity around carbon rights, transparent benefit-sharing models, and nested accounting frameworks will ensure environmental and social credibility. Lastly, REDD+ initiatives should be aligned with India's broader development goals.

Integration into national programs such as the Green India Mission, CAMPA, etc., is the key to achieving synergistic outcomes. It should also support biodiversity conservation, water security, and rural employment, ensuring that REDD+ is not only a climate tool but also a vehicle for ecological and socio-economic transformation.

References

Ahonen, H. M., Kessler, J., Michaelowa, A., Espelage, A., & Hoch, S. (2022). Governance of fragmented compliance and voluntary carbon markets under the Paris Agreement. *Politics and Governance, 10*(1), 103–115. https://doi.org/10.17645/pag.v10i1.4759

Angelsen, A. (2017). REDD+ as result-based aid: General lessons and bilateral agreements of Norway. *Review of Development Economics, 21*(2), 237–264. https://doi.org/10.1111/rode.12271

Angelsen, A., Hermansen, E. A. T., & Rajão, R. (2018). Results-based payment. In A. Angelsen, C. Martius, V. de Sy, H. T. Nguyen, & L. Sills (Eds.), *Transforming REDD+: Lessons and new directions*, (pp. 41–52). Center for International Forestry Research (CIFOR).

Atela, J. O., Minang, P. A., Quinn, C. H., & Duguma, L. A. (2015). Implementing REDD+ at the local level: Assessing the key enablers for credible mitigation and sustainable livelihood outcomes. *Journal of Environmental Management, 157*, 238–249. https://doi.org/10.1016/j.jenvman.2015.04.015

Awono, A., Somorin, O. A., Atyi, R. E., & Levang, P. (2014). Tenure and participation in local REDD+ projects: Insights from southern Cameroon. *Environmental Science & Policy, 35*, 76–86. https://doi.org/10.1016/j.envsci.2013.01.017

Barman, D., Pandey, R., Sharma, A., Singh, M., Arshad, A., Tamang, M., Shukla, G., Panwar, P., & Chakravarty, S. (2025). Forest degradation: Its impact and conservation efforts in India. In G. Shukla, A. Manohar K., A. Raj Kizha, P. Panwar, & S. Chakravarty (Eds.), *Forest degradation and management* (Forestry Sciences, Vol. 87). Cham: Springer. https://doi.org/10.1007/978-3-031-84055-5_21

Bekele, M. B., & Zewdie, S. Z. (2018). REDD+ MRV implementation in Ethiopia: Review of the context, framework and progress. *CAB Reviews: Perspectives in Agriculture, Veterinary Science, Nutrition and Natural Resources, 13*(062). https://pdfs.semanticscholar.org/d662/30a7e549492c0d220f1aa55d18228a2b1452.pdf

Bhatt, H., Jugran, H. P., & Pandey, R. (2024). Cultural ecosystem services nexus with Socio-Cultural attributes and traditional ecological knowledge for managing community forests of Indian western Himalaya. *Ecological Indicators, 166*, 1123.

Bhatt, H., & Jugran, H.P. (2024). Community-managed forests and their effectiveness in SDG implications in the Western Himalayan region. In: S. Tripathi, R. Bhadouria, S. C. Garkoti (Eds.), *Warming Mountains*. Cham: Springer. https://doi.org/10.1007/978-3-031-62197-0_22

Bhattacharya, P., & Mehra, S. (2021). REDD+ in the Indian context: Planning and implementation scenario. In A. Kaushik, C. P. Kaushik, & S. D. Attri (Eds.), *Climate resilience and environmental sustainability approaches* (pp. 53–75). Springer. https://doi.org/10.1007/978-981-16-0902-2_4

Bhattarai, N. (2024). *Studies on addressing challenges and enhancing community engagement in REDD+ implementation in Nepal and India* (Doctoral dissertation). Graduate School of Environmental Science, Hokkaido University, Sapporo, Japan. https://eprints.lib.hokudai.ac.jp/dspace/bitstream/2115/91772/1/Nabin_Bhattarai.pdf

Bhattarai, N., Karky, B. S., Avtar, R., Thapa, R. B., & Watanabe, T. (2023). Are countries ready for REDD+ payments? REDD+ readiness in Bhutan, India, Myanmar, and Nepal. *Sustainability, 15*(7), 6078. https://doi.org/10.3390/su15076078

Botero Tabares, A. M. (2025). *Environmental integrity of tropical forest activities under the Paris Agreement Article 6.2*. University of Eastern Finland. https://erepo.uef.fi/bitstreams/877b5060-7076-4ff9-8828-cb0234f1abb2/download

Caldecott, J., Mahaningtyas, A., Howard, B., Williams, D., & Lincoln, P. (2018). *Third independent review of the Indonesia–Norway cooperation on reducing greenhouse gas emissions from REDD+*. LTS International & Chr. Michelsen Institute (CMI). https://open.cmi.no/cmi-xmlui/handle/11250/2568523

Chand, H. B., Singh, S., Kumar, A., Kewat, A. K., Bhatt, R., & Bohara, R. (2021). Comprehensive Overview of REDD+ in India: Status, Opportunities and Challenges. *Grassroots Journal of Natural Resources, 4*(3), 185–200. https://doi.org/10.33002/nr2581.6853.040314

Corbera, E., Martin, A., & Springate-Baginski, O. (2020). Sowing the seeds of sustainable rural livelihoods? An assessment of participatory forest management through REDD+ in Tanzania. *Land Use Policy, 97*, Article 104793. https://doi.org/10.1016/j.landusepol.2017.09.037

Dhungana, S., Poudel, M., & Bhandari, T. S. (Eds.). (2018). *REDD+ in Nepal: Experiences from the REDD Readiness Phase*. REDD Implementation Centre, Ministry of Forests and Environment, Government of Nepal. https://www.icimod.org/himaldoc

Dugasseh, F. A., & Andersen, M. S. (2024). Non-carbon benefits of REDD+ implementation and sustainable emission reductions—A review. *Forests, Trees and Livelihoods, 33*(4), 299–318. https://doi.org/10.1080/14728028.2024.2383737

Duchelle, A. E., Seymour, F., Brockhaus, M., & Angelsen, A. (2019). *Forest-based climate mitigation: Lessons from REDD+ implementation*. Center for International Forestry Research (CIFOR). https://helda.helsinki.fi/bitstreams/e5958414-8d13-4534-b57a-a3a0c1fbbab6/download

Duchelle, A. E., Simonet, G., Sunderlin, W. D., & Wunder, S. (2018). What is REDD+ achieving on the ground? *Current Opinion in Environmental Sustainability, 32*, 134–140. https://doi.org/10.1016/j.cosust.2018.07.001

Dunlop, T., & Corbera, E. (2016). Incentivizing REDD+: How developing countries are laying the groundwork for benefit-sharing. *Environmental Science & Policy, 63*, 44–54. https://doi.org/10.1016/j.envsci.2016.04.018

Duraisami, M., Singh, R., & Chaliha, S. (2022). *Roadmap for scaling trees outside forests in India: Learnings from select states on policy incentives, enabling conditions, and barriers*. Working Paper, World Resources Institute India. https://wri-india.org/sites/default/files/roadmap-scaling-trees-outside-forests-India-learnings-select-states-policy-incentives.pdf

Datta, M. S. (2009, July). *Convergence or divergence: Institutional challenges for promoting market-based instruments in developing countries—A case study of carbon offset markets*. Paper presented at the International Sustainable Development Research Conference, Utrecht, Netherlands. ERP Environment.

FSI. (2021). *India State of Forest Report 2021*. Forest Survey of India, Ministry of Environment, Forest and Climate Change, Government of India. https://fsi.nic.in/isfr2021

Gilani, H. R., Yoshida, T., & Innes, J. L. (2017). A Collaborative Forest Management user group's perceptions and expectations on REDD+ in Nepal. *Forest Policy and Economics, 80*, 27–33. https://doi.org/10.1016/j.forpol.2017.03.004

GoI. (2022). *India's updated first nationally determined contribution under Paris Agreement (2021–2030)*. Government of India Submission to UNFCCC.

Gorain, S., Malakar, A., Dutta, S., & Das, S. (2025). Assessing forest cover, carbon stocks and fire dynamics in India and economic valuation of forest carbon in Asia: A meta-analysis. *Discover Forestry, 1*(16). https://doi.org/10.1007/s44415-025-00014-3

Grassi, G., House, J., Dentener, F., Federici, S., Den Elzen, M., & Penman, J. (2017). The key role of forests in meeting climate targets requires science for credible mitigation. *Nature Climate Change, 7*(3), 220–226. https://doi.org/10.1038/nclimate3227

Griscom, B. W., Adams, J., Ellis, P. W., Houghton, R. A., Lomax, G., Miteva, D. A., et al. (2017). Natural climate solutions. *Proceedings of the National Academy of Sciences, 114*(44), 11645–11650. https://doi.org/10.1073/pnas.1710465114

Gunawardena, M. A., Lokupitiya, E., & Gunawardena, P. (2024). Land degradation neutrality and carbon neutrality: Approaches, synergies, and challenges. *Frontiers in Forests and Global Change, 7*, 1398864. https://doi.org/10.3389/ffgc.2024.1398864

Harris, N. L., Gibbs, D. A., Baccini, A., Birdsey, R. A., DeBruin, S. P., Farina, M., & Walker, W. S. (2021). Global maps of twenty-first century forest carbon fluxes. *Nature Climate Change, 11*(3), 234–240. https://doi.org/10.1038/s41558-020-00976-6

Hazarika, B., & Bhattacharjee, N. (2022). Population pressure and its impact on forest resources in North East India. *Journal of Applied School Psychology, 6*(4), 4245–4255.

Helland, J., & Mæstad, O. (2015). *Experiences with results-based payments in Norwegian development aid.* Chr. Michelsen Institute (CMI). https://open.cmi.no/cmi-xmlui/handle/11250/2475179

Ickowitz, A., Sills, E., & de Sassi, C. (2017). Estimating smallholder opportunity costs of REDD$^+$: A pantropical analysis from households to carbon and back. *World Development, 95*, 15–26. https://doi.org/10.1016/j.worlddev.2017.02.022

IPCC. (2021). *Climate change 2021: The physical science basis.* Contribution of Working Group I to the Sixth Assessment Report of the Intergovernmental Panel on Climate Change. Cambridge University Press. https://www.ipcc.ch/report/ar6/wg1/

Jagger, P., Brockhaus, M., Duchelle, A. E., Gebara, M. F., Lawlor, K., Resosudarmo, I. A. P., & Sunderlin, W. D. (2014). Multi-level policy dialogues, processes, and actions: Challenges and opportunities for national REDD+ safeguards measurement, reporting, and verification (MRV). *Forests, 5*(9), 2136–2162. https://doi.org/10.3390/f5092136

Johnson, B. A., & Scheyvens, H. (2016). Characteristics of the remote sensing data used in the proposed UNFCCC REDD+ forest reference emission levels (FRELs). *International Archives of the Photogrammetry, Remote Sensing and Spatial Information Sciences, XLI-B8*, 669–675. https://doi.org/10.5194/isprs-archives-XLI-B8-669-2016

Joshi, R., Singh, H., Chhetri, R., Poudel, S. R. & Rijal, S. (2021). Carbon sequestration potential of community forests: A comparative analysis of soil organic carbon stock in community managed forests of Far-Western Nepal. *Eurasian Journal of Soil Science, 10*(2), 96–104. https://doi.org/10.18393/ejss.825066.

Kaul, M., Dadhwal, V. K., & Mohren, G. M. J. (2009). Land use change and net carbon flux in Indian forests. *Forest Ecology and Management, 258*(2), 100–108. https://doi.org/10.1016/j.foreco.2009.03.049

Kengoum Djiegni, F., Pham, T. T., & Ntirumenyerwa Mihigo, B.- P. (2024). *REDD+ benefit sharing mechanisms in the Democratic Republic of the Congo: Legal and institutional frameworks, policy implementation, and project experiences.* Center for International Forestry Research (CIFOR). https://doi.org/10.17528/cifor-icraf/009119

Khatun, K., Gross-Camp, N., Corbera, E., Martin, A., Ball, S., & Massao, G. (2015). When participatory forest management makes money: Insights from Tanzania on governance, benefit sharing, and implications for REDD+. *Environment and Planning a: Economy and Space, 47*(10), 2097–2112. https://doi.org/10.1177/0308518X15595899

Kim, J. (2018). Identifying and overcoming constraints on establishing feasible national REDD+ measurement, reporting and verification systems [Doctoral dissertation, University of Leeds]. White Rose eTheses Online. https://etheses.whiterose.ac.uk/id/eprint/23677/1/Kim_J_PhD_Geography_2018.pdf

Kishwan, J. (2023). REDD+ and its implementation in the Himalayan region: Policy issues. In P. Ghosh (Ed.), *Carbon management for promoting local livelihoods in the Hindu Kush Himalayan (HKH) region* (pp. 93–106). Elsevier. https://www.sciencedirect.com/science/article/pii/B9780443194153000062

Lawlor, K., Myers Madeira, E., Blockhus, J., & Ganz, D. J. (2013). Community participation and benefits in REDD+: A review of initial outcomes and lessons. *Forests, 4*(2), 296–318. https://doi.org/10.3390/f4020296

Lofts, K., Sarmiento Barletti, J. P., & Larson, A. M. (2021). *Lessons towards rights-responsive REDD+ safeguards from a literature review.* Working Paper 280, CIFOR.

Lyster, R. (2011). REDD+, transparency, participation and resource rights: The role of law. *Environmental Science & Policy, 14*(2), 118–126. https://doi.org/10.1016/j.envsci.2010.11.008

Mbano-Mweso, N. (2023). *Gender and REDD+ governance in Malawi: Enhancing women's right to participation.* In A. O. Jegede (Ed.), Implementing REDD+ in Africa (pp. xx–xx). Springer, Cham. https://doi.org/10.1007/978-3-031-39397-6_7

McDermott, C. L., & Ituarte-Lima, C. (2016). Safeguarding what and for whom? The role of institutional fit in shaping REDD+ in Mexico. *Ecology and Society, 21*(1), 9. https://doi.org/10.5751/ES-08088-210109

McDermott, C., Vira, B., & Walcott, J. (2022). *The evolving governance of REDD+: From Market Mechanism to Performance-Based Aid.* Working Paper 40, University of Oxford. https://researchportal.helsinki.fi/files/209083341/ws40_1.pdf

Meinshausen, M., Lewis, J., McGlade, C., et al. (2022). Realization of Paris Agreement pledges may limit warming just below 2 °C. *Nature, 604*(7905), 304–309. https://doi.org/10.1038/s41586-022-04553-z

Mitchell, A. L., Rosenqvist, A., & Mora, B. (2017). Current remote sensing approaches to monitoring forest degradation in support of countries measurement, reporting and verification (MRV) systems for REDD+. *Carbon Balance and Management, 12*, 9. https://doi.org/10.1186/s13021-017-0078-9

MoEFCC. (2022). *Updated First Nationally Determined Contribution under the Paris Agreement.* Ministry of Environment, Forest and Climate Change, Government of India.

Moonen, P. C. J., Verbist, B., Schaepherders, J., Bwama Meyi, M., Van Rompaey, A., & Muys, B. (2016). Actor-based identification of deforestation drivers paves the road to effective REDD+ in DR Congo. *Land Use Policy, 58*, 123–132. https://doi.org/10.1016/j.landusepol.2016.07.019

Morita, K., & Matsumoto, K. (2023). Challenges and lessons learned for REDD+ finance and its governance. *Carbon Balance and Management, 18*, 8. https://doi.org/10.1186/s13021-023-00228-y

Nautiyal, S., Singh, H., Kamboj, S., Kumar, A., Pal Panwar, V., & Bala, N. (2022). Carbon regulating services of a 30-year-old reclaimed limestone mine area in Foothills of Himalaya. *Indian Forester, 148*(3), 338–343. https://doi.org/10.36808/if/2022/v148i3/168187

Nayak, R., Karanth, K. K., Dutta, T., Defries, R., Karanth, K. U., & Vaidyanathan, S. (2020). Bits and pieces: Forest fragmentation by linear intrusions in India. *Land Use Policy, 99*, Article 104619. https://doi.org/10.1016/j.landusepol.2020.104619

Neeff, T. (2017). *From reference levels to results reporting: REDD+ under the UNFCCC.* Food and Agriculture Organization of the United Nations (FAO). https://openknowledge.fao.org/handle/20.500.14283/i7352en

Pan, Y., Birdsey, R. A., Fang, J., Houghton, R., Kauppi, P. E., Kurz, W. A., & Hayes, D. (2011). A large and persistent carbon sink in the world's forests. *Science, 333*(6045), 988–993. https://doi.org/10.1126/science.1201609

Pandey, D. (2021). MRV issues in REDD+ in the Indian context. In J. Kishwan (Ed.), *Managing India's forests in a changing climate: Emerging concepts and their operationalization* (pp. 270–280). USAID (Tetra Tech).

Parrotta, J., Mansourian, S., & Grima, N. (2022). Forests, climate, biodiversity and people: Assessing a decade of REDD+. In *Global assessment report on biodiversity and ecosystem services (Technical Report).* U.S. Forest Service. https://research.fs.usda.gov/treesearch/download/66157.pdf

Pasgaard, M., Sun, Z., Müller, D., & Mertz, O. (2016). Challenges and opportunities for REDD+: A reality check from perspectives of effectiveness, efficiency and equity. *Environmental Science & Policy, 63*, 161–169. https://doi.org/10.1016/j.envsci.2016.05.021

Paudel, N. S., Vedeld, P. O., & Khatri, D. B. (2015). Prospects and challenges of tenure and forest governance reform in the context of REDD+ initiatives in Nepal. *Forest Policy and Economics, 52*, 1–8. https://doi.org/10.1016/j.forpol.2014.12.009

Pham, T. T., Moeliono, M., Yuwono, J., Dwisatrio, B., & Gallo, P. (2021). REDD+ finance in Brazil, Indonesia and Vietnam: Stakeholder perspectives between 2009–2019. *Global Environmental Change, 70*, Article 102330. https://doi.org/10.1016/j.gloenvcha.2021.102330

Phelps, J., Friess, D. A., & Webb, E. L. (2012). Win–win REDD+ approaches belie carbon–biodiversity trade-offs. *Biological Conservation, 154*, 53–60. https://doi.org/10.1016/j.biocon.2012.03.003

Rakatama, A., Pandit, R., Ma, C., & Iftekhar, S. (2017). The costs and benefits of REDD+: A review of the literature. *Forest Policy and Economics, 75*, 103–111. https://doi.org/10.1016/j.forpol.2016.10.006

Riamit, S., & Tauli-Corpuz, V. (2012). *Indigenous peoples' perspectives and activities in monitoring, reporting, and indicators development for REDD+ and a review of the MRV concepts, tools and instruments*. Forest Carbon Partnership Facility. https://www.forestcarbonpartnership.org/system/files/documents/Indigenous%20Peoples%20Perspective%20on%20Community%20based%20MRV%20for%20Social%20and%20Enviromental%20Standards.pdf

Sandker, M., Neeff, T., Todd, K., & Poultouchidou, A. (2022). *From reference levels to results: REDD+ reporting by countries*. Food and Agriculture Organization of the United Nations (FAO). https://openknowledge.fao.org/items/3ff44092-746a-4bad-8a6f-edc7e9212845

Schroeder, H., Di Gregorio, M., Brockhaus, M., & Pham, T. T. (2020). Policy learning in REDD+ donor countries: Norway, Germany and the UK. *Global Environmental Change, 63*, Article 102106. https://doi.org/10.1016/j.gloenvcha.2020.102106

Silori, C. S., Frick, S., Luintel, H., & Poudyal, B. H. (2013). Social safeguards in REDD+: A review of existing initiatives and challenges. *Journal of Forest and Livelihood, 11*(2), 66–78. https://doi.org/10.3126/jfl.v11i2.8619

Skutsch, M. M. (2017). *The evolution of international policy on REDD+*. Oxford Research Encyclopedia of Climate Science. https://doi.org/10.1093/acrefore/9780190228620.001.0001/acrefore-9780190228620-e-43

Smith, R., Guevara, O., Wenzel, L., & Dudley, N. (2019). Ensuring co-benefits for biodiversity, climate change and sustainable development. In W. Leal Filho, J. Barbir, & R. Preziosi (Eds.), *Handbook of climate change and biodiversity* (pp. xx–xx). Cham: Springer. https://doi.org/10.1007/978-3-319-98681-4_9

Sud, R., Sharma, J. V., & Bansal, A. K. (2012). *International REDD+ architecture and its relevance for India.* Ministry of Environment, Forest and Climate Change. https://moef.gov.in/uploads/2019/06/redd-bk5.pdf

Tan, X. C., Wang, Y., Gu, B. H., Kong, L. S., & Zeng, A. (2022). Research on the national climate governance system toward carbon neutrality—A critical literature review. *Fundamental Research, 2*(3), 384–391. https://doi.org/10.1016/j.fmre.2022.03.010

Trivedi, A. (2025). Implementing REDD+ safeguards and protecting the rights of tribal and forest communities in India: With special emphasis on the principle of free, prior, and informed consent. *Jindal Global Law Review*. https://doi.org/10.1007/s41020-025-00264-4

UNEP. (2023). *Emissions Gap Report 2023: Broken record – Temperatures hit new highs, yet world fails to cut emissions (again)*. United Nations Environment Programme. https://www.unep.org/resources/emissions-gap-report-2023

UNFCCC. (2022). *REDD+ infographic: A holistic framework for climate action in the forest sector* [Infographic]. https://redd.unfccc.int/files/redd__infographic.pdf

UNFCCC. (2023). *Lima REDD+ Information Hub.*United Nations Framework Convention on Climate Change. https://redd.unfccc.int/

UN-REDD Programme. (2020, September 4). Indonesia receives results-based payments from Green Climate Fund. UN-REDD Newsroom. https://www.un-redd.org/news/indonesia-receives-results-based-payments-green-climate-fund

Vijge, M. J., & Gupta, A. (2014). Framing REDD+ in India: Carbonizing and centralizing Indian forest governance? *Environmental Science & Policy, 38*, 17–27. https://doi.org/10.1016/j.envsci.2013.10.032

Vijge, M. J., Brockhaus, M., Di Gregorio, M., & Muharrom, E. (2016). Framing national REDD+ benefits, monitoring, governance and finance: A comparative analysis of seven countries. *Global Environmental Change, 39*, 57–68. https://doi.org/10.1016/j.gloenvcha.2016.04.013

Vijitharan, S., Tsusaka, T. W., & Sasaki, N. (2024). Evaluating REDD+ progress and results-based incentives in Sri Lanka: A comparative analysis across Southeast and South Asian countries. *Trees, Forests and People, 13*, Article 100481.

Voigt, C., Nemitz, D., & Ferreira, F. (2025). The Paris agreement and the importance of the Warsaw framework for REDD+ (WFR). *Review of European, Comparative & International Environmental Law, 34*(1), 45–59. https://doi.org/10.1111/reel.12590

Wong, G. Y., Luttrell, C., Loft, L., Yang, A., Pham, T. T., Naito, D., et al. (2019). Narratives in REDD+ benefit sharing: Examining evidence within and beyond the forest sector. *Climate Policy, 19*(8), 1038–1051. https://doi.org/10.1080/14693062.2019.1618786

Yang, H., Song, M., Son, H., Kim, R., & Choi, E. (2025). Evaluating REDD+ readiness: High-potential countries based on MRV capacity. *Forests, 16*(1), 67. https://doi.org/10.3390/f16010067

Chapter 18
Urban Green Spaces and Blue-Green Infrastructure: Pathways to Carbon Neutrality and Building Sustainable Cities

Apurva Malik and Hukum Singh

Abstract Rapid urbanization has disrupted urban ecosystems, fragmented natural landscapes, and intensified challenges such as the urban heat island effect, air pollution, and waterlogging, all of which affect human well-being. Hence, addressing these issues necessitates integrating nature-based solutions, such as urban green spaces (UGS) and Blue-Green Infrastructure (BGI), into urban planning. These solutions are pivotal in advancing cities toward carbon neutrality, enhancing resilience, and ensuring long-term sustainability. UGS and BGI facilitate urban cooling, efficient water management, and habitat creation, enhancing overall urban environmental health. The BGI, such as bioswales and vegetative rain gardens, improves water retention, reduces flood risks, and enhances ecological connectivity and aesthetic value. BGI helps to mitigate environmental risks and foster well-being by promoting stress reduction, physical activity, social interaction, and community cohesion. This chapter highlights case studies such as Singapore's vertical gardens, Copenhagen's flood-resilient parks, and India's Pashan Lake restoration, emphasizing the adaptability and success of these solutions in diverse urban contexts. However, implementing UGS and BGI faces challenges such as land-use conflicts, financing limitations, and ensuring equitable access. Addressing these barriers requires robust policy frameworks, innovative technologies, and active community engagement. Aligning ecological, social, and economic objectives with UGS and BGI systems represents transformative pathways for building sustainable, livable, and resilient cities.

Keywords Urban green spaces · Sustainable urban planning · Urban resilience · Blue-green infrastructure · Urban heat island effect · Air pollution mitigation

A. Malik (✉) · H. Singh
ICFRE-Forest Research Institute, Dehradun, Uttarakhand, India
e-mail: apurva.malik14@gmail.com

H. Singh (ed.), *Forestry for a Carbon-Neutral and Sustainable Future*, Sustainability Solutions, https://doi.org/10.1007/978-3-032-06427-1_18

18.1 Introduction

Rapid urbanization has significantly degraded urban ecosystems by fragmenting natural landscapes and triggering a cascade of environmental and ecological challenges across multiple scales in metropolitan regions, including the urban heat island effect, worsening air pollution, and frequent waterlogging (Grimm et al., 2008; Wu, 2014). These challenges demand immediate solutions as cities increasingly face the impacts of climate change, water scarcity, and declining public health (Hoek et al., 2013). Cities, covering only 2–3% of the Earth's surface, consume 75% of global energy and contribute 80% of CO_2 emissions, generating vast amounts of waste and polluting air and water resources. It is estimated that 68% of the global population will reside in urban areas by 2050, further intensifying these issues (Gu & Dupre, 2021).

Integrating nature-based solutions, such as urban green spaces (UGS) and Blue-Green Infrastructure (BGI), into urban planning has become a crucial strategy for addressing these pressing issues. These solutions offer sustainable approaches by integrating natural ecosystems with urban development, enabling the efficient management of resources, reducing environmental footprints, and enhancing public well-being (Steffen et al., 2015). This approach supports cities in transitioning toward carbon neutrality, resilience, and long-term sustainability, addressing the need for innovative solutions to ensure the viability of urban areas for future generations (Singh, 2024b).

Urban Green Spaces and Blue-Green Infrastructure, encompassing a network of natural, semi-natural, and designed environmental features, is defined as a strategically planned system that integrates green spaces (such as parks, street trees, and green roofs) and blue elements (including rivers, wetlands, and rain gardens) (Kassim et al., 2013). BGI plays a central role in enhancing urban ecosystems and addressing the adverse effects of urbanization. Its role in providing vital ecosystem services, such as climate regulation, air purification, and biodiversity support, underscores its importance in fostering resilient cities (Bisht et al., 2024). Moreover, BGI facilitates urban cooling, efficient water management, and habitat creation, enhancing the overall ecological health of the urban areas (Dorst et al., 2019). For instance, bioswales and vegetative rain gardens improve water retention, reduce flood risks, and enhance ecological connectivity and aesthetic value.

Besides ecological benefits, BGI addresses socio-economic and cultural dimensions by offering recreational spaces, promoting public well-being, and fostering eco-cultural landscapes (Singh, 2024a). Its multifunctional approach creates inclusive urban environments, balancing the needs of nature and people while enhancing cities' adaptive capacity to climate-induced stresses (Toxopeus & Polzin, 2021).

Meanwhile, BGI addresses water management issues, reduces flood risks, and promotes groundwater recharge, all while contributing to carbon sequestration and emissions reductions. The integration of such solutions within urban planning aligns with the United Nations Sustainable Development Goals (SDGs), particularly Goals

11 (Sustainable Cities and Communities) and 13 (Climate Action), and underpins the transition to net-zero emissions by enhancing urban sustainability and liveability.

Moreover, the co-benefits of these infrastructures extend beyond environmental advantages. These nature-based solutions foster green economies and create jobs, supporting social equity and economic resilience. Their adoption in cities worldwide helps illustrate the scalability and adaptability of these systems in various climatic and socio-economic contexts.

However, achieving their full potential requires overcoming land-use conflicts, developing effective financing mechanisms, and ensuring equitable access. Leveraging interdisciplinary approaches and innovative technologies may optimize the design, implementation, and maintenance of green and blue-green infrastructures to create carbon–neutral urban landscapes (Joshi et al., 2021).

This chapter aims to unveil the multidimensional role of Urban Green Spaces (UGS) and Blue-Green Infrastructure (BGI) in achieving climate-resilient, low-carbon urban landscapes through integrated ecological and urban design approaches (Pakhira et al., 2024). While existing literature highlights the ecological and socio-economic benefits of green infrastructure (Kabisch et al., 2017), few studies have offered a comparative assessment of successful implementations across regions. This chapter distinguishes itself by presenting a cross-continental synthesis of case studies, reflecting how different cities, from Singapore to Copenhagen and Ahmedabad, have adapted UGS and BGI strategies to address region-specific urban stressors. The novelty lies in aligning urban ecological planning with climate adaptation, SDGs, and carbon mitigation across varied climatic zones and socio-political contexts. By exploring case-based patterns and synthesizing outcomes around key themes such as climate resilience, carbon neutrality, equity, and multi-functionality, this work provides actionable insights for international researchers, urban planners, and policy designers.

18.2 Urban Green Spaces: A Tool for Sustainability

Urban green spaces are defined as all urban lands covered by vegetation, including natural and semi-natural areas such as parks, urban forests, wetlands, grasslands, and private gardens, irrespective of size and function. These spaces are vital for ecological, social, and economic sustainability (Malik et al., 2024). They can serve as a health-promoting setting for all urban community members. Urban green space systems include protecting existing green spaces, creating new spatial forms, and restoring and maintaining connectivity among diverse green spaces (Kong et al., 2010).

Urban green spaces play a pivotal role in addressing climate change through carbon sequestration. Parks and urban forests are significant carbon sinks, with trees storing carbon in their biomass and soil. Studies show that urban forests in U.S. cities sequester approximately 22.8 million tons of carbon annually (Nowak & Crane, 2002). In contrast, roadside plantations in tropical regions capture about 20–30 kg

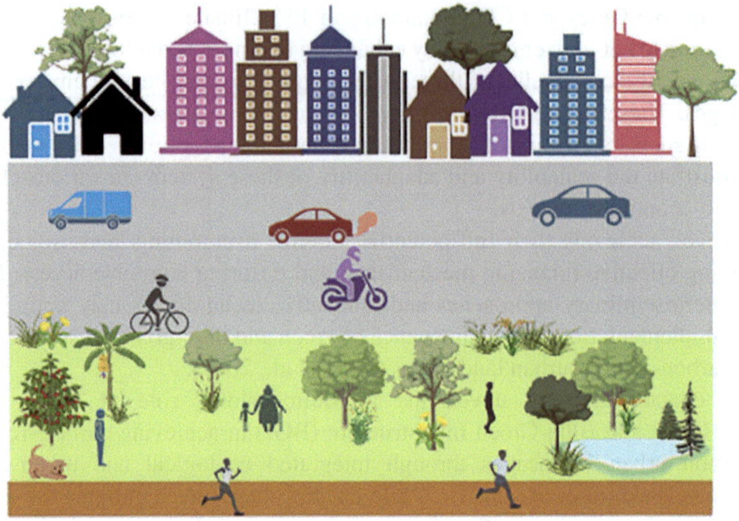

Fig. 18.1 Pictorial representation of Urban Green Spaces (UGS)

of carbon per tree annually (Baro et al., 2015). Green roofs sequester up to 375 g of carbon per square meter annually, while community gardens increase soil carbon storage by up to 20% (Pearson & Hodgkin, 2010). Wetlands are particularly effective at storing 2–3 times more carbon than terrestrial ecosystems, primarily due to their anaerobic soils (Mitsch et al., 2013) (Fig. 18.1).

Additionally, urban green spaces play a crucial role in biodiversity protection by providing habitats for flora and fauna, as well as maintaining ecological connectivity. Urban biodiversity corridors enhance species movement, and the planting of native species boosts habitat resilience (Tzoulas et al., 2007). Ecosystem services encompass air purification, water management, microclimate regulation, and the provision of aesthetic and cultural enrichment. These spaces also offer areas for relaxation, recreation, and social interaction, promoting emotional well-being and reducing stress (Rawat & Singh, 2024). Accessible and equitably distributed green spaces enhance public health by promoting active lifestyles and improving air quality, particularly in densely populated urban areas (Zhang & Qian, 2024). Inner-city green spaces are vital for improving air quality by absorbing pollutant gases and particulates, which are responsible for respiratory infections. Green spaces also help reduce the energy costs of cooling buildings effectively (Chauhan & Singh, 2024).

Urban green spaces significantly contribute to sustainable urban development through their multifaceted roles. They mitigate climate change, enhance biodiversity, deliver vital ecosystem services, and improve social and health outcomes. A well-integrated network of green spaces is essential for creating resilient and liveable cities that align with global sustainability objectives.

18.3 Blue-Green Infrastructure (BGI): the Nexus of Water and Green Systems

Blue-Green Infrastructure (BGI) exemplifies a forward-thinking urban planning model that intertwines water management with ecological preservation. Holistically addressing environmental, social, and economic challenges through BGI offers a sustainable pathway for cities to adapt to climate change and enhance the quality of urban life. The strategic implementation of BGI can transform cities into resilient ecosystems, striking a balance between development and the natural environment.

Blue-Green Infrastructure integrates a transformative approach to urban planning that merges water management systems (blue) with natural and vegetative or green spaces to create multifunctional landscapes that enhance urban resilience, ecological balance, and community well-being. As a strategic framework, BGI emphasizes the creation of multifunctional landscapes addressing urban challenges, including climate change adaptation, flood management, biodiversity loss, and declining public health (Walsh et al., 2016). Its implementation ensures a sustainable coexistence of natural ecosystems and urban environments. Blue-green infrastructure can significantly contribute to the urban environment, offering multiple benefits that include improved water supply, enhanced flood mitigation, increased terrestrial biodiversity, urban cooling, increased resilience to climate change effects, and support for urban agriculture, ultimately leading to improved human well-being. Blue-green infrastructure is also used to ensure controlled flooding in concave green areas or detention basins (Liu et al., 2014) (Fig. 18.2).

BGI is a nature-based solution that improves urban functionality while preserving environmental integrity. BGI focuses on sustainable water resource management by integrating green areas (such as parks and urban forests) with blue elements (including rivers and wetlands) to create a cohesive urban planning approach. This integration supports efficient water cycling, carbon storage, and biodiversity. Its multifunctional nature provides diverse benefits, including climate regulation, water management, recreation, and enhanced air quality, contributing to more liveable and

Fig. 18.2 Pictorial representation of Blue-Green Infrastructure (BGI)

sustainable cities. Ecosystem restoration aims to revitalize degraded urban landscapes, improve ecological connectivity, and strengthen cities' natural capacity to sustain wildlife. BGI also enhances urban resilience by developing adaptive systems that can withstand extreme weather events, such as floods and droughts. Additionally, it emphasizes community involvement to foster inclusive urban spaces, ensure fair access to green and blue infrastructure, promote public health, and enhance the overall quality of life (Hansen et al., 2015; Kabisch et al., 2017; Mell et al., 2013).

Blue-green infrastructure can be classified as wetlands, which serve as natural water filtration systems and carbon sinks. They absorb excess rainfall, reduce urban flooding, and provide habitats for a diverse range of species. For example, the Cheonggyecheon Stream Restoration in Seoul transformed a paved-over stream into a thriving wetland that manages floods, revitalizes biodiversity, and offers recreational benefits. Bioswales are vegetated, shallow ditches designed to capture, filter, and slowly release stormwater runoff. They help improve water quality by trapping pollutants and enhancing groundwater recharge (Vega, 2019). Another type of BGI is Rain Gardens, which are landscaped depressions filled with vegetation that absorb and treat stormwater. They prevent waterlogging, reduce surface runoff, and support native plant species, such as those found in Chicago's Millennium Park (Davis et al., 2009).

BGI offers comprehensive solutions for urban resilience, particularly in mitigating the impacts of climate change, such as flood mitigation and Water Management. BGI systems reduce the risk of urban flooding by enhancing water retention. Features like wetlands and bioswales act as natural reservoirs, slowing down water flow during heavy rains (Gupta et al., 2024). Urban areas prone to flooding, such as Copenhagen, have adopted permeable pavements and green roofs as part of their Building Green Infrastructure (BGI) strategies, significantly reducing flood risks (Mabrouk et al., 2024). In contrast, rain gardens and bioswales capture and purify stormwater, decreasing the burden on urban drainage systems. Wetlands and riparian zones act as natural sponges, regulating water availability during dry spells and storing excess rainwater during storms. Integrating these systems into urban planning can simultaneously address water scarcity and pollution (Fletcher et al., 2013).

Beyond its immediate ecological functions, BGI promotes urban cooling through green roofs, parks, and tree-lined streets, which mitigate the urban heat island effect and enhance thermal comfort. Additionally, BGI promotes biodiversity conservation by connecting fragmented habitats and facilitating the migration of species, thereby preserving genetic diversity. It also provides social and health benefits through accessible green and blue spaces that encourage physical activity, mental well-being, and social cohesion.

18.4 Co-Benefits of Urban Green and Blue-Green Systems

Urban BGI systems are gaining popularity globally and are essential for achieving carbon neutrality by acting as carbon sinks, conserving biodiversity, and aligning urban planning efforts with global sustainability goals. These systems improve air and water quality, reduce noise pollution, and mitigate the impacts of extreme weather events, thereby lowering environmental health risks in cities. Moreover, BGI promotes well-being by reducing stress, encouraging physical activity, fostering social connections, and strengthening community ties. Addressing challenges such as climate change, public health, and resource efficiency, BGI exemplifies a multifaceted approach to enhancing urban ecosystems (World Health Organization, 2023).

18.4.1 Urban Heat Island (UHI) Effect Mitigation

Heat-absorbing materials, such as asphalt and concrete, have replaced vegetation in urban areas, making them warmer than surrounding vegetated rural areas, which contributes to the Urban Heat Island (UHI) effect. Urban green spaces mitigate the Urban Heat Island (UHI) effect through shading and evapotranspiration, thereby cooling the air and reducing localized temperatures. Similarly, blue-green systems, such as ponds and wetlands, contribute to evaporative cooling, thereby enhancing the combined cooling effect. Studies show that a tree-lined street can reduce local temperatures by 2–3 °C compared to a treeless area (Bowler et al., 2010). These systems create localized microclimates that alleviate heat stress, particularly during heat waves, thereby improving urban thermal comfort and reducing energy demands for cooling.

18.4.2 Improved Air Quality, Enhanced Water Quality, and Flood Management

Green spaces improve air quality through trees and vegetation, which act as natural air filters, capturing particulate matter (PM10 and PM2.5), absorbing nitrogen and sulfur oxides (NOx and SOx), and reducing carbon monoxide (CO) levels, thereby also lowering ozone levels. Nowak et al. (2014) reported that urban forests can remove up to 711,000 metric tons of air pollution annually in the U.S. alone. Studies also revealed that a mature tree can absorb up to 20 kg of CO_2 annually while simultaneously producing oxygen (Nowak et al., 2006). Additionally, roadside vegetation reduces vehicular emissions, contributing to healthier urban air. Blue-green systems, such as bioswales, rain gardens, and wetlands, play a crucial role in water management. These systems filter stormwater, removing pollutants such as heavy metals, nutrients, and sediments before reaching waterways. They also prevent river nutrient runoff by improving infiltration, which is vital in urban areas prone to waterlogging. This dual benefit ensures cleaner air and water, addressing urban environmental challenges while enhancing public health (Berland et al., 2017).

18.4.3 Biodiversity Conservation and Carbon Sequestration

Urban green and blue spaces provide essential habitats for wildlife, supporting biodiversity even in densely populated cities. These areas serve as ecological corridors, allowing species to migrate and maintain genetic diversity. For example, urban wetlands support bird and aquatic species, while green roofs and vertical gardens create habitats for pollinators, such as bees and butterflies (Tzoulas et al., 2007).

Green infrastructure helps in energy conservation by reducing the need for artificial cooling and heating. Green roofs and walls help insulate buildings and reduce energy consumption. Trees and plants also act as carbon sinks, storing CO_2 in their biomass and soils. Urban wetlands can sequester up to 2–3 times more carbon per unit area than terrestrial systems (Mitsch et al., 2013).

18.4.4 Improved Mental and Physical Health and Socio-Economic Benefits

Green and blue spaces offer recreational opportunities, reduce stress, and enhance well-being. Parks, urban forests, and community gardens provide venues for recreation, relaxation, and social interaction, thereby improving both mental and physical well-being. Green spaces reduce stress, encourage physical activity, and foster community cohesion. Similarly, blue-green spaces, such as waterfront parks and wetlands, add aesthetic and recreational value, promoting activities like birdwatching, kayaking, and fishing. These spaces also boost property values and stimulate local economies by attracting tourists and enhancing urban appeal (Haq, 2011).

18.5 India's Initiatives in Implementing Urban Green Spaces (UGS) and Blue-Green Infrastructure (BGI)

India intends to maintain ecological balance, promote climate resilience, and enhance urban sustainability. The organization has undertaken significant policy-driven initiatives as part of its broader strategy to promote the concepts of Urban Green Spaces (UGS) and Blue-Green Infrastructure (BGI). Some key schemes and initiatives aimed at integrating nature-based solutions are:

18.5.1 Green Highway (Plantation, Transplantation, Beautification and Maintenance) Policy, 2015

The Green Highway Policy, launched by the Ministry of Road Transport and Highways in New Delhi, aims to create eco-friendly highway corridors through collaborative efforts involving communities, farmers, private entities, NGOs, and government bodies. The policy seeks to mitigate air pollution and dust by planting trees and shrubs along national highways, serving as natural sinks for air pollutants, and preventing soil erosion on embankment slopes. Key objectives include enhancing carbon sequestration, curbing vehicular emissions, and improving the visual appeal of highways (Garg et al., 2016). The initiative mandates the allocation of 1% of the project cost for highway greening and emphasizes the use of native plant species to boost local biodiversity. By November 2024, approximately 465.42 lakh plants had been planted along National Highways as part of the Green Highways (Plantation, Transplantation, Beautification and Maintenance) Policy, 2015. The NHAI has also released updated progress reports and revised Standard Operating Procedures (SOPs) to guide and enhance plantation activities.

18.5.2 Nagar Van Yojana (NVY), 2020

This scheme was launched in 2020 by the Ministry of Environment, Forest, and Climate Change, Union Government of India, aiming to enhance urban living conditions by expanding green cover. It targets the development of 600 city forests and 400 city gardens nationwide to mitigate urban pollution, enhance carbon sequestration, and promote biodiversity enrichment. The initiative focuses on establishing "Nagar Van" (Urban Forests) spanning 10–50 hectares in cities and towns in collaboration with municipal bodies. A key objective is to conserve forested areas within or on the outskirts of urban centres, preventing degradation and encroachment. Additionally, it seeks active involvement from local communities and various agencies in developing biodiverse green spaces that deliver social and environmental benefits (Singh, 2024). Financial support is provided by the Union Government of India for both the creation and maintenance of these forests, with a strong emphasis on public engagement to promote collective responsibility for urban ecosystems.

As of August 2023, 385 Nagar Van projects have been approved nationwide, with 173 of them sanctioned in the 2021–22 fiscal year across 26 states. The scheme offers financial support of ₹4 lakhs per hectare for the development and upkeep of these urban forests. Notably, the initial goal of establishing 100 Nagar Vans within the first 100 days of the initiative was exceeded, reaching 111 sanctioned projects. The NVY has an estimated budget allocation of ₹895 crores for the period spanning from 2020–21 to 2024–25.

18.5.3 Blue-Green Master Plan for Delhi

The Delhi Development Authority (DDA) has introduced an innovative Blue-Green Master Plan to integrate water bodies and green spaces, thereby enhancing the city's climate resilience. This comprehensive plan emphasizes the restoration of lakes, the development of bioswales, and the implementation of natural stormwater management systems. Under the proposed "Green–Blue Policy," water bodies, including rivers, canals, ponds, wetlands, and floodplains, along with their adjacent lands, will be revitalized. The DDA intends to integrate approximately 50 major drains across the city by eliminating pollution sources, combining mechanized and natural treatment systems, and strictly prohibiting waste dumping. Additionally, the Master Plan for Delhi 2041 (MPD-2041) outlines a strategic vision for sustainable urban development, incorporating lessons from previous 20-year plans. Key focus areas include flood control, water management through blue-green corridors, and the creation of recreational spaces around restored water bodies.

18.5.4 Smart Cities Mission 2015

The Smart Cities Mission aims to foster sustainable urban development by integrating green and blue infrastructure into city planning, enhancing liveability and resilience. Several cities, such as Pune and Surat, have launched notable projects under this initiative. Pune focuses on stormwater management through blue-green infrastructure (BGI) and green corridors, while Surat emphasizes urban afforestation and biodiversity parks to mitigate the heat island effect.

The mission's core objective is to drive economic growth and improve urban quality of life by transforming existing neighbourhoods through retrofitting and redevelopment while creating new greenfield areas to accommodate the rising population. Adopting smart technologies and data-driven solutions can help cities enhance their infrastructure, service delivery, and governance, ensuring greater inclusivity. Key elements include promoting mixed land use, expanding housing opportunities, developing open spaces, and improving pedestrian-friendly transit options. Additionally, smart governance models and identity-based development further contribute to building vibrant, inclusive cities for the future. Over 8000 diverse, multi-sectoral projects have been initiated across 100 cities, with a total investment of approximately ₹1.6 lakh crore. Remarkably, more than 90% of these projects—worth ₹1,45,312 crores—have already been completed. 75% of projects have been completed in 75 cities, with 17 cities achieving 100% completion. Financially, the Central Government allocated ₹48,000 crores to the mission, of which ₹46,787 crores has been disbursed, with over 90% of the funds already utilised.

18.5.5 Urban Flood Mitigation and Water Conservation Initiative (2024)

India recently launched an initiative with a $300 million investment to mitigate urban flooding and conserve water in seven major cities. Key objectives include reducing flood risks, promoting water conservation, restoring ecosystems, and enhancing inter-departmental coordination. The initiative emphasizes blue-green infrastructure (BGI) by constructing wetlands, restoring water bodies, and expanding urban green spaces to manage stormwater and reduce surface runoff. Key features include upgrading drainage systems, installing permeable pavements, and using technology for real-time monitoring and early warning systems. Urban green spaces will help mitigate the effects of heat islands and increase groundwater recharge. Capacity building and public engagement are integral to its success. Expected outcomes are improved urban resilience, sustainable water resources, enhanced liveability, and biodiversity conservation. Challenges such as coordination, funding, and community participation will be addressed through task forces, public–private partnership (PPP) models, and participatory planning. This initiative marks a significant step towards climate-resilient urban development.

18.6 Successful Implementation of the Concept of UGS and BGI

The global implementation of UGS and BGI strategies reveals a diversity of innovative urban design practices tailored to specific environmental challenges. Key functional themes, including flood management, urban cooling, ecosystem restoration, social engagement, and multi-functionality, have grouped the case studies from Europe, Asia, and North America. These cases are not only reflective of localized best practices but also provide a comparative lens to assess the performance, adaptability, and replicability of BGI strategies. The observed patterns suggest that climate-responsive design, strong community engagement, and policy integration are recurring enablers of success, irrespective of the region. Some of the examples are:

18.6.1 Copenhagen: Flood-Resilient Blue-Green Infrastructure Systems

Copenhagen, Denmark, is a leading example of flood-resilient urban planning through innovative Blue-Green Infrastructure (BGI) solutions. The city's Cloudburst Management Plan incorporates cutting-edge features, such as permeable pavements, bioswales, and retention basins, to effectively mitigate the impacts of intense rainfall. Multifunctional parks, such as Superkilen and Tasinge Plads, further demonstrate the integration of recreational spaces with stormwater management systems. These initiatives have significantly bolstered Copenhagen's resilience to climate-induced flooding while fostering a model of sustainable and adaptive urban living (Winter, 2019).

18.6.2 Rotterdam, Netherlands: Water Plazas

Rotterdam exemplifies excellence in adaptive water management through pioneering Blue-Green Infrastructure (BGI) initiatives. The city's water plazas are ingenious, multifunctional spaces that serve as vibrant recreational areas during dry periods and as essential floodwater storage during heavy rainfall. These dynamic solutions mitigate urban flooding risks, foster community engagement, and elevate the city's aesthetic appeal. Rotterdam's forward-thinking approach underscores the importance of integrating water-sensitive urban designs in addressing climate challenges and enhancing urban resilience effectively (Dai et al., 2018).

18.6.3 New York City, USA: The High Line

The High Line is a pioneering example of repurposing urban green space. An abandoned elevated railway was transformed into a lush, linear park spanning 1.45 miles, providing recreational opportunities while serving as a hub for biodiversity, featuring native plants that attract pollinators and birds. The park contributes to urban cooling, mitigating the urban heat island effect, and enhances community engagement through public art installations and cultural events. Additionally, its design exemplifies sustainable urban revitalization, inspiring cities globally to reimagine underutilized infrastructure (Black & Richards, 2020).

18.6.4 Singapore: Integrating Green Roofs and Vertical Gardens

Singapore is globally recognized for its exemplary incorporation of green infrastructure into urban development. Through initiatives such as the "Skyrise Greenery Incentive Scheme," the city-state actively encourages the integration of green roofs and vertical gardens into building designs. Iconic landmarks like Marina Bay Sands and Changi Airport feature verdant landscapes that mitigate the urban heat island effect while enhancing their aesthetic appeal. The Park Connector Network also interlinks parks and green spaces, fostering biodiversity and ecological connectivity. These forward-thinking measures have effectively lowered urban temperatures and enriched urban biodiversity, setting a benchmark for sustainable urban planning (Malik, 2024).

18.6.5 Ahmedabad: Sabarmati Riverfront Development Project

The Sabarmati Riverfront in Ahmedabad integrates Blue-Green Infrastructure principles by transforming the polluted riverbanks into a vibrant urban space. This project incorporates landscaped gardens, walking paths, and community spaces while improving water quality and managing urban flooding. The riverfront has enhanced recreational opportunities and boosted biodiversity along the river corridor, demonstrating how ecological restoration can coexist with urban development (Mahadevia & Sharma, 2023).

18.6.6 Pune: Pashan Lake Revitalization

Pashan Lake, a notable urban wetland in Pune, has undergone extensive revitalization efforts aimed at restoring its ecological integrity. Key initiatives included dredging to enhance water retention, planting native vegetation to support local flora and fauna, and establishing buffer zones to mitigate the impact of urban runoff. These measures have improved the lake's hydrological capacity, fostered biodiversity, and transformed it into a valuable recreational hub for the community. This project exemplifies effective and sustainable wetland management in urban environments (Mitra et al., 2022).

While Copenhagen and Rotterdam emphasize stormwater retention and flood resilience through bioswales and adaptive landscapes, Singapore showcases vertical greening and biodiversity corridors to combat heat stress and urban biodiversity loss. Meanwhile, India's Pashan Lake restoration and the Sabarmati Riverfront demonstrate hybrid approaches that integrate ecological restoration with public amenity development. These cases illustrate how BGI serves as a flexible urban planning tool that varies in scale and form, yet maintains its core objectives of resilience, ecosystem service enhancement, and community well-being.

18.7 Challenges in Implementation and Management

UGS and Blue-Green Infrastructure (BGI) development and maintenance face numerous challenges that require careful planning and innovative solutions. These challenges are often interlinked with social, economic, and environmental factors, as detailed below:

18.7.1 Land-Use Conflicts and Urban Space Constraints

The ever-increasing demand for urban land, driven by population growth and industrial development, often leads to conflicts over land use. Allocating space for green and blue-green systems competes with other priorities, such as housing, commercial developments, and infrastructure projects. Additionally, in densely populated cities, the limited availability of land poses significant challenges in implementing expensive green or BGI projects. Balancing development needs with environmental preservation remains a key hurdle (Kabisch et al., 2015).

18.7.2 Financing and Maintenance of Green-BGI Systems

Urban green and blue-green systems require significant initial investment for planning, design, and implementation. Securing consistent funding can be challenging, especially for municipalities operating under budget constraints. Moreover, the ongoing maintenance of these systems, such as ensuring vegetation health, cleaning water bodies, and maintaining permeable surfaces, requires dedicated resources and expertise. Inadequate financial planning often results in deteriorated systems that fail to deliver the intended benefits (Depietri et al., 2012).

18.7.3 Environmental and Structural Challenges in Management

Urban Blue-Green Infrastructure (BGI) presents both challenges and benefits. Routine maintenance, such as grass cutting, generates green waste requiring sustainable disposal. Constructed wetlands (CWs) enhance habitats and improve water quality, but they can also become breeding grounds for mosquitoes and attract large bird populations, thereby increasing disease risks for nearby communities. Tree roots near sidewalks often cause damage to infrastructure, lifting and cracking surfaces as they seek moisture and nutrients (Wong et al., 1988). Addressing these issues demands innovative design, proactive maintenance, and community engagement to balance ecological benefits with urban needs (Shah et al., 2024).

18.7.4 Equity and Accessibility Issues

Ensuring equitable distribution and accessibility of green spaces and BGI systems is another critical challenge. In many urban areas, these spaces are disproportionately concentrated in affluent neighbourhoods, leaving marginalized and low-income communities with limited or no access. This inequitable distribution exacerbates social disparities and denies vulnerable populations the benefits of urban greenery, such as improved health, recreational opportunities, and enhanced climate resilience. Lack of cohesive policies and inter-agency coordination often hinders the integrated implementation of BGI systems.

A limited public understanding of the benefits of green and blue-green systems can result in resistance to their implementation. Adapting designs to local climatic conditions and urban needs requires advanced technical expertise, which may not always be readily available (Wolch et al., 2014).

18.8 Policy and Planning Approaches

Urban Green Spaces (UGS) and Blue-Green Infrastructure (BGI) are vital for sustainable urban development. Policy and planning approaches are critical in integrating these systems into urban landscapes and aligning ecological, social, and economic goals.

18.8.1 Incorporating BGI into Urban Master Plans

Integrating Blue-Green Infrastructure (BGI) with urban planning frameworks ensures that sustainable practices are embedded in long-term development strategies, supporting their seamless implementation and effectiveness. This approach emphasizes aligning urban planning with ecological principles, prioritizing land-use policies that accommodate green and blue spaces alongside traditional infrastructure. Urban master plans must prioritize BGI as a foundational element, emphasizing its multifunctionality in addressing urban challenges such as flooding, air pollution, and biodiversity loss. Tools like Geographic Information Systems (GIS) and spatial mapping can help planners identify suitable areas for BGI projects, ensuring optimal connectivity and ecological functionality. Cities like Singapore have successfully incorporated green roofs and vertical gardens into urban development policies, showcasing the potential of embedding BGI in planning frameworks (Tan, 2006).

Comprehensive policies should also establish regulations for land-use zoning, green connectivity, and urban resilience to support long-term sustainability goals (Kabisch et al., 2015).

18.8.2 Leveraging Nature-Based Solutions for Sustainable Development Goals (SDGs)

BGI aligns closely with the United Nations' Sustainable Development Goals (SDGs), particularly SDG 11 (Sustainable Cities and Communities), SDG 13 (Climate Action), SDG 15 (Life on Land), and SDG 6 (Clean Water and Sanitation). Nature-based solutions promote ecosystem-based approaches, as BGI contributes to carbon sequestration, improved air and water quality, and biodiversity conservation. It also addresses urban challenges while enhancing ecosystem services. For instance, urban wetlands and bioswales help manage stormwater while supporting aquatic habitats, and urban forests enhance carbon sequestration while providing shade and recreation opportunities. Policy frameworks that align nature-based solutions (NbS) with the Sustainable Development Goals (SDGs) can unlock funding opportunities, foster innovation, and promote inclusive urban development (Cohen-Shacham et al., 2016).

18.8.3 Public–Private Partnerships and Community Engagement

The successful implementation of BGI systems often depends on the collaboration between governments, private entities, and local communities. Public–private partnerships (PPPs) can secure the financial and technical resources required for large-scale projects, such as constructing green roofs or rehabilitating urban wetlands. For example, Melbourne's "Green Your Laneway" initiative brought together government agencies, private stakeholders, and local communities to transform urban alleys into green corridors, showcasing the power of collective action (Bush et al., 2021). Community engagement is equally critical, fostering a sense of ownership and stewardship among residents. Initiatives like community gardens, citizen science programs, and participatory planning processes empower communities to contribute actively to BGI projects, ensuring their relevance and sustainability. Moreover, businesses that invest in urban greenery often benefit from enhanced brand value and improved corporate social responsibility (CSR) outcomes (Foster et al., 2011).

18.9 Future Directions and Innovations

Cities increasingly adopt integrated urban planning strategies that align green and blue-green infrastructure (BGI) with carbon neutrality goals. Urban forests, carbon offset initiatives, and water-sensitive urban designs are becoming central to city master plans. Policies now emphasize mixed-use developments that blend green spaces with residential and commercial areas, optimizing land use and advancing sustainability objectives.

Technological innovations are transforming the management of urban green and BGI systems. Innovative irrigation systems with sensors optimize water usage by monitoring soil moisture and weather conditions. IoT-enabled green infrastructure provides real-time data on air quality, water flow, and ecosystem health. Modular vertical gardens and green roofs enhance scalability, while algae-based bio-curtains and solar-integrated green walls maximize carbon sequestration and energy efficiency.

Advanced tools, such as LiDAR and satellite imagery, are being leveraged to monitor vegetation health, carbon storage, and heat mitigation. Social outcomes are assessed through community feedback and participatory approaches to ensure accessibility, equity, and public satisfaction. Establishing standardized metrics for ecosystem services and integrating them into policy frameworks enhances the effectiveness and accountability of green and blue-green infrastructure (BGI) systems, supporting resilient and sustainable urban futures.

18.10 Conclusion

The integration of urban green spaces, Blue-Green infrastructure (BGI), and nature-based solutions represents pivotal strategies in addressing the multifaceted challenges of urbanization, climate change, sustainability, and carbon neutrality. This approach can mitigate environmental hazards, enhance biodiversity, and foster healthier urban living. These approaches not only strengthen the resilience of urban environments but also improve the quality of life for residents by providing access to nature, protecting biodiversity, and reducing environmental hazards such as air pollution and noise. The successful implementation of these solutions, as demonstrated in case studies such as Singapore's vertical gardens, Copenhagen's flood-resilient parks, and Indian initiatives like the restoration of Pashan Lake, showcases their adaptability across diverse urban contexts. However, land-use conflicts, financing constraints, and equitable access must be addressed through informed policy-making, community engagement, and innovative approaches. Embracing these solutions and aligning them with global sustainability goals may help cities mitigate the impacts of climate change while enhancing the quality of urban life for present and future generations.

References

Baro, F., Haase, D., Gómez-Baggethun, E., & Frantzeskaki, N. (2015). Mismatches between ecosystem services supply and demand in urban areas: A quantitative assessment in five European cities. *Ecological Indicators, 55*, 146–158.

Berland, A., Shiflett, S. A., Shuster, W. D., Garmestani, A. S., Goddard, H. C., Herrmann, D. L., & Hopton, M. E. (2017). The role of trees in urban stormwater management. *Landscape and Urban Planning, 162*, 167–177.

Black, K. J., & Richards, M. (2020). Eco-gentrification and who benefits from urban green amenities: NYC's high Line. *Landscape and Urban Planning, 204*, Article 103900.

Bowler, D. E., Buyung-Ali, L., Knight, T. M., & Pullin, A. S. (2010). Urban greening to cool towns and cities: A systematic review of the empirical evidence. *Landscape and Urban Planning, 97*(3), 147–155.

Bisht, S., Singh, H., & Kumar, A. (2024). Ecosystem services and maintaining their flow in urban environments: Nurturing sustainable flow in urban environments. In *Urban forests, climate change and environmental pollution: Physio-biochemical and molecular perspectives to enhance urban resilience* (pp. 447–467). Cham: Springer Nature Switzerland.

Bush, J., Ashley, G., Foster, B., & Hall, G. (2021). Integrating green infrastructure into urban planning: Developing Melbourne's green factor tool. *Urban Planning, 6*(1), 20–31.

Chauhan, A., & Singh, H. (2024). Biological insights into the adaptive responses of medicinal plants to changing urban environments. In *Urban forests, climate change and environmental pollution: Physio-biochemical and molecular perspectives to enhance urban resilience* (pp. 73–92). Cham: Springer Nature Switzerland.

Cohen-Shacham, E., Walters, G., Janzen, C., & Maginnis, S. (2016). Nature-based solutions to address global societal challenges. IUCN: Gland. *Switzerland, 97*, 2016–2036.

Dai, L., Wörner, R., & van Rijswick, H. F. (2018). Rainproof cities in the Netherlands: Approaches in Dutch water governance to climate-adaptive urban planning. *International Journal of Water Resources Development, 34*(4), 652–674.

Davis, A. P., Hunt, W. F., Traver, R. G., & Clar, M. (2009). Bioretention technology: Overview of current practice and future needs. *Journal of Environmental Engineering, 135*(3), 109–117.

Depietri, Y., Renaud, F. G., & Kallis, G. (2012). Heat waves and floods in urban areas: A policy-oriented review of ecosystem services. *Sustainability Science, 7*, 95–107.

Dorst, H., Van der Jagt, A., Raven, R., & Runhaar, H. (2019). Urban greening through nature-based solutions–Key characteristics of an emerging concept. *Sustainable Cities and Society, 49*, Article 101620.

Fletcher, T. D., Andrieu, H., & Hamel, P. (2013). Understanding, management and modelling of urban hydrology and its consequences for receiving waters: A state of the art. *Advances in Water Resources, 51*, 261–279.

Foster, J., Lowe, A., & Winkelman, S. (2011). The value of green infrastructure for urban climate adaptation. *Center for Clean Air Policy, 750*(1), 1–52.

Grimm, N. B., Faeth, S. H., Golubiewski, N. E., Redman, C. L., Wu, J., Bai, X., & Briggs, J. M. (2008). Global change and the ecology of cities. *Science, 319*(5864), 756–760.

Gu, D., Andreev, K., & Dupre, M. E. (2021). Major trends in population growth around the world. *China CDC Weekly, 3*(28), 604.

Gupta, U., Shukla, S., Bhardwaj, L. K., Rath, P., Upadhyay, D., Sharma, B., … & Singh, H. (2024). Unraveling urban plant strategies for heavy metal uptake and detoxification. In *Urban forests, climate change and environmental pollution: Physio-biochemical and molecular perspectives to enhance urban resilience* (pp. 93–119). Cham: Springer Nature Switzerland.

Hansen, R., Frantzeskaki, N., McPhearson, T., Rall, E., Kabisch, N., Kaczorowska, A., & Pauleit, S. (2015). The uptake of the ecosystem services concept in planning discourses of European and American cities. *Ecosystem Services, 12*, 228–246.

Haq, S. M. A. (2011). Urban green spaces and an integrative approach to sustainable environment. *Journal of Environmental Protection, 2*(5), 601–608.

Hoek, G., Krishnan, R. M., Beelen, R., Peters, A., Ostro, B., Brunekreef, B., & Kaufman, J. D. (2013). Long-term air pollution exposure and cardio-respiratory mortality: A review. *Environmental Health, 12*, 1–16.

Joshi, R., Singh, H., Chhetri, R., Poudel, S. R., & Rijal, S. (2021). Carbon sequestration potential of community forests: A comparative analysis of soil organic carbon stock in community managed forests of Far-Western Nepal. *Eurasian Journal of Soil Science, 10*(2), 96–104. https://doi.org/10.18393/ejss.825066.

Kabisch, N., Qureshi, S., & Haase, D. (2015). Human–environment interactions in urban green spaces—A systematic review of contemporary issues and prospects for future research. *Environmental Impact Assessment Review, 50*, 25–34.

Kabisch, N., Stadler, J., Korn, H., & Bonn, A. (2017). Nature-Based solutions for societal goals under climate change in urban areas–Synthesis and Ways Forward. In *Nature-based solutions to climate change adaptation in urban areas: Linkages between science, policy and practice* (pp. 323–336).

Kassim, H., Peterson, J., Bauer, M. W., Connolly, S., Dehousse, R., Hooghe, L., & Thompson, A. (2013). *The European Commission of the twenty-first century*. OUP Oxford.

Kong, F., Yin, H., Nakagoshi, N., & Zong, Y. (2010). Urban green space network development for biodiversity conservation: Identification based on graph theory and gravity modeling. *Landscape and Urban Planning, 95*(1–2), 16–27.

Liu, W., Holst, J., & Yu, Z. (2014). Thresholds of landscape change: A new tool to manage green infrastructure and social–economic development. *Landscape Ecology, 29*, 729–743.

Mabrouk, M., Han, H., Mahran, M. G. N., Abdrabo, K. I., & Yousry, A. (2024). Revisiting urban resilience: A systematic review of multiple-scale urban form indicators in flood resilience assessment. *Sustainability, 16*(12), 5076.

Mahadevia, D., & Sharma, R. (2023). *Urban Planning and Its Discontents: Practice in Contemporary India*. Taylor & Francis.

Malik, A. (2024). Smart green housing: Vertical farming. In *Recent advancements in sustainable agricultural practices: Harnessing technology for water resources, irrigation and environmental management* (pp. 133–149). Singapore: Springer Nature Singapore.

Malik, A., Singh, H., Kumar, K., Kumari, G., & Sojitra, A. (2024). Understanding the impact of climate change and environments on urban forests: Insights from morphological, physiological, and molecular perspectives. In *Urban forests, climate change and environmental pollution: Physio-biochemical and molecular perspectives to enhance urban resilience* (pp. 161–183). Cham: Springer Nature Switzerland.

Mell, I. C., Henneberry, J., Hehl-Lange, S., & Keskin, B. (2013). Promoting urban greening: Valuing the development of green infrastructure investments in the urban core of Manchester. *UK. Urban Forestry & Urban Greening, 12*(3), 296–306.

Mitra, P., Haldar, A., & Banerjee, P. (2022). Public participation in restoration and sustainable use of wetland ecosystem services in India. In *Handbook of research on monitoring and evaluating the ecological health of wetlands* (pp. 280–303).

Mitsch, W. J., Bernal, B., Nahlik, A. M., Mander, Ü., Zhang, L., Anderson, C. J., & Brix, H. (2013). Wetlands, carbon, and climate change. *Landscape Ecology, 28*, 583–597.

Nowak, D. J., & Crane, D. E. (2002). Carbon storage and sequestration by urban trees in the USA. *Environmental Pollution, 116*(3), 381–389.

Nowak, D. J., Crane, D. E., & Stevens, J. C. (2006). Air pollution removal by urban trees and shrubs in the United States. *Urban Forestry & Urban Greening, 4*(3–4), 115–123.

Nowak, D. J., Hirabayashi, S., Bodine, A., & Greenfield, E. (2014). Tree and forest effects on air quality and human health in the United States. *Environmental Pollution, 193*, 119–129.

Pakhira, R., Biswas, S., & Singh, H. (2024). Urbanization and climate change: Impacts, adaptation and mitigation strategies. In *Urban forests, climate change and environmental pollution: Physio-biochemical and molecular perspectives to enhance urban resilience* (pp. 47–72). Cham: Springer Nature Switzerland.

Pearson, D., & Hodgkin, K. (2010). The role of community gardens in urban agriculture. In *Community Gardens Conference: Promoting sustainability, health and inclusion in the city* (pp. 99–105). COGS University of Canberra.

Rawat, A., & Singh, H. (2024). Role of urban forests in modulating urban climate. In *Urban forests, climate change and environmental pollution: Physio-biochemical and molecular perspectives to enhance urban resilience* (pp. 351–371). Cham: Springer Nature Switzerland.

Shah, A. M., Liu, G., Nawab, A., Li, H., Xu, D., Yeboah, F. K., & Zhang, L. (2024). Sustainability and resilience interface at typical urban green and blue infrastructures: Costs, benefits, and impacts assessment. *Frontiers in Sustainable Cities, 6*, 1453829.

Singh, H. (2024a). *Urban forests, climate change and environmental pollution: Physio-biochemical and molecular perspectives to enhance urban resilience*. Springer Nature. https://doi.org/10.1007/978-3-031-67837-0

Singh, H. (2024b). Climate change and forests: Biological perspectives. In *Forests and climate change: Biological perspectives on impact, adaptation, and mitigation strategies* (pp. 1–17). Singapore: Springer Nature Singapore.

Steffen, W., Richardson, K., Rockström, J., Cornell, S. E., Fetzer, I., Bennett, E. M., & Sörlin, S. (2015). Planetary boundaries: Guiding human development on a changing planet. *Science, 347*(6223), 1259855.

Tan, K. W. (2006). A greenway network for Singapore. *Landscape and Urban Planning, 76*(1–4), 45–66.

Toxopeus, H., & Polzin, F. (2021). Reviewing financing barriers and strategies for urban nature-based solutions. *Journal of Environmental Management, 289*, Article 112371.

Tzoulas, K., Korpela, K., Venn, S., Yli-Pelkonen, V., Kaźmierczak, A., Niemela, J., & James, P. (2007). Promoting ecosystem and human health in urban areas using Green Infrastructure: A literature review. *Landscape and Urban Planning, 81*(3), 167–178.

Vega, O. M. (2019). *Green infrastructure in the City of Vancouver: Performance monitoring of stormwater tree trenches and bioswales* (Doctoral dissertation, University of British Columbia).

Walsh, C. J., Booth, D. B., Burns, M. J., Fletcher, T. D., Hale, R. L., Hoang, L. N., & Wallace, A. (2016). Principles for urban stormwater management to protect stream ecosystems. *Freshwater Science, 35*(1), 398–411.

World Health Organization. (2023). *Regional Office for Europe (2017) Urban green space interventions and health: a review of impacts and effectiveness.*

Winter, A. K. (2019). The green city citizen: Exploring the ambiguities of sustainable lifestyles in Copenhagen. *Environmental Policy and Governance, 29*(1), 14–22.

Wolch, J. R., Byrne, J., & Newell, J. P. (2014). Urban green space, public health, and environmental justice: The challenge of making cities 'just green enough.' *Landscape and Urban Planning, 125*, 234–244.

Wong, T. W., Good, J. E. G., & Denne, M. P. (1988). Tree root damage to pavements and kerbs in the city of Manchester. *Arboricultural Journal, 12*(1), 17–34.

Wu, J. (2014). Urban ecology and sustainability: The state-of-the-science and future directions. *Landscape and Urban Planning, 125*, 209–221.

Zhang, F., & Qian, H. (2024). A comprehensive review of the environmental benefits of urban green spaces. *Environmental Research, 118837.*

Chapter 19
Plant Growth-Promoting Rhizobacteria (PGPR): A Biological Approach to Boost the Carbon Sequestration Potential of Forestry Species

Aradhna Kumari, Munmun Dash, Sanjib Kumar Sahoo, Santosh Kumar Singh, and P. K. Mishra

Abstract As global carbon emissions continue to rise, effective strategies for enhancing forest productivity and accelerating carbon sequestration are crucial. Plant growth-promoting rhizobacteria (PGPR) are one of the biological approaches that stimulate plant growth, improve nutrient uptake, enhance disease resistance, and promote soil health, offering an eco-friendly and sustainable measure to boost forest establishment and development. Studies have demonstrated that PGPR may increase root biomass, facilitate efficient nutrient cycling, and promote plant stress tolerance, thereby contributing to enhanced carbon fixation in forest ecosystems. Additionally, the symbiotic relationships between PGPRs and tree species improve soil carbon storage by influencing microbial communities and soil organic matter dynamics. This chapter examines the role of PGPRs in enhancing the growth and carbon sequestration capacity of forestry species, highlighting the need for further research into their application in forest management practices. Moreover, this chapter inferred that utilising PGPRs could provide an innovative and cost-effective solution for enhancing the carbon sequestration potential of forests, contributing to climate change mitigation efforts.

A. Kumari
College of Agriculture, Jawaharlal Nehru Krishi Vishwa Vidyalaya, Ganj Basoda, Vidisha, MP, India

M. Dash · S. K. Sahoo
Odisha University of Agriculture and Technology, Bhubaneswar, Odisha, India

S. K. Singh (✉)
Department of Soil Science, Dr Rajendra Prasad Central Agricultural University, Pusa, Samastipur, Bihar, India
e-mail: santoshraupusa@gmail.com

P. K. Mishra
College of Agriculture, Jawaharlal Nehru Krishi Vishwavidyalaya, Powarkheda, Madhya Pradesh, India

H. Singh (ed.), *Forestry for a Carbon-Neutral and Sustainable Future*, Sustainability Solutions, https://doi.org/10.1007/978-3-032-06427-1_19

Keywords Carbon fixation · Disease resistance · Nutrient cycling · Stress tolerance · Siderophore

19.1 Introduction

The impacts of global climate change are becoming increasingly severe, necessitating urgent measures to adapt and mitigate its adverse effects across various sectors (Stocker et al., 2015; Kumari et al., 2022, 2025b; Patni et al., 2022; Verma et al., 2024a; Verma et al., 2024b; Arumugam et al., 2025). Simultaneously, the global political system has struggled to establish binding international agreements to reduce Greenhouse Gas (GHG) emissions, a challenge evident after the UN meetings in Copenhagen in 2009 and Warsaw in 2013. Human emissions are currently at levels previously predicted as a worst-case scenario in the IPCC's 3rd report released in 2001 (Lucas, 2021). This report had warned that emissions would reach these levels by now if no action were taken, and that prediction has come true. The headline of the IPCC's March 30, 2014, report emphasised that "Warming of the climate system is unequivocal, and since the 1950s, many of the observed changes are unprecedented over decades to millennia".

The Earth's atmosphere and oceans have become hotter; there has been a reduction in snow and ice cover, sea levels have risen, and the levels of greenhouse gases have increased, indicating the need for urgent action to address this challenge. Hence, understanding the basic facts and assessing realistic mitigation options for both the short and long term is crucial. The global average concentration of CO_2, the primary greenhouse gas caused by human activities, is now approaching 400 parts per million (ppm). The rate of increase in atmospheric CO_2 over the past century has been unprecedented in the last 22,000 years (Bush, 2019). If GHG emissions continue at current rates, CO_2 levels could reach 500 ppm by 2050 and 800 ppm by 2100 (van Vuuren et al., 2015). The persistence of current patterns is not limited by the availability of fossil fuels but rather by international accords or a major economic downturn. The continued release of greenhouse gases will lead to further global temperature increases and alterations across all components of the climate system. It is becoming more evident that addressing climate change will necessitate significant and long-term cuts in greenhouse gas emissions (Kumari et al., 2024).

Recent research in India emphasises the role of reforestation on agricultural lands in reducing CO_2 emissions (Kumari et al., 2018; Jagadesh et al., 2024a). Specifically, using marginal agrarian land for forest restoration holds significant economic and environmental potential (Jagadesh et al., 2024a, 2024b; Chazdon et al., 2020). The results suggest that planting fast-growing trees on less productive agricultural lands is crucial for carbon absorption (Kumari et al., 2017; Pérez-Cruzado et al., 2012). However, another study indicates the need for specific public incentives to encourage large-scale plantations (Enters et al., 2003). Economic analyses show that the costs associated with reforesting agricultural lands are a significant barrier for landowners, although these costs are lower than those of other methods for reducing CO_2 emissions (Bustamante et al., 2014).

Additionally, another study highlighted the need to improve methodologies for assessing the climate impact of forest biomass used for energy (Brandão et al., 2019). It noted that previous research had failed to consider the use of forest carbon stocks when evaluating greenhouse gas (GHG) emissions in the absence of energy demand. Without such demand, forests can continue growing or be used for timber production. To accurately assess carbon sequestration, it is essential to evaluate the carbon stored in forest production, considering non-harvesting scenarios and the potential use of forest biomass for bioenergy (Jagadesh et al., 2024a; Raison, 2024). Forests play a key role globally, offering essential benefits such as fiber, pharmaceuticals, recreation, and wildlife habitats (Jagadesh et al., 2024a). However, global population growth poses a threat to forests, as harvesting rates often exceed regeneration rates (Alam et al., 2024). Low seedling survival rates and inadequate early growth at reforested sites impede effective forest regeneration (Singh et al., 2013).

An economical and eco-friendly method to enhance the productivity of newly planted forest stands is to inoculate seedlings with root-associated plant growth-promoting microorganisms during the nursery phase. Traditionally, in forestry, this has been limited to mycorrhizal fungi inoculation to improve seedling performance in areas lacking suitable fungal symbionts (Hagh-Doust et al., 2022). However, plant root systems, including mycorrhizae, are heavily populated by symbiotic soil microorganisms, with bacteria being especially prevalent (Martin et al., 2017). Certain root-associated bacteria have demonstrated the ability to enhance tree seedling growth and/or mycorrhizal colonization in both short-term controlled and field studies (Compant et al., 2010), suggesting that such bacteria may play a significant role among the various soil biological factors influencing seedling growth. The term "plant growth-promoting rhizobacteria" (PGPR) refers to naturally occurring soil bacteria that promote plant growth when applied to seeds or roots (Saeed et al., 2021). Inoculating forest seedlings with bacteria before outplanting is an inexpensive and straightforward process, making PGPR a promising and cost-effective solution for enhancing plantation establishment (Kumari et al., 2025a). Additionally, Santoyo et al. (2021) have discovered that some PGPR strains can colonize internal plant tissues. The more intimate relationship between these endophytic bacteria and plant cells could potentially enhance the effectiveness of PGPR, as opposed to rhizosphere associations, where microorganisms are physically distinct from the root system.

19.2 PGPR: Types and Mechanism of Action

A comprehensive understanding of the mechanisms by which PGPR operates is essential for optimizing rhizosphere processes that significantly impact plant productivity. PGPR action mechanisms are typically categorized into direct and indirect

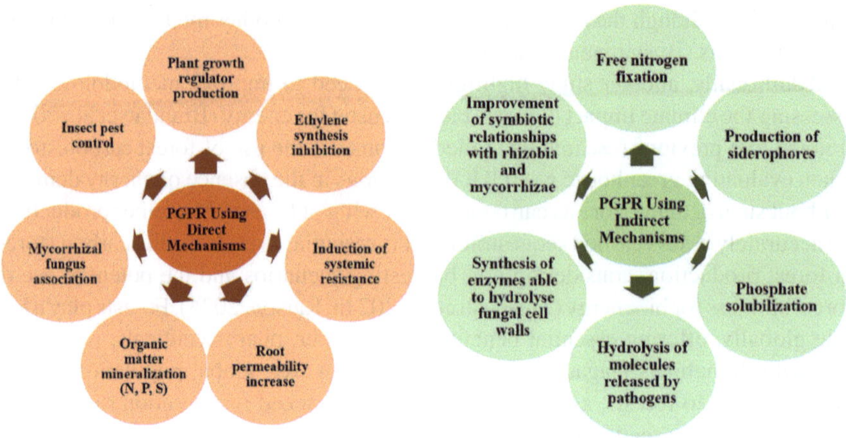

Fig. 19.1 PGPR using direct and indirect mechanisms

mechanisms. While the distinction between these two categories is not always clear-cut, indirect mechanisms generally occur outside the plant. In contrast, direct mechanisms occur within the plant and directly influence its metabolism. Direct mechanisms involve the plant's defence metabolic processes, which respond to signals from the bacteria affecting the plant's system. Direct mechanisms include those that alter the balance of plant growth regulators. This can occur when microorganisms release growth regulators that are integrated into the plant or when the microorganisms absorb plant-released hormones. Another key direct mechanism is the induction of systemic resistance to plant pathogens and protection against high salinity stress, which enhances the plant's ability to adapt (Jangra et al., 2024). In contrast, indirect mechanisms involve processes outside the plant, such as improving nutrient availability, inhibiting harmful microorganisms, and facilitating free nitrogen fixation in the rhizosphere, thereby enhancing nitrogen supply to the plant (Fig. 19.1).

19.3 Role of PGPR in Carbon Sequestration

Soil microbial activities are influenced by elevated temperatures and carbon dioxide levels, either directly or indirectly. At higher temperatures, microbial activity tends to increase, creating a positive feedback loop that exacerbates climate change, whereas reduced moisture conditions can have the opposite effect (Sofi et al., 2016). The aggregation of ecosystem carbon is maintained by balancing plant productivity and heterotrophic respiration from the decomposition of soil organic matter (SOM) (Schlesinger and Andrews, 2000). Numerous studies have shown that elevated CO_2 (eCO_2) enhances plant growth and photosynthetic carbon input into soils. These increased carbon inputs stimulate microbial growth, thereby expanding soil microbial communities under elevated CO_2. This may accelerate SOM decomposition and result in net soil carbon losses. Elevated CO_2 also induces rhizosphere priming effects, which amplify SOM decomposition linked to microbial activity (Niehus et al. 2017). PGPR's enzymatic activity also contributes to SOM breakdown in soil. Soil moisture is a critical factor in microbial-driven climate change processes. In drought or water-stressed soils, microbial activity increases due to the lower water levels and the introduction of oxygen into previously anaerobic soils (Zibilske and Bradford 2007). Peatlands and wetlands are significant carbon reservoirs (Ward et al., 2007), and the accelerated degradation of stable and resistant organic matter under dry conditions could have substantial implications for the global carbon cycle.

19.4 Mechanisms of PGPR-Driven Carbon Sequestration

19.4.1 Improved Plant Growth

PGPRs enhance plant growth by producing plant hormones (like auxins, cytokinins, and gibberellins) that enhance root development and increase plant biomass (Arumugam et al., 2025; Dash et al., 2025; Kumari et al., 2025a). More biomass means greater carbon fixation in plant tissues. There is a diverse range of PGPR-induced changes in plants, and the promotion of growth is most likely a result of a complex combination of multiple pathways that affect both plant development and nutrition (Bharti et al., 2016; Kumari et al., 2025a). PGPR positively affects plant growth directly and indirectly (Weyens et al., 2009; Asad et al., 2019). Plant growth under the influence of rhizobacteria is a multigeneous process specific to the bacteria and plants that are individually participating. This "additive hypothesis" is a complex phenomenon involving a cumulative effect of changes in the expression of various genes, ultimately influencing the global plant multifactor metabolic system (Bharti et al., 2016). Direct mechanisms of plant growth stimulation by bacteria rely on facilitating the uptake of nutrients and synthesizing or regulating the hormonal status of plants (Fig. 19.1) (Kong and Glick, 2017; Backer et al., 2018). Indirect mechanisms of PGPR influence plant growth and comprise a range of mechanisms that prevent or suppress plant diseases (Goswami et al., 2016; Asad et al., 2019).

19.4.2 Enhanced Soil Organic Carbon Storage

PGPRs stimulate root growth and microbial activity in the rhizosphere, which can lead to increased root exudation (Kumari et al., 2023, 2025a). These exudates, including sugars and organic acids, contribute to the soil organic carbon pool, a significant component of the soil carbon sequestration process (Fig. 19.2).

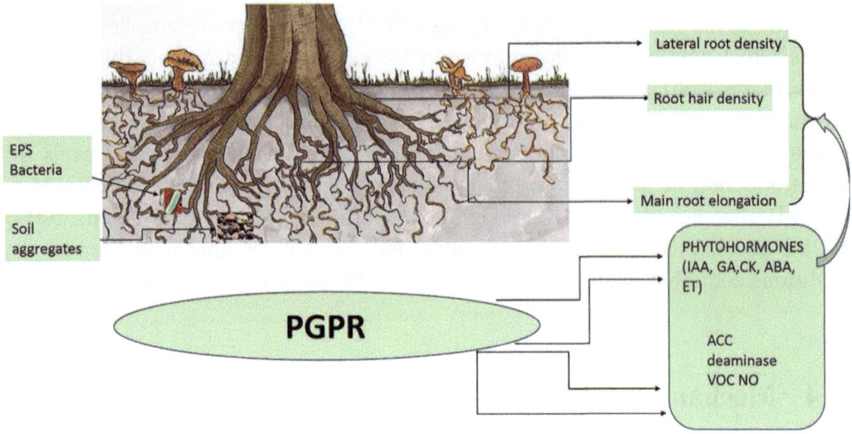

Fig. 19.2 PGPR interaction with forest trees' roots

19.4.3 Increased Soil Microbial Activity

PGPRs can enhance microbial activity in the soil, particularly those involved in the breakdown of organic matter and the formation of stable soil organic carbon (Arumugam et al., 2025; Dash et al., 2025). This process helps lock carbon in inorganic forms in the soil, making it less likely to return to the atmosphere. The mutualistic interaction between existing microorganisms and plants can be further enhanced by introducing nanoparticles (NPs) (Ameen et al., 2021). Phosphate-solubilizing bacteria, such as *Pseudomonas sp.*, can enhance phosphorus availability in the soil by solubilizing it from nanoparticles (NPs), which in turn promotes root colonization and overall plant growth (De Souza-Torres et al., 2021). The presence of NPs alongside PGPR may also increase the abundance of certain microbial groups in the soil, contributing to overall improvements in soil health (Gautam et al., 2025). This increase in beneficial bacteria has been attributed to the enhanced organic matter content in the soil and its decomposition, driven by the interaction between NPs and the microbial inoculant (Ahmad et al., 2022; De Souza-Torres et al., 2021; Semwal et al., 2025).

19.4.4 Influence on Root Development and Soil Carbon Storage

PGPRs can significantly enhance root development and soil carbon sequestration. These beneficial microorganisms interact with plant roots to enhance nutrient uptake, stimulate root growth, and improve soil structure, promoting carbon storage in soil systems (Jagadesh et al., 2024c; Kumari et al., 2023). PGPRs influence root development through several mechanisms, including the production of phytohormones, nitrogen fixation, and the solubilization of essential nutrients (Dash et al., 2025). Auxins, gibberellins, and cytokinins produced by PGPRs stimulate root elongation, branching, and overall root biomass. Enhanced root systems enable plants to access deeper soil layers, thereby increasing nutrient absorption and enhancing resilience to abiotic stress (Arumugam et al., 2025; Kumari et al., 2021).

Additionally, PGPRs facilitate the solubilization of phosphorus and other nutrients that are otherwise unavailable to plants (Arumugam et al., 2025; Kumari et al., 2025a). By making these nutrients accessible, PGPRs indirectly support the formation of dense, well-developed root systems, further improving plant health and productivity. The enhanced root systems promoted by PGPRs contribute significantly to soil carbon sequestration. As roots grow, they release exudates, such as sugars, organic acids, and amino acids, into the rhizosphere. These exudates serve as a primary carbon source for soil microbial communities, stimulating their activity and promoting the accumulation of biomass. Active microbial communities, in turn, contribute to the formation of soil organic matter, which acts as a stable carbon reservoir.

19.4.5 Nutrient Cycling and Soil Health Enhancement

Several studies have demonstrated that certain beneficial soil bacteria, known as phosphate-solubilising bacteria (PSB), can convert insoluble forms of phosphorus, such as hydroxyapatite, di-calcium phosphate, and rock phosphate, into soluble forms that plants can utilise (Kumari et al., 2025a; Turan et al., 2012). PSBs use various mechanisms, including the release of enzymes and acids, to solubilise phosphorus. Inoculating plants with PSB boosts growth and significantly enhances crop yield. Key bacterial genera with phosphate-solubilising abilities include Serratia, Alcaligenes, Rhizobium, Acinetobacter, Pseudomonas, Bacillus, and others. Symbiotic nitrogen fixation occurs when plants, particularly legumes, form a mutually beneficial relationship with BNF bacteria like Rhizobium, which enables nitrogen fixation in nodules on plant roots (Ojuederie et al., 2019).

Furthermore, though somewhat, non-symbiotic nitrogen-fixing bacteria, such as Cyanobacteria and Azotobacter, contribute to nitrogen fixation. Approximately 80% of the nitrogen available to plants comes from BNF, while the remaining 20% is from non-symbiotic sources. Free-living nitrogen-fixing bacteria, such as *Acetobacter diazotrophicus* and *Azoarcus*, play a significant role in agriculture but do not form symbiotic relationships with plant roots. PGPRs enhance carbon and nitrogen mineralisation (Sahoo et al., 2023). PGPR produces siderophores, compounds with a strong affinity for iron, which enable it to acquire iron. These siderophores help bacteria obtain Fe^{3+} (ferric ions) from the soil, which is the form plants can absorb. PGPR enhances plant growth and improves iron availability, both of which are essential for proper plant development.

19.5 Application of PGPR in Forestry

The application of PGPR in forestry is depicted in Table 19.1.

19.6 Benefits of PGPR in Forestry Species

PGPRs play a crucial role in helping plant hosts adapt to various adverse climatic conditions, including cold temperatures, drought, temperature fluctuations, and pathogen infections. In addition to enhancing the plant's resistance to various biotic and abiotic stresses, these symbiotic microbes play a crucial role in regulating the plant's defence mechanisms (Kumar et al., 2022). Inoculation with Binucleate Rhizoctonia (BnR) significantly enhanced the early growth of Scots pine seedlings in nitrogen-deficient nursery soil. It showed increased root length and a reduction in root width compared to the non-inoculated controls [Gronberg et al., 2006]. The C4 perennial succulent xerohalophyte *Haloxylon ammodendron*, known for its exceptional resistance to both salt and drought stress, was studied by He et al. (2018), who identified a novel salt-tolerant bacterial strain, *Pseudomonas sp. M30-35*, in the rhizosphere of this plant. The strain's tolerance to abiotic stress and its plant growth-promoting (PGP) capabilities are associated with 34 genes in *Pseudomonas sp. M30-35*. Additionally, inoculation with the PGPR strain *Bacillus subtilis* induced significant alterations in gene expression in *Acacia gerrardii*, including the activation of transcription factors that regulate stress-related genes, highlighting its potential for mitigating the effects of salinity and drought stress (Lara-Chavez et al., 2015). PGPRs significantly enhance a plant's ability to survive and thrive in adverse conditions by promoting resilience.

19.7 Synergies with Other Sustainable Practices

The role of PGPR, particularly *Rhizobium*, includes the production of phytohormones (Swarnalakshmi et al., 2020), solubilization of inorganic phosphorus (Turan et al., 2012), release of siderophores (Sarwar et al., 2020), enhancing soil enzymatic activities, leading to improved nutrient cycling (Sahoo et al., 2022a), and the suppression of plant pathogens (Wang et al., 2021). PGPR directly enhances plant growth by improving nutrient acquisition, particularly through the solubilization of inorganic phosphates and the production of iron-chelating siderophores, thereby boosting phosphorus and iron uptake. Several bacterial species from the genera *Bacillus*, *Rhizobium*, and *Pseudomonas* are among the most effective phosphate-solubilizing microbes (Rawat et al., 2021). Additionally, non-symbiotic nitrogen fixers, such as Azotobacter, have also been reported to exhibit phosphate solubilization (Ahmed et al., 2006). The phosphate-solubilizing activity of *Rhizobium* (e.g., *Rhizobium/*

Table 19.1 Role of PGPRs in forestry species

PGPRs	Role	Results	References
Proteobacteria, Actinobacteria and Firmicutes	IAA and Siderophore production	Increased iron in the plants	Flores-Nunez et al., (2018)
Pseudomonas and Bacillus sp.	Endophytic colonization	Increase in dry weight	Chanway et al., (2000)
Pseudomonas putida DSM291 (P), Bacillus subtilis strain FzB24 (B), and Enterobacter cloacae (E)	seed germination traits and early growth	Improved seed germination and early growth	Rostamikia et al., (2024)
Paenibacillus sp. RRB2, Azospirillumbrasilense RRAK5, Bacillus subtilis subsp. subtilis RRD8, Burholdriakururiensis RRAK1, and Pseudomonas stutzeri RRAN2	plant growth-promoting (PGP) activity	Biomass production	Radhapriya et al, (2018)
Gram-positive cocci, Gram-negative bacilli, Gram-positive non-endospore-forming bacilli, and Gram-positive endospore-forming bacilli	IAA production, Siderophore production, Phosphorus solubilization	Increased siderophores production and P solubilization	Garcia-Villaraco et al., (2024)
Pseudomonas protegens DA1.2 and Pseudomonas sp. 4CH	tree seedling growth	Boost chlorophyll content, N balance index, and carbon storage	Nazarov et al., (2023)
R. qingshengii CPT4 and R. erythropolis CPT9	Phytohormones production	Increase shoot dry mass, plant growth	Heredia-Acuña et al., (2019)
P. gessardii	phosphate solubilization	phosphate solubilization	Heredia-Acuña et al., (2019)
Bacillus subtillis	Carbon storage	Increase net photosynthetic assimilation	Jang et al., (2017)

(continued)

Table 19.1 (continued)

PGPRs	Role	Results	References
Pseudomonas fluorescens	Carbon storage	Increased chlorophyll content, plant height, leaf area, stem diameter, dry matter production, and net photosynthetic rate	Heydari et al., (2023)
Pseudomonas aeruginosa RRALC3	PGP activities	IAA production, Siderophore production, N mineralization, and increased carbon content	Radhapriya et al., (2015)

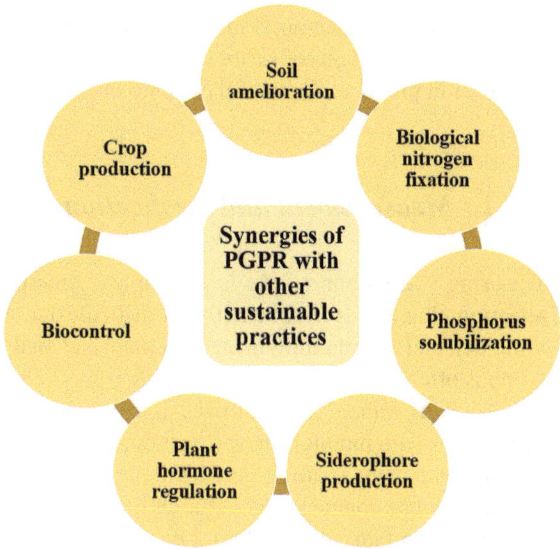

Fig. 19.3 Synergies of PGPRs with other sustainable practices

Bradyrhizobium) is linked to the production of 2-ketogluconic acid, which aids in phosphate solubilization. Various PGPRs have been shown to interact with both C3 and C4 plants, significantly boosting their yield (Ojuederie et al., 2019). PGPR biofertilizers can potentially increase crop yield and supplement or even reduce the need for urea-N (Athul et al., 2022; Sahoo et al., 2022b). Notably, *Azospirillum*, *Burkholderia*, and *Herbaspirillum* are commonly found in the rice rhizosphere, while strains of *Bacillus subtilis* are widely recognised for their disease-suppressing and antibiotic-producing properties (Kokalis-Burelle et al. 2006) (Fig. 19.3).

19.8 Challenges and Limitations in Adoption of PGPR

Forest soils play a crucial role in the global carbon cycle, as forests cover approximately 4.1 billion hectares worldwide. These ecosystems are estimated to store approximately 1240 Pg of carbon, with around 536 Pg in vegetation and up to 704 Pg in soil (O'Dell, 2022). Forests are responsible for storing more than 70% of global soil organic carbon (SOC), and forest soils are thought to contain about 43% of the ecosystem's carbon within the top meter of soil (Ahmed, 2018). However, this substantial carbon stock is vulnerable to depletion due to factors such as reduced biomass returning to the soil, alterations in soil moisture and temperature, and variations in the decomposability of soil organic matter, which depend on the carbon-to-nitrogen ratio and lignin content (Cotrufo et al. 2022). Human activities, such as converting forests into agricultural land, can lead to a 20–25% reduction in soil organic carbon stocks. Deforestation releases approximately 1.6–1.7 Pg of carbon annually, accounting for roughly 20% of global human-caused emissions (Ciais et al.,

2014). While significant opportunities exist to enhance carbon storage and sequestration potential in the soils of various ecosystems, several challenges make this difficult to achieve in practice. Some of these challenges include:

19.8.1 Measurement and Verification

Measuring soil carbon stocks is a complex, time-consuming, and costly process. Detecting changes as small as 10% is difficult due to sampling errors, small-scale variability, and uncertainties in measurement and analysis. The annual increase in soil carbon is often minimal, typically ranging from 0.25 to 1.0 tons per hectare. Furthermore, tracking minor gains or losses in soil carbon at different scales is challenging due to issues with monitoring, verification, sampling, and depth (Kumar et al., 2020). Even if small changes are detected, linking them to specific land management practices or land use changes is not straightforward. Additionally, the soil's capacity to sequester and retain carbon is finite, eventually reaching a steady state after a specific period.

19.8.2 Carbon Pools

Sequestered carbon exists in the soil in various pools, each with distinct residence times within the ecosystem. These include.

- **Passive, recalcitrant, or refractory pool**: This carbon has a very long residence time, ranging from decades to thousands of years.
- **Active, labile, or fast pool**: Carbon in this pool is more readily decomposed, with a residence time from one day to a year.
- **Slow, stable, or humus pool**: Carbon in this pool has a longer turnover time due to slower decomposition, with a typical residence time ranging from one to ten years.

19.8.3 Permanence

A significant challenge for soil carbon sequestration is the non-permanence of sequestered carbon, which can be released back into the atmosphere through decomposition or mineralization. This means that sequestered carbon is considered a short-term solution for removing carbon from the atmosphere. The rate at which carbon is lost depends on various factors, including climate, land use, and management practices.

19.8.4 Separation

It is difficult to distinguish the carbon sequestered in soil due to management interventions from that which occurred naturally. The principle of separation requires distinguishing between carbon sequestered as a result of human actions, such as land management, and that which would have naturally occurred. Methods are needed to differentiate carbon captured through human management from naturally sequestered carbon (Joshi et al., 2021).

There are established practices and strategies that effectively increase soil carbon stocks across various terrestrial ecosystems. Most of these approaches enhance carbon storage in biomass through photosynthesis, promoting the build-up of below-ground and soil carbon by increasing organic matter deposition. Dhaliwal et al. (2019) suggested that soil organic carbon levels can be improved by increasing the input of organic matter, modifying the decomposability of this matter, incorporating organic material into deeper soil layers, and improving the physical protection of soil aggregates or the formation of organo-mineral complexes.

19.9 Future Perspectives and Research Directions

Future research should focus on identifying specific PGPR strains or combinations that can address ongoing agricultural challenges, such as the reliance on harmful pesticides and fertilisers and farming in soils with low fertility. Additionally, PGPR applications may contribute to the development of functional foods that offer health benefits beyond essential nutrition.

19.10 Conclusion

Plant growth-promoting rhizobacteria (PGPRs) are one of the most effective and environmentally friendly methods for enhancing agricultural productivity. In the dynamic agricultural context, producing adequate plant biomass is crucial. A wide variety of microorganisms present in the soil surrounding plant roots, as well as at the root interface, have the potential to boost plant growth and increase biomass production. PGPRs play a crucial role in rhizosphere engineering due to their ability to promote plant growth and development. Over the past two to three decades, numerous PGPR strains have been identified, studied, and applied to support optimal growth in various plant species under both normal and stress conditions. Additionally, PGPRs can neutralise various contaminants and manage different plant diseases. It is expected that PGPRs will eventually replace synthetic growth regulators, herbicides, and chemical fertilisers, which can harm sustainable agriculture. Furthermore, research into the mechanisms of PGPR-mediated phyto-stimulation will help develop more effective rhizobacterial strains that can perform well in diverse agroecological environments.

References

Ahmad, F., Ahmad, I., Aqil, F., Ahmed Wani, A., & Sousche, Y. S. (2006). Plant growth promoting potential of free-living diazotrophs and other rhizobacteria isolated from Northern Indian soil. *Biotechnology Journal: Healthcare Nutrition Technology, 1*(10), 1112–1123.

Ahmad, H. M., Fiaz, S., Hafeez, S., Zahra, S., Shah, A. N., Gul, B., et al. (2022). Plant growth-promoting rhizobacteria eliminate the effect of drought stress in plants: A review. *Frontiers in Plant Science, 13.* https://doi.org/10.3389/fpls.2022.875774.

Ahmed, I. U. (2018). Forest soil C: stock and stability under global change. *New Perspectives in Forest Science, 37.*

Alam, N., Ullah, Z., Ahmad, B., Ali, A., & Syed, K. (2024). Population growth poses a significant threat to forest ecosystems: A case study from the Hindukush-Himalayas of Pakistan. *PLoS ONE, 19*(11), Article e0302192.

Ameen, F., Alsamhary, K., Alabdullatif, J. A., & ALNadhari, S. (2021). A review on metal-based nanoparticles and their toxicity to beneficial soil bacteria and fungi. *Ecotoxicology and Environmental Safety, 213*, 112027.

Arumugam, T., Jagadesh, M., Hegde, V., Rane, J., Srinivasarao, C., Krishnaveni, A., Singh, S.K., Kumari, A., Dash, M., Garg, V.K. (2025). Improvement of sugarcane crop quality by using some PGPRs under biotic and abiotic stresses. In: Verma et al. (Eds.), *Sugarcane Cultivation and Management*. Canada: Apple Academic Press/CRC Press (Taylor and Francis group). ISBN9781003504122.

Athul, P. P., Patra, R. K., Sethi, D., Panda, N., Mukhi, S. K., Padhan, K., & Pattanayak, S. K. (2022). Efficient native strains of rhizobia improved nodulation and productivity of French bean (*Phaseolus vulgaris* L.) under rainfed condition. *Frontiers in Plant Science, 13*, 1048696.

Backer, R., Rokem, J. S., Ilangumaran, G., Lamont, J., Praslickova, D., Ricci, E.,... & Smith, D. L. (2018). Plant growth-promoting rhizobacteria: context, mechanisms of action, and roadmap to commercialization of biostimulants for sustainable agriculture. *Frontiers in Plant Science, 9*, 1473.

Bharti, N., Pandey, S. S., Barnawal, D., Patel, V. K., & Kalra, A. (2016). Plant growth promoting rhizobacteria*Dietzia natronolimnaea modulates the expression of stress responsive genes providing protection of wheat from salinity stress. Scientific Reports, 6*(1), 34768.

Brandão, M., Kirschbaum, M. U., Cowie, A. L., & Hjuler, S. V. (2019). Quantifying the climate change effects of bioenergy systems: Comparison of 15 impact assessment methods. *GCB Bioenergy, 11*(5), 727–743.

Bush, M. J. (2019). *Climate change and renewable energy: How to end the climate crisis.* Springer Nature.

Bustamante, M., Robledo-Abad, C., Harper, R., Mbow, C., Ravindranat, N. H., Sperling, F., & Smith, P. (2014). Co-benefits, trade-offs, barriers and policies for greenhouse gas mitigation in the agriculture, forestry and other land use (AFOLU) sector. *Global Change Biology, 20*(10), 3270–3290.

Chanway, C. P., Shishido, M., Nairn, J., Jungwirth, S., Markham, J., Xiao, G., & Holl, F. B. (2000). Endophytic colonization and field responses of hybrid spruce seedlings after inoculation with plant growth-promoting rhizobacteria. *Forest Ecology and Management, 133*(1–2), 81–88.

Chazdon, R. L., Lindenmayer, D., Guariguata, M. R., Crouzeilles, R., Benayas, J. M. R., & Chavero, E. L. (2020). Fostering natural forest regeneration on former agricultural land through economic and policy interventions. *Environmental Research Letters, 15*(4), Article 043002.

Ciais, P., Sabine, C., Bala, G., Bopp, L., Brovkin, V., & House, J. I. (2014). Carbon and other biogeochemical cycles. In *Climate Change 2013: The Physical Science Basis. Contribution of Working Group I to the Fifth Assessment Report of the Intergovernmental Panel on Climate Change Change* (pp. 465–570). Cambridge University Press.

Compant, S., Van Der Heijden, M. G., & Sessitsch, A. (2010). Climate change effects on beneficial plant–microorganism interactions. *FEMS Microbiology Ecology, 73*(2), 197–214.

Cotrufo, M. F., & Lavallee, J. M. (2022). Soil organic matter formation, persistence, and functioning: A synthesis of current understanding to inform its conservation and regeneration. *Advances in Agronomy, 172*, 1–66.

Dash, M., Tripathy, S., Jagadesh, M., Boopathi, N., Kumari, A., Singh, S. K., & Jaiswal, A. (2025). Role of rhizomicrobiome in in-situ and ex-situ conservation of plant community. In *Rhizomicrobiome in Sustainable Agriculture and Environment* (pp. 229–244). Academic Press.

De Souza-Torres, A., Govea-Alcaide, E., Gomez-Padilla, E., Masunaga, S. H., Effenberger, F. B., Rossi, L. M., & Jardim, R. F. (2021). Fe3O4 nanoparticles and Rhizobium inoculation enhance nodulation, nitrogen fixation and growth of common bean plants grown in soil. *Rhizosphere, 17*, Article 100275.

Dhaliwal, S. S., Naresh, R. K., Mandal, A., Singh, R., & Dhaliwal, M. K. (2019). Dynamics and transformations of micronutrients in agricultural soils as influenced by organic matter build-up: A review. *Environmental and Sustainability Indicators, 1*, Article 100007.

Enters, T., Durst, P. B., & Brown, C. (2003). What does it take to promote forest plantation development? Incentives for tree-growing in countries of the Pacific rim. *Unasylva, 54*(1), 11–18.

Flores-Nunez, V. M., Amora-Lazcano, E., Rodríguez-Dorantes, A., Cruz-Maya, J. A., & Jan-Roblero, J. (2018). Comparison of plant growth-promoting rhizobacteria in a pine forest soil and an agricultural soil. *Soil Research, 56*(4), 346–355.

Garcia-Villaraco, A., Ramos Solano, B., Gutierrez-Mañero, F. J., & Lucas, J. A. (2024). Deciphering the structural and functional diversity of rhizobacteria from stone pine inoculated with plant growth promoting rhizobacteria (PGPR) before and after transplanted into degraded agricultural soil. *Soil Systems, 8*(2), 39.

Gautam, K., Singh, H., & Sinha, A. K. (2025). Nanotechnology in plant nanobionics: Mechanisms, applications, and future perspectives. *Advanced Biology, 9*(4), 2400589.

Goswami, D., Thakker, J. N., & Dhandhukia, P. C. (2016). Portraying mechanics of plant growth promoting rhizobacteria (PGPR): a review. *Cogent Food & Agriculture, 2*(1), 1127500.

Gronberg, H., Kaparakis, G., & Sen, R. (2006). Binucleate Rhizoctonia (Ceratorhiza spp.) as non-mycorrhizal endophytes alter *Pinus sylvestris* L. seedling root architecture and affect growth of rooted cuttings. *Scandinavian Journal of Forest Research, 21*(6), 450–457.

Hagh-Doust, N., Färkkilä, S. M., Moghaddam, M. S. H., & Tedersoo, L. (2022). Symbiotic fungi as biotechnological tools: Methodological challenges and relative benefits in agriculture and forestry. *Fungal Biology Reviews, 42*, 34–55.

He, A. L., Niu, S. Q., Zhao, Q., Li, Y. S., Gou, J. Y., Gao, H. J., & Zhang, J. L. (2018). Induced salt tolerance of perennial ryegrass by a novel bacterium strain from the rhizosphere of a desert shrub Haloxylon ammodendron. *International Journal of Molecular Sciences, 19*(2), 469.

Heredia-Acuña, C., Almaraz-Suarez, J. J., Arteaga-Garibay, R., Ferrera-Cerrato, R., & Pineda-Mendoza, D. Y. (2019). Isolation, characterization and effect of plant-growth-promoting rhizobacteria on pine seedlings (*Pinus pseudostrobus* Lindl.). *Journal of Forestry Research, 30*(5), 1727–1734.

Heydari, M., Hajinia, S., Jafarian, N., Karamian, M., Mosa, Z., Asgharzadeh, S., & Prévosto, B. (2023). Synergistic use of biochar and the plant growth-promoting rhizobacteria in mitigating drought stress on oak (*Quercus brantii* Lindl.) seedlings. *Forest Ecology and Management, 531*, 120793.

Jagadesh, M., Dash, M., Kumari, A., Singh, S. K., Verma, K. K., Kumar, P., Bhatt, R. & Sharma, S. K. (2024c). Revealing the hidden world of soil microbes: Metagenomic insights into plant, bacteria, and fungi interactions for sustainable agriculture and ecosystem restoration. *Microbiological Research*, 127764.

Jagadesh, M., Dash, M., Singh, S. K., Kumari, A., & Verma, K. K. (2024b). Carbon footprints in agriculture: Challenges, mitigation strategies, and future prospects. In *Carbon footprint assessments: Case studies & best practices* (pp. 17–43). Cham: Springer Nature Switzerland.

Jagadesh, M., Dash, M., Singh, S.K., Kumari, A., Garg, V.K., Jaiswal, A. (2024a). Carbon sequestration potential of forests and forest soils and their role in climate change mitigation. In H. Singh (Ed.), *Forests and climate change*. Singapore: Springer. https://doi.org/10.1007/978-981-97-3905-9_16

Jang, J. H., Woo, S. Y., Kim, S. H., Khaine, I., Kwak, M. J., Lee, H. K., & Lee, W. Y. (2017). Effects of increased soil fertility and plant growth-promoting rhizobacteria inoculation on biomass yield, energy value, and physiological response of poplar in short-rotation coppices in a reclaimed tideland: A case study in the Saemangeum area of Korea. *Biomass and Bioenergy, 107*, 29–38.

Jangra, A., Kumar, K., Maikhuri, S., Bhandari, M. S., Pandey, S., Singh, H., & Barthwal, S. (2024). Unveiling stress-adapted endophytic bacteria: Characterizing plant growth-promoting traits and assessing cross-inoculation effects on Populus deltoids under abiotic stress. *Plant Physiology and Biochemistry, 210*, 108610.

Joshi R, Singh H, Chhetri R, Poudel SR and Rijal S 2021. Carbon sequestration potential of community forests: A comparative analysis of soil organic carbon stock in community managed forests of Far-Western Nepal. *Eurasian Journal of Soil Science, 10*(2), 96–104. https://doi.org/10.18393/ejss.825066.

Kokalis-Burelle, N., Kloepper, J. W., & Reddy, M. S. (2006). Plant growth-promoting rhizobacteria as transplant amendments and their effects on indigenous rhizosphere microorganisms. *Applied Soil Ecology, 31*(1-2), 91–100.

Kumar, A., Dwivedi, G. K., Tewari, S., Jaipaul, Sah, V. K., Singh, H., Kumar, P., Kumar, N., & Kausal, R. (2020). Soil organic carbon pools under *Terminalia chebula* Retz. based agroforestry system in Himalayan foothills, India. *Current Science, 118*, 1098–1103. https://doi.org/10.18520/cs/v118/i7/1098-1103.

Kumari, A., Dash, M., Singh, S. K., Jagadesh, M., Mathpal, B., Mishra, P. K., Pandey, S. K., & Verma, K. K. (2023). Soil microbes: A natural solution for mitigating the impact of climate change. *Environmental Monitoring and Assessment, 195*, 1436.

Kumari, A., Dash, M., Jagadesh, M., & Singh, S. K. (2025b). Impact of climate change on food crops cultivation in the present consequence of global warming. In *Climate-Smart Agricultural Technologies: Approaches for Field Crops Production Systems* (pp. 1–16). Singapore: Springer Nature Singapore.

Kumari, A., Singh, S. K., Singh, A. K., & Khan, I. (2017). Physiological evaluation of drought tolerance in Poplar *(Populus deltoids L.)* for different drought levels. *Journal of AgriSearch, 4*(2), 128–132.

Kumari, A., Singh, A. K., Singh, S. K., & Khan, I. M. (2018). Morphological, physiological and biochemical responses of poplar plants to drought stress. *Journal of AgriSearch, 5*(3), 184–189.

Kumar, V., Jain, L., Chaudhary, S., & Soni, R. (2022). Endophytes as potential plant growth promoters in forestry: Recent advances and perspectives. *Microbes in Agri-Forestry Biotechnology*, 241–262.

Kumari, A., Lakshmi, G. A., Krishna, G. K., Patni, B., Prakash, S., Bhattacharyya, M., & Verma, K. K. (2022). Climate change and its impact on crops: A comprehensive investigation for sustainable agriculture. *Agronomy, 12*(12), 3008.

Kumari, A., Singh, S. K., & Rai, G. K. (2021). *Physiological and biochemical response of plants under drought stress. Plant Abiotic Stress Tolerance: Physiochemical and Molecular Avenues.* Deepika Book Agency: Delhi, India, pp 77–92.

Kumari, G., Kotiyal, P. B., Singh, H., Kumar, M., Kumar, N., Malik, A., … & Singh, S. (2024). Predicting future climate change effects on biotic communities: A species distribution modeling approach. In *Forests and climate change: Biological perspectives on impact, adaptation, and mitigation strategies* (pp. 137–168). Singapore: Springer Nature Singapore.

Kumari, A., Singh, S. K., Al-Tawaha, A. R. M., Thangadurai, D., Panigatti, S., & Sangeetha, J. (2025a). Effect of mycorrhizal inoculation on citrus seedling growth and nutrient uptake. In J. Sangeetha, A. R. M. Al Tawaha, D. Thangadurai (Eds.), *Microbial fertilizer technology for sustainable crop production.* AAP/CRC Press, Taylor and Francis group.

Lara-Chavez, A., Lowman, S., Kim, S., Tang, Y., Zhang, J., Udvardi, M., & Mei, C. (2015). Global gene expression profiling of two switchgrass cultivars following inoculation with *Burkholderia phytofirmans* strain PsJN. *Journal of Experimental Botany, 66*(14), 4337–4350.

Lucas, A. (2021). Risking the earth Part 1: Reassessing dangerous anthropogenic interference and climate risk in IPCC processes. *Climate Risk Management, 31*, Article 100257.

Martin, F. M., Uroz, S., & Barker, D. G. (2017). Ancestral alliances: plant mutualistic symbioses with fungi and bacteria. *Science, 356*(6340), eaad4501.

Nazarov, A., Chetverikov, S., Chetverikova, D., Tuktarova, I., Ivanov, R., Urazgildin, R., & Kudoyarova, G. (2023). Microbial preparations combined with humic substances improve the quality of tree planting material needed for reforestation to increase carbon sequestration. *Sustainability, 15*(9), 7709.

Niehus, R., Picot, A., Oliveira, N. M., Mitri, S., & Foster, K. R. (2017). The evolution of siderophore production as a competitive trait. *Evolution, 71*(6), 1443–1455.

O'Dell, A. (2022). *Scotland's blue carbon: the contribution from seaweed detritus.* Doctoral dissertation, University of the Highlands and Islands

Ojuederie, O. B., Olanrewaju, O. S., & Babalola, O. O. (2019). Plant growth promoting rhizobacterial mitigation of drought stress in crop plants: Implications for sustainable agriculture. *Agronomy, 9*(11), 712.

Patni, B., Bhattacharyya, M., Kumari, A., & Purohit, V. K. (2022). Alarming influence of climate change and compromising quality of medicinal plants. *Plant Physiology Reports*, 1–10.

Pérez-Cruzado, C., Mohren, G. M., Merino, A., & Rodríguez-Soalleiro, R. (2012). Carbon balance for different management practices for fast growing tree species planted on former pastureland in southern Europe: A case study using the CO_2 Fix model. *European Journal of Forest Research, 131*, 1695–1716.

Radhapriya, P., Ramachandran, A., Anandham, R., & Mahalingam, S. (2015). *Pseudomonas aeruginosa* RRALC3 enhances the biomass, nutrient and carbon contents of *Pongamia pinnata* seedlings in degraded forest soil. *PLoS ONE, 10*(10), Article e0139881.

Radhapriya, P., Ramachandran, A., & Palani, P. (2018). Indigenous plant growth-promoting bacteria enhance plant growth, biomass, and nutrient uptake in degraded forest plants. *3 Biotech, 8*, 1–10.

Raison, R. J. (2024). A review of the impacts of sustainable harvesting, non-harvest management and wildfire on net carbon emissions from Australian native forests. *Australian Forestry*, 1–19.

Rawat, P., Das, S., Shankhdhar, D., & Shankhdhar, S. C. (2021). Phosphate-solubilizing microorganisms: Mechanism and their role in phosphate solubilization and uptake. *Journal of Soil Science and Plant Nutrition, 21*(1), 49–68.

Rostamikia, Y., Teimouri, M., & Jafari, F. (2024). Improvement of seed germination and early growth of Caucasian whortleberry (*Vaccinium arctostaphylos* L.) by plant growth promoting rhizobacteria and cold stratification.

Saeed, Q., Xiukang, W., Haider, F. U., Kučerik, J., Mumtaz, M. Z., Holatko, J., & Mustafa, A. (2021). Rhizosphere bacteria in plant growth promotion, biocontrol, and bioremediation of contaminated sites: A comprehensive review of effects and mechanisms. *International Journal of Molecular Sciences, 22*(19), 10529.

Sahoo, S. K., Mishra, K. N., Panda, N., Padhan, K., Mohanty, S., Kumar, K., & Sethi, D. (2022a). Nine years of integrated nutrient management practices on soil microbial activities in a cereal-based cropping system. *International Journal of Plant and Soil Science, 34*, 1234–1242.

Sahoo, S. K., Mishra, K. N., Panda, N., Sethi, D., & Dey, P. (2023). Carbon and nitrogen mineralization kinetics in soil under differential nutrient management regimes in inceptisols of coastal India. *Communications in Soil Science and Plant Analysis, 54*(20), 2782–2799.

Sahoo, S. K., Mishra, K. N., Panda, N., Panda, R. K., Padhan, K., Mohanty, S., & Sethi, D. (2022b). System productivity and nutrient recoveries as influenced by nine years of long-term INM practices under acidic inceptisols of India. In *Biological Forum* (Vol. 14, No. 3, pp. 1036–1040).

Santoyo, G., Urtis-Flores, C. A., Loeza-Lara, P. D., Orozco-Mosqueda, M. D. C., & Glick, B. R. (2021). Rhizosphere colonization determinants by plant growth-promoting rhizobacteria (PGPR). *Biology, 10*(6), 475.

Sarwar, S., Khaliq, A., Yousra, M., Sultan, T., Ahmad, N., & Khan, M. Z. (2020). Screening of siderophore-producing PGPRs isolated from groundnut (*Arachis hypogaea* L.) rhizosphere and their influence on iron release in soil. *Communications in Soil Science and Plant Analysis, 51*(12), 1680–1692.

Semwal, A., Gautam, K., Bhatt, M., Yadav, R., Barthwal, S., & Singh, H. (2025). Green synthesis and application of metal nanoparticles (MNPs): A sustainable strategy for protecting wood from fungal and termite decay. *Next Materials, 9*, 100962

Singh, M., Kumari, A., & Verma, K. K. (2013). Physiological, growth, and biomass attributes in *Populus deltoides* L. (clones G–48 and Kranti) influenced by water stress. *Arboriculture & Urban Forestry (AUF), 39*(5), 226–230.

Sofi, J. A., Lone, A. H., Ganie, M. A., Dar, N. A., Bhat, S. A., Mukhtar, M., Dar, M. A., & Ramzan, S. (2016). Soil microbiological activity and carbon dynamics in the current climate change scenarios: A review. *Pedosphere, 26*(5), 577–591.

Stocker, T. F., Dahe, Q., Plattner, G. K., & Tignor, M. (2015, September). IPCC workshop on regional climate projections and their use in impacts and risk analysis studies. In *workshop report* (Vol. 15, p. 18). Bern, Switzerland: Intergovernmental Panel on Climate Change.

Swarnalakshmi, K., Yadav, V., Tyagi, D., Dhar, D. W., Kannepalli, A., & Kumar, S. (2020). Significance of plant growth promoting rhizobacteria in grain legumes: Growth promotion and crop production. *Plants, 9*(11), 1596.

Turan, M., Gulluce, M., von Wirén, N., & Sahin, F. (2012). Yield promotion and phosphorus solubilization by plant growth–promoting rhizobacteria in extensive wheat production in Turkey. *Journal of Plant Nutrition and Soil Science, 175*(6), 818–826.

van Vuuren, D. P., van Sluisveld, M., & Hof, A. F. (2015). *Implications of long-term scenarios for medium-term targets (2050)*. PBL Netherlands Environmental Assessment Agency.

Verma, K. K., Joshi, A., Song, X. P., Singh, S., Kumari, A., Arora, J., Singh, S. K., Solanki, M. K., Seth, C. S., & Li, Y. R. (2024a). Synergistic interactions of nanoparticles and plant growth promoting rhizobacteria enhancing soil-plant systems: A multigenerational perspective. *Frontiers in Plant Science, 15*, 1376214.

Verma, K. K., Song, X. P., Kumari, A., Jagadesh, M., Singh, S. K., Bhatt, R., Singh, M., Seth, C.S. & Li, Y. R. (2024b). Climate change adaptation: Challenges for agricultural sustainability. *Plant, Cell & Environment.*

Wang, H., Liu, R., You, M. P., Barbetti, M. J., & Chen, Y. (2021). Pathogen biocontrol using plant growth-promoting bacteria (PGPR): Role of bacterial diversity. *Microorganisms, 9*(9), 1988.

Ward, S. E., Bardgett, R. D., McNamara, N. P., Adamson, J. K., & Ostle, N. J. (2007). Long-term consequences of grazing and burning on northern peatland carbon dynamics. *Ecosystems, 10*(7), 1069–1083.

Weyens, N., Van Der Lelie, D., Taghavi, S., Newman, L., & Vangronsveld, J. (2009). Exploiting plant–microbe partnerships to improve biomass production and remediation. *Trends in Biotechnology, 27*(10), 591–598.

Chapter 20
Insight into Ecosystem-Based Solutions and Green Credit for Climate Change Mitigation and Sustainable Future

Prateek Srivastava and D. R. Vickram

Abstract Ecosystem-based Solutions (EbS) harness natural processes to address climate change, bolster resilience, and advance sustainable development. Leveraging the functions of EbS such as carbon sequestration, flood regulation, and soil stabilization, delivers both environmental and socio-economic benefits. Incentive frameworks, such as India's Green Credit Programme (GCP) and comparable global schemes, are catalyzing private and public investment in ecosystem stewardship. Closely related Nature-based Solutions (NbS) further align biodiversity conservation with human well-being, reinforcing the interconnected goals of ecological health and community resilience. Effective EbS deployment rests on several core principles: enhancing ecosystem and social resilience, fostering multi-sectoral, decentralized governance, integrating adaptive management to address dynamic environmental conditions, and blending rigorous scientific data with local knowledge through participatory decision-making. Prioritizing the minimization of trade-offs and the maximization of benefits ensures that ecological integrity and human livelihoods are advanced in tandem. Despite their promise, EbS encounters obstacles including limited funding, fragmented governance structures, low public awareness, and technical knowledge gaps. To embed EbS within national and international climate strategies, robust monitoring and evaluation systems, targeted capacity building, and innovative financing models are essential. Future policy directions should emphasize multi-stakeholder collaboration, expanded investments in research and education, and scalable funding mechanisms. Overcoming these barriers, EbS and NbS may serve as transformative pathways toward a more resilient, equitable, and sustainable future.

Keywords Ecosystem · Sustainability · Nature-based solutions · Climate change mitigation

P. Srivastava (✉)
Amrut Nature Solutions Private Limited, Indore, Madhya Pradesh, India
e-mail: prateek.ecology@gmail.com

D. R. Vickram
Hindalco Industries Limited, Ranchi, Jharkhand, India

H. Singh (ed.), *Forestry for a Carbon-Neutral and Sustainable Future*, Sustainability Solutions, https://doi.org/10.1007/978-3-032-06427-1_20

20.1 Introduction

Ecosystem-based solutions (EbS) are innovative, nature-centred approaches that harness the benefits of biodiversity and ecosystem services to address critical societal challenges, including climate change mitigation and adaptation. As the impacts of climate change intensify globally, EbS is gaining recognition for its ability to provide sustainable and resilient pathways to reduce greenhouse gas emissions, enhance carbon sequestration, and support the adaptive capacity of ecosystems. At its core, EbS involves the conservation, sustainable management, and restoration of natural or modified ecosystems to enhance their natural functions, such as carbon storage, water filtration, and climate regulation. For instance, afforestation, reforestation, agroforestry, and wetland restoration are prime examples of EbS that contribute to both environmental and socio-economic goals (Joshi et al., 2025). These strategies align with key international frameworks, such as the United Nations Sustainable Development Goals (SDGs) and the Paris Agreement, demonstrating their significance in achieving global climate targets (Seddon et al., 2020). One of the defining attributes of EbS is its holistic approach. Unlike conventional technological solutions that often address specific issues in isolation, EbS acknowledges the interconnectedness of ecosystems, human well-being, and climate systems. For example, restoring mangrove forests not only enhances carbon sequestration but also protects coastal communities from storm surges and supports biodiversity by providing habitats for various species (Narayan et al., 2016). Such multifaceted benefits make EbS an economically viable option for climate action, as it reduces the need for costly infrastructure and repairs while enhancing ecosystem services. Moreover, EbS solutions often promote inclusive community participation by integrating traditional knowledge and local practices, thereby fostering stewardship and ownership among communities (Kumar & Singh, 2020).

Another critical advantage of EbS lies in its capacity for long-term impact. Ecosystems, when managed sustainably, continue to provide services that grow in value over time. For instance, reforestation projects not only act as carbon sinks but also improve soil fertility, increase water retention, and support pollination services crucial for agriculture. These cascading benefits ensure that EbS projects are not only climate-resilient but also economically sustainable. Additionally, EbS supports biodiversity conservation, which is essential for maintaining ecological balance and preventing species extinction. Biodiverse ecosystems are more resilient to disturbances, ensuring their ability to function effectively under changing climatic conditions (IPCC, 2019).

Despite its numerous benefits, the implementation of EbS faces challenges. Land-use conflicts, lack of funding, inadequate policy frameworks, and insufficient stakeholder engagement can hinder the adoption of EbS strategies. Moreover, the benefits of EbS, though significant, often take time to materialize, making them less appealing compared to immediate technological fixes. Overcoming these barriers requires integrated planning, cross-sectoral collaboration, and investments in capacity-building to ensure the scalability and effectiveness of EbS. Policymakers and practitioners must also address the issue of equitable benefit distribution, ensuring that vulnerable and marginalized communities derive tangible benefits from EbS initiatives (Kabisch et al., 2017).

20.1.1 Green Credit Mechanisms

Green credit mechanisms are innovative financial frameworks that integrate environmental considerations into lending and investment decisions, fostering a shift toward sustainable development. These mechanisms incentivize individuals, businesses, and governments to adopt eco-friendly practices and discourage activities that harm the environment. Aligning financial support with sustainability goals may guide green credit systems in contributing to the global effort to combat climate change, protect ecosystems, and promote green growth (Wang & Zhi, 2016).

The fundamental principle of green credit mechanisms is environmental accountability. Financial institutions assess the social and ecological impacts of projects before granting loans or investments. Businesses seeking financial support must demonstrate adherence to sustainability standards, such as reducing carbon emissions, conserving biodiversity, or minimizing waste. This ensures that credit flows toward projects aligned with international frameworks, such as the United Nations Sustainable Development Goals (SDGs) and the Paris Agreement (OECD, 2021). One of the key roles of green credit mechanisms is to redirect financial flows toward sustainable sectors, such as renewable energy, sustainable agriculture, and pollution control technologies. For instance, renewable energy projects, including solar and wind farms, often receive preferential financing terms, which encourages their adoption and reduces reliance on fossil fuels. Similarly, green credit systems support sustainable agriculture practices, such as organic farming and agroforestry, which contribute to soil health, water conservation, and carbon sequestration (Srivastava & Singh, 2024).

20.1.1.1 India's Green Credit Programme (GCP)

Green credits represent units of incentive provided to individuals, communities, and organizations that engage in activities delivering positive environmental impacts. The Green Credit Programme (GCP), recently introduced in India as part of the Lifestyle for Environment (LiFE) initiative, exemplifies this approach (Government of India, 2023). The GCP encourages voluntary environmental actions across diverse sectors, including tree plantation, water conservation, sustainable agriculture, waste management, and air pollution reduction (Drishti IAS, 2024). The GCP aims to enhance environmental sustainability while fostering community engagement and responsibility by creating a market-based mechanism for these activities. The operational framework of green credit mechanisms typically involves a registration process where participants submit their environmental projects for verification. Upon successful evaluation by designated agencies, participants receive certificates of green credit that can be traded on a dedicated platform (BanQu, 2024). This trading aspect allows for flexibility and encourages further investment in sustainable practices by providing a financial return on environmental efforts.

The primary objectives of green credit mechanisms include promoting sustainable practices, enhancing ecological health, and supporting community livelihoods. By incentivizing a wide range of environmentally friendly activities, these mechanisms aim to create synergies between economic development and ecological conservation. For instance, tree plantation initiatives not only contribute to carbon sequestration but also improve local biodiversity and enhance soil health (Murray et al., 2019). Moreover, green credit mechanisms provide opportunities for various stakeholders, ranging from grassroots individuals to large corporations, to engage in environmental stewardship. This inclusivity fosters a sense of ownership among participants and encourages collective action towards sustainability goals (Government of India, 2023). Aligning economic incentives with environmental outcomes, the green credit mechanisms may effectively mobilize resources for conservation efforts while addressing pressing environmental challenges.

While both green credits and carbon credits aim to incentivize environmental sustainability actions, they differ significantly in focus and scope. Carbon credits primarily target the reduction of greenhouse gas emissions through specific projects, such as renewable energy initiatives or carbon capture technologies (Koshish India, 2024). In contrast, green credits encompass a broader range of activities beyond carbon emissions reduction, including water management and ecosystem restoration.

Additionally, the eligibility criteria for earning these credits vary. Green credits are accessible to a wide range of stakeholders engaged in eco-friendly actions, from individuals to organizations. In contrast, carbon credits are typically earned by entities that reduce emissions below a predetermined baseline (Koshish India, 2024). This distinction enables green credit mechanisms to promote a range of environmental actions while fostering community participation in sustainability efforts.

While green credit mechanisms present significant opportunities for promoting sustainability, several challenges must be addressed. The effective implementation of green credit programs necessitates robust regulatory frameworks to ensure transparency and accountability in credit generation and trading processes (Koshish India, 2024). Without clear guidelines and oversight, there is a risk that the system will be misused or exploited. For green credit mechanisms to succeed, widespread awareness and participation among stakeholders are essential. Educational campaigns play a crucial role in informing communities about the benefits of adopting environmentally friendly practices and participating in credit programs (Drishti IAS, 2024). Establishing effective monitoring and verification systems is crucial for accurately assessing the impact of activities that generate green credits. Accurate data collection will ensure that credits reflect genuine environmental benefits and maintain the integrity of the trading system (Indian Express, 2023).

The potential for greenwashing poses a serious challenge to the integrity of the Green Credit Initiative. Companies may exploit the system by purchasing credits without making substantive changes to their operations or investing in genuine sustainability initiatives (Koshish India, 2024). This practice can mislead consumers and stakeholders about a company's environmental impact, ultimately eroding public trust in sustainability claims. For instance, if businesses focus solely on purchasing credits as a means to offset their environmental footprint without implementing internal measures to reduce pollution or resource consumption, they may be perceived as engaging in superficial sustainability efforts. This behavior not only undermines the goals of the Green Credit Initiative but also detracts from meaningful progress toward climate action.

The success of the Green Credit Initiative is also contingent upon market dynamics, including price discovery and liquidity of credits. Establishing fungibility between different types of green credits can be challenging due to variations in environmental benefits associated with each credit type. For example, credits generated from biodiversity conservation may not be directly comparable to those from water conservation efforts, complicating trading mechanisms. Market volatility poses another concern; fluctuations in credit prices could create uncertainty for businesses engaged in this market. If prices are too low, there may be insufficient incentive for companies to invest in sustainable practices (BanQu, 2024).

20.1.1.2 Global Green Credit Mechanisms

An exemplary implementation of green credit systems can be observed in China. The country introduced Green Credit Guidelines in 2012, mandating banks to incorporate environmental and social risk assessments into their lending processes. This policy has led to increased investments in renewable energy, energy efficiency, and environmental restoration projects. Between 2013 and 2019, green loans in China experienced significant growth, demonstrating the effectiveness of such mechanisms in promoting sustainability (Xu et al., 2021). Similarly, the European Investment Bank (EIB) has adopted a green financing strategy that prioritizes climate action and environmental sustainability, reflecting the global relevance of green credit frameworks (OECD, 2021). Green credit mechanisms also play a pivotal role in fostering innovation. Companies investing in eco-friendly technologies, such as electric vehicles, energy-efficient systems, and waste management solutions, often benefit from favorable credit terms. This reduces financial barriers and encourages businesses to develop and implement innovative solutions to environmental challenges. Additionally, these mechanisms contribute to raising awareness among stakeholders about the importance of integrating sustainability into economic activities.

Despite their benefits, the implementation of green credit systems faces several challenges. A key issue is the lack of standardization in defining and measuring the environmental impact of projects, which can lead to inconsistencies and potential "greenwashing." For example, companies may overstate the ecological benefits of their projects to qualify for green financing. Additionally, limited awareness among businesses and financial institutions about green credit opportunities can hinder their adoption. Regulatory frameworks in many regions are still evolving, necessitating stronger policies and incentives to scale up green credit systems effectively.

20.1.2 Nature-Based Solutions (NbS)

Nature-based solutions (NbS) are actions that leverage the power of nature to address societal challenges, offering benefits to both biodiversity and human well-being. These solutions involve the sustainable management, protection, and restoration of ecosystems to achieve long-term outcomes, such as mitigating climate change, reducing disaster risk, ensuring water security, and improving public health. The International Union for Conservation of Nature (IUCN) defines NbS as "actions to protect, sustainably manage and restore natural or modified ecosystems that address societal challenges effectively and adaptively, simultaneously providing human well-being and biodiversity benefits" (IUCN, 2011).

NbS frameworks are built on the principles of sustainability, adaptability, and inclusivity. They emphasize ecosystem services such as carbon sequestration, water filtration, and pollination and their role in creating resilient and sustainable systems. For example, green infrastructure projects, such as urban forests, green roofs, and constructed wetlands, are Nature-Based Solutions (NbS) approaches that address urban challenges, including heatwaves, air pollution, and flooding while enhancing biodiversity. Similarly, agricultural practices such as agroforestry and regenerative farming integrate NbS by improving soil health, increasing carbon storage, and enhancing resilience to climate change impacts (Cohen-Shacham et al., 2019). One of the most widely adopted frameworks for NbS is the United Nations' Sustainable Development Goals (SDGs), particularly Goals 13 (Climate Action), 14 (Life Below Water), and 15 (Life on Land). These goals emphasize the interconnectedness of ecosystems, climate resilience, and human well-being, providing a global roadmap for integrating Nature-Based Solutions (NbS) into policy and practice. Additionally, the European Union has developed an NbS framework to guide research, implementation, and scaling, with a focus on ecosystem restoration, climate adaptation, and socio-economic development (Raymond et al., 2017).

While NbS offers transformative potential, its implementation requires collaborative governance, cross-sectoral integration, and careful monitoring to ensure equitable outcomes. Integrating NbS into mainstream climate and development policies is crucial for achieving a sustainable and resilient future. Nature-based solutions (NbS) are increasingly recognized as critical tools for addressing climate change and promoting sustainable development. The term encompasses a range of strategies that utilize natural processes and ecosystems to tackle societal challenges such as climate change, biodiversity loss, and disaster risk reduction. While the concept shares similarities with other approaches, such as ecosystem-based adaptation (EbA) and ecosystem services, it is distinct in its emphasis on achieving multiple co-benefits, including human well-being and biodiversity conservation, through ecosystem-centered actions (Cohen-Shacham et al., 2019). Unlike traditional gray infrastructure solutions, NbS leverages the dynamic capacities of natural systems, offering resilience and adaptability to environmental changes. Frameworks for NbS implementation are designed to guide their integration into climate and development policies. Effective NbS implementation requires clarity in terminology to avoid misuse or "green washing," where projects labelled as NbS fail to deliver genuine environmental or social benefits. To ensure accountability, frameworks emphasize evidence-based decision-making, stakeholder engagement, and rigorous monitoring and evaluation processes. Integrating NbS into global and local strategies enables policymakers to harness the synergistic benefits of nature, addressing the climate crisis while fostering sustainable and equitable development.

20.2 The Role of Ecosystems in Carbon Sequestration

Ecosystems play a pivotal role in carbon sequestration by acting as natural sinks that absorb and store atmospheric carbon dioxide (CO_2), thereby mitigating climate change. Forests, wetlands, grasslands, and marine ecosystems are among the most effective carbon sinks, each contributing uniquely to carbon storage and reduction of emissions. Forest ecosystems, for example, store vast amounts of carbon in their biomass and soil. Tropical rainforests, such as the Amazon, are particularly significant, accounting for approximately one-third of the carbon stored in terrestrial ecosystems (Pan et al., 2011). Trees capture CO_2 through photosynthesis and store it in their trunks, roots, and leaves, while forest soils sequester organic carbon over extended periods. Sustainable forest management practices, such as afforestation and reforestation, can enhance this sequestration potential.

Wetland ecosystems, including peatlands and mangroves, are among the most carbon-dense environments. Peatlands, which cover only about 3% of the Earth's land surface, store twice as much carbon as all the world's forests combined (Parish et al., 2008). This is because the waterlogged conditions in wetlands slow down the decomposition of organic matter, leading to the accumulation of carbon-rich peat. Mangroves, on the other hand, are coastal ecosystems that not only sequester carbon in their biomass but also trap organic carbon in their sediments. These ecosystems provide dual benefits by acting as carbon sinks and protecting coastal areas from erosion and extreme weather events. Grasslands and savannas also contribute significantly to carbon sequestration, primarily through their soil. These ecosystems, while less conspicuous than forests, store substantial amounts of carbon underground. Grassland soils are remarkably resilient to disturbances, making them reliable carbon reservoirs. Practices such as rotational grazing and restoring degraded grasslands can enhance their carbon storage capacity (Conant et al., 2017). Marine ecosystems, including seagrasses, salt marshes, and coral reefs, contribute to what is known as "blue carbon." Seagrasses, for instance, can sequester carbon at rates up to 35 times faster than tropical rainforests due to their ability to trap organic matter in underwater sediments (Fourqurean et al., 2012). Protecting and restoring these ecosystems is crucial, as their degradation not only releases stored carbon but also diminishes their ability to act as sinks.

20.3 Key Principles Underlying EbS

Ecosystem-based solutions (EbS) are guided by several key principles that ensure their effectiveness in addressing environmental challenges, particularly in the context of climate change. These principles emphasize the importance of integrating ecological health with socio-economic considerations, promoting resilience, and fostering sustainable practices.

20.3.1 Promoting Resilience of Ecosystems and Societies

Promoting resilience in ecosystems and societies is a fundamental principle of Ecosystem-Based Solutions (EbS). This principle focuses on enhancing the capacity of both natural systems and human communities to withstand, adapt to, and recover from environmental changes and disturbances, particularly those induced by climate change. Resilience is not merely about bouncing back from adverse events; it involves the ability to transform and evolve in response to changing conditions (IUCN, 2011).

Resilience, in ecological terms, refers to an ecosystem's capacity to absorb disturbances while maintaining its essential structure, processes, and functions. This concept encompasses various aspects such as biodiversity, ecosystem services, and Adaptive Capacity. Diverse ecosystems are generally more resilient because they can better withstand stresses and recover from disturbances (Singh, 2024). High biodiversity contributes to functional redundancy, where multiple species can perform similar roles within an ecosystem, ensuring stability even when some species are lost. Healthy ecosystems provide vital services, such as water purification, soil fertility, and carbon sequestration that enhance resilience. For instance, wetlands act as buffers against floods and storms, while forests stabilize soils and improve water retention. To promote resilience in ecosystems, several strategies can be employed. Some of these include Restoration of Degraded Ecosystems, Sustainable Management Practices, Protecting Natural Habitats, and Integrating Traditional Knowledge. Revitalizing degraded ecosystems is essential for restoring their functionality and resilience. For example, reforestation projects can restore carbon sinks while enhancing biodiversity and ecosystem services. Restoration efforts should prioritize native species that are well-adapted to local conditions, as these species are more likely to thrive and contribute to the ecosystem's stability. Implementing sustainable land-use practices helps maintain ecosystem health. Practices such as agroforestry, sustainable forestry, and conservation agriculture can enhance soil health and biodiversity while reducing vulnerability to climate impacts. Conserving existing natural habitats is crucial for maintaining the resilience of ecosystems. Protected areas serve as refuges for biodiversity, helping to preserve ecological processes that support resilience against climate change. Incorporating traditional ecological knowledge into management practices can enhance resilience by leveraging local insights into ecosystem dynamics and sustainable practices. Engaging indigenous communities in conservation efforts fosters a sense of ownership and ensures that management strategies align with local needs.

20.3.2 Multi-sectoral Approaches

EbS advocates for multi-sectoral approaches that recognize the interdependencies among various sectors, including agriculture, forestry, water management, and urban planning. This principle emphasizes the need for collaboration across different sectors to achieve comprehensive solutions that address multiple environmental and social challenges simultaneously. This principle acknowledges that ecosystems provide essential services that underpin the functions of diverse sectors and that effective management must involve collaboration across these sectors to enhance resilience and sustainability.

Multi-sectoral approaches involve integrating efforts from various sectors, to address environmental challenges holistically. This integration is crucial because ecosystems often span multiple jurisdictions and affect various social groups differently. For instance, the health of a watershed impacts agriculture, fisheries, and urban water supply, necessitating coordinated management efforts across these sectors.

Ecosystems provide essential services that underpin various economic sectors. For instance, agriculture relies on healthy soils, clean water, and pollination services provided by natural ecosystems. Similarly, forestry and fisheries depend on robust ecological systems to sustain their productivity. Recognizing these interdependencies is crucial for developing integrated management strategies that enhance both ecosystem health and economic resilience. Climate change presents multifaceted challenges that single-sector interventions cannot effectively address. For example, managing water resources requires collaboration between agricultural practices, urban planning, and conservation efforts to ensure sustainable use and distribution. A multi-sectoral approach enables a more comprehensive understanding of these challenges, facilitating the development of solutions that consider the needs and impacts across various sectors. Multi-sectoral strategies promote resilience by diversifying responses to environmental pressures. Integrating efforts across sectors may help communities to better adapt to changes and recover from disturbances. For example, combining sustainable agricultural practices with watershed management can enhance food security while also improving water quality and increasing water availability.

Key Components of Multi-Sectoral Approaches include Collaboration Across Sectors, Engagement of Multiple Stakeholders, Addressing Conflicting Priorities, Spatial Complexity, and Integration of Knowledge Systems. Fostering cooperation among different sectors enables stakeholders to align their goals and resources, thereby achieving common objectives. A multi-stakeholder approach is crucial for developing policies that consider the diverse needs and perspectives of various groups, including local communities, government agencies, non-governmental organizations (NGOs), and the private sector. Engaging these stakeholders ensures that diverse viewpoints are represented in decision-making processes, leading to more equitable and effective outcomes. Different sectors may have conflicting priorities or mandates that can hinder effective ecosystem management. A multi-sectoral approach helps identify these conflicts early on and develop strategies to reconcile them. For example, balancing agricultural expansion with conservation efforts requires understanding the trade-offs involved and finding solutions that satisfy both agricultural productivity and ecological health. Ecosystems often operate at different scales, meaning that broader regional or national policies can influence local initiatives. For example, restoring floodplains in a locality to reduce flooding risks requires coordinated river management upstream. Understanding this spatial complexity is crucial for designing effective, multi-sectoral strategies that consider how actions in one area can impact others. Multi-sectoral approaches benefit from integrating scientific knowledge with traditional ecological knowledge. This combination allows for a more comprehensive understanding of ecosystem dynamics and enhances the effectiveness of management strategies. Local communities often possess valuable insights into ecosystem functions and sustainable practices that can inform policy decisions (Abdelmagied & Mpheshea, 2020).

20.3.3 Decentralized Management

Effective ecosystem management is increasingly recognized as a necessity for sustainable development, particularly in the face of climate change and biodiversity loss. A key principle underpinning successful ecosystem management is decentralization to the lowest appropriate level, which allows local communities to take ownership of their resources. This approach promotes greater efficiency, equity, and accountability in the management of ecosystems. Moreover, local knowledge and practices are critical for understanding specific ecological contexts and ensuring that management strategies are culturally appropriate and effective.

Decentralization involves transferring authority and responsibility from central governments to local or regional entities. This shift is grounded in several theoretical and practical considerations involving.

20.3.3.1 Local Knowledge Utilization

Local communities possess unique knowledge about their ecosystems, which has been developed over generations through interaction with their environment. This traditional ecological knowledge (TEK) includes insights into species behavior, seasonal changes, and sustainable resource management practices (Berkes, 2009). Decentralizing management allows policies to be tailored to reflect this knowledge, leading to more effective and culturally relevant solutions (Meynen & Doornbos, 2004).

20.3.3.2 Improved Efficiency

Decentralized systems can operate more efficiently by reducing bureaucratic layers often associated with centralized governance. Local management allows for quicker decision-making and adaptation to changing conditions. For example, decentralized water management systems can respond more rapidly to local needs compared to centralized systems that may be slow to react due to distance and administrative hurdles (FAO, 1996).

20.3.3.3 Enhanced Equity

Decentralization promotes equity by giving marginalized communities a voice in decision-making regarding resource management. When local stakeholders participate in governance processes, it fosters a sense of ownership and accountability (Ribot, 2002). This participatory approach ensures that the benefits of ecosystem services are distributed equitably among community members rather than being concentrated in the hands of a few.

20.3.3.4 Increased Accountability

Local management structures can enhance accountability by establishing clear lines of responsibility among community members. When communities manage their resources, they are more likely to hold each other accountable for sustainable practices (Agrawal & Ostrom, 1999). Furthermore, decentralized systems often utilize transparent mechanisms—such as public meetings or community-led initiatives to track resource use and decision-making processes.

A few associated case studies supporting decentralized management are:

Case Study 1

Community-Based Natural Resource Management (CBNRM)

CBNRM initiatives empower local communities to manage their natural resources sustainably. Namibia's Community-Based Natural Resource Management Programme is a prominent example of successful CBNRM implementation. Established in 1996, this program allows local communities to manage wildlife resources within designated conservancies. Communities are granted user rights over wildlife populations, providing incentives for conservation through eco-tourism and sustainable hunting (Jones & Murphree, 2001). As a result, wildlife populations have increased significantly in these areas, while local communities have benefited economically (Fig. 20.1). In Brazil, extractive reserves were created to allow local communities to sustainably harvest non-timber forest products while conserving biodiversity in the Amazon rainforest. These reserves provide Indigenous peoples with land rights, enabling them to continue traditional harvesting practices that support their livelihoods without resorting to environmentally damaging activities, such as logging or cattle ranching (Schwartzman et al., 2000). The Communal Areas Management Programme for Indigenous Resources (CAMPFIRE) in Zimbabwe exemplifies how Community-Based Natural Resource Management (CBNRM) can enhance community livelihoods while promoting conservation. Through CAMPFIRE, local communities manage wildlife resources and receive a share of the revenue generated from tourism and hunting activities. This program has led to increased wildlife populations and improved community welfare (Murphree, 1993).

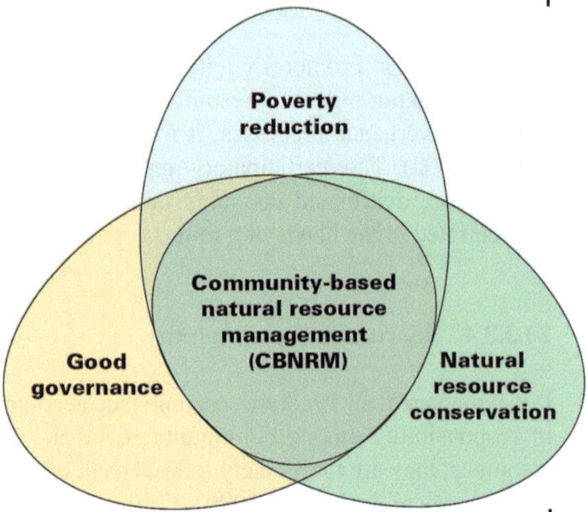

Fig. 20.1 CBNRM management system (Resource Africa, 2024)

CBNRM seeks to transfer authority over natural resources from central governments to local communities. This empowerment enables communities to take control of their resources and make decisions that align with their needs and values (Ribot, 2002). The approach promotes the sustainable use of natural resources by creating incentives for communities to manage their resources wisely. CBNRM promotes practices that conserve ecosystems while providing livelihoods, enabling communities to benefit economically from their resource management efforts (Berkes et al., 2001a, 2001b). CBNRM emphasizes participatory governance structures that involve community members in decision-making processes. This participatory approach fosters transparency, accountability, and inclusiveness in resource management (Agrawal & Ostrom, 1999). Recognizing the value of traditional ecological knowledge is crucial for effective resource management. CBNRM incorporates local customs, practices, and knowledge systems into management strategies, ensuring that they are culturally appropriate and context-specific (Berkes et al., 2001a, 2001b).

Community-Based Natural Resource Management (CBNRM) represents a transformative approach to natural resource governance that empowers local communities while promoting environmental sustainability. However, challenges such as institutional barriers, capacity constraints, conflicts over resource use, and market access must be addressed to ensure the long-term success of CBNRM initiatives. Continued support from governments and NGOs is crucial for enhancing the capacity of local communities and fostering environments that enable effective resource management.

Case Study 2

Integrated Water Resources Management (IWRM)

Integrated Water Resources Management (IWRM) is a holistic approach to managing water resources that emphasizes the interconnectedness of water, land, and related resources. Established as a framework to address the challenges of water scarcity, pollution, and ecosystem degradation, Integrated Water Resources Management (IWRM) seeks to maximize economic and social welfare while ensuring the sustainability of vital ecosystems (Global Water Partnership [GWP], 2000). According to GWP (2000), IWRM is defined as *a process that promotes the coordinated development and management of water, land, and related resources to maximize resultant economic and social welfare equitably without compromising the sustainability of vital ecosystems* (Fig. 20.2).

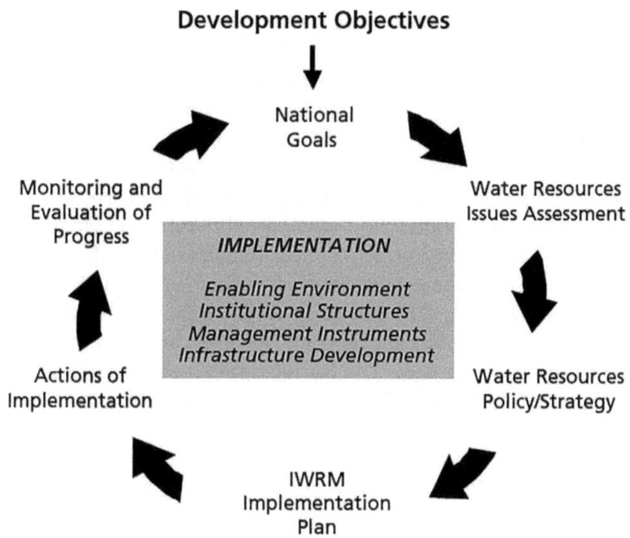

Fig. 20.2 Stages in IWRM planning and implementation (UNDESA, 2005)

IWRM recognizes that water resources are interdependent and must be managed in a way that considers their multiple uses—domestic, agricultural, industrial, and ecological. This integrated perspective is crucial for addressing competing demands on water resources (Agarwal et al., 2000). Effective Integrated Water Resources Management (IWRM) involves the participation of all stakeholders, including local communities, government agencies, and private sector actors. Engaging users in decision-making processes fosters transparency and accountability (GWP, 2000). IWRM acknowledges the central role of women in water management. Women often bear the brunt of water scarcity and play a key role in managing household water supplies (GWP, 2000; United Nations [UN], 2013). Water is recognized as an economic good with value in all its competing uses. This principle encourages efficient allocation and use of water resources (GWP, 2000).

Case Study 3

Decentralized Forestry Management

Decentralized forestry management has emerged as a vital approach to sustainable resource management, particularly in developing countries where local communities depend heavily on forest resources for their livelihoods. In Burkina Faso, the control over fuelwood management has been devolved to local committees, allowing local stakeholders to make informed decisions based on their knowledge of local conditions. This article examines the implementation of decentralized forestry management in Burkina Faso, highlighting its impacts, successes, and challenges.

In Burkina Faso, local committees play a crucial role in decentralized forestry management. These committees are typically composed of community members who are directly affected by forest management practices. They are responsible for making decisions regarding the sustainable use of forest resources, including fuelwood harvesting and conservation measures. The establishment of these committees has led to significant improvements in forest management outcomes due to several factors, including local knowledge and expertise, as well as enhanced accountability (FAO, 1996). One notable example of decentralized forestry management in Burkina Faso is the Nazinon Forest Management Project. Initiated in 1985, this project aimed to combat deforestation and promote sustainable fuelwood production for Ouagadougou, the capital city (FAO, 1996). The project involved establishing Forest Management Groups (FMGs) composed of local community members who were trained in sustainable forest management practices. The Nazinon project successfully implemented a system of selective logging that allowed for sustainable fuelwood production while minimizing environmental impact.

The establishment of FMGs fostered greater community involvement in forest management decisions. Community members actively participated in monitoring forest conditions and enforcing regulations related to harvesting practices (Ribot et al., 2006). The Nazinon project contributed to biodiversity conservation efforts in Burkina Faso by promoting sustainable logging practices and protecting critical habitats within the forest.

20.3.4 Adaptive Management

Adaptive management (AM) is increasingly recognized as a foundational principle of Ecosystem-Based Solutions (EbS), which aim to address complex environmental challenges by integrating ecological, social, and economic considerations. This approach acknowledges that ecosystems are dynamic and subject to change, necessitating flexible management strategies that can be adjusted in response to ongoing monitoring and evaluation of ecological conditions.

Adaptive management is defined as a systematic approach to resource management that emphasizes learning through experimentation and iterative decision-making (Holling, 1978; Williams & Hummelbrunner, 2010). This process involves several key components.

20.3.4.1 Monitoring and Evaluation

Continuous monitoring of ecological conditions is essential for adaptive management. It provides the data necessary to assess the effectiveness of management actions and to identify when adjustments are needed (Cumming et al., 2015).

20.3.4.2 Flexibility in Management Strategies

Adaptive management promotes flexibility in decision-making, allowing managers to adjust their strategies in response to new information or changing conditions. This flexibility is crucial for addressing the inherent uncertainties associated with ecosystem dynamics (Allen et al., 2011).

20.3.4.3 Learning from Outcomes

Adaptive management is a dynamic, learning-based approach that emphasizes iterative decision-making informed by both successes and failures (Holling & Meffe, 1996). It is particularly suited to complex, unpredictable ecosystems where traditional control-oriented methods often fall short (Gunderson et al., 2016). Focusing on ecological resilience and integrating stakeholder perspectives may enable adaptive management responses to ecological changes while balancing trade-offs across ecosystem services (Cumming et al., 2015).

In practice, this approach has been applied successfully across various sectors. The U.S. Forest Service uses adaptive strategies to reconcile timber production with biodiversity conservation through continuous ecological monitoring (Williams & Hummelbrunner, 2010). Similarly, the Everglades restoration project employs hydrological assessments to guide real-time interventions (Zedler & Kercher, 2005). In fisheries, adaptive regulations such as catch limits based on ongoing stock assessments—support ecological sustainability and community livelihoods (Hilborn & Walters, 1992). These examples underscore the utility of adaptive management in fostering resilient, evidence-based environmental governance.

20.3.5 Minimizing Trade-Offs and Maximizing Benefits

Ecosystem-based solutions (EbS) aim to reconcile the often-conflicting objectives of development and conservation by minimizing trade-offs and maximizing benefits for both ecosystems and human communities. This principle encourages decision-makers to consider the long-term implications of their actions on ecosystem health and the provision of services. Aligning development initiatives with ecological sustainability may avoid unintended negative impacts on natural systems, ultimately fostering resilience and enhancing the well-being of communities.

Trade-offs in ecosystem management refer to the compromises that must be made when pursuing multiple objectives, such as economic development, biodiversity conservation, and social equity. These trade-offs can manifest in various ways:

20.3.5.1 Resource Allocation

When resources are allocated to one ecosystem service, such as agricultural production, it may come at the expense of another service, such as water quality regulation (Rodriguez et al., 2006). For example, intensive farming practices can lead to nutrient runoff that degrades water quality, illustrating a trade-off between agricultural productivity and ecosystem health.

20.3.5.2 Spatial and Temporal Trade-Offs

Spatial trade-offs occur when the benefits derived from one area are diminished by the use of another area for a different purpose. Temporal trade-offs involve decisions that prioritize short-term gains over long-term sustainability (Zheng et al., 2019). For instance, logging in a forest may provide immediate economic benefits but can compromise long-term carbon storage and habitat provision.

20.3.5.3 Stakeholder Conflicts

Different stakeholders may have competing interests regarding the provision of ecosystem services. For instance, conservationists may advocate for protected areas that limit land use, while local communities may depend on those lands for livelihoods (Friman et al., 2022). Balancing these interests requires careful negotiation and consideration of trade-offs.

EbS seeks to maximize benefits by promoting synergies among ecosystem services while minimizing trade-offs. Several strategies can facilitate this process. Effective EbS requires integrated planning that considers the interconnectedness of various ecosystem services. Adoption of a landscape approach that encompasses multiple land uses, such as agriculture, forestry, and conservation, enables decision-makers to identify opportunities for synergy (Costanza et al., 2014). Additionally, implementing NbS that harnesses ecological processes may help decision-makers achieve win–win outcomes that benefit both people and nature. Engaging local communities in decision-making processes is essential for identifying trade-offs and maximizing benefits. Local stakeholders possess valuable knowledge about their ecosystems and can contribute to the development of solutions that align with their needs and values (Berkes et al., 2001a, 2001b). Participatory approaches foster ownership of management strategies and enhance compliance with conservation measures.

Case Study 1

Land Use Planning in Bolivia

Bolivia, one of the world's most biodiverse countries, faces mounting pressures from deforestation, land-use change, and climate impacts (GEF, 2018). In response, integrated approaches have emerged to align agricultural development with biodiversity conservation. A key example is the incorporation of agrobiodiversity into land-use planning, particularly in highland regions where smallholder farmers cultivate diverse maize landraces critical for food security (Cochrane and Souza, 2018). Studies in Valle Alto demonstrate the viability of intercropping systems, such as maize with peach, that sustain biodiversity while enhancing farmer income. Participatory land-use planning enables communities to balance agricultural expansion with conservation by identifying cultivation zones and protecting critical habitats (Pérez et al., 2013). Incentive-based mechanisms—such as training in agroecological practices and reciprocal watershed agreements—further encourage sustainable land use (Friman et al., 2022). These efforts have improved ecosystem services, strengthened climate resilience, and empowered local communities through inclusive decision-making processes.

Case Study 2

Wetland Restoration Projects

Wetland restoration projects exemplify the potential of Ecosystem-Based Solutions (EbS) to deliver ecological and community benefits while minimizing trade-offs. Two key case studies—Fishlake Wetland Restoration in the UK and Sembakkam Lake Restoration in India—demonstrate how integrating ecological goals with community engagement enhances both biodiversity and ecosystem services. In Fishlake, a 60-ha site was restored to enhance flood resilience, improve wildlife habitat, and increase public access, supported by stakeholder engagement and hydrological interventions (Sheffield University, n.d.). Similarly, the Sembakkam Lake project addressed urban degradation through pollution control and eco-restoration, with local communities actively involved in planning and monitoring (Nature Conservancy India, 2022). Both projects report improved water quality, species diversity, and public participation. These examples highlight the effectiveness of participatory restoration models in aligning ecological functions with socio-economic needs. They offer replicable frameworks for urban and peri-urban wetland conservation, emphasizing the value of inclusive, nature-based approaches in climate adaptation and ecosystem resilience.

20.3.6 *Incorporating Best Available Science and Local Knowledge*

The implementation of Ecosystem-Based Solutions (EbS) requires a comprehensive understanding of ecological dynamics and socio-economic contexts. This understanding is best achieved by integrating the best available scientific evidence with traditional knowledge from local communities. Utilizing a wide range of knowledge sources guides decision-makers in developing culturally relevant and effective solutions that address both environmental and social challenges.

Scientific evidence serves as a critical foundation for the effective implementation of EbS. It provides data-driven insights into ecosystem functions, service provision, and the impacts of human activities on natural systems. For instance, a review by Murray et al. (2019) emphasizes that restoring coastal wetlands not only enhances biodiversity but also provides essential ecosystem services such as storm protection and carbon sequestration. Such scientific findings inform policymakers about the potential benefits of restoration initiatives, guiding resource allocation and management strategies. Moreover, scientific research helps identify best practices for ecosystem management. For example, studies have shown that adaptive management approaches, characterized by ongoing monitoring and iterative decision-making, can significantly improve resource management outcomes. Grounding EbS in robust scientific evidence may provide insights into better understanding the complexities of ecosystems and making informed decisions that promote sustainability while addressing community needs.

20.3.6.1 Incorporating Traditional Ecological Knowledge (TEK)

While scientific evidence is vital, incorporating traditional knowledge from local communities is equally important in developing effective EbS. TEK encompasses the insights, practices, and cultural beliefs that indigenous peoples and local communities have developed over generations (Berkes et al., 2001a, 2001b). This knowledge provides valuable context regarding local ecosystems, including species interactions, seasonal changes, and sustainable resource use practices. Engaging local communities in the decision-making process fosters a sense of ownership over natural resources and enhances compliance with conservation measures (Friman et al., 2022). For instance, in many regions worldwide, local stakeholders have successfully implemented sustainable land management practices based on their traditional knowledge, leading to improved ecological outcomes (Berkes et al., 2001a, 2001b). By integrating TEK with scientific research, decision-makers can develop solutions that are not only ecologically sound but also culturally relevant and socially acceptable.

20.3.6.2 Synergies Between Scientific Evidence and Traditional Knowledge

The combination of scientific evidence and traditional knowledge creates synergies that enhance the effectiveness of EbS. This integrative approach allows for a more comprehensive understanding of ecosystem dynamics while addressing the socioeconomic realities faced by local communities. For example, collaborative projects that involve scientists working alongside local stakeholders can lead to innovative solutions that leverage both forms of knowledge. Additionally, integrating these knowledge systems can help identify trade-offs between different ecosystem services. For instance, understanding local agricultural practices through TEK can inform scientists about how specific farming techniques impact soil health and biodiversity (Rodriguez et al., 2006). This information enables decision-makers to develop strategies that maximize benefits across multiple ecosystem services while minimizing negative impacts.

20.3.7 Participatory Governance

Participatory governance is an essential framework for ensuring that all stakeholders have a voice in ecosystem management decisions. This principle emphasizes transparency, accountability, and inclusiveness in governance processes, recognizing that diverse perspectives contribute to more equitable outcomes. Engaging local communities in decision-making may foster participatory governance, which in turn promotes a sense of ownership and responsibility toward ecosystem conservation, ultimately leading to more sustainable and effective management practices.

Participatory governance involves the active engagement of various stakeholders including local communities, indigenous peoples, government agencies, and non-governmental organizations in the decision-making processes related to resource management. This approach is grounded in the recognition that local knowledge and experiences are invaluable for understanding ecological dynamics and addressing environmental challenges (Berkes et al., 2001a, 2001b). Incorporating diverse perspectives into governance frameworks might help to enhance the legitimacy and effectiveness of management decisions. One of the primary benefits of participatory governance is its ability to promote transparency and accountability. When stakeholders are involved in decision-making processes, they are more likely to hold each other accountable for their actions. This increased scrutiny can lead to better compliance with regulations and conservation measures (Friman et al., 2022). Furthermore, transparency in governance processes fosters trust among stakeholders, which is crucial for building collaborative relationships necessary for effective ecosystem management.

Engaging local communities in decision-making not only empowers them but also enhances their commitment to sustainable resource management. When communities have a say in how their natural resources are managed, they are more likely to adopt practices that align with conservation objectives (Ribot et al., 2006). For instance, community-based resource management initiatives have shown that involving local stakeholders can lead to improved ecological outcomes while simultaneously addressing social and economic needs (Berkes et al., 2001a, 2001b). Participatory governance also allows for the integration of TEK into management strategies. TEK encompasses the insights and practices developed by indigenous peoples and local communities over generations, providing a rich understanding of local ecosystems (Berkes et al., 2001a, 2001b).

Several case studies illustrate the effectiveness of participatory governance in ecosystem management. For example, in Brazil's Atlantic Forest region, community-led initiatives have successfully restored degraded landscapes while enhancing biodiversity. These projects involved residents in planning and implementation processes, leading to increased community ownership and improved ecological outcomes. Another example is found in the La Campana–Peñuelas Biosphere Reserve in Chile, where participatory governance has been enhanced through transdisciplinary approaches that engage diverse stakeholders in assessing the reserve's performance (Peredo et al., 2020). This initiative has demonstrated how collaborative efforts can address environmental conflicts while promoting social equity.

20.4 Integrating EbS into Climate Policy

Incorporating EbS into national and international climate policies offers a strategic pathway to achieve sustainability and resilience in addressing climate change. Despite their critical importance, EbS have often been undervalued or overlooked in policy frameworks. Integrating them into climate policy involves recognizing their contributions, embedding them into economic systems, and ensuring their protection through innovative governance mechanisms (Costanza et al., 1997). The integration of EbS into climate policy is essential for developing effective strategies that address the challenges posed by climate change while promoting sustainable development.

At the national level, integrating ecosystem services into climate policy necessitates that governments align environmental objectives with socioeconomic goals. Policies can be designed to incentivize practices that enhance ecosystem services, such as reforestation, conservation agriculture, and wetland restoration. For instance, Payment for Ecosystem Services (PES) schemes reward landowners or communities for maintaining or improving ecosystem services. Costa Rica's PES program is a notable example, where landowners receive payments for reforestation and forest conservation, contributing to the country's climate mitigation goals (Pagiola, 2008). Integrating ecosystem services into climate policy helps identify synergies and trade-offs between conservation objectives and development goals. For instance, maintaining healthy wetlands can provide flood protection while also enhancing biodiversity and improving water quality (Murray et al., 2019).

Benefits of Integrating Ecosystem Services into Climate Policy

1. Enhanced Resilience: Incorporating ecosystem services into climate policy can strengthen community resilience to climate-related impacts. For example, protecting coastal ecosystems such as mangroves can reduce vulnerability to storm surges and flooding while providing habitat for marine species (Barbier et al., 2011).
2. Cost-Effectiveness: Nature-based solutions that leverage ecosystem services often prove to be more cost-effective than traditional engineering approaches. For instance, using wetlands for water purification or flood prevention can be less expensive than constructing artificial treatment facilities or levees (Maes et al., 2012).
3. Improved Stakeholder Engagement: Integrating ecosystem services into climate policy fosters greater stakeholder engagement by highlighting the value of local ecosystems to communities. When communities recognize the benefits provided by nearby natural resources, they are more likely to participate in conservation efforts (Berkes et al., 2001a, 2001b).

Case Study 1

Mexico City Water Resilience Project

The Mexico City Water Resilience Project is a comprehensive initiative designed to enhance the city's capacity to manage its water resources in the face of climate change and urbanization challenges. As one of the largest megacities in the world, Mexico City faces significant water-related challenges, including water supply shortages, pollution, and vulnerability to flooding. The project seeks to address these challenges through a combination of innovative strategies, community engagement, and the integration of ecosystem services into water management practices (WWF, 2023).

One of the project's core objectives is to align with national and local climate policies, such as Mexico's General Law on Climate Change and the country's Nationally Determined Contributions (NDCs) under the Paris Agreement (GCF, 2023). Integrating ecosystem services into water management decisions may assist the project to enhance the adaptive capacity of local communities while ensuring sustainable resource use. The implementation of the Mexico City Water Resilience Project involved several key strategies. The project promotes the installation of RWHS in households, particularly targeting vulnerable groups such as single mothers, Indigenous people, and individuals with disabilities. These systems capture rainwater from rooftops and store it for domestic use, providing families with a reliable source of water for several months each year (World Economic Forum, 2021). To date, over 13,000 households have benefited from this initiative. Community Engagement: Engaging local communities in decision-making processes is central to the project's success. Training programs are provided for residents on the operation and maintenance of RWHS, which not only empowers them but also promotes equitable distribution of water-related tasks between genders (World Economic Forum, 2021). This participatory approach fosters a sense of ownership over water resources and encourages sustainable practices. Integrated Water Management: The project emphasizes an integrated approach to water management that combines green infrastructure with traditional engineering solutions. For example, optimizing constructed wetlands alongside RWHS can enhance socio-hydrological resilience by improving water quality and increasing storage capacity during heavy rainfall events (BGS, 2023). This holistic strategy recognizes the interconnectedness of urban water systems and ecosystem services.

The Mexico City Water Resilience Project has yielded several positive outcomes. Early assessments indicate improvements in water security for participating households, reducing their vulnerability to climate-related impacts (GCF, 2023). Additionally, the emphasis on community engagement has strengthened social cohesion around shared resource management goals.

Case Study 2

Lamu Port Development in Kenya

The Lamu Port Development project is a significant infrastructure initiative aimed at transforming Lamu County, Kenya, into a key transport and logistics hub for East Africa and the Horn of Africa. As part of the larger Lamu Port South Sudan Ethiopia Transport

(LAPSSET) corridor program, this project seeks to enhance regional connectivity and economic growth while integrating an understanding of ecosystem services into its planning and implementation processes.

The Lamu Port project is designed to accommodate some of the world's largest ships, with a capacity to handle approximately 23.9 million tonnes of cargo by 2030 (Kenya Ports Authority [KPA], 2023). The initiative aligns with Kenya's Vision 2030 development plan, which aims to position the country as a competitive hub for transport and logistics in the region (Orf, 2023). Recognizing the potential impacts on local ecosystems, the project emphasizes the importance of integrating ecosystem services into its planning framework to ensure sustainable development. One of the primary objectives of the Lamu Port project is to minimize adverse environmental impacts while maximizing social and economic benefits for local communities. Before construction, comprehensive social and ecological impact assessments (ESIAs) were conducted to evaluate potential effects on marine ecosystems, mangroves, and local livelihoods (FAO, 2023). These assessments informed decision-making processes and helped identify strategies for mitigating negative impacts associated with infrastructure development.

Conducting comprehensive assessments is an essential step in engaging local communities in decision-making processes and implementing effective mitigation measures. As Kenya continues to pursue ambitious infrastructure projects under the LAPSSET corridor program, lessons learned from the Lamu Port project will be valuable for future endeavors seeking to harmonize development objectives with ecological sustainability.

20.5 Challenges to Implementing EbS

While EbS present promising approaches to addressing climate change and promoting environmental sustainability, several barriers hinder their widespread adoption and successful implementation. These challenges include limited funding, complex governance structures, and inadequate public awareness, each of which poses significant obstacles to scaling up EbS at local, national, and global levels. Understanding and addressing these barriers is crucial for overcoming implementation gaps and ensuring that EbS can fully deliver its potential in mitigating the impacts of climate change.

20.5.1 Funding Constraints

One of the primary challenges to implementing EbS is the lack of sufficient and consistent funding. Many EbS initiatives, such as ecosystem restoration or sustainable land management, require long-term financial commitments. Yet, they often compete with short-term, high-return projects that may be perceived as more economically viable. Funding for EbS is usually fragmented and can be challenging to secure, especially for initiatives in low-income countries or marginalized communities (Mandle et al., 2021). Furthermore, the financial benefits of EbS, such as improved ecosystem services and climate resilience, often accrue over a long period, making it harder to attract private investment or government funding. The challenge is compounded by the lack of dedicated financial instruments that prioritize ecosystem-based solutions in global financial markets.

20.5.2 Governance and Institutional Challenges

The complexity of governance structures presents another significant challenge to implementing EbS. Effective ecosystem management often requires collaboration across multiple levels of government (local, regional, and national) as well as between various stakeholders, including community groups, private sectors, and non-governmental organizations. However, fragmented and siloed governance structures can impede coordination and decision-making, particularly in countries with weak institutional frameworks or unclear land tenure rights. This lack of coordination can lead to delays in project implementation, inefficiency, or conflict over land and resource use.

Additionally, local communities often face challenges in accessing decision-making processes despite being the primary beneficiaries of EbS projects. Ensuring inclusive governance is crucial for the success of these initiatives. The participation of local stakeholders, particularly indigenous communities and farmers, is essential in the design and management of ecosystem-based solutions. Without active engagement from those who directly interact with ecosystems, the risk of failure or unintended consequences increases.

20.5.3 Public Awareness and Education

A significant barrier to the widespread adoption of EbS is the limited public awareness and understanding of the importance of ecosystem services. Many communities, especially in urban areas, may not recognize the value of ecosystems in providing critical services such as flood regulation, water purification, and climate stabilization. The lack of awareness can result in limited support for policies that prioritize environmental conservation and sustainable practices, making it challenging to implement EbS at the scale needed for significant climate impact (Gupta et al., 2015).

Furthermore, misconceptions about the costs and benefits of EbS can hinder their adoption. Some policymakers and businesses may view ecosystem-based approaches as too costly or impractical compared to engineered solutions. As a result, there may be resistance to integrating EbS into national or local climate policies, even when they offer long-term, cost-effective benefits. Increasing public awareness through education campaigns, media outreach, and community engagement can help shift perceptions and build broader support for EbS.

20.5.4 Scientific and Technical Barriers

Another challenge is the limited scientific knowledge and technical capacity to design and implement EbS effectively. While there is growing evidence supporting the effectiveness of EbS, knowledge gaps remain in understanding the specific ecological, social, and economic conditions under which these solutions work best. EbS often requires integrated approaches that combine environmental, social, and economic data, which can be complex and resource-intensive to collect and analyze. Additionally, there may be a lack of local expertise to implement these solutions effectively, particularly in regions where ecosystem management is underdeveloped.

20.5.5 Monitoring and Evaluation

Effective monitoring and evaluation are crucial for ensuring the success of EbS projects. However, tracking the impacts of EbS on ecosystem services and climate resilience can be challenging due to the complex and often long-term nature of these solutions. Without proper monitoring, it is challenging to evaluate the effectiveness of EbS and make necessary adjustments to enhance outcomes. The lack of standardized metrics and methodologies for assessing ecosystem-based projects further complicates this issue, resulting in a scarcity of reliable data to inform policy decisions (Mandle et al., 2021).

20.6 Future Directions: Policy Recommendations for Sustainable Implementation

As the urgency of addressing climate change intensifies, integrating Ecosystem-Based Solutions (EbS) and green credit systems into climate strategies becomes paramount. To enhance this integration, several actionable policy recommendations can be proposed. These recommendations aim to create a supportive framework that encourages investment in EbS while ensuring that green credit systems effectively contribute to climate mitigation and adaptation efforts.

First and foremost, governments should establish clear regulatory frameworks that define and standardize EbS and green credit systems. This includes developing comprehensive guidelines for the issuance of green bonds and carbon credits ensuring that projects meet rigorous environmental and social criteria. Standardization will enhance transparency and credibility, attracting more investors to these financial instruments (Flammer, 2021). Additionally, creating a centralized registry for green credits can facilitate tracking and trading, thereby increasing market efficiency and trust among stakeholders (Böhringer et al., 2022). Furthermore, it is essential to promote public–private partnerships (PPPs) to leverage resources and expertise in implementing EbS. Governments can incentivize collaboration between public entities, private companies, and non-governmental organizations (NGOs) to design and execute projects that deliver both ecological and economic benefits. For instance, PPPs can facilitate the development of large-scale reforestation or wetland restoration projects, which can be funded through green bonds while simultaneously generating carbon credits (Peters-Stanley et al., 2013). To foster collaboration, stakeholders can share risks and rewards, ultimately leading to more successful and sustainable outcomes. Capacity building is another critical area for policy intervention. Governments should invest in training programs and resources to enhance the knowledge and skills of local communities, practitioners, and policymakers regarding EbS and green credit systems. This includes providing technical assistance for project design, implementation, and monitoring, as well as educating stakeholders about the benefits and opportunities associated with these approaches (Bennett et al., 2016). Empowering local communities to engage in EbS initiatives not only enhances project effectiveness but also ensures that the benefits are equitably distributed.

Additionally, integrating EbS into national and local climate action plans is crucial for ensuring that these solutions are prioritized in climate strategies. Policymakers should conduct comprehensive assessments of the potential contributions of EbS to climate goals, including carbon sequestration, biodiversity enhancement, and community resilience. Governments can create synergies between ecological and economic objectives, ultimately leading to more holistic and effective climate strategies by incorporating EbS into climate action frameworks (Díaz et al., 2015).

It is crucial to establish monitoring and evaluation mechanisms to assess the effectiveness of EbS and green credit systems continuously. Policymakers should develop indicators and metrics that measure the ecological, social, and economic impacts of these initiatives. Regular reporting and feedback loops will enable stakeholders to learn from successes and challenges, fostering adaptive management and continuous improvement.

20.7 Conclusion

Integrating Ecosystem-Based Solutions (EbS) with green credit systems presents a compelling strategy for addressing climate change and biodiversity loss. EbS through conservation, restoration, and sustainable management of ecosystems—supports carbon sequestration and enhances climate resilience. Coupled with green credit mechanisms, such as carbon credits and green bonds, they offer financial incentives to scale sustainable practices. However, barriers including limited funding, weak governance, and low public awareness must be addressed. Strengthening regulatory frameworks, ensuring financial transparency, and promoting community-level capacity building are essential. Incorporating EbS and green credit frameworks into national climate strategies can bridge ecological goals with economic incentives. Ultimately, valuing nature's role in climate resilience is key to achieving long-term sustainability and a carbon–neutral future.

References

Abdelmagied, M., & Mpheshea, M. (2020). *Ecosystem-based adaptation in the agriculture sector—A nature-based solution (NbS) for building the resilience of the food and agriculture sector to climate change*. Rome, FAO. https://openknowledge.fao.org/server/api/core/bitstreams/c98872f8-0d2c-47aa-b916-ea8c3743df00/content

Resource Africa. (2024, April 2). *The commons and the theory behind community-based natural resource management*. Resource Africa. https://www.resourceafrica.net/the-commons-and-the-theory-behind-community-based-natural-resource-management/

Agarwal, A., Narain, S., & Singh, S. (2000). *Water pollution in India*. Centre for Science and Environment.

Agrawal, A., & Ostrom, E. (1999). Collective action, property rights, and decentralization in resource use in India and Nepal. *Politics & Society, 27*(4), 485–514.

Allen, C. R., & Gunderson, L. H., et al. (2011). Pathology and resilience in social-ecological systems. In F. S. Chapin (Ed.), *Principles of ecosystem stewardship: Sustainability through resource management* (pp. 37–51). Springer.

BanQu. (2024). *India's Green Credit Programme: What it is & how to prepare*. https://www.banqu.co/blog/indias-green-credit-programme-what-it-is-how-to-prepare

Barbier, E. B., Hacker, S. D., Kennedy, C., Koch, E. W., Stier, A. C., & Silliman, B. R. (2011). The value of estuarine and coastal ecosystem services. *Ecological Monographs, 81*(2), 169–193.

Berkes, F. (2009). Evolution of co-management: Role of knowledge generation, bridging organizations and social learning. *Journal of Environmental Management, 90*(5), 1692–1702.

Berkes, F., Folke, C., & Gadgil, M., et al. (2001a). Traditional ecological knowledge, biodiversity, resilience and sustainability. In F. Berkes (Ed.), *Navigating social-ecological systems: Building resilience for complexity and change* (pp. 271–287). Cambridge University Press.

Berkes, F., Hughes, T. P., Steneck, R. S., & Wilson, J. A. (2001b). Globalization of conservation: A perspective from Canada. *Environmental Conservation, 28*(2), 102–112.

BGS (British Geological Survey). (2023). *Understanding the resilience of Mexico City to future water crises*. https://www.bgs.ac.uk/news/understanding-the-resilience-of-mexico-city-to-future-water-crises/

Böhringer, C., Fischer, C., Rosendahl, K. E., & Rutherford, T. F. (2022). Potential impacts and challenges of border carbon adjustments. *Nature Climate Change, 12*(1), 22–29.

Cochrane, T. A., & Souza, R. C. (2018). Agrobiodiversity conservation in smallholder systems: Evidence from Bolivia. *Agricultural Systems, 165*(1), 1–10.

Cohen-Shacham, E., Andrade, A., Benessaiah, K., Bernhardt, J. R., & Gollner, S.(2019). *Nature-based solutions to address global societal challenges*. The Nature Conservancy.

Conant, R. T., Cerri, C. E., Osborne, B. B., & Paustian, K. (2017). Grassland management impacts on soil carbon stocks: A new synthesis. *Ecological Applications, 27*(2), 662–668.

Costanza, R., d'Arge, R., De Groot, R., Farber, S., Grasso, M., Hannon, B., Limburg, K., Naeem, S., O'neill, R. V., Paruelo, J., Raskin, R. G., & Van Den Belt, M. (1997). The value of the world's ecosystem services and natural capital. *Nature, 387*(6630), 253–260.

Costanza, R., de Groot, R., Farber, S., Grasso, M., Hannon, B., Limburg, K., Farber, S., & Ricketts, T. (2014). Changes in the global value of ecosystem services. *Global Environmental Change, 26*(1), 152–158.
Cumming, G. S., Allen, C. R., & Angeler, D. G. (2015). Understanding the role of adaptive capacity in managing social-ecological systems. *Ecology and Society, 20*(1), 12.
Díaz, S., Demissew, S., Carabias, J., Joly, C., Lonsdale, M., Ash, N., Larigauderie, A., Adhikari, J. R., Arico, S., Báldi, A., Bartuska, A., & Zlatanova, D. (2015). The IPBES conceptual framework—Connecting nature and people. *Current Opinion in Environmental Sustainability, 14*, 1–16.
Drishti IAS. (2024). *Green credit program*. https://www.drishtiias.com/daily-updates/daily-news-analysis/green-credit-program-1
FAO (Food and Agriculture Organization). (1996). *Decentralization and forestry: A review of experiences*. FAO Forestry Paper No. 132.
FAO (Food and Agriculture Organization). (2023). *The economics of ecosystem services of the Tana River Basin*. https://www.fao.org/fishery/ar/openasfa/ce0a2be4-d4e0-42d6-8555-5a0152f3de41
FAO. (1996). *Participatory forest management: A strategy for sustainable forest management*. Food and Agriculture Organization of the United Nations.
Flammer, C. (2021). Corporate green bonds. *Journal of Financial Economics, 142*(2), 499–516.
Fourqurean, J. W., Duarte, C. M., Kennedy, H., Marbà, N., Holmer, M., Mateo, M. A., Apostolaki, E. T., Kendrick, G. A., Krause-Jensen, D., McGlathery, K. J. & Serrano, O. (2012). Seagrass ecosystems as a globally significant carbon stock. *Nature Geoscience, 5*(7), 505–509.
Friman, E. L., Galvin, K. A., & Karambiri, H. (2022). Community forestry management implementation in Burkina Faso: Gendered dynamics and customary complexity. *International Journal of the Commons*.
GCF (Green Climate Fund). (2023). *Enhancing resilience of vulnerable families in Mexico city to climate change impacts on water*. https://www.greenclimate.fund/document/enhancing-resilience-vulnerable-families-mexico-city-impacts-climate-change-water-0
Global Environment Facility [GEF]. (2018). *Biodiversity conservation GEFI & II projects*. https://www.gefieo.org/sites/default/files/documents/lb-case-study-bolivia.pdf
Global Water Partnership [GWP]. (2000). *Integrated water resources management*. https://www.gwp.org/en/GWP-CEE/about/what-is-iwrm/
Government of India. (2023). *Green Credit Program (GCP) and the Ecomark scheme*. https://pib.gov.in/PressReleaseIframePage.aspx?PRID=1967476
Gunderson, L., Allen, C., & Holling, C. S. (2016). *Foundations of ecological resilience*. Island Press.
Gupta, G., Parihar, S. S., Ahirwar, N. K., Snehi, S. K., & Singh, V. (2015). Plant growth promoting rhizobacteria (PGPR): Current and future prospects for development of sustainable agriculture. *Journal of Microbial & Biochemical Technology, 7*(2), 096–102.
GWP. (2000). *Integrated water resources management*. Global Water Partnership.
Hilborn, R., & Walters, C. J. (1992). *Quantitative fisheries stock assessment: Choice, dynamics and uncertainty*. Chapman & Hall.
Holling, C. S. (1978). *Adaptive environmental assessment and management*. Wiley.
Holling, C. S., & Meffe, G. K. (1996). Command and control and the pathology of natural resource management. *Conservation Biology, 10*(2), 328–337.
Koshish India. (2024). *Green credit programme: What it is & why it's important*. https://www.koshishindia.in/blogs/green-credit-programme
IUCN. (2011). https://portals.iucn.org/library/efiles/documents/2011-064.pdf
Jones, B., & Murphree, M. (2001). The evolution of community-based wildlife management in Africa. In D. Hulme & M. Murphree (Eds.), *African wildlife & livelihoods: The promise and performance of community conservation* (pp. 1–22). James Currey.
Joshi, R., Sharma, B., Singh, H., Dhakal, N., Ayer, S., & Maraseni, T. (2024). Poplar plantation as an agroforestry approach: Economic benefits and its role in carbon sequestration in North India. *Journal of Resources and Ecology, 15*(4), 880–888.

Kabisch, N., Korn, H., Stadler, J., & Bonn, A. (2017). *Nature-based solutions to climate change adaptation in urban areas: Linkages between science, policy and practice*. Springer Nature.

Kenya Ports Authority [KPA]. (2023). *Lamu port overview*. https://www.kpa.co.ke/lamu-port/

Kumar, M., & Singh, H. (2020). Agroforestry as a nature-based solution for reducing community dependence on forests to safeguard forests in rainfed areas of India. In *Nature-based solutions for resilient ecosystems and societies* (pp. 289–306). Singapore: Springer Singapore.

Maes, J., Paracchini, M. L., & Zulian, G. (2012). Mapping ecosystem services for policy support. *Environmental Science & Policy, 19–20*(1), 1–12.

Mandle, L., Shields-Estrada, A., Chaplin-Kramer, R., Mitchell, M. G., Bremer, L. L., Gourevitch, J. D., Hawthorne, P., Johnson, J. A., Robinson, B. E., Smith, J. R., Sonter, L. J., & Ricketts, T. H. (2021). Increasing decision relevance of ecosystem service science. *Nature Sustainability, 4*(2), 161–169.

Meynen, W., & Doornbos, M. (2004). Decentralizing natural resource management: A recipe for sustainability and equity. *European Journal of Development Research*.

Murphree, M. (1993). Communal areas management programme for indigenous resources (CAMPFIRE) in Zimbabwe. In D. Hulme & M. Murphree (Eds.), *African wildlife & livelihoods: The promise and performance of community conservation* (pp. 43–64). James Currey.

Murray, B. C., Pendleton, L. H., & Jenkins, W. A. (2019). *Coastal blue carbon: Methods for assessing carbon stocks in Mangroves Seagrasses*. The Nature Conservancy.

Narayan, S., Beck, M. W., Reguero, B. G., Losada, I. J., Van Wesenbeeck, B., Pontee, N., Sanchirico, J. N., Ingram, J. C., Lange, G. M., & Burks-Copes, K. A. (2016). The effectiveness, costs and coastal protection benefits of natural and nature-based defences. *PLoS One, 11*(5), e0154735.

Nature Conservancy India. (2022). *An integrated approach for urban wetland restoration—A case study of Sembakkam Lake*. https://www.tncindia.in/content/dam/tnc/nature/en/documents/india/CoP27SembakkamReport.pdf

Pagiola, S. (2008). Payments for environmental services in Costa Rica. *Ecological Economics, 65*(4), 712–724.

Pan, Y., Birdsey, R. A., Fang, J., Houghton, R., Kauppi, P. E., Kurz, W. A., Phillips, O. L., Shvidenko, A., Lewis, S. L., Canadell, J. G., Ciais, P., & Hayes, D. (2011). A large and persistent carbon sink in the world's forests. *Science, 333*(6045), 988–993.

Parish, F., Sirin, A. A., Charman, D., Joosten, H., Minaeva, T. Y., & Silvius, M. (2008). Assessment on peatlands, biodiversity and climate change.

Peredo Parada, A., Winkler, H., & Hauck, J. (2020). Enhancing participatory governance in biosphere reserves through co-production of knowledge. *Frontiers in Environmental Science, 8*(1), 1–14.

Pérez, M. R., Becerra, J. C., & Gutiérrez, J. A. (2013). Land use planning as a tool for agrobiodiversity conservation: A case study from Bolivia. *Environmental Management, 52*(6), 1355–1369.

Raymond, C. M., Frantzeskaki, N., Kabisch, N., Berry, P., Breil, M., Nita, M. R., Geneletti, D., & Calfapietra, C. (2017). A framework for assessing and implementing the co-benefits of nature-based solutions in urban areas. *Environmental Science & Policy, 77*, 15–24.

Ribot, J. C., Agrawal, A., & Larson, A. M. (2006). Recentralizing while decentralizing: How national governments reappropriate forest resources. *World Development, 34*(11), 1864–1886.

Ribot, J. C. (2002). Democratic decentralization of natural resources: Institutional choice and discretionary power in African conservation. *Public Administration & Development, 22*(1), 53–65.

Rodriguez, J. P., Beard Jr., K. H., Bennett, E. M., Cummings, G. S., & Peterson, G. D. (2006). Trade-offs across space, time and ecosystem services. *Ecology and Society, 11*(1), 28.

Schwartzman, S., Moreira, A., & Nepstad, D.(2000). Protected areas in the Brazilian Amazon: a new approach to conservation. In D. Pollock et al. (Eds.), *The Amazon rainforest* (pp. 157–174). Yale University Press.

Seddon, N., Chausson, A., Berry, P., Girardin, C. A., Smith, A., & Turner, B. (2020). Understanding the value and limits of nature-based solutions to climate change and other global challenges. *Philosophical Transactions of the Royal Society B, 375*(1794), 20190120.

Sheffield University. (n.d.). *Case study: Wetland restoration at fishlake.* https://www.sheffield.ac.uk/doncatchment/projects/fishlake-wetland-restoration

Singh, H. (2024). *Urban forests, climate change and environmental pollution: Physio-biochemical and molecular perspectives to enhance urban resilience.* Springer Nature. https://doi.org/10.1007/978-3-031-67837-0

Srivastav, A., & Singh, H. (2024). Agroforestry adoption in the eastern region of Uttar Pradesh, India: A way to mitigate climate change. In *Forests and climate change: Biological perspectives on impact, adaptation, and mitigation strategies* (pp. 445–467). Singapore: Springer Nature Singapore.

UNDESA. (2005). *Integrated Water Resources Management (IWRM).* https://www.un.org/waterforlifedecade/iwrm.shtml

United Nations. (2013). *Water security & the global water agenda.* United Nations.

Wang, Y., & Zhi, Q. (2016). The role of green finance in environmental protection: Two aspects of market mechanism and policies. *Energy Procedia, 104*, 311–316.

Williams, B. K., & Hummelbrunner, R. (2010). *Systems concepts in action: A practitioner's toolkit.* Stanford University Press.

World Economic Forum. (2021). *How improving water security is tackling the gender gap in Mexico city.* https://www.weforum.org/stories/2021/09/water-security-gender-gap-mexico-city/

WWF (World Wildlife Fund). (2023). *Case studies on integrating ecosystem services and climate resilience in infrastructure development: Lessons for advocacy.* https://files.worldwildlife.org/wwfcmsprod/files/Publication/file/qopcq0me_Case_Studies_on_Integrating_Ecosystem_Services_and_Climate_Resilience_in_Infrastructure_Development_Lessons_for_Advocacy.PDF

Xu, Z., Peng, J., Qiu, S., Liu, Y., Dong, J., & Zhang, H. (2021). Responses of spatial relationships between ecosystem services and the Sustainable Development Goals to urbanization. *Science of the Total Environment, 850*, Article 157868.

Zedler, J. B., & Kercher, S. (2005). Wetland resources: Status, trends, ecosystem services. In D. J. Patil et al.(Eds.), *Wetlands ecology* (pp. 31–56). Springer Netherlands.

Zheng, Z. Y., Wang, Y. Z., Liu, Y. H., & Li, Y. F. (2019). Trade-offs among ecosystem services under different land use scenarios: A case study from China. *Land Use Policy, 87*(1), 104029.

Chapter 21
Forest-Based Climate Solutions in the Indian Himalayas: A Community-Centric Pathway to Carbon Neutrality

Harshit Pant, Aditi Mishra, and Paras Upadhyaya

Abstract The Indian Himalayan Region (IHR) is a globally significant ecological zone that plays a critical role in India's efforts toward carbon neutrality. This chapter examines the multifaceted role of community-based forest governance, particularly through Van Panchayats (VPs), in promoting forest-based climate mitigation and ecosystem resilience. Drawing on empirical assessments conducted in 35 VPs of the Hawalbagh development block in Uttarakhand, the study quantifies carbon sequestration potential, evaluates forest fire mitigation strategies, and assesses the viability of pine needle-based bioeconomy models. Based on estimates from the Forest Survey of India, the total forest area managed by these Van Panchayats (VPs) is approximately 881.62 ha, which sequesters an estimated 5,289.72 tons of CO_2 annually. Furthermore, the collection and utilization of pine needle biomass significantly reduced fire incidences while providing livelihood opportunities to local communities, particularly women. The production of over 11,900 eco-friendly items offset an estimated 2.18 tons of CO_2, demonstrating the tangible benefits of circular bioeconomy practices. The chapter emphasizes the importance of integrating Traditional Ecological Knowledge (TEK), participatory governance, and modern monitoring tools to scale ecosystem-based adaptation (EbA) and nature-based solutions (NbS). Despite institutional and policy challenges, the study highlights the potential of VPs to serve as decentralized, inclusive, and scalable platforms for advancing India's Nationally Determined Contributions (NDCs) and achieving long-term climate goals through sustainable forest management.

Keywords Carbon neutrality · Indian Himalayan region · Forest-based climate solutions · Community forest management · Ecosystem-based adaptation · Traditional ecological knowledge

H. Pant (✉) · A. Mishra · P. Upadhyaya
Centre for Socio-economic Development, G.B. Pant National Institute of Himalayan Environment, Kosi-Katarmal, Almora, Uttarakhand, India
e-mail: harshit.pantj@gbpihed.nic.in

H. Singh (ed.), *Forestry for a Carbon-Neutral and Sustainable Future*, Sustainability Solutions, https://doi.org/10.1007/978-3-032-06427-1_21

21.1 Introduction

The Indian Himalayan Region (IHR) is a biodiversity hotspot and a critical carbon sink, making it central to India's climate action and sustainable development goals (Singh & Kumar, 2022). Forest-based climate solutions in this region integrate traditional knowledge, community engagement, and modern science to promote carbon neutrality, support local livelihoods, and enhance ecosystem resilience (Saikia et al., 2021). The Indian Himalayan Region (IHR) represents one of the most ecologically significant and environmentally sensitive zones in the world. Stretching over 2,500 km across 11 Indian states from Jammu & Kashmir in the west to Arunachal Pradesh in the east, the region is home to diverse flora and fauna, pristine forests, and vital watersheds. These forests are not only biodiversity hotspots but also critical carbon sinks that play a substantial role in mitigating the impacts of climate change. As India moves toward achieving its ambitious climate commitment of reaching net-zero carbon emissions by 2070, the IHR assumes a pivotal position in this national and global endeavour. Forests in the IHR sequester large volumes of atmospheric carbon dioxide, regulate hydrological cycles, and provide ecosystem services vital for the livelihoods of millions of mountain dwellers (Bhatt et al., 2024a, b, Hussain et al., 2024). However, these forests are increasingly under threat from climate change-induced stressors, unsustainable land-use practices, and infrastructural development. The urgency to preserve and restore these ecosystems has never been greater and has been reported by many researchers across the region (Dhyani et al., 2021; Elegbeleye et al., 2025; Erdoğdu et al., 2025; Gabric, 2023; Gadgil et al., 1993; Gupta et al., 2024; Haq et al., 2023; IPCC, 2018; Karki & Adhikari, 2015; Kc et al., 2022; Kumar et al., 2022; Kumari et al., 2024a; Longbottom et al., 2014; Makundi, 1997; Malhi et al., 2002; Meetei et al., 2023; Negi et al., 2023; Panthi, 2011; Prentice et al., 2001; Rajbhandari et al., 2024; Rao et al., 2003, 2022; Rist et al., 2010; Sahana Florence & Mishra, 2024; Sharma et al., 2024; Singh, 2008; Tiwari & Sunny, 2024; Ontl et al., 2020).

Empirical evidence from India's State of Forest Reports indicates a substantial increase in forest carbon stocks, from 6.07 gigatons in 1994 to 7.20 gigatons in 2021, reflecting an annual sequestration rate of approximately 48.5 million tonnes of carbon. This enhancement in carbon storage capacity can be attributed to policy-driven afforestation initiatives, improved forest management practices, and the restoration of degraded ecosystems. Notably, India's Third National Communication to the UNFCCC (2023) quantifies the climate-regulating function of its forests, reporting that they functioned as a net sink, offsetting 20.3% of the nation's gross greenhouse gas emissions in 2019. This finding underscores the critical role of terrestrial ecosystems in national climate change mitigation strategies. In alignment with its updated Nationally Determined Contributions (NDCs), India has committed to creating an additional carbon sink of 2.5–3.0 gigatons of CO_2 equivalent through the expansion of forest and tree cover by 2030. (Nayak & Srivastava, 2021). This target, grounded in the principles of nature-based solutions, represents a significant component of India's climate action framework, integrating both mitigation and adaptation objectives while addressing sustainable development priorities. The implementation of this commitment necessitates continued investment in forest conservation, sustainable management practices, and robust monitoring systems to ensure the permanence and additionality of carbon stocks. These measures are particularly crucial in light of projected climate change impacts on forest ecosystems, including altitudinal shifts in vegetation zones and changes in species composition, as observed in Himalayan forests.

21.2 The Indian Himalayan Region (IHR): Ecological and Socioeconomic Significance

21.2.1 Ecological Diversity

The IHR spans approximately 533,000 km^2, covering about 16.2% of India's total land area. It encompasses a vast altitudinal gradient from foothills at 200 m to high-altitude peaks exceeding 7,000 m. This range creates a unique mosaic of ecosystems, ranging from subtropical forests to alpine meadows, which support over 10,000 species of plants and 1,100 species of vertebrates (Mishra et al., 2024). Many of these are endemic, rare, or threatened, making the IHR a global biodiversity hotspot.

The forests of the region contribute significantly to India's forest carbon stock. According to the India State of Forest Report (ISFR, 2021), the total forest and tree cover in the IHR is approximately 38% of its geographical area much higher than the national average of 24%. These forests serve as vital carbon sinks, absorbing large amounts of CO_2 annually.

21.2.2 Socioeconomic Dependence on Forests

Forests in the IHR provide direct and indirect benefits to over 50 million people. They offer timber, fuelwood, fodder, non-timber forest products (NTFPs), and ecosystem services, including soil conservation, water regulation, and climate buffering (Sati, 2023a, 2023b). Traditional agroforestry systems and community-managed forests are embedded in local culture and livelihoods.

Yet, the economic dependence on forest resources often places pressure on these ecosystems, leading to overharvesting and degradation. The dual role of forests as both a source of livelihood and a tool for carbon sequestration, makes sustainable management a complex yet essential endeavor (Negi, 2022, Bhatt et al., 2024a).

21.3 Climate Change and Its Impact on Himalayan Forests

21.3.1 Ecological Impacts

Climate change is manifesting rapidly in the Himalayas, often at a rate faster than the global average. Rising temperatures, altered precipitation patterns, glacial retreat, and increasing frequency of extreme weather events have begun to disrupt forest ecosystems (Kumari et al., 2024b). Shifts in vegetation zones, reduced regeneration of specific tree species, and increased vulnerability to pests and diseases are among the significant observed impacts (Negi, 2022).

A study by the Indian Institute of Remote Sensing (IIRS) reveals that tree lines are shifting upward, and the area under temperate broadleaf forests is declining. This has implications for species composition, habitat quality, and the long-term potential for carbon sequestration.

21.3.2 Socioeconomic Impacts, Equity, and Forest-Dependent Livelihoods

Climate-induced forest degradation affects the livelihoods of local communities who rely heavily on forests. Decline in NTFP availability, reduced agricultural productivity due to erratic rainfall, and increased fuelwood scarcity are affecting the socio-economic stability of mountain populations. Moreover, forest degradation exacerbates disaster risks, such as landslides, flash floods, and droughts, which are becoming more frequent in the region. Socio-economic equity is another critical aspect of SFM, particularly in the IHR, where many communities rely on forests for subsistence and income (Das & Mallick, 2024). Non-timber forest products (NTFPs), including medicinal plants, wild foods, and fodder, serve as economic safety nets, especially during agricultural downturns (Peerzada et al., 2022). However, access to these resources is often unequal, with disparities based on caste, gender, and land ownership (Sheikh et al., 2022). Community-based governance models, such as Joint Forest Management (JFM) and Van Panchayats, have played a pivotal role in addressing socio-ecological inequities by institutionalizing local participation and promoting equitable benefit-sharing mechanisms (Dhyani & Dhyani, 2020).

Ecological Sustainability in Sustainable Forest Management

Ecological sustainability is a fundamental principle of Sustainable Forest Management (SFM), particularly in biodiverse and fragile regions like the Indian Himalayan Region (IHR). Forests in this region provide essential ecological services, including biodiversity conservation, water regulation, soil stabilization, and climate mitigation (Kumar, 2023). However, their vulnerability to erosion, landslides, and climate change necessitates careful management. Biodiversity conservation is fundamental to ecological sustainability, necessitating the protection not only of individual species but also of entire ecosystems. This includes ensuring habitat connectivity, maintaining diverse forest structures, and adopting landscape-scale conservation approaches (Negi et al., 2019; Kotwal et al., 2008). The Bhopal-India (B-I) framework's Criteria and Indicators (C&I) method helps monitor forest health through measurable factors like species diversity, natural regeneration, and soil quality (Kotwal et al., 2008). Despite these tools, uncertainties remain regarding how biodiversity responds to human interventions, prompting the need for adaptive management strategies that incorporate ongoing ecological monitoring (Sati, 2023a, 2023b).

Community-managed forests in the IHR have shown greater biodiversity and resilience compared to government-managed areas, highlighting the value of traditional practices such as rotational grazing, multi-species planting, and sacred grove conservation (Måren et al., 2014; Rao, 2024). These approaches align with the principle of intergenerational equity, which underscores the need to preserve forest ecosystems for future generations by sustaining their ecological adaptability and resilience (Xu et al., 2019).

Forest-Based Climate Solutions

Emerging bioeconomic approaches, such as value-added NTFP processing and improved market access, offer new livelihood opportunities, particularly for women and youth (Pandey et al., 2016). However, achieving accurate socio-economic equity requires policy interventions that recognize customary rights, secure land tenure, and provide targeted support for marginalized groups (Gautam et al., 2008).

21.3.3 Sustainable Forest Management (SFM) and Participatory Forest Management (PFM)

SFM is a cornerstone of climate mitigation and adaptation in forested regions. In the IHR, SFM must be tailored to the diverse ecological zones and cultural contexts. The principles of SFM involve maintaining forest productivity, enhancing carbon stocks, preserving biodiversity, and ensuring socio-economic benefits for forest-dependent communities. Van Panchayats (VPs) in Uttarakhand and Joint Forest Management Committees (JFMCs) across the IHR have demonstrated the efficacy of community-based forest governance. These institutions empower local communities to take responsibility for forest protection, fire management, and sustainable harvesting. However, challenges remain in ensuring equitable participation, capacity building, and legal recognition at various capacites in different forest management systems (Rawal et al., 2025).

21.3.4 Forest Monitoring and Planning

Modern technological tools are playing an increasingly vital role in advancing the principles and practice of Sustainable Forest Management (SFM). The integration of remote sensing technologies, Geographic Information System (GIS)-based forest inventories, and standardized carbon accounting frameworks has significantly enhanced the capacity to monitor and manage forest ecosystems. These tools enable precise assessments of forest cover change over time, estimation of above- and belowground biomass, and tracking of carbon stock dynamics all of which are essential for understanding forest health and productivity. In the context of global climate commitments, such technological integration is particularly relevant to mechanisms like REDD+ (Reducing Emissions from Deforestation and Forest Degradation), which require robust data to quantify carbon sequestration and emissions reductions. Moreover, the accurate monitoring enabled by these systems supports forest certification processes and facilitates participation in emerging carbon markets and trading schemes, thereby linking local forest conservation efforts to international climate finance. As noted by Liang et al. (2025), the deployment of such tools not only improves transparency and accountability in forest governance but also strengthens the scientific foundation for policy formulation and resource allocation in climate mitigation strategies.

21.3.5 Agroforestry, Silvicultural Systems, Afforestation and Reforestation Initiatives

Agroforestry, an age-old practice in the Himalayas, involves integrating trees with crops and livestock to enhance agricultural productivity (Devi et al., 2023). It improves soil fertility, biodiversity, and income diversification while sequestering carbon both above and below ground (Pant et al., 2025). The National Agroforestry Policy (2014) and the State Agroforestry Missions in Himachal Pradesh, Uttarakhand, and Sikkim provide institutional support for promoting agroforestry as a climate-smart land-use option. Models such as alder-based agroforestry in Nagaland or *Grewia optiva*-based systems in Uttarakhand have shown promising results in terms of both carbon sequestration and support for rural livelihoods. National programs, such as the Green India Mission (GIM), the Compensatory Afforestation Fund Management and Planning Authority (CAMPA), and the National Afforestation Programme (NAP), are funding large-scale reforestation and ecosystem restoration in the IHR (Devi et al., 2023; Ozukum et al., 2023). State governments have also launched region-specific reforestation efforts, such as the "Catch the Rain" campaign in Himachal Pradesh and spring-shed restoration in Uttarakhand. However, these efforts must ensure species diversity, ecological appropriateness, and community involvement to avoid the pitfalls of monoculture plantations and short-term afforestation drives.

21.3.6 Limitations and Challenges in Forest Carbon Sequestration: Implications for Long-Term Storage Strategies

While forests serve as critical carbon sinks, their future carbon sequestration potential is constrained by multiple stressors, including nutrient depletion, atmospheric pollution, and climate change-induced shifts in ecological conditions (Singh et al., 2024; Beedlow et al., 2004; He et al., 2025; Liu et al., 2023). These factors may reduce forest productivity and carbon uptake efficiency over time, necessitating a change in management strategies toward enhancing ecosystem resilience alongside carbon storage. Long-term carbon retention depends not only on aboveground biomass but also on belowground soil carbon pools, which are vulnerable to degradation (Schimel et al., 2001; Houghton, 2003). Protecting carbon-rich soils, particularly in old-growth and high-biomass forests, must therefore be a priority in climate mitigation policies.

21.4 Climate-Smart Forest Management: Balancing Carbon and Ecological Resilience

Forest managers are increasingly adopting climate-adaptive practices that integrate carbon conservation with broader ecological objectives. Vulnerability assessments and long-term adaptation planning are now recognized as essential tools for sustaining carbon stocks while maintaining ecosystem functionality (Pant & Pant, 2025). In the Indian Himalayan Region (IHR), mixed temperate and subtropical forests demonstrate high carbon sequestration efficiency. However, their storage capacity varies with elevation, precipitation regimes, and anthropogenic pressures (Kumari et al., 2022). Community-based forest management models such as those implemented in Tripura and Uttarakhand have proven effective in enhancing carbon stocks through participatory conservation and sustainable use (Deb et al., 2021).

21.4.1 Threats to Forest Carbon Sinks: Degradation and Land-Use Pressures

Despite their sequestration potential, forests in India face persistent degradation due to shifting cultivation, infrastructure expansion, and unregulated grazing, particularly in ecologically sensitive regions like the IHR (Borah et al., 2025). These disturbances not only release stored carbon but also exacerbate biodiversity decline, soil erosion, and hydrological disruptions (Patra et al., 2022). Additionally, increasing forest fires and habitat fragmentation further diminish carbon retention capacity, underscoring the need for enhanced protective measures. The cooperatively governed forest institutions like Van Panchayats and Cooperative Forest Societies promote sustainable forest management, biodiversity conservation, and local livelihoods. However, they face challenges related to governance autonomy and integration with state-led programs like Joint Forest Management, which can dilute traditional community control and affect forest sustainability and socioeconomic outcomes (Bhatt & Pant, 2024)

21.5 Ecosystem-Based Adaptation (EbA) and Co-benefits for VPs

EbA is increasingly recognized as a vital approach to climate resilience in the IHR, where forest ecosystems act as natural buffers against climate-induced risks such as landslides, floods, and droughts. VPs, which are unique to Uttarakhand and part of the central Himalayas, are central to the implementation and scaling of EbA in this region. Restoring and sustainably managing forest landscapes, VPs contribute to a range of EbA outcomes, including the reforestation of degraded and landslide-prone slopes, the restoration of watersheds, the establishment of vegetative buffers near glacial lakes and river systems, and the creation of biodiversity corridors (Bhatt & Pant, 2024). These interventions not only enhance the adaptive capacity of local ecosystems but also protect critical services such as water regulation, soil stabilization, and habitat connectivity. VPs, with their deep-rooted institutional presence and TEK, are ideally positioned to operationalize EbA at the grassroots level, linking climate adaptation with decentralized governance (Bhatt & Pant, 2023).

The co-benefits of EbA strategies implemented through VPs are far-reaching (Pant & Pant, 2025). These include carbon sequestration through native species plantations, reduction in forest fire risk via pine needle removal, and sustainable livelihood generation through eco-friendly product development and regulated NTFP harvesting. As custodians of over 12% of forest area in some Himalayan districts, VPs have also become key platforms for promoting climate-smart agroforestry, community-led soil and water conservation, and bioeconomy models that reduce emissions and increase resilience. Importantly, EbA practices through VPs are being increasingly mainstreamed into SAPCCs, supported by national programs such as the Green India Mission and the NAFCC (Bhatt & Pant, 2024). However, to fully realize the potential of this synergy, there is a need for capacity-building, technical support, and financial mechanisms such as Payments for Ecosystem Services (PES) and REDD+ readiness. By bridging policy, science, and traditional knowledge, Van Panchayats can serve as effective institutional vehicles for delivering low-cost, scalable, and community-driven EbA interventions, thereby making meaningful contributions to India's carbon neutrality and climate adaptation objectives (Table 21.1).

Table 21.1 Roadmap to carbon neutrality in the Indian Himalayan Region (IHR)

Action area	Key strategies	Carbon benefit	Co-benefits
Agroforestry and mixed land use	Integration of trees with crops and livestock; promotion of organic farming	Increased carbon sequestration in soils and biomass	Improved livelihoods, soil fertility, and agrobiodiversity
Community forestry	Local governance through Van Panchayats; participatory forest management	Enhanced carbon stock, reduced forest degradation	Strengthened social capital, equity, and climate resilience
Forest restoration	Reforestation, assisted regeneration, native species plantation	Recovery of carbon sinks; improved adaptation capacity	Water regulation, pollination, and ecosystem services
Fire and watershed management	Fire prevention measures, pine needle removal, watershed and spring conservation	Avoided emissions from wildfires; stabilized landscapes	Disaster risk reduction, water security, and ecosystem protection
Ecosystem-based adaptation (EbA)	Restoration of degraded slopes, riparian buffers, and biodiversity corridors	Enhanced ecosystem carbon storage and climate buffering	Biodiversity conservation, disaster resilience, food and water security
Traditional ecological knowledge (TEK) and indigenous knowledge systems (IKS)	Use of traditional land, water, and forest management practices; sacred groves; and seasonal calendars	Sustained low-carbon resource use; climate-adaptive behavior	Cultural continuity, ecological ethics, intergenerational knowledge
Policy integration and PES	Incentives for conservation (e.g., REDD+, PES), carbon market access, MRV systems	Monetized ecosystem services; verified emission reductions	Economic diversification, institutional capacity building
Green energy integration	Deployment of solar, biogas, hydrogen, and improved cookstoves	Reduced household and fossil fuel emissions	Clean energy access, health improvements, and women's empowerment

21.6 Traditional Ecological Knowledge (TEK) and Its Integration in Sustainable Forest Management

Indigenous communities in the IHR possess a deep understanding of ecological knowledge regarding sustainable land use, seasonal changes, and resource management (Bhatt et al., 2024a). Practices such as rotational grazing, sacred groves, seasonal harvesting of NTFPs, and traditional fire control mechanisms reflect adaptive strategies that have evolved over centuries. Incorporating TEK into formal forest governance can enhance the relevance, cultural legitimacy, and resilience of climate adaptation interventions. Indigenous Knowledge Systems (IKS) have long played a significant role in promoting sustainable forest management in the Himalayas (Bhatt et al., 2024a). Practices such as seasonal harvesting, sacred grove preservation, and rotational grazing have maintained ecological balance for generations (Negi et al., 2022). Formal recognition and integration of these practices into modern forest policies can enhance both ecological sustainability and socio-economic equity, ensuring that management strategies are both scientifically sound and culturally appropriate. By combining traditional wisdom with adaptive management, forest governance in the IHR can better address the dual challenges of conservation and livelihood security. The critical role of Himalayan women in Uttarakhand's circular bioeconomy and livelihood generation, highlighting their sustainable resource use, traditional knowledge, and eco-friendly practices also safeguards the forest conservation through their immense support. By actively participating in REDD+ initiatives, they strengthen community forestry institutions like Van Panchayats, contributing to forest conservation, carbon sequestration, and climate resilience. Their efforts empower local communities, enhance socioeconomic resilience, and promote biodiversity conservation through sustainable value chains and green entrepreneurship, driving environmental sustainability and inclusive economic growth in the region (Pant & Rawal, 2025).

21.7 Policy Framework and Institutional Support

India's forest and climate policy landscape provides a strong foundation for promoting carbon neutrality in the IHR. National frameworks such as the NAPCC, GIM, and National Agroforestry Policy emphasize ecosystem restoration, community engagement, and sustainable land use. At the state level, SAPCCs aim to localize these goals, with several Himalayan states recognizing the role of VPs and TEK in forest-based mitigation and adaptation. However, challenges persist in terms of institutional coordination, adequate funding, and operational convergence between forest, climate, and rural development departments. Strengthening policy implementation through capacity-building, cross-sectoral integration, and formal inclusion of VPs in carbon governance frameworks will be essential to realize their full climate potential.

21.7.1 National Climate Commitments and Forest Policy Integration: Policy Landscape and Institutional Challenges for Forest-Based Climate Solutions in the IHR

India's national climate agenda, particularly its Nationally Determined Contributions (NDCs) under the Paris Agreement, commits to creating an additional carbon sink of 2.5–3.0 billion tonnes of CO_2-equivalent through forest and tree cover by 2030. Achieving meaningful progress toward climate mitigation targets necessitates the effective alignment of forest management policies with broader climate action strategies. This integration is particularly critical in forest-rich and ecologically sensitive regions such as the Indian Himalayan Region (IHR), where forests serve as both carbon sinks and biodiversity reservoirs. As emphasized by Gorain et al. (2025), synergizing forest governance frameworks with national and subnational climate agendas enhances the potential for co-benefits, including carbon sequestration, ecosystem resilience, and livelihood support for forest-dependent communities. Ensuring such policy coherence requires adaptive, region-specific approaches that recognize the ecological complexity and socio-cultural diversity of Himalayan landscapes. The strategic relevance of this alignment is further underscored by empirical evidence from forest carbon estimates in the IHR (Table 21.2), which illustrate the region's substantial contribution to national climate goals through natural climate solutions. Therefore, harmonizing sustainable forest management (SFM) with climate resilience planning in these areas is not only desirable but imperative for achieving India's long-term environmental commitments.

A robust policy framework exists to support forest-based climate action in the IHR, yet gaps in implementation, coordination, and financing continue to limit its full potential. Key national instruments, such as the National Forest Policy (1988), emphasize ecological stability, forest conservation, and principles of people's participation, which align well with community-based models like Van Panchayats (Stevens, 2022). The National Agroforestry Policy (2014) promotes tree-based farming systems, particularly relevant for hill agriculture and degraded lands. The GIM of 2014 further strengthens this agenda by prioritizing ecosystem restoration and livelihood enhancement through community engagement. Additionally, the NAFCC has financed ecosystem-based adaptation (EbA) projects in Himalayan states, recognizing their climate vulnerability (Kumar et al., 2025).

Despite this supportive policy architecture, cross-sectoral convergence remains weak, with limited real-time monitoring and underfunded mandates at the state level. SAPCCs aligned with the NAPCC provide state-specific roadmaps. For instance, Uttarakhand's SAPCC highlights the role of ecosystem resilience, reforestation, and the integration of TEK; Himachal Pradesh focuses on REDD+ readiness and catchment treatment; while Sikkim and Arunachal Pradesh emphasize alpine ecosystem conservation and climate-smart agriculture. However, the implementation of these plans is often hindered by institutional fragmentation, technical capacity gaps, and limited availability of localized ecological data.

Table 21.2 IOOI framework for carbon neutrality policy and institutional programs in the Indian Himalayan Region (IHR)

IOOI element	General policy/institutional frameworks	Specific programmatic examples in the IHR
Input	• National Action Plan on Climate Change (NAPCC) and State Action Plans (SAPCCs) • Bureau of Energy Efficiency (BEE) • Carbon Credit Trading Scheme (CCTS) • Ecosystem-Based Adaptation (EbA) integrated into forest, agriculture, and water missions • Recognition of TEK and Indigenous Knowledge Systems in community resource planning • Van Panchayats as decentralized governance platforms for forest management • PES (Payments for Ecosystem Services), REDD+, and climate finance mechanisms	• Ladakh's Carbon Neutral Action Plan (CNAP) and Uttarakhand's SAPCC prioritizing EbA and TEK • VP-based forest monitoring, pine needle removal, and biodiversity registers • Community-based vulnerability assessments using local knowledge • Partnerships with ICIMOD, TERI, and GBPIHED for technical support • Renewable energy procurement, eco-certification for NTFPs, and watershed diagnostics
Output	• Implementation of EbA interventions: slope stabilization, watershed restoration, buffer plantations • Integration of TEK into local climate adaptation and resource management practices • Strengthening Van Panchayat governance for forest protection and carbon stock enhancement • Capacity building on MRV, carbon accounting, and nature-based enterprises	• Assisted Natural Regeneration (ANR) and native species plantations by VPs • Pine needle biomass collection and product diversification (e.g., briquettes, paper items) • Restoration of springs and degraded slopes using EbA approaches • Local documentation of TEK through biodiversity registers and participatory mapping • Skill development in green livelihoods, including eco-tourism and biomass use
Outcome	• Increased community-led carbon sequestration and emission reduction practices • Broader application of TEK for adaptive forest, water, and land use planning • Institutional strengthening of VPs for integrated landscape management • Reduced exposure to climate risks through EbA and NbS implementation	• Reduced fire incidences due to pine litter removal and controlled grazing • Improved forest health and ecosystem connectivity • Enhanced adaptive capacity of communities via EbA-based restoration • Increased participation of women and marginalized groups in climate action • Strengthened synergy between formal policy and traditional institutions

(continued)

Table 21.2 (continued)

IOOI element	General policy/institutional frameworks	Specific programmatic examples in the IHR
Impact	• Accelerated contribution to India's net-zero emissions targets • Resilient socio-ecological systems in high-altitude and mid-hill zones • Sustainable rural economies built on ecosystem services and TEK • Preservation of biodiversity and carbon-rich landscapes under community stewardship	• Sustained regional carbon sinks managed by VPs • Institutionalization of EbA and TEK in climate governance • Improved food-water-energy nexus in mountain areas • Long-term ecological and social resilience through inclusive, place-based strategies

On the global and regional front, initiatives such as REDD+ pilots in Mizoram, Himachal Pradesh, and Uttarakhand are laying the foundation for carbon finance and emission reductions from deforestation. Institutions like ICIMOD are fostering transboundary cooperation at the pan-Himalayan level, while GEF and UNDP-supported projects are advancing biodiversity conservation and community-based climate resilience in the region (Bhardwaj et al., 2024; Bhattarai, 2024; Trivedi, 2025). These collaborations present opportunities for regional knowledge exchange, innovation, and policy harmonization, but require scaling and integration with local governance systems.

Notably, the Uttarakhand SAPCC acknowledges the climate mitigation and adaptation potential of VPs, yet they continue to face structural and institutional challenges. These include limited financial autonomy, bureaucratic delays in approving plantation or harvesting activities, and a lack of institutional convergence between forest, rural development, and climate departments. Addressing these challenges through capacity-building, decentralized planning, and integrated policy frameworks is essential to harness the full potential of community-managed forests in India's pathway to carbon neutrality (Chaturvedi et al., 2022).

Recent efforts by the Uttarakhand Forest Department, such as capacity-building workshops, GIS-based forest monitoring, and VP federations, are steps toward strengthening this system. However, sustained investment, legal safeguards, and decentralized governance are essential to scale up their carbon neutrality potential.

21.7.2 A Roadmap to Carbon Neutrality in the Indian Himalayan Region

Achieving carbon neutrality in the IHR requires a multidimensional, community-centered, and ecosystem-based strategy that aligns ecological restoration with rural development and climate mitigation goals. The region's vast forest cover, fragile ecosystems, and high dependence on natural resources offer both opportunities and vulnerabilities in the face of climate change. A strategic roadmap must therefore integrate scientific innovation, traditional knowledge, inclusive governance, and scalable policy frameworks. A targeted, inclusive, and science-driven roadmap is essential to align the Himalayan forest sector with India's carbon neutrality goals (Chaudhry & Tambe, 2023). Central to this roadmap is the strengthening of community-based forest management. Institutions such as VPs, JFMCs, and traditional indigenous councils must be legally empowered to take on enhanced roles in forest governance. Financial incentives, technical training, and capacity-building initiatives should be extended to local communities to promote sustainable harvesting of forest products, effective forest fire management, and participatory monitoring (Bhatt et al., 2024b). Integrating TEK and community-led resource mapping into official planning processes can ensure that management practices are both culturally grounded and ecologically sound along with the institutional involvement in grassroot level decision making (Upadhyaya & Pant, 2025).

Equally important is scaling up ecosystem restoration and climate-resilient afforestation. Restoration efforts must prioritize native species, promote ecological corridors, and adopt landscape-level planning to restore degraded forests and enhance biodiversity. In glacial-fed catchments, combining forest restoration with watershed development can secure critical water sources and support ecosystem resilience (Joshi et al., 2021). Climate-smart afforestation practices tailored to high-altitude ecological dynamics should be implemented in degraded forest zones and buffer areas to bolster carbon sequestration and prevent future degradation.

To incentivize conservation and generate sustainable livelihoods, the roadmap emphasizes the promotion of carbon finance and nature-based enterprises. Community-managed forests should be linked to voluntary carbon markets through robust MRV systems. Simultaneously, it is essential to strengthen support for forest-based enterprises that can contribute to both livelihood enhancement and forest conservation. These include the development and diversification of Non-Timber Forest Product (NTFP) value chains, processing and marketing of medicinal and aromatic plants, and the promotion of community-led ecotourism initiatives. Such enterprises not only generate income but also incentivize sustainable resource use and conservation stewardship at the grassroots level. Targeted interventions must prioritize the inclusion and empowerment of women, tribal groups, and other marginalized communities who are often the primary custodians of traditional ecological knowledge and heavily dependent on forest resources for their subsistence. Strengthening institutional capacities, improving market access, and facilitating access to credit and technical training are critical for unlocking the full potential of these nature-based enterprises. By aligning these initiatives with broader sustainable development and climate adaptation goals, forest-based livelihoods can become a cornerstone of inclusive green growth in ecologically sensitive regions such as the Indian Himalayas (Peerzada et al., 2021; Siegner et al., 2021). Transparent mechanisms for benefit-sharing will ensure the equitable distribution of revenues and enhance local ownership of conservation outcomes (Table 21.3).

Table 21.3 Suggested policy and institutional frameworks for developing the roadmap to carbon neutrality in IHR

Framework/ institution	Main role/function	Key impact on carbon neutrality
National Action Plan on Climate Change	Sets national climate missions and targets	Guides sectoral and regional climate action
Bureau of Energy Efficiency (BEE)	Administers the carbon market in India, playing a central role in setting emission reduction targets, issuing carbon credits, and ensuring regulatory compliance under market-based mechanisms such as the Perform, Achieve, and Trade (PAT) scheme and the Indian Carbon Market (ICM) framework	Operationalizes emission reduction
NSCICM (National Steering Committee for Indian Carbon Market)	The National Steering Committee for the Implementation of the Indian Carbon Market (NSCICM) oversees the governance of the carbon market, providing strategic direction, recommending operational procedures, setting emission reduction targets, and ensuring alignment with national climate objectives	Ensures market integrity and compliance
Carbon Credit Trading Scheme (CCTS)	Enables trading of carbon credits, compliance and offset	Incentivizes emission reductions
SAPCCs	Localizes climate action at the state level	Tailors solutions to regional needs
PES Schemes	Incentivizes ecosystem conservation by communities	Aligns economic and climate benefits
Regional Action Plans (e.g., Ladakh)	Pilots' neutral strategies in sensitive regions	Demonstrates scalable models

Integrating forest and climate policy across governance levels is also crucial. Achieving meaningful policy coherence in forest-based climate solutions necessitates strengthened coordination among various governmental departments and institutions responsible for managing forests, biodiversity, disaster risk reduction, and climate change. Fragmentation in governance often leads to overlapping mandates, inefficient resource use, and conflicting priorities that undermine long-term sustainability goals. Integrative frameworks and cross-sectoral dialogue are essential to ensure that policies related to forest conservation, ecosystem services, and climate adaptation are mutually reinforcing rather than working in isolation. In particular, the convergence of forest management with climate resilience planning, biodiversity action plans, and disaster risk mitigation strategies is critical for addressing the complex, interlinked challenges facing ecologically fragile regions such as the IHR. Establishing institutional mechanisms for interdepartmental collaboration through joint task forces, shared data systems, and co-financing models can greatly enhance the effectiveness and impact of forest-based climate initiatives (Bhatt & Pant, 2024; Stevens & Satterfield, 2024).

SAPCCs should be localized through climate-resilient village development planning in Himalayan blocks, ensuring that forest-based climate solutions are tailored to local vulnerabilities and capacities. Establishing inclusive science-policy-community platforms will bridge the gap between research and practice, fostering adaptive governance rooted in evidence and lived experience (Sugandha et al., 2025).

Finally, enhanced monitoring, research, and data systems are foundational to effective climate action in the forest sector. Investments in remote sensing technologies, LiDAR mapping, and community forest inventories will enable accurate assessments of carbon stocks and effective land-use tracking (Gupta et al., 2022). Regional research hubs should be developed in partnership with institutions like the G.B. Pant Institute, ICIMOD, and state forest departments to drive innovation. At the grassroots level, participatory data generation through citizen science initiatives and community biodiversity registers (CBRs) can democratize forest monitoring and build a shared sense of stewardship. Together, these strategies provide a comprehensive pathway for leveraging Himalayan forests as a cornerstone of India's carbon neutrality and climate resilience agenda.

21.7.3 Rationale of the Present Study

VPs the community-based forest governance institutions operating across Uttarakhand and other parts of the IHR, play a vital role in integrating local forest management with national and global climate goals, particularly those related to carbon neutrality. As decentralized entities with statutory authority over village-level forest governance, VPs function as crucial intermediaries between state forest departments and local communities. Their participatory governance approach empowers villagers to manage, protect, and sustainably utilize forest resources, fostering a sense of stewardship and collective responsibility.

One of the most significant contributions of VPs toward carbon neutrality is their capacity to enhance forest carbon stocks. Through community-driven afforestation programs, natural regeneration of degraded lands, and effective protection measures against overexploitation, VPs contribute to increased carbon sequestration. Traditional ecological knowledge embedded in their management practices such as rotational grazing, seasonal restrictions on lopping, and the conservation of sacred groves plays a crucial role in maintaining healthy forest ecosystems that function as effective carbon sinks. In numerous instances, VPs have transformed barren or degraded landscapes into dense, multi-species forest patches, thereby directly contributing to climate mitigation through enhanced biomass accumulation and carbon storage (Bhatt et al., 2024a; Negi, 2022).

In addition to ecological restoration, VPs also support sustainable livelihoods that reduce reliance on environmentally damaging practices. By regulating the extraction of fuelwood, fodder, and NWFPs, they help curb forest degradation while ensuring that the subsistence needs of forest-dependent communities are met. Furthermore, several VPs have begun integrating climate-resilient practices such as agroforestry, medicinal plant cultivation, and community-based ecotourism. These initiatives not only diversify local income sources but also reinforce ecosystem integrity and long-term sustainability.

Given the growing emphasis on NbS and community-driven approaches in climate policy frameworks, there is an urgent need to evaluate and document the role of Van Panchayats in contributing to carbon neutrality. This study aims to fill that gap by assessing the ecological and socio-economic impacts of VP-managed forests, with a focus on their carbon sequestration potential and livelihood implications in the context of climate resilience in the IHR.

21.7.4 Forest-Based Carbon Neutrality in Uttarakhand: The Role of Van Panchayats

Crucially, VPs embody a participatory governance model that enhances local stewardship and accountability, both of which are essential for long-term climate action (Rawal & Jugran, 2024). This bottom-up approach ensures that forest conservation measures are socially acceptable, culturally appropriate, and ecologically informed. The participatory monitoring of forest health, carbon stock assessments, and forest fire prevention by VP members supports robust climate mitigation frameworks.

However, their full potential remains constrained by challenges such as limited technical capacity, inadequate financial support, and overlapping jurisdictions with state agencies. Addressing these constraints through policy support, integration into national carbon accounting frameworks (such as REDD+), and capacity-building initiatives can amplify the role of VPs in India's climate strategy.

In summary, Van Panchayats are key actors in forest-based climate solutions, with the potential to significantly contribute to carbon neutrality by combining local ecological knowledge, sustainable resource management, and community-driven governance. Empowering and scaling up the VP model could be a transformative step toward inclusive, low-carbon rural development in the Himalayas.

21.7.5 Role of Van Panchayats (VPs) in Achieving Carbon Neutrality Through Forest Resource Management

Uttarakhand, located in the Central Indian Himalayan Region, is endowed with abundant forest resources, with approximately 45.44% of its geographical area under forest cover (FSI, 2021). Owing to its extensive forested landscapes, rich biodiversity, and community-centric conservation practices, the state holds a strategically important position in India's pursuit of carbon neutrality. A key feature of Uttarakhand's forest governance is the long-standing tradition of community-based forest management through VPs.

Established under the Kumaun Panchayat Forest Rules of 1931, Van Panchayats represent one of the earliest and most enduring models of decentralized forest governance in India. These institutions legally empower local communities to manage, conserve, and sustainably utilize designated forest areas. As of 2023, Uttarakhand had 12,089 functioning VPs, collectively managing nearly 16% (5,446,964 ha) of the state's total forest area. However, recent records indicate a decline in their number to approximately 11,217 by 2024, with the area under VP jurisdiction reduced to around 4,520,000 ha (MoEFCC, 2018; UKFD, 2024). Despite this reduction, Van Panchayats continue to represent one of the most extensive and influential community forestry systems in the country, offering valuable insights into participatory natural resource management and climate-responsive forest governance.

21.7.6 Van Panchayats as Enablers of Carbon Sequestration

VPs play a pivotal role in advancing carbon neutrality at the grassroots level through a range of community-led forest management practices. Their efforts in forest conservation and regeneration, including regular patrolling, fire prevention, and control of illegal activities, significantly curb emissions from deforestation and degradation. Many VPs have also undertaken afforestation and enrichment planting using native species, thereby enhancing long-term carbon sequestration in both biomass and soil. In addition, the sustainable harvesting of NTFPs and regulated fuelwood collection reduce extraction pressure on core forests while supporting subsistence livelihoods and promoting a low-emission rural economy. Collectively, these practices contribute to the increased above-ground biomass, improved soil organic carbon, and enhanced ecosystem functions such as biodiversity conservation, water regulation, and climate resilience, positioning VPs as vital actors in India's carbon neutrality pathway.

21.7.7 Community Engagement and Traditional Ecological Knowledge (TEK)

Van Panchayats rely on deep-rooted traditional ecological knowledge systems that guide resource use based on seasonal availability, ecological thresholds, and social norms. Practices such as rotational grazing, seasonal harvesting bans, and opening/closure of the VP area and Sacred groves preservation are examples of TEK that align with carbon sequestration goals and climate adaptation (Bhatt et al., 2024a, b). Local norms also ensure social accountability, wherein rule-breakers face community-imposed sanctions, leading to high compliance with conservation norms, a challenge that formal forest departments often struggle to achieve (Rawal & Jugran, 2024).

21.7.8 Quantifying Carbon Benefits

Although systematic carbon accounting in Van Panchayat-managed areas remains in its early stages, emerging evidence from pilot studies indicates their significant potential in enhancing forest carbon stocks. Research conducted by Dangwal et al. (2022), Bhatt et al. (2024a, 2024b), and others has demonstrated that community-managed forests under Van Panchayats tend to store considerably higher levels of biomass and carbon compared to open-access or degraded forest areas. These findings underscore the ecological efficacy of participatory forest governance in promoting sustainable land use and carbon sequestration. Given their institutional structure, social legitimacy, and capacity for ecosystem restoration, Van Panchayats present a promising entry point for implementing REDD+ initiatives in the Indian Himalayan Region. Integrating VPs into REDD+ frameworks could not only contribute to national climate mitigation goals but also provide co-benefits in terms of livelihood support, biodiversity conservation, and community empowerment.

To fully harness the carbon mitigation potential of VPs, several enabling steps are essential. First, capacity building of VP members in carbon stock estimation through simplified and community-friendly MRV protocols is crucial. This should be complemented by the integration of remote sensing technologies with ground-level forest inventories, ensuring accurate and scalable carbon assessments. Moreover, linking Van Panchayats to voluntary carbon markets or PES schemes could provide tangible economic incentives for conservation, thereby monetizing their ecological contributions. Promoting mixed-species plantations and reviving traditional agroforestry systems within Van Panchayat jurisdictions would not only enhance biodiversity and soil health but also improve long-term carbon sequestration. At the policy level, establishing district-level carbon baselines and formally including VPs' forests in state-level GHG inventories would strengthen climate accountability and enable targeted investments in nature-based solutions. Together, these measures can position Van Panchayats as credible actors in both local livelihood enhancement and global climate mitigation efforts.

21.7.9 Policy Support and Institutional Challenges

The Uttarakhand State Action Plan on Climate Change (SAPCC) acknowledges the potential of community forest management in enhancing climate resilience and carbon sequestration (Sharma et al., 2017). However, Van Panchayats face several structural and policy barriers:

- Limited financial autonomy and over-reliance on government grants.
- Bureaucratic delays in the approval of plantation activities or harvesting plans.
- Lack of convergence between the forest, climate, and rural development departments.

Recent efforts by the Uttarakhand Forest Department, such as capacity-building workshops, GIS-based forest monitoring, and Van Panchayat federations, are steps toward strengthening this system. However, sustained investment, legal safeguards, and decentralized governance are essential to scale up their carbon neutrality potential.

The VPs represent a unique model of community-managed forest governance in the IHR with significant potential to contribute to India's carbon neutrality goals. However, their role in carbon neutrality remains under-assessed and largely excluded from formal climate accounting frameworks. Existing studies indicate that forests managed by Van Panchayats generally exhibit higher biomass stocks compared to degraded or open-access forest areas; however, comprehensive and systematic data on their carbon sequestration potential remain limited (Bhatt & Pant, 2024; Nagahama et al., 2022). This study addresses that gap by estimating the carbon neutrality potential of VP forests in the Hawalbagh block of Almora district while also examining community perspectives on forest-based livelihoods and fire mitigation. The findings aim to inform the integration of VPs into carbon finance mechanisms, REDD+ strategies, and state-level GHG inventories.

21.8 Methodology

The present study employed a quasi-experimental research design to assess the carbon neutrality potential of community-managed forests under VPs in the Hawalbagh block of Almora district, Uttarakhand. The study combined quantitative estimation of forest-based carbon sequestration with qualitative insights from community members to provide a holistic understanding of forest-climate-livelihood linkages. The carbon sequestration potential was estimated using secondary data on forest areas managed by 35 Van Panchayats, employing standardized carbon assessment protocols in accordance with the guidelines provided by the Forest Survey of India (FSI, 2021). In parallel, household-level surveys and focus group discussions were conducted to capture community perceptions on forest-based livelihoods, fire mitigation practices, and the adoption of sustainable forest resource use. A specific focus was placed on the collection and utilization of pine needles (Pirul), with an emphasis on evaluating their role in reducing forest fire risk, generating supplementary rural income, and contributing to carbon offset efforts through biomass utilization and avoided emissions. Integrating spatial forest data with participatory inputs offers grounded evidence on how Van Panchayats can function as effective institutional platforms for achieving localized carbon neutrality and advancing nature-based climate solutions.

One intervention site (with pine needle collection and product-making activities) at the Rural Technology Complex (RTC) of GBPNIHE was undertaken as per the research-based amenities, and to estimate the carbon neutrality of pine needle-based products. As part of a broader initiative supported by the National Mission on Himalayan Studies (NMHS), a pine needle-based paper processing unit was established at the Regional Training Centre (RTC) and implemented between 2016 and 2020 (Negi et al., 2021). The intervention aimed to demonstrate the ecological and socio-economic benefits of utilizing dry pine needles (Pirul), which are highly flammable and contribute significantly to recurring forest fires in the middle Himalayan zone. At the RTC, collected pine needles were processed into a range of paper-based products, including wedding cards, file folders, carry bags, and handcrafted decorative items. Alongside production, the facility functioned as a training centre for local communities, with a particular focus on empowering women from marginalized and economically disadvantaged groups. By imparting skills in pine needle collection, processing, and product development, the initiative promoted value addition and provided alternative livelihood opportunities, reducing community dependence on ecologically harmful extraction practices.

To assess the intervention's effectiveness, field-based research was carried out in nearby Van Panchayats within the Hawalbagh developmental block. A mixed-methods approach was employed, incorporating structured vegetation surveys, fire incident records from the Forest Department, and systematic monitoring of biomass conditions. In addition, qualitative data were collected through semi-structured interviews and focus group discussions with trained beneficiaries to capture local perceptions of ecological and livelihood changes. The carbon neutrality contribution of the project was evaluated by estimating avoided carbon emissions both from reduced forest fire incidents due to needle removal and from the substitution of conventional, more carbon-intensive products with pine needle-based alternatives. The findings offer compelling empirical evidence that strategic Pirul management not only mitigates wildfire risk but also supports low-carbon rural development and inclusive green economy pathways. The intervention thus exemplifies a scalable model that integrates community-based forest management with climate mitigation, sustainable livelihoods, and ecosystem resilience in the Indian Himalayan Region.

Primary data on pine needle (Pirul) availability, collection practices, processing methods, and product development were systematically collected through participatory rural appraisal (PRA) techniques, including focus group discussions (FGDs) and direct field observations following Bhatt et al. (2024a). These engagements involved a diverse group of stakeholders, with a particular emphasis on women's self-help groups and local beneficiaries who were directly involved in the intervention. Participants were trained in the collection and sorting of pine litter, ensuring quality control and sustainability in raw material extraction. Additionally, hands-on training sessions were conducted to build local capacity in the production of value-added products such as file folders, eco-friendly carry bags, decorative handicrafts, and pine needle-based biomass briquettes. These products served not only as sustainable alternatives to conventional materials but also as potential sources of supplementary income for forest-dependent households. To ensure analytical rigor, triangulation of primary field data with secondary sources including relevant academic literature, government policy documents, and climate action reports was undertaken. This process allowed for a more nuanced understanding of the intervention's relevance and alignment within the broader context of India's climate mitigation goals, forest governance frameworks, and community-based natural resource management strategies. The integrated approach provided critical insights into the potential of pine needle utilization as a nature-based solution that addresses both environmental and socio-economic challenges in the Indian Himalayan Region.

21.9 Results

The results highlight the significant role of Van Panchayats in reducing forest fire incidents and enhancing carbon sequestration through community-based forest management. Simultaneously, strong community interest in pine needle-based enterprises, particularly with high levels of women's participation, points to the viability of integrating sustainable livelihoods with ecological restoration. Together, these findings underscore the potential of decentralized, inclusive, and nature-based approaches in advancing carbon neutrality in the Himalayan region.

21.9.1 Role of VPs in Achieving Carbon Neutrality Through Forest Resource Management: Community Perceptions and Potential for Scalable Forest-Based Solutions

The role of community participation in forest conservation and climate mitigation is increasingly evident through localized data and lived experiences. In the Hawalbagh development block of Uttarakhand, a study conducted across 35 Van Panchayats (VPs) revealed notable outcomes in both ecological and socio-economic dimensions. Approximately 59% of respondents reported a decline in forest fire incidents following targeted community interventions such as fire line creation and seasonal monitoring. This reflects not only the effectiveness of decentralized forest governance but also the heightened awareness and vigilance. These ecological and socio-economic outcomes reflect the synergistic benefits of community stewardship and sustainable resource use that have been fostered through participatory management (Fig. 21.1).

Equally significant is the growing interest in nature-based enterprises that link livelihood generation with forest sustainability. The study found that 67% of respondents expressed a willingness to participate in scaling up pine needle-based enterprises, underscoring the dual potential of such initiatives to simultaneously reduce fire-prone biomass and generate value-added products and rural employment opportunities. This strong community interest reflects the perceived socio-economic and ecological benefits of Pirul utilization, positioning it as a viable strategy for integrated forest fire management and livelihood enhancement in the Himalayan context.

The findings also highlight the significant involvement of women in forest-based livelihood activities. Notably, 96% of the respondents acknowledged the active

(a) (b)

Fig. 21.1 The perception-based community assessment at village level **a** mixed forest VP and agriculture at Balta village (Almora), **b** perception-based assessment survey at Manaun village (Almora)

participation of women in pine needle-based product development, indicating a strong gendered dimension to local climate action (Fig. 21.1). This engagement not only enhances household income but also contributes to carbon neutrality efforts through product diversification and biomass conversion. Transforming fire-prone forest residues, particularly pine needles, into value-added products such as biomass briquettes, handicrafts, and eco-friendly packaging materials enables women-led initiatives to contribute meaningfully to forest fire risk reduction while advancing sustainable resource utilization. This approach not only mitigates ecological hazards but also fosters inclusive green entrepreneurship, empowering marginalized communities especially women with alternative livelihood opportunities rooted in climate-resilient practices. This highlights the crucial role of women in climate-resilient forest economies and positions them as key stakeholders in both ecological restoration and low-carbon development pathways.

The analysis of 35 Van Panchayats (VPs) within the Hawalbagh development block of Almora district reveals a cumulative forest area of approximately 881.62 ha, with an estimated annual carbon sequestration potential of 5,289.72 tons of CO_2. This estimation is based on the Forest Survey of India's (FSI, 2021) standardized coefficient of 6.0 tons of CO_2 sequestered per hectare per year for moderately dense forests. Among the surveyed VPs, Raun, Khoont, and Bimola demonstrated the highest carbon sequestration capacities, collectively contributing nearly 30% of the total estimated offset. This variation can be attributed to differences in forest area, vegetative density, and management intensity across individual panchayats. Importantly, even smaller VPs such as Jyula, Dal, and Baari contributed meaningfully to local carbon neutrality goals, underscoring the cumulative ecological value of decentralized forest governance structures.

These findings reinforce the critical role that community-managed forests play in advancing regional climate mitigation strategies. They highlight the potential of Van Panchayats to function as effective carbon sinks when sustainably managed and protected through participatory approaches. The aggregated impact of small-scale, village-level interventions further illustrates the scalability of nature-based solutions rooted in traditional ecological knowledge and localized stewardship. This evidence supports the case for strengthening Van Panchayats as institutional anchors within India's broader forest-based carbon neutrality agenda, while promoting policies that enhance their capacity, legal recognition, and access to carbon financing mechanisms (Fig. 21.2).

Given the widespread issue of pine needle accumulation in Chir pine-dominated forests an ecological challenge closely linked to increased forest fire incidence there is a compelling rationale for integrating bioeconomy models into community forest management plans. The findings from this study highlight the potential of nature-based, community-driven interventions that transform fire-prone biomass into economically valuable products, such as briquettes, paper goods, and handicrafts. Such approaches not only mitigate ecological risks but also align with the broader goals of sustainable rural development. The demonstrated community willingness to participate in both fire prevention and enterprise development signals a high degree of local readiness to co-own and co-implement climate-resilient forest strategies. This dual commitment ecological and entrepreneurial suggests that local forest-dependent communities can be key partners in operationalizing low-carbon development at the grassroots level. Scaling up these integrated bioeconomy models across other Himalayan districts could significantly enhance forest health, strengthen carbon sequestration capacities, and improve rural livelihoods three pillars critical to India's long-term strategy for achieving carbon neutrality.

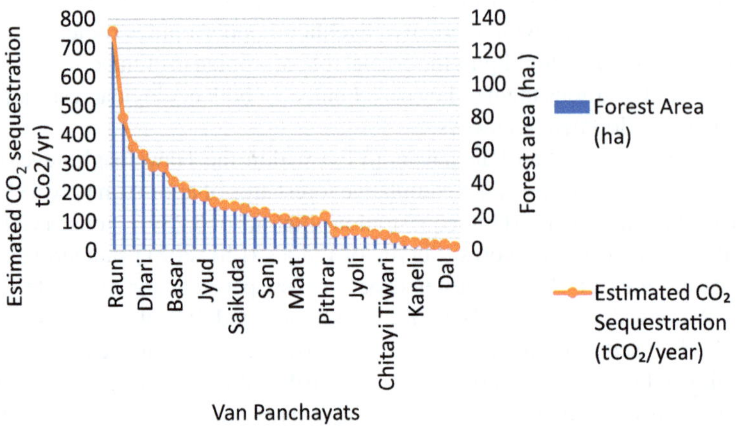

Fig. 21.2 Role of van panchayats (VPs) in local carbon neutrality at the forest level at Almora district (Uttrakhand)

21.9.2 Achieving Carbon Neutrality Through Pine Needle-Based Products and Forest Fire Mitigation

The widespread presence of chir pine (*Pinus roxburghii*) in the mid-altitude zones of the Indian Himalayan Region, particularly in Uttarakhand, presents both ecological challenges and opportunities. Pine needle litter (*Pirul*), if left unmanaged, accumulates in large quantities on the forest floor, becoming a major fire hazard during dry months. Forest fires not only degrade biodiversity and ecosystems but also result in significant carbon emissions, undermining efforts toward carbon neutrality. In response, innovative utilization of pine needles for eco-friendly product development has emerged as a dual-benefit strategy: mitigating carbon emissions from forest fires and promoting sustainable rural livelihoods.

The collection of dry pine needle litter one of the primary contributors to surface fuel loads in Chir pine-dominated, community-managed forests holds significant potential for reducing wildfire risk and the associated carbon emissions. By systematically removing this flammable biomass, community members contribute directly to fire prevention while simultaneously creating opportunities for value addition. The harvested pine needles are processed into a diverse range of eco-friendly products, including file covers, carry bags, envelopes, ornaments, biomass briquettes, and decorative items. These products not only serve as sustainable alternatives to conventional materials but also reflect the potential for circular bioeconomy models in forest-dependent rural economies. Over the course of the intervention, a total of 11,910 such items were produced, demonstrating both the scalability of pine needle utilization and its role in fostering low-carbon, livelihood-oriented development pathways (Fig. 21.3). This not only prevents the natural decomposition or combustion of the biomass, which would otherwise release CO_2, but also replaces conventional plastic and paper products with biodegradable, low-emission alternatives. The program has actively engaged local women and rural households, generating an income of ₹1,53,967 through institutional sales to entities like ICAR, the Forest Department, and the Governor's House. This demonstrates a sustainable, inclusive model for promoting green livelihoods while supporting mitigation objectives. Additionally, the utilization of pine cones, bark, and small branches in the construction of check dams, contour barriers, and other soil conservation structures exemplifies the principles of NbS. These low-cost, locally sourced materials enhance a range of critical ecosystem services, including soil stabilization, improved water infiltration and retention, and the preservation of microhabitats essential for biodiversity. By reducing erosion, controlling runoff on sloped terrain, and fostering vegetation regrowth, such interventions play a vital role in restoring ecological functionality in degraded forest landscapes. Importantly, these co-benefits extend beyond immediate mitigation outcomes, contributing to long-term ecosystem resilience and climate adaptation. Embedding NbS into community forest management strategies not only reinforces the ecological integrity of Himalayan forest ecosystems but also ensures that mitigation and adaptation goals are achieved in an inclusive, cost-effective, and contextually appropriate manner (Table 21.4).

(a) wedding card of first author **(b) file folder for carbon-neutral Ladakh**

(c) pine-needle rakhi and earrings **(d) pine-needle bio-briquette**

Fig. 21.3 Pine needle-based products: outcomes and dissemination

This integrated approach contributes to India's carbon neutrality agenda on multiple fronts by harmonizing ecological restoration with social and economic development. First and foremost, it facilitates emission reduction through the systematic removal of dry pine needles from forest floors, thereby significantly reducing the frequency, intensity, and spread of forest fires one of the major episodic sources of carbon emissions in the Himalayan landscape. Forest fires not only release vast amounts of stored carbon into the atmosphere but also degrade biomass, soil quality, and biodiversity (Sharma & Pant, 2017; Pant & Mishra, 2025). By addressing this risk through community-led collection efforts, the intervention prevents carbon loss and safeguards carbon stocks.

In parallel, the collected biomass is converted into a variety of value-added products, supporting rural livelihoods while substituting carbon-intensive materials. This supports circular economy principles and reduces the carbon footprint of household and artisanal production systems. Moreover, the initiative empowers local

Table 21.4 Role of pine products towards carbon neutrality contribution in Indian Himalayan Region

Component	Details	Carbon neutrality contribution
Raw material	Pine needle litter (*Pirul*) collected from forests (e.g., Hawalbagh Block, Almora, Uttarakhand)	Reduces fuel load in forests, lowering wildfire risk and associated emissions
Activities undertaken	Collection, processing, and conversion of pine needles into paper-based and decorative products	Prevents decomposition and combustion of biomass, reducing CO_2 release
Products developed	File covers, folders, bags, envelopes, notepads, wedding cards, ornaments, bio-briquettes, etc. (Total = 11,910 items)	Substitutes plastic/paper products with biodegradable, low-emission alternatives
Beneficiary engagement	Local women and rural households are involved in the collection and production	Promotes inclusive green livelihoods and community engagement in mitigation
Economic impact	Income generated: ₹1,53,967 from product sales to institutions (ICAR, Forest Dept., VPKAS, Governor's House)	Incentivizes sustainable biomass removal, aligning with carbon offset pathways
Forest fire prevention	Reduction of dry litter on forest floors curbs fire intensity and frequency	Avoids episodic high-emission wildfire events
Use of tree parts	Pine cones, bark, and branches are used in decorative items and check dams for soil conservation	Enhances ecosystem resilience and multifunctional land use
Environmental co-benefits	Reduced plastic use, improved soil and water conservation, and habitat protection	Strengthens adaptive capacity and ecosystem-based climate action
Scalability potential	Applicable across pine-dominated landscapes in mid-altitude Himalayas	Can significantly contribute to forest-based emission reduction at scale
Policy relevance	Aligned with NMHS, Green India Mission, REDD+, and carbon market readiness strategies	Supports India's NDCs and carbon neutrality goal by 2070

communities especially women and marginalized groups by creating green jobs, enhancing skills, and strengthening institutional capacity, thus embedding climate action within inclusive development. By integrating fire prevention, sustainable enterprise, and forest ecosystem management, this model exemplifies how localized NbS can generate synergistic benefits for mitigation, adaptation, and socio-economic resilience core pillars of the carbon neutrality framework.

Secondly, by preventing wildfires, the initiative supports carbon sequestration and retention, helping preserve above-ground biomass and soil carbon stocks essential for long-term climate stability. Thirdly, the program has created sustainable livelihoods, especially for women and marginalized communities, demonstrating how climate mitigation can be effectively linked with inclusive rural development. Lastly, the initiative promotes a circular bioeconomy by transforming pine needle waste into value-added products such as briquettes, paper items, and handicrafts, offering low-emission alternatives to plastic and paper, and advancing a zero-waste, community-driven climate solution.

Furthermore, additional components of the pine tree such as cones, bark, and branches were creatively repurposed for multiple ecological and economic uses. These included the crafting of decorative items and artificial ornaments, as well as their integration into soil and water conservation structures such as check dams. Such nature-based interventions contribute directly to enhancing ecosystem resilience by reducing erosion, retaining moisture, and supporting habitat restoration. In contrast, pine needles were utilized for the production of biochar and subsequently processed into biomass briquettes, offering a sustainable and low-emission energy alternative for rural households.

The carbon mitigation co-benefits of these interventions, while significant, remain largely under-quantified. Nevertheless, the integrated model demonstrates high potential for replication across pine-dominated mid-altitude landscapes of the IHR. By reducing the frequency and intensity of forest fires, lowering regional carbon emissions, and creating community-based green enterprises, this approach supports both climate mitigation and sustainable development. When linked to PES schemes and emerging carbon market mechanisms, such community-led initiatives could meaningfully contribute to India's NDCs under the Paris Agreement. Furthermore, this model aligns closely with key national strategies, including the GIM and REDD+ readiness programs, thereby reinforcing India's commitment to achieving net-zero emissions by 2070 through inclusive, ecosystem-based climate solutions.

The pine needle-based livelihood and forest management interventions collectively demonstrate a tangible contribution toward local carbon neutrality. Through the production of approximately 11,910 eco-friendly items, including file covers, folders, wedding cards, ornaments, and bio-briquettes, an estimated 2.18 tons of CO_2 emissions have been offset. These products substitute conventional plastic and paper materials, thereby reducing both manufacturing emissions and post-use waste. The use of pine needle bio-briquettes, even at a modest 1 kg per unit, further curbs emissions by displacing traditional firewood and preventing the open burning of highly flammable forest litter (Table 21.5). Beyond the immediate product-level benefits, the systematic collection of pine needles plays a crucial ecological role by significantly reducing surface fuel loads, thereby lowering the intensity and frequency of forest fires a frequently overlooked yet critical source of episodic carbon emissions in Himalayan forests. This intervention directly contributes to emission avoidance while improving forest health and resilience. In parallel, the repurposing of pine biomass including branches, cones, and bark for the construction of soil and water conservation structures, such as check dams, enhances ecosystem functionality. These structures aid in stabilizing fragile hill slopes, reducing erosion, and improving soil moisture retention, which indirectly supports long-term carbon storage and forest regeneration.

Table 21.5 Estimated carbon offset potential of pine needle-based products (based on FSI, 2021)

Component	Details	Estimated CO_2 offset per unit (kg)	Total units	Total CO_2 offset (kg)	CO_2 offset (tons)
Paper-based products	File covers, folders, bags, envelopes, notepads	0.10	9,000	900.0	0.90
Decorative products	Wedding cards, ornaments, and handmade items	0.15	2,000	300.0	0.30
Bio-briquettes	Pine needle briquettes used as a fuel substitute (1 kg each)	0.25	910	227.5	0.23
Forest fire risk reduction	Litter removal impact (per hectare, indirect effect)	500.00	1	500.0	0.50
Plastic substitution	Avoided plastic use in bags, folders, and packaging	0.05	3,000	150.0	0.15
Soil and water conservation	Use of pine branches, cones, and check dams for erosion and moisture control	100.00	1	100.0	0.10

Collectively, these nature-based interventions offer a multidimensional solution that addresses biodiversity conservation, disaster risk reduction, and climate change mitigation within an integrated framework. They also generate socio-economic co-benefits by strengthening community livelihoods-particularly for women and economically vulnerable rural households-through green enterprise development and participatory forest management (Pant, 2025). Importantly, this model demonstrates high replicability across similar pine-dominated mountain landscapes in the IHR, offering a scalable pathway for operationalizing India's commitments to nature-based climate action and inclusive, ecosystem-based development.

21.10 Discussion

The findings of this study underscore the critical role that decentralized, community-driven forest management systems particularly Van Panchayats can play in advancing India's pathway to carbon neutrality. By integrating traditional ecological knowledge with scientific approaches such as standardized carbon accounting, these local institutions are uniquely positioned to address both climate mitigation and adaptation objectives in a holistic manner. The pilot initiative on pine needle utilization serves as a compelling example of how low-cost NbS can deliver multifaceted co-benefits: reducing wildfire risk, preventing episodic carbon emissions, enhancing forest health, and generating livelihood opportunities for rural and marginalized communities.

Despite these promising outcomes, several challenges persist. The absence of standardized MRV systems, limited access to carbon finance mechanisms, and fragmented institutional coordination continue to constrain the scalability and long-term sustainability of such initiatives. Additionally, while India's forest policy frameworks including the National Forest Policy and REDD+ strategy are progressive in vision, they often lack operational mechanisms that are tailored to the unique socio-ecological contexts of mountain regions. Addressing these gaps requires a paradigm shift toward landscape-level forest governance that is both inclusive and adaptive.

This study advocates for the mainstreaming of climate-responsive, community-based forest governance models supported by enabling policy environments and robust financial frameworks such as REDD+ and PES. Such a shift would not only accelerate India's progress toward achieving its NDCs under the Paris Agreement but also enhance ecological resilience, foster green livelihoods, and promote social equity across the fragile Himalayan ecosystems arguably among the most climate-vulnerable regions in the country.

21.10.1 The Central Role of Forests in the Indian Himalayan Region (IHR)

The IHR serves as a linchpin in India's climate mitigation strategy, owing not only to its extensive forest cover and rich biodiversity but also to its complex and dynamic socio-ecological systems. The forests of the IHR play a vital role in carbon sequestration, biodiversity conservation, hydrological regulation, and sustaining the livelihoods of over 50 million people across diverse ethnic and ecological landscapes. Notably, the region has exhibited a substantial increase in carbon stocks, rising from 6.07 gigatons in 1994 to 7.20 gigatons in 2021, highlighting the impact of policy-driven afforestation initiatives, community-based forest governance, and improved forest management practices (FSI, 2021). These forests currently offset more than 20% of India's gross greenhouse gas emissions, underscoring their indispensable contribution to meeting national climate targets and advancing India's NDCs under the Paris Agreement. Given the region's ecological sensitivity and strategic importance, the IHR must be prioritized within national frameworks for climate action, forest conservation, and sustainable development (Pant et al, 2024).

21.10.2 Community-Based Forest Governance: Strengths and Challenges

VPs and JFMCs have emerged as effective models for decentralized forest governance in the IHR. These institutions draw on IKS and foster local participation to enhance forest protection, fire prevention, and the sustainable use of natural resources. Empirical studies suggest that community-managed forests often exhibit greater biodiversity, ecological resilience, and regeneration potential compared to those managed solely by state authorities. Traditional practices such as rotational grazing, sacred grove conservation, and multi-species plantation not only promote ecological sustainability but also embody principles of intergenerational equity and cultural stewardship. However, the long-term viability of these models is constrained by several institutional and socio-political barriers. These include limited financial autonomy, bureaucratic hurdles in plantation and harvesting approvals, and poor coordination across forest, climate, and rural development departments.

Furthermore, social inequities based on caste, gender, and land ownership can impact equitable access to forest resources and decision-making processes. To unlock the full potential of community forest governance, there is a pressing need for legal empowerment, capacity building, and policy convergence at all levels, including local, state, and national. Strengthening these institutions with sustained financial support and inclusive governance mechanisms will be essential for realizing their role in advancing carbon neutrality, biodiversity conservation, and climate resilience in the IHR.

21.10.3 Integrating Traditional Knowledge and Modern Science

The integration of TEK and IKS with scientific forest management is a recurring theme in the IHR. Indigenous practices, such as seasonal harvesting, preserving sacred groves, and rotational farming, have maintained ecological balance for generations. Formal recognition of these practices within policy and management frameworks can enhance ecological sustainability and socio-economic equity. Adaptive management that combines traditional wisdom with modern tools, such as remote sensing, GIS-based inventories, and carbon accounting, can enhance MRV for carbon credits and ecosystem services.

21.10.4 Ecosystem-Based Adaptation (EbA) and Nature-Based Solutions

VPs in the Indian Himalayan Region offer a compelling example of how Ecosystem-Based Adaptation (EbA) and Nature-Based Solutions (NbS) can be effectively implemented at the grassroots level to advance carbon neutrality goals. Through community-led forest governance, VPs promote the conservation, restoration, and sustainable use of local ecosystems, thereby enhancing both ecological resilience and carbon sequestration. Key EbA practices adopted by VPs include the removal of pine needle litter to reduce forest fire risk, the plantation of native and climate-resilient species, and the integration of traditional agroforestry systems, all of which contribute to reducing the vulnerability of forest-dependent communities to climate extremes. Simultaneously, these actions serve as NbS by leveraging ecosystem functions, such as biomass accumulation, soil stabilization, and watershed protection, to deliver co-benefits for climate mitigation and adaptation. Moreover, VPs play a vital role in promoting community stewardship, incorporating TEK into forest management, and fostering inclusive green livelihoods through the value addition of NTFPs and pine-based enterprises. Despite their potential, the effectiveness of VPs in scaling EbA and NbS remains constrained by limited technical capacity, a lack of formal carbon accounting systems, and insufficient access to climate finance mechanisms, such as REDD+ or PES. Strengthening institutional support, enhancing participatory monitoring frameworks, and aligning VP initiatives with national programs such as the GIM and SAPCC will be critical to unlocking their full potential (Kumar et al., 2025). When adequately supported, VPs can become powerful local actors in delivering low-cost, scalable, and culturally grounded solutions that bridge climate mitigation, adaptation, and sustainable development in the fragile Himalayan ecosystem.

21.10.5 Policy and Institutional Frameworks: Opportunities for Convergence

India's national climate commitments, including the creation of an additional carbon sink of 2.5–3.0 gigatons of CO_2-equivalent by 2030, hinge on effective policy integration and institutional support. Key frameworks such as the National Forest Policy, National Agroforestry Policy, Green India Mission, and REDD+ provide the backbone for forest-based climate solutions. However, the lack of cross-sectoral convergence, real-time monitoring, and adequate funding at the state level hampers progress.

State-specific SAPCCs have tailored approaches; for instance, Uttarakhand emphasizes ecosystem resilience and TEK, Himachal Pradesh focuses on REDD+ readiness, and Sikkim prioritizes alpine ecosystem preservation. Despite their comprehensive scope, implementation remains inconsistent due to institutional fragmentation and a lack of localized data.

21.10.6 Quantifying and Monetizing Carbon Benefits

Systematic carbon accounting in VPs remains at a nascent stage, yet it is a critical prerequisite for integrating these ecosystems into emerging carbon markets and PES schemes. Empirical studies and pilot assessments indicate that VP-managed forests typically maintain higher biomass and carbon stock levels compared to open-access or degraded forestlands, primarily due to community enforcement, regulated extraction, and ecological restoration practices. This positions Van Panchayat forests as promising candidates for REDD+ programs and participation in voluntary carbon markets, where verified emissions reductions can be monetized. However, realizing this potential requires a set of enabling interventions. First, training local communities in simplified carbon stock estimation protocols can enhance grassroots participation in carbon accounting. Second, integrating remote sensing technologies with field-based forest inventories can improve the accuracy and scalability of carbon assessments. Third, establishing district-level carbon baselines and incorporating VP forest data into state-level GHG inventories would ensure formal recognition of their climate mitigation contributions. By institutionalizing these steps, Van Panchayats could transition from informal conservation actors to recognized climate service providers, thereby unlocking new streams of finance and incentives for sustainable forest management and community development.

21.10.7 Socio-economic Equity and Livelihood Diversification

Forest-dependent communities in the IHR rely on timber, fuelwood, fodder, and NTFPs for their livelihoods. While these resources provide economic safety nets, access is often unequal. Community-based governance models have helped address some disparities through institutionalized participation and the sharing of benefits. Emerging bioeconomic approaches such as value-added NTFP processing and improved market access offer new livelihood opportunities, especially for women and youth. However, achieving accurate equity requires policy interventions that recognize customary rights, secure land tenure, and support marginalized groups. Women play a vital role in advancing carbon neutrality through sustainable practices, resource management, and community leadership. Their engagement in renewable energy, forest conservation, and eco-friendly livelihoods significantly contributes to reducing carbon emissions, enhancing climate resilience, and achieving national and global targets for a low-carbon, sustainable future which can be enhanced through the modern technological interventions for ground based forest resource management (Pant & Bhatt, 2023).

21.10.8 Addressing Threats and Enhancing Resilience

Persistent threats, including shifting cultivation, infrastructure expansion, unregulated grazing, and increasing forest fires, continue to degrade forests and reduce their carbon retention capacity. Climate change exacerbates these challenges, causing shifts in vegetation zones, reduced regeneration, and increased vulnerability to pests and diseases. Effective forest management must therefore prioritize both carbon storage and ecosystem resilience, employing adaptive strategies informed by vulnerability assessments and long-term planning.

21.10.9 Challenges and Opportunities

Achieving carbon neutrality through VP management in the Himalayan region is a promising yet complex endeavor, marked by a dynamic interplay of opportunities and challenges. The VPs, rooted in participatory governance, offer a decentralized framework for sustainable forest management that aligns closely with climate mitigation objectives. These community institutions manage vast tracts of forested landscapes that not only serve as biodiversity reservoirs but also act as significant carbon sinks. Their ability to integrate traditional TEK, enforce localized resource-use norms, and promote nature-based enterprises, such as pine needle utilization, agroforestry, and NTFP-based livelihoods, places them at the forefront of grassroots climate solutions. However, despite this potential, several structural and operational challenges hinder their effectiveness in contributing to carbon neutrality. These include limited awareness and capacity among VP members to conduct carbon stock assessments, the lack of simplified MRV protocols, and inadequate linkage with carbon finance mechanisms such as REDD+ or voluntary carbon markets. Many Van Panchayats also suffer from poor institutional coordination, absence of digitized forest records, and ambiguous legal tenure rights, which weaken their ability to attract long-term conservation investments.

Furthermore, growing anthropogenic pressures, encroachment, overgrazing, and the increasing frequency of climate-induced disturbances, such as forest fires, landslides, and drying springs, further strain their ecological capacity. Nonetheless, there exists a clear opportunity to transform Van Panchayats into climate-resilient institutions by mainstreaming them into state climate action plans, providing targeted training in carbon accounting, facilitating access to PES, and enhancing convergence with programs like the GIM, National Adaptation Fund for Climate Change, and CAMPA. With the right mix of policy support, scientific guidance, and financial incentives, Van Panchayats can evolve into robust actors in India's carbon neutrality pathway, linking forest conservation, rural development, and climate justice in the fragile Himalayan ecosystem.

21.10.10 Roadmap for Integrating EbA and Community Governance for Climate-Resilient Himalayan Forests

A targeted, inclusive, and science-driven roadmap is essential for aligning the Himalayan forest sector with India's carbon neutrality objectives. Central to this vision is the strengthening of community-based forest management, particularly through the legal and financial empowerment of institutions like Van Panchayats, alongside participatory planning and capacity-building. Equally important is the scaling up of ecosystem restoration, emphasizing the use of native species plantations, ecological corridors, and climate-resilient afforestation in degraded and disaster-prone landscapes (Fig. 21.2). To ensure long-term sustainability, the roadmap must also promote carbon finance mechanisms and nature-based enterprises, enabling community forests to access voluntary carbon markets and generate green livelihoods. Effective climate action further requires the integration of forest, climate, and rural development policies, with localized implementation of Sustainable Agricultural and Rural Development (SAPCC) plans. Ultimately, robust MRV systems that integrate remote sensing technologies with community-based inventories are crucial for monitoring progress and ensuring accountability in carbon governance.

The framework in Fig. 21.4 is structured across six key pillars: Policy and Regulatory Foundation, Institutional Mechanisms, Community and Stakeholder Engagement, Technical and Programmatic Interventions, Regional and Transboundary Cooperation, and MRV. It highlights the alignment of national and subnational policies (e.g., SAPCCs, NAPCC, Ladakh CNAP), the central role of VPs and community institutions, and the deployment of nature-based solutions such as agroforestry, springshed management, and renewable energy. The diagram emphasizes the integration of traditional and scientific knowledge, the engagement of women and indigenous groups, and the need for robust MRV systems. Lead institutions, including the MoEFCC, NITI Aayog, ICIMOD, TERI, and local bodies, play a collaborative role in operationalizing this roadmap.

VPs represent a scalable, inclusive, and ecologically sound model for advancing carbon neutrality in the Himalayas. Investing in these community institutions and integrating them into national and subnational climate strategies may accelerate India's transition toward a net-zero future while promoting biodiversity conservation, ecosystem stability, and sustainable rural development.

21.11 Conclusion and Recommendations

Forests of the IHR are indispensable for India's climate mitigation, biodiversity conservation, and sustainable development, and are central to India's climate mitigation efforts and its trajectory towards carbon neutrality. As dynamic socio-ecological systems, they play a vital role not only in sequestering atmospheric carbon dioxide

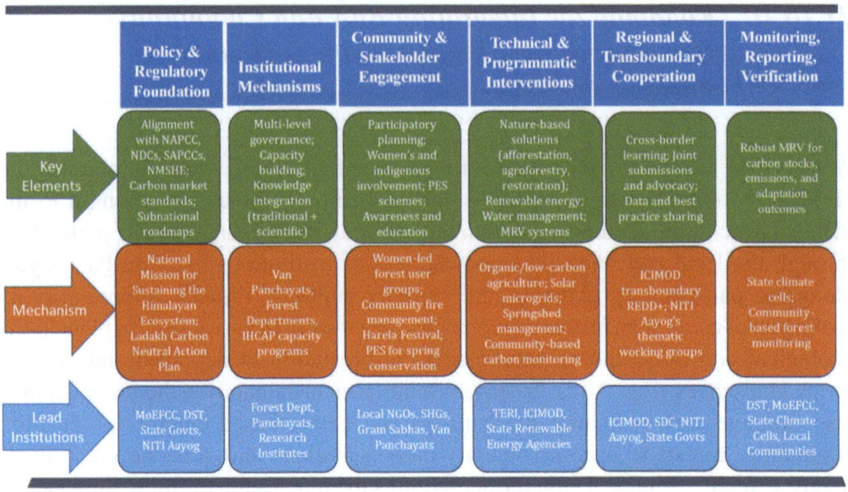

Fig. 21.4 Integrated framework for forest-based carbon neutrality in the Indian Himalayan region

but also in preserving biodiversity, regulating hydrology, reducing disaster risks, and supporting the livelihoods of millions of forest-dependent communities. However, the accelerating pace of climate change, coupled with pressures from unsustainable development, population growth, and governance challenges, poses a threat to the ecological integrity and long-term viability of these forests. To safeguard their functions, it is critical to adopt an integrated strategy that combines conservation, restoration, sustainable land use, community participation, and innovative financing mechanisms. Community-led models, integration of traditional knowledge, and innovative nature-based solutions offer promising pathways for achieving carbon neutrality. However, realizing this potential requires overcoming institutional, technical, and socio-economic barriers through coordinated policy action, investment in capacity building, and empowerment of local communities.

Community-led institutions, such as Van Panchayats in Uttarakhand, offer a proven model of decentralized forest governance that aligns well with India's climate goals. These institutions have demonstrated their capacity to sustainably manage forests, prevent degradation, and restore degraded ecosystems through local participation, TEK and inclusive decision-making. By integrating Van Panchayats into carbon credit mechanisms and PES schemes, and enhancing their institutional capacities in carbon monitoring, adaptive silviculture, and ecological restoration, they can emerge as key actors in forest-based climate mitigation (Box 1). Additionally, encouraging women and youth participation, promoting mixed-species plantations, and establishing district-level GHG inventories will further strengthen their contribution to carbon neutrality. Nature-based solutions anchored in traditional land-use systems such as agroforestry, rotational farming, and community forestry can be scaled up across the IHR to improve carbon sequestration while enhancing ecological resilience. These approaches improve soil health, diversify rural incomes, and reduce vulnerability to climate shocks. Simultaneously, conservation of natural carbon sinks, especially old-growth forests, sacred groves, and biodiversity-rich alpine zones, must be prioritized to preserve existing carbon stocks and ecosystem services.

Box 1: Policy recommendations for van panchayats and carbon neutrality in the IHR

Thematic area	Policy recommendations
Legal and institutional support	• Grant statutory recognition and strengthen tenure rights of VPs • Allocate dedicated budgets for VP operations
Capacity building and technical support	• Train VP members in carbon stock estimation, MRV, and EbA practices • Facilitate participatory forest monitoring and planning
Climate finance and carbon markets	• Enable access to REDD+, PES, and voluntary carbon markets • Establish VP-level or local, regional aggregation platforms for carbon crediting
Ecosystem-based adaptation and TEK	• Integrate EbA and traditional ecological knowledge into SAPCCs and forest working plans • Promote native species plantations and biodiversity corridors
Cross-sectoral policy convergence	• Align VP activities with rural employment (MGNREGA), organic farming, and watershed programs • Establish inter-departmental coordination platforms
Nature-based enterprises and bioeconomy	• Promote the development of pine needle-based and invasive species-based products and value chains • Support women- and youth-led green enterprises
Monitoring, reporting and verification	• Combine remote sensing with community forest inventories for MRV • Integrate VP forest data into state-level GHG inventories

Innovative initiatives such as the use of pine needle (Pirul)-based products represent low-cost, high-impact solutions for reducing forest fire risks while generating rural livelihoods and supporting bio-based circular economies. These localized interventions, when combined with modern technology for fire early warning systems, remote sensing, and climate monitoring, can drastically reduce emissions from forest fires and biomass loss, contributing to India's net-zero goal. From a policy perspective, aligning national frameworks such as the National Agroforestry Policy, Green India Mission, and REDD+ strategy with SAPCCs and local governance structures is vital for convergence and scale. Moreover, investing in climate inventories, data transparency, and carbon MRV systems at the state and district levels will support evidence-based forest management and enable participation in international climate finance mechanisms. Achieving carbon neutrality in the Indian Himalayas requires a multi-dimensional and inclusive approach, one that harmonizes scientific innovation with grassroots action, traditional knowledge with adaptive management, and ecological integrity with sustainable livelihoods. Building on existing institutional frameworks, empowering communities, and ensuring policy coherence across scales may guide India in transforming its Himalayan forests into resilient carbon sinks, paving the way for a just, climate-secure, and ecologically sustainable future.

Abbreviations

Abbreviations	Detail
BEE	Bureau of Energy Efficiency
CAMPA	Compensatory Afforestation Fund Management and Planning Authority
CBR	Community Biodiversity Register
CCTS	Carbon Credit Trading Scheme
FGD	Focused Group Discussion
GEF	Global Environment Facility
GHG	Greenhouse Gases
GIM	Green India Mission
EbA	Ecosystem-Based Adaptations
ICIMOD	International Centre for Integrated Mountain Development
IOOI	Input–Output–Outcome–Impact
IHR	Indian Himalayan Region
IPCC	Intergovernmental Panel on Climate Change
IHCAP	Indian Himalayas Climate Adaptation Programme
ISFR	India State of Forest Report
IKS	Indigenous Knowledge Systems
JFMCs	Joint Forest Management Committees
MRV	Monitoring, Reporting, and Verification
MGNREGA	Mahatma Gandhi National Rural Employment Guarantee Act
NbS	Nature-based Solutions
NITI aayog	National Institution for Transforming India
NAFCC	National Adaptation Fund for Climate Change
NAPCC	National Action Plan on Climate Change

(continued)

(continued)

Abbreviations	Detail
NMSHE	National Mission for Sustaining the Himalayan Ecosystem
NMHS	National Mission on Himalayan Studies
NSCICM	National Steering Committee on Implementation of Carbon Market
NTFPs	Non-Timber Forest Products
NWFPs	Non-wood Forest Products
PES	Payment for Ecosystem Services
PRA	Participatory Rural Appraisal
REDD+	Reducing Emissions from Deforestation and Forest Degradation
SAPCCs	State Action Plans on Climate Change
SDC	State Data Centre
TEK	Traditional Ecological Knowledge
TERI	The Energy and Resources Institute
UNDP	United Nations Development Program

References

Beedlow, P. A., Tingey, D. T., Phillips, D. L., Hogsett, W. E., & Olszyk, D. M. (2004). Rising atmospheric CO_2 and carbon sequestration in forests. *Frontiers in Ecology and the Environment, 2*(6), 315–322.

Bhardwaj, D. R., Salve, A., Kumar, J., Kumar, A., Sharma, P., & Kumar, D. (2024). Biomass production and carbon storage potential of agroforestry land use systems in high hills of north-western Himalaya: An approach towards natural based climatic solution. *Biomass Conversion and Biorefinery, 14*(15), 18079–18092.

Bhatt, H., & Pant, H. J. (2023). Uttarakhand's forest resource's: Nurturing a green economy for sustainable tomorrow. *ENVIS Newsletter on Himalayan Ecology, 20*(3), 6.

Bhatt, H., & Pant, H.J. (2024). Community-managed forests and their effectiveness in SDG implications in the Western Himalayan Region. In *Warming mountains: Implications for livelihood and sustainability* (pp. 435–458). Cham: Springer Nature Switzerland. https://doi.org/10.1007/978-3-031-62197-0_22. ISBN 978-3-031-62196-3

Bhatt, H., Pant, H. J., & Pandey, R. (2024a). Cultural ecosystem services nexus with socio-cultural attributes and traditional ecological knowledge for managing community forests of Indian Western Himalaya. *Ecological Indicators*, *166*, 1123.

Bhatt, H., Verma, D., Jugran, H. P., Kumar, V., & Pandey, R. (2024b). Community perception about ecosystem services assessment from forests managed under different regimes in Hindu Kush Himalaya. *Environmental and Sustainability Indicators*, 100463.

Bhattarai, N. (2024). *Studies on addressing challenges and enhancing community engagement in REDD+ implementation in Nepal and India* (Doctoral dissertation, 北海道大学).

Borah, D., Rout, J., & Nooruddin, T. (2025). Impacts of land use changes in Indian Himalayas: A socio-economic and ecological perspective. In *People and mountain environments: Interconnectedness for sustainable development in the Himalayas* (pp. 115–147). Springer Nature Switzerland.

Chaturvedi, R. K., Strappason, A., & Garg, A. (2022). Forests, trees and agroforestry: Their roles in India's sustainable development and climate action.
Chaudhry, P., & Tambe, S. (2023). Investigation of the Indian Forest sector's performance during the last three and half decades. *Ecological Questions, 34*(3), 1–10.
Dangwal, B., Rana, S. K., Negi, V. S., & Bhatt, I. D. (2022). Forest restoration enhances plant diversity and carbon stock in the sub-tropical forests of western Himalaya. *Trees, Forests and People, 7*(100201). https://doi.org/10.1016/j.tfp.2022.100201
Das, A., & Mallick, P. H. (2024). Exploring livelihood dependency on provisioning ecosystem services in a protected tropical forest area of eastern India: Keys to sustainable forest management. *Environment, Development and Sustainability*. https://doi.org/10.1007/s10668-024-04933-7
Deb, D., Jamatia, M., Debbarma, J., Ahirwal, J., Deb, S., & Sahoo, U. K. (2021). Evaluating the role of community-managed forest in carbon sequestration and climate change mitigation of Tripura, India. *Water, Air, & Soil Pollution, 232*(5), 166.
Devi, N. B., Lepcha, N. T., Bhutia, P. T., Rocky, P., Sahoo, U. K., Pandey, R., & Nath, A. J. (2023). Biodiversity and ecosystems services of the agroforestry systems of the Himalayan region: An overview. In *Agroforestry for sustainable intensification of agriculture in Asia and Africa* (pp. 487–513).
Dhyani, S., Murthy, I. K., Kadaverugu, R., Dasgupta, R., Kumar, M., & Adesh Gadpayle, K. (2021). Agroforestry to achieve global climate adaptation and mitigation targets: Are South Asian countries sufficiently prepared? *Forests, 12*(3), 303.
Dhyani, S., & Dhyani, D. (2020). Local socio-economic dynamics shaping forest ecosystems in Central Himalayas. In *Socio-economic and eco-biological dimensions in resource use and conservation: Strategies for sustainability* (pp. 31–60).
Elegbeleye, I., Oguntona, O., & Elegbeleye, F. (2025). Green hydrogen: Pathway to net zero green house gas emission and global climate change mitigation. *Hydrogen, 6*(2), 29.
Erdoğdu, A., Dayi, F., Yanik, A., Yildiz, F., & Ganji, F. (2025). Innovative solutions for combating climate change: Advancing sustainable energy and consumption practices for a greener future. *Sustainability, 17*(6), 2697.
Forest Survey of India (FSI). (2021). *India state of forest report 2021*. Ministry of Environment, Forest and Climate Change, Government of India.
Gabric, A. J. (2023). The climate change crisis: A review of its causes and possible responses. *Atmosphere, 14*(7), 1081.
Gadgil, M., Berkes, F., & Folke, C. (1993). Indigenous knowledge for biodiversity conservation. *Ambio, 22*(2/3), 151–156. http://www.jstor.org/stable/4314060
Gautam, A. P., Karmacharya, M. B., & Karna, B. K. (2008). Community forestry, equity and sustainable livelihoods in Nepal. In *Proceedings of the 12th Biennial conference of the international association for the study of the commons (IASC) held during* (pp. 14–18).
Gorain, S., Malakar, A., Dutta, S., & Das, S. (2025). Assessing forest cover, carbon stocks and fire dynamics in India and economic valuation of forest carbon in Asia: A meta-analysis. *Discover Forests, 1*(1), 1–29.
Gupta, A., Srivastava, P. K., Satish, K. V., Chauhan, A., & Pandey, P. C. (2022). Challenges and future possibilities toward Himalayan forest monitoring. In *Advances in remote sensing for forest monitoring* (pp. 329–350).
Gupta, H., Janju, S., Mahajan, A., Singh, C., Sharma, S., & Prajapati, A. (2024). Forests and agroforestry: Nature-based solutions for climate change mitigation. In *Forests and climate change: Biological perspectives on impact, adaptation, and mitigation strategies* (pp. 421–443). Springer Nature Singapore.
Haq, S. M., Pieroni, A., Bussmann, R. W., Abd-ElGawad, A. M., & El-Ansary, H. O. (2023). Integrating traditional ecological knowledge into habitat restoration: Implications for meeting forest restoration challenges. *Journal of Ethnobiology and Ethnomedicine, 19*(1), 33.

He, J., Zhou, Y., Pandey, R., Stringer, L. C., Cao, Y., Bhatt, H., Luo, R., Wang, S., & Li, T. (2025). Characteristics and framework for assessing supply and demand relationship for ecosystem services using a trade-off and synergy lens. *Land Degradation & Development*.

Houghton, R. A. (2003). Why are estimates of the terrestrial carbon balance so different? *Global Change Biology, 9*(4), 500–509.

Hussain, S., Sharma, S., Bhatti, R. C., & Singh, A. N. (2024). Sustainability in the Indian Himalayan Region: Understandings the ecosystem services, climate change impacts, land use shifts and their threats. In *The Himalayas in the Anthropocene: Environment and development* (pp. 33–57). Springer Nature Switzerland.

IPCC. (2018). Summary for policymakers. In V. Masson-Delmotte, P. Zhai, H.-O. Pörtner, D. Roberts, J. Skea, P. R. Shukla, A. Pirani, W. Moufouma-Okia, C. Péan, R. Pidcock, S. Connors, J. B. R. Matthews, Y. Chen, X. Zhou, M. I. Gomis, E. Lonnoy, T. Maycock, M. Tignor, & T. Waterfield (Eds.), *Global warming of 1.5°C. An IPCC special report on the impacts of global warming of 1.5°C above pre-industrial levels and related global greenhouse gas emission pathways, in the context of strengthening the global response to the threat of climate change, sustainable development, and efforts to eradicate poverty* (pp. 3–24). Cambridge University Press. https://doi.org/10.1017/9781009157940.001

ISFR. (2021). India State of Forest Report, Forest Survey of India. ISBN 978-81-950073-1-8

Joshi, R., Singh, H., Chhetri, R., Poudel, S. R., & Rijal, S. (2021). Carbon sequestration potential of community forests: A comparative analysis of soil organic carbon stock in community managed forests of Far-Western Nepal. *Eurasian Journal of Soil Science, 10*(2), 96–104. https://doi.org/10.18393/ejss.825066

Karki, M., & Adhikari, J. R. (2015). Integrating indigenous, local and modern knowledge for sustainable conservation and management of forest ecosystems in Nepal. *Forestry Nepal: Gateway to Forestry Information in Nepal.*

Kc, K. B., Tzadok, E., & Pant, L. (2022). Himalayan ecosystem services and climate change driven agricultural frontiers: A scoping review. *Discover Sustainability, 3*(1), 35.

Kotwal, P. C., Omprakash, M. D., Gairola, S., & Dugaya, D. (2008). Ecological indicators: Imperative to sustainable forest management. *Ecological Indicators, 8*(1), 104–107.

Kumar, M. (2023). Forestry policies and practices to promote climate change adaptation in the Indian Western Himalayan states. *Climate change adaptation, risk management and sustainable practices in the Himalaya* (pp. 65–87). Springer International Publishing.

Kumar, D., Thakur, C. L., Bhardwaj, D. R., Sharma, N., Sharma, P., & Sankhyan, N. (2022). Biodiversity conservation and carbon storage of Acacia catechu Willd. Dominated northern tropical dry deciduous forest ecosystems in north-western Himalaya: Implications of different forest management regimes. *Frontiers in Environmental Science, 10*, 981608.

Kumar, S., Verma, P., Kanyal, H. S., Jha, P. C., & Rai, J. (2025). Initiatives for challenges faced by developed countries and India on green growth and sustainable development in the World. In Demystifying emerging trends in green technology (pp. 219–231).

Kumari, R., Kumar, A., Saikia, P., & Khan, M. L. (2022). Vulnerability assessment of the Indian Himalayan forests in terms of biomass production and carbon sequestration potential in changing climatic conditions. *Handbook of climate change mitigation and adaptation* (pp. 147–161). Springer International Publishing.

Kumari, Y., Garima, Kumari, N., Mehta, A., Walia, K., Devi, M., Kumari, R., Rana, D., & Verma, S. (2024a). Empowering climate change resilience through forests and agroforestry. In *Forests and climate change: Biological perspectives on impact, adaptation, and mitigation strategies* (pp. 489–512). Springer Nature Singapore.

Kumari, G., Kotiyal, P. B., Singh, H., Kumar, M., Kumar, N., Malik, A., ... & Singh, S. (2024b). Predicting future climate change effects on biotic communities: A species distribution modeling approach. In *Forests and climate change: Biological perspectives on impact, adaptation, and mitigation strategies* (pp. 137–168). Singapore: Springer Nature Singapore.

Liang, S., Ziegler, A. D., Reich, P. B., Zhu, K., Wang, D., Jiang, X.,... & Zeng, Z. (2025). Climate mitigation potential for targeted forestation after considering climate change, fires, and albedo. *Science Advances, 11*(15), eadn7915.

Liu, H., Zhang, L., Ma, Q., Zhao, W., & Chen, Y. (2023). From tree to forest: Multiple carbon sink constraints. *The Innovation, 4*(4).

Longbottom, T. L., Townsend-Small, A., Owen, L. A., & Murari, M. K. (2014). Climatic and topographic controls on soil organic matter storage and dynamics in the Indian Himalaya: Potential carbon cycle–climate change feedbacks. *Catena, 119*, 125–135.

Makundi, W. R. (1997). Global climate change mitigation and sustainable forest management—The challenge of monitoring and verification. *Mitigation and Adaptation Strategies for Global Change, 2*, 133–155.

Malhi, Y., Meir, P., & Brown, S. (2002). Forests, carbon and global climate. *Philosophical Transactions of the Royal Society of London. Series A: Mathematical, Physical and Engineering Sciences, 360*(1797), 1567–1591.

Måren, I. E., Bhattarai, K. R., & Chaudhary, R. P. (2014). Forest ecosystem services and biodiversity in contrasting Himalayan forest management systems. *Environmental Conservation, 41*(1), 73–83.

Meetei, K. B., Tsopoe, M., Giri, K., Mishra, G., Verma, P. K., & Rohatgi, D. (2023). Climate-resilient pathways and nature-based solutions to reduce vulnerabilities to climate change in the Indian Himalayan Region. In *Climate change in the Himalayas* (pp. 89–119). Academic Press.

Mishra, R., Khurana, R., Mohanty, A., & Agarwal, P. (2024). Emerging threats to mountain ecosystem services under changing climate. In *Warming mountains: Implications for livelihood and sustainability* (pp. 25–43). Springer Nature Switzerland.

MoEFCC. (2018). *Draft national forest policy*. The Ministry of Environment, Forest and Climate Change.

Nagahama, K., Tachibana, S., & Rakwal, R. (2022). Critical aspects of people's participation in community-based forest management from the case of Van Panchayat in Indian Himalaya. *Forests, 13*(10), 1667.

Nayak, S., & Srivastava, E. (2021). Approaches, extent and issues at the global level. *Environment and Sustainable Development: Perspectives and Issues.*

Negi, G. C. S. (2022). Trees, forests and people: The Central Himalayan case of forest ecosystem services. *Trees, Forests and People, 8*, Article 100222.

Negi, V. S., Pathak, R., Rawal, R. S., Bhatt, I. D., & Sharma, S. (2019). Long-term ecological monitoring on forest ecosystems in Indian Himalayan Region: Criteria and indicator approach. *Ecological Indicators, 102*, 374–381.

Negi, V. S., Pathak, R., Dhyani, V., Durgapal, M., Joshi, R. K., & Bhatt, I. D. (2022). Land restoration in the Himalayan Region: Steps towards biosphere integrity. *Land Use Policy, 121*, Article 106317.

Negi, V. S., Pathak, R., Thakur, S., Joshi, R. K., Bhatt, I. D., & Rawal, R. S. (2023). Scoping the need of Mainstreaming indigenous knowledge for sustainable use of bioresources in the Indian Himalayan region. *Environmental Management, 72*(1), 135–146.

Negi, G. C. S., Chauhan, D. S., Bisht, D. D., Pant, H., Arya, S. C., Deorari, M., Rawat, D. S., & Sundriyal, R. C. (2021). Technical manual on use of Pine needles for making Environment friendly products and avoidance of forest fire. GBPNIHE (G.B. Pant National Institute of Himalayan Environment). ISBN: 9789355937155.

Ontl, T. A., Janowiak, M. K., Swanston, C. W., Daley, J., Handler, S., Cornett, M., Hagenbuch, S., Handrick, C., Mccarthy, L., & Patch, N. (2020). Forest management for carbon sequestration and climate adaptation. *Journal of Forestry, 118*(1), 86–101. https://doi.org/10.1093/jofore/fvz062

Ozukum, L., Feroze, S. M., & Roy, A. (2023). Alder-based farming system of Nagaland: Valuating the eco-services from nitrogen fixing trees. *Indian Journal of Agricultural Economics, 78*(3), 420–431.

Pandey, A. K., Tripathi, Y. C., & Kumar, A. (2016). Non timber forest products (NTFPs) for sustained livelihood: Challenges and strategies. *Research Journal of Forestry, 10*(1), 1–7.

Pant, H. (2025). Ritual as resilience: Celebration of Harela festival. *LEISA INDIA* June, 16–19.

Pant, H., & Mishra, A. (2025). Role of local community in the forest fire prevention and management: Insights from Uttarakhand State. In H. Singh, & N. Kumar (Eds.), *Forest fire and climate change*. Springer, Cham. https://doi.org/10.1007/978-3-031-89967-6_13

Pant, H., Upadhyaya, P., & Ghosh, P. (2025). Advancing agroforestry research in the Indian Himalayas: a systematic review of knowledge trends, research gaps, and circular economy pathways. *Agroforestry Systems, 99*(8), 253.

Pant, H. J., & Bhatt, H. (2023). Relationship of ecosystem services with community managed forests. In M. K. Arya (Eds.), *Biodiversity, environment and ecosystem services* (pp. 77–88). Delhi (India): Discovery Publishing House. ISBN 978-81-959169-3-1.

Pant, M., & Pant, H. (2025). Translational dynamics of social-ecological management system of community forestry: prospects for ecosystem-based adaptation in community-managed forests of Uttarakhand. In: Khan, Y. I., Goswami, M., Nautiyal, S. (Eds.), *Ecosystem-based approaches for resilience building in Himalayan landscapes. Disaster resilience and green growth*. Springer, Singapore. https://doi.org/10.1007/978-981-95-2007-7_14

Pant, H. J., & Rawal, R. (2025). Role of Himalayan women in circular bioeconomy & livelihood generation. In J. Joshi, H. V. Pant, S. Vinwal, N. Bisht (Eds.), *Harmony in heritage: women, folk culture and sustainability in Uttarakhand*. IBS Pvt. Ltd. ISBN 978-93-6729-069-9.

Pant, H. J., Bhatt, H., Bisht, D. S., & Bisht, D. (2024). Impact of changing climate pattern on forest resources and socio-economic dynamics of a marginalized village in Kumaon Himalayan region. In *Himalayan ecology: Special focus on Millets* (pp. 63–66). EIACP- GBPNIHE. ISBN 978-93-5967-660-9.

Panthi, B. B. (2011). Additional benefits of community managed forest: A case study of Champadevi community forest. *Nepal Journal of Science and Technology, 12*, 127–132.

Patra, S., Kumar, A., & Saikia, P. (2022). Deforestation and forests degradation impacts on livelihood security and climate change: Indian initiatives towards its mitigation. *Environmental degradation: Challenges and strategies for mitigation* (pp. 371–392). Springer International Publishing.

Peerzada, I. A., Chamberlain, J., Reddy, M., Dhyani, S., & Saha, S. (2021). Policy and governance implications for transition to NTFP-based bioeconomy in Kashmir Himalayas. *Sustainability, 13*(21), 11811.

Peerzada, I. A., Islam, M. A., Chamberlain, J., Dhyani, S., Reddy, M., & Saha, S. (2022). Potential of NTFP based bioeconomy in livelihood security and income inequality mitigation in Kashmir Himalayas. *Sustainability, 14*(4), 2281.

Prentice, I. C., Farquhar, G. D., Fasham, M. J. R., Goulden, M. L., Heimann, M., Jaramillo, V. J., Kheshgi, H. S., Le Quéré, C., Scholes, R. J., Wallace, D. W., & Archer, D. (2001). The carbon cycle and atmospheric carbon dioxide. In *Climate change 2001: The scientific basis, Intergovernmental panel on climate change*.

Rajbhandari, S., Winyuchakrit, P., Pradhan, B. B., Chaichaloempreecha, A., Pita, P., & Limmeechokchai, B. (2024). Thailand's net-zero emissions by 2050: Analysis of economy-wide impacts. *Sustainability Science, 19*(1), 189–202.

Rao, N. N. (2024). *Rural development: Community forest management and sustainable development*. Laxmi Book Publication.

Rao, K. S., Semwal, R. L., Maikhuri, R. K., Nautiyal, S., Sen, K. K., Singh, K., Chandrasekhar, K., & Saxena, K. G. (2003). Indigenous ecological knowledge, biodiversity and sustainable development in the central Himalayas.

Rao, K. S., Semwal, R. L., Ghoshal, S., Maikhuri, R. K., Nautiyal, S., & Saxena, K. G. (2022). Participatory active restoration of communal forests in temperate Himalaya, India. *Restoration Ecology, 30*(1), Article e13486.

Rawal, R., & Jugran, H. P. (2024). Impact of climate change and mitigation plans for building climate-resilient village ecosystem: Challenges and adaptive strategies. In S. Tripathi, R. Bhadouria, & S. C. Garkoti (Eds.), *Warming mountains*. Springer. https://doi.org/10.1007/978-3-031-62197-0_10

Rawal, R., Pant, H., & Ghildiyal, S. (2025). An ecosystem-based assessment of resilient traditional restoration practices towards biodiversity conservation and ecosystem restoration in high altitudes of Darma Valley in Uttarakhand Himalayan Region. In: Khan, Y. I., Goswami, M., Nautiyal, S. (Eds.), *Ecosystem-based approaches for resilience building in Himalayan landscapes. Disaster resilience and green growth.* Springer, Singapore. https://doi.org/10.1007/978-981-95-2007-7_12

Rist, L., Shaanker, R. U., Milner-Gulland, E. J., & Ghazoul, J. (2010). The use of traditional ecological knowledge in forest management: An example from India. *Ecology and Society, 15*(1).

Sahana Florence, P., & Mishra, A. (2024). Traditional ecological knowledge repository in the Indian Himalayas: An overview. In *Addressing the climate crisis in the Indian Himalayas: Can traditional ecological knowledge help?* (pp. 293–311).

Saikia, P., Nag, A., Kumari, R., Kumar, A., & Khan, M. L. (2021). Forest degradation prevention through nature-based solutions: An Indian perspective. In *Handbook of ecological and ecosystem engineering* (pp. 333–352).

Sati, V. P. (2023a). *Sustainable forest management in the Himalaya.* Springer.

Sati, V. P. (2023b). Ecosystem services valuation and payment for livelihood sustainability in the Indian Central Himalayan region. *Journal of Resources and Ecology, 14*(3), 468–478.

Schimel, D. S., House, J. I., Hibbard, K. A., Bousquet, P., Ciais, P., Peylin, P., Braswell, B. H., Apps, M. J., Baker, D., Bondeau, A., & Canadell, J. (2001). Recent patterns and mechanisms of carbon exchange by terrestrial ecosystems. *Nature, 414*(6860), 169–172.

Sharma, S., & Pant, H. (2017). Vulnerability of Indian Central Himalayan forests to fire in a warming climate and a participatory preparedness approach based on modern tools. *Current Science*, 2100–2105.

Sharma, S., Joshi, R., Pant, H., & Dhyani, P. P. (2017). *Climate change & North-West Himalaya: prioritization of agriculture based livelihood actions.* GB Pant National Institute of Himalayan Environment and Sustainable Development, Kosi-Katarmal, Almora (p. 27).

Sharma, R., Singh, S., Huria, A., Chauhan, A., & Barman, T. (2024). Sustainable land management practices for healthy ecosystem services: A cornerstone of biodiversity sustenance in the Indian Northwestern Himalaya. In *Sustainable land management in India* (pp. 235–252). Springer.

Sheikh, N., Patra, S., Kumar, A., & Saikia, P. (2022). Indian forests: Sustainable uses and its role in livelihood security. In *Land degradation neutrality: Achieving SDG 15 by forest management* (pp. 437–452). Springer Nature Singapore.

Siegner, M., Panwar, R., & Kozak, R. (2021). Community forest enterprises and social enterprises: The confluence of two streams of literatures for sustainable natural resource management. *Social Enterprise Journal, 17*(4), 584–603.

Singh, P. P. (2008). Exploring biodiversity and climate change benefits of community-based forest management. *Global Environmental Change, 18*(3), 468–478.

Singh, H., & Kumar, M. (2022). Climate change and its impact on Indian Himalayan forests: Current status and research needs. In *Climate change: Impacts, responses and sustainability in the Indian Himalaya* (pp. 223–242). Cham: Springer International Publishing.

Singh, M., Singh, H., Kumar, A., Kumar, N., Kumar, M., Barthwal, S., & Thakur, A. (2024). Soil nitrogen availability determines the CO fertilization effect on tree species (*Neolamarckia cadamba*): Growth and physiological evidence. *Environmental Sustainability, 7*(1), 53–60.

Stevens, M. P. (2022). *Exploring complexity in changing practices of care: A mixed methods inquiry into rights, relations, and knowledge in protected area conservation* (Doctoral dissertation, University of British Columbia).

Stevens, M., & Satterfield, T. (2024). Many mountain paths: Perceiving change in the management of community forests in the Hindu Kush Himalaya, Uttarakhand, India. *Environment and Planning E: Nature and Space, 7*(3), 1142–1169.

Sugandha, Phartiyal, M., & Jugran, H. P. (2025). Scoping of forest resources in light of climate change implications in Western Himalayan Region. In *Climate change impact on Himalayan biodiversity.* Cham: Springer Nature Switzerland. https://doi.org/10.1007/978-3-031-77149-1_5. ISBN 978-3-031-77149-1.

Tiwari, C., & Sunny, A. (2024). To the ones who listen to the mountains: Evaluating the role of traditional ecological knowledge in conservation efforts in the Himalayas. In *Learning 'from' and 'with' the locals: Traditional knowledge systems for environmental sustainability in the Himalayas* (pp. 3–22). Springer Nature Switzerland.

Trivedi, A. (2025). Implementing REDD+ safeguards and protecting the rights of tribal and forest communities in India: With special emphasis on the principle of free, prior, and informed consent. *Jindal Global Law Review*, 1–38.

UKFD. (2024). *Uttrakhand Forest Department, Uttarakhand Approved Forest Panchayat Amendment Rules.*

Upadhyaya, P., & Pant, H. J. (2025). Role of NIHE in pursuit of innovative technologies for monitoring, conservation, and sustainable resource management in challenging Himalayan landscapes. *ENVIS Bulletin Himalayan Ecology, 32*, 102–118. ISSN 2455-6815

Xu, J., Badola, R., Chettri, N., Chaudhary, R. P., Zomer, R., Pokhrel, B., Hussain, S. A., Pradhan, S., & Pradhan, R. (2019). Sustaining biodiversity and ecosystem services in the Hindu Kush Himalaya. In *The Hindu Kush Himalaya assessment: Mountains, climate change, sustainability and people* (p. 127-1).

Chapter 22
Mitigation Co-benefit of Ecosystem-Based Adaptation Measures

Deependra Joshi, Rajeev Joshi, Keshav Prasad Khanal, Top Bahadur Khatri, Santosh Mani Nepal, Buddi Sagar Poudel, Raju Sapkota, Binod Thapa, Digambar Singh Dahal, and Bhola Dhakal

Abstract Ecosystem-based Adaptation (EbA) is a nature-based solution to climate change adaptation. As it focuses on the conservation, restoration, and sustainable management of natural resources, the EbA project could contribute significantly to climate change mitigation as one of its co-benefits. Thus, quantification of CO_2 emission reduction and sequestration is essential. The EbA II project, funded by the Global Environmental Facility (GEF) and implemented by the Ministry of Forests and Environment, Nepal, and the United Nations Environment Programme, has been implementing EbA measures in the three hill and mountain districts of Nepal: Dolakha, Salyan, and Achham. This chapter aims to quantify the CO_2 sequestration and emission reduction resulting from the implementation of EbA measures in these districts. This findings demonstrates that the EbA II project contributes to reducing approximately 194,053 tons of CO_2 annually, and within the project's 5-year period, it contributes to lowering nearly one million tons of CO_2. This paper provides evidence to position Ecosystem-Based Adaptation as a crucial approach to addressing climate change adaptation and mitigation. Therefore, the findings are vital for informing the government's future climate change-related project design and implementation.

D. Joshi · K. P. Khanal · T. B. Khatri · S. M. Nepal · R. Sapkota · B. Thapa · D. S. Dahal · B. Dhakal
Ecosystem Based Adaptation-EbA II Project, Ministry of Forests and Environment, Climate Change Management Division, Govt. of Nepal, Kathmandu, Nepal

R. Joshi (✉)
College of Natural Resource Management, Faculty of Forestry, Agriculture and Forestry University, Katari, Udayapur, Nepal
e-mail: joshi.rajeev20@gmail.com; rajeevjoshi@afu.edu.np

College of Economics and Management, Northwest A&F University, Yangling District, Xianyang, Shaanxi, China

B. S. Poudel
Department of National Parks and Wildlife Conservation, Kathmandu, Nepal

H. Singh (ed.), *Forestry for a Carbon-Neutral and Sustainable Future*, Sustainability Solutions, https://doi.org/10.1007/978-3-032-06427-1_22

Keywords Climate change · Ecosystem-based adaptation · Carbon emissions · EbA co-benefits

22.1 Introduction

The project "Catalyzing Ecosystem Restoration for Climate Resilient Natural Capital and Rural Livelihoods in Degraded Forests and Rangelands of Nepal" (EbA II) is funded by the Global Environment Facility (GEF)/Least Developed Country Fund (LDCF). This project was launched in May 2018 and is expected to close by December 2024. The project focuses on decreasing the vulnerability of communities to the impacts of climate change. It aims to increase the capacity of local communities to adapt to climate change by implementing Ecosystem-based Adaptation measures in degraded forests and rangelands in the mid-hills (Salyan and Achham Districts) and high mountain areas in Dolakha District (UNEP, 2021).

Ecosystem-based adaptation (EbA) measures refer to the use of ecosystem services and biodiversity to help people adapt to the adverse effects of climate change (CBD, 2009). These measures have the potential to provide several mitigation benefits in addition to their primary function of enhancing resilience to climate change impacts. The EbA measures can contribute to climate change mitigation by enhancing carbon sequestration and reducing carbon emissions from deforestation and forest degradation. Many ecosystems, such as forests and wetlands, store significant amounts of carbon (Adhikari et al., 2009). Conserving and restoring these ecosystems with implementing various EbA measures may help sequester carbon dioxide from the atmosphere, reducing greenhouse gas concentrations and mitigating climate change (Singh, 2024). Similarly, minimizing and reducing deforestation by protecting and managing forests can help mitigate deforestation and forest degradation, which are significant sources of carbon emissions (Gregersen et al., 2007; Singh et al., 2023; Garima et al., 2024). Conserving these carbon sinks helps mitigate climate change through ecosystem-based adaptation (EbA) (Kumar et al., 2021). Similarly, sustainable land management practices, such as agroforestry and sustainable agricultural practices, can also enhance soil health and reduce the need for chemical fertilizers, leading to lower emissions of nitrous oxide, a potent greenhouse gas (Srivastava & Singh, 2024).

Ecosystem-based Adaptation (EbA) measures have been widely recognized as cost-effective, nature-based solutions that deliver multiple co-benefits, including climate change mitigation (IPCC, 2021; Seddon et al., 2020). While global and regional studies have highlighted the theoretical potential of EbA in reducing greenhouse gas emissions and enhancing carbon sinks (Griscom et al., 2017; UNEP, 2023), a critical gap remains in empirical evidence from the Himalayas, one of the world's most climate-vulnerable regions. The novelty of this study lies in quantifying the mitigation co-benefits of EbA interventions implemented under the EbA II project in Nepal, which integrates forest restoration, afforestation, and sustainable forest management in hill and mountain landscapes. Specifically, this study aims

Table 22.1 Forest area of EbA II project intervention sites

Project sites	Achham	Salyan	Dolakha	Total
Total forest restoration area (ha)	348	873	172	1393
Total sites	35	18	21	74
Sustainable forest management (ha)	10,000			

Data source EbA II project report

to (i) estimate the annual and cumulative carbon sequestration and emission reduction resulting from EbA interventions, and (ii) demonstrate how such interventions contribute to national climate goals and global carbon neutrality pathways. This study advances the global discourse on scaling nature-based solutions for simultaneous adaptation and mitigation, thereby informing climate policy and investment priorities at regional and international levels by providing robust estimates of carbon benefits.

22.2 Materials and Methods

22.2.1 Study Area

The study sites of this research were located in two hill districts (Salyan and Achham) and one high mountain district (Dolakha) of Nepal. The project covers 18 wards of 10 municipalities. The total forest area coverage under this study is presented in Table 22.1. The total forest area reforested due to the EbA II project intervention is 1,393 ha, and the total forest area under sustainable management is approximately 1,000 ha, comprising 132 Community forests.

22.2.2 Baseline Data

Calculating the carbon benefits of forest conservation involves estimating the amount of carbon stored in the forest and the emissions that would occur if the forests were to be deforested or degraded. We used a secondary source of information for the carbon emission calculation, and no direct measurement of carbon stock change was conducted. We reviewed the relevant literature available on carbon stock changes and carbon sequestration rates of similar forest types at the EbA II project sites.

There is no specific baseline data available on forest carbon stock from BbA II project sites. We used the data provided by DFRS (DFRS, 2015) for the baseline carbon stock of the project sites. We used the total annual carbon stock of plantation forests and the annual increment of carbon stocks in sustainably managed forests as the annual baseline carbon increment due to the project intervention. Here's a general outline of the method we used to calculate the carbon impact of forest conservation.

22.2.3 Calculate Carbon Stock

Carbon stock calculations for plantation forests and rehabilitated degraded forests were made based on the assumption of the percentage change in carbon stock due to improved forest management. For the new plantation, the year-wise carbon stock was calculated up to the rotation age (20 years). We reviewed studies from similar forest types and ecological zones to estimate the forest carbon stock and annual carbon sequestration of plantation and sustainably managed forests. A survey conducted in the Synagja district of the middle hill of western Nepal at an elevation from 979 to 1320 m above sea level (masl) on Carbon Sequestration of Community Forest (Anup, 2012) found that the total carbon stock annual carbon sequestration rate and total CO_2 mitigation potential as 122.29, 0.45 and 1.64 ton ha^{-1}, respectively. Similarly, a study conducted on the Hill Sal Forests of Nepal found that the carbon sequestration rate of the Hill Sal forest was 2.6 Mg Ton ha^{-1} yr^{-1} (Thapa-Magar & Shrestha, 2015). Another study conducted in a community forest in Nepal found that the average annual carbon increment of the community forests was 2.19 tons ha^{-1} (Shrestha et al., 2013). Similarly, an assessment conducted by ICIMOD on forest carbon stock and carbon sequestration rates at the ICIMOD Knowledge Park at Godavari, Nepal, found that the carbon sequestration rate of 2.65 Ton C ha^{-1} yr^{-1} (Karki et al., 2016). Different studies have shown similar data on baseline carbon stock and annual carbon sequestration rates in the middle hills of Nepal, and these data do not significantly deviate from the DFRS data. We used the data provided by DFRS (2015) for the baseline carbon stock of the project sites.

22.2.4 Estimate the Emissions Reduction and Carbon Sequestration

The potential emissions reduction resulting from deforestation or forest degradation was determined by calculating the difference between the current carbon stock (baseline) and the potential carbon emissions that would occur if the forest were to be converted to another use (e.g., agriculture or urban development). The carbon stock report of Forest Resource Assessment (DFRS, 2015) indicated a Carbon stock of plantation forest as 79.43 Ton Carbon ha^{-1}: We assume 79.43 Ton C ha^{-1} of plantation forest after year 10 of plantation. We used factors of 0.2, 0.3, 0.4, and up to 0.9, with a value of 79.43 t C for the year from the second to the ninth year of plantation.

22.2.5 *Time Horizon*

Carbon sequestration can be a long-term process, so understanding the timeframe is essential for accurate calculations. We assume that full forest growth can be achieved in 20 years of plantation or conservation.

22.2.6 *Calculation of the Carbon Storage Potential of Plantation Forests and Natural Forests*

Carbon Calculation of Plantation Forests

The carbon storage potential of plantation forests was estimated, predicting how much carbon these forests would store over time. The calculation can be complex, and the value may vary based on factors such as tree species, age, site conditions, and forest management practices. However, we assume that the forest will reach its full growth in 20 years and use the same carbon stock quantity for planted forests as for the natural forests at the study sites. For this, we utilize the Forest Resource Assessment (DFRS, 2015) data on carbon stock in middle mountain forests.

Enhance Carbon Sequestration Due to SFM

We used the present-day tree carbon sequestration rate of 0.5 ha^1 yr^1 (IPCC default value, Tier 1) for the Tropical moist forest. The IPCC value is 3 tons ha^{-1} above-ground biomass and 25% of above-ground biomass as below-ground biomass. Therefore, the total carbon stock of a 1 ha forest would be 3 tons of above-ground biomass and 25% of 3 tons of above-ground carbon stock as below-ground carbon stock. Therefore, the total carbon stock of one hectare of forests would be 3.75 tons (3 + 3 * 25% (AGB + BGB) = 3.75 ton ha^{-1} yr^{-1}). We assume a 5% increase in carbon biomass after five years of sustainable management. Therefore, the enhanced carbon sequestration after 5 years would be 3.9375 ($3.75 + 0.05 \times 3.75 = 3.9375$). We used factors of 0.2. 0.4, 0.6, and 0.8 of this value (3.9375) as annual carbon increment for the first, second, third, and fourth years of sustainable forest management. After year 5, we assume an annual carbon increment of 10% until the rotation age of 20 years. Details of Carbon Emission reduction and carbon sequestration are presented in Table 22.2.

22.3 Results and Discussion

Annual carbon sequestration resulting from the establishment of new plantations and the application of sustainable forest management practices is presented in Table 22.2. The yearly carbon sequestration resulting from the plantation of 1,393 ha of forest was found to be 116,178 tons. Similarly, annual carbon sequestration from SFM of

Table 22.2 Calculation of Annual CO_2 emission reduction/sequestration

Item/description	Year 1	Year 2	Year 3	Year 4	Year 5	Year 6	Year 7	Year 8	Year 9	Year 10	Year 11
Carbon sequestration in plantation forest[a] (Ton)		22,129	33,194	44,258	55,323	66,388	77,452	88,517	99,581	110,646	116,178
Enhance carbon sequestration due to sustainable management[b] (Ton/ha)	7,875	15,750	23,625	31,500	39,375	39,769	40,166	40,568	40,974	41,384	41,797
Total carbon sequestration (Ton)	7,875	37,879	56,819	75,758	94,698	106,156	117,619	129,085	140,555	152,030	157,976

Item/description	Year 12	Year 13	Year 14	Year 15	Year 16	Year 17	Year 18	Year 19	Year 20	Annual	Total
Carbon sequestration in plantation forest[a] (Ton)	121,987	128,087	134,491	141,215	148,276	155,690	163,475	171,648	180,231	116,178	2,058,766
Enhance carbon sequestration due to sustainable management[b] (Ton/ha)	42,215	42,637	43,064	43,494	43,929	44,369	44,812	45,261	45,713	44,364	758,278
Total carbon sequestration (Ton)	164,203	170,724	177,555	184,710	192,206	200,059	208,287	216,909	225,944	194,053	2,817,045

Unit: Ton Carbon

[a] Plantation area 1393 ha; Calculation/*Source* Assume 79.43 Ton Carbon per ha of plantation forest after year 5 of plantation

[b] Total area of SFM: 132 CFs with approximate area of 10,000 ha; Calculation/*Source* Present tree carbon sequestration rate/ha/yr (IPCC) for Tropical moist forest $= 3 + 3 * 25\%$ (AGB + BGB) $= 3.75$ ton/ha/yr; assume to increase by 5% in 5 years. So, enhance carbon sequestration after 5 years $= 3.75 + 0.05 * 3.75 = 3.9375$

10,000 ha was found to be 44,364 tons, and the total carbon sequestration of both plantation and SFM practices was 194,053 tons.

The results of this study highlight the substantial potential of both new forest plantations and sustainable forest management (SFM) practices in enhancing carbon sequestration, a critical step in mitigating climate change. The establishment of 1,393 ha of new plantations resulted in an annual carbon sequestration of 116,178 tons, demonstrating the effectiveness of afforestation and reforestation in capturing atmospheric carbon. This finding aligns with previous studies that emphasize the role of new plantations in increasing forest carbon stocks, especially in regions where deforestation and forest degradation have previously occurred (Joshi et al., 2020, 2021, 2023).

Furthermore, the implementation of SFM practices across 10,000 ha contributed an additional 44,364 tons of carbon sequestration annually (Table 22.2). This reinforces the importance of maintaining and enhancing the productivity and health of existing forests through sustainable practices (Marchi et al., 2018). SFM not only enhances carbon uptake but also ensures long-term forest resilience, biodiversity conservation, and the delivery of ecosystem services, making it a crucial component of climate action strategies (Kumar et al., 2022).

Combined, these two approaches yielded a total annual carbon sequestration of 194,053 tons. This significant figure highlights the synergistic impact of integrating afforestation and reforestation efforts with the sustainable management of existing forests. The results also emphasize the potential of forest-based carbon sequestration strategies in contributing to national and global carbon neutrality goals.

However, several factors could influence the outcomes, including the species composition of the plantations, the age and growth rate of the trees, local climatic conditions, and the specific management practices applied (Pawson et al., 2013). Future research could explore these factors in more depth to optimize carbon sequestration outcomes (Onwuka & Adu, 2024). Additionally, expanding the geographic scope of such initiatives and incorporating community engagement and indigenous knowledge could further enhance the effectiveness and sustainability of these practices (Thompson et al., 2020).

Tables 22.3 present the annual and total CO_2 emission reductions and sequestrations resulting from project implementation. It shows that 711,527 tons of CO_2 emissions can be reduced and sequestered through the project interventions of the EbA II project. Within the project's 5-year period, a total of 1,001,108 tons of CO_2 can be sequestered or emissions reduced due to the project intervention. Table 22.4 and Fig. 22.1 depict the annual CO_2 sequestration/remission reduction from the EbA II project. The EbA II project sequestered approximately 1 million tons of CO_2 over 5 years.

This study provides compelling evidence of the significant role that ecosystem-based adaptation (EbA) interventions play in mitigating climate change while enhancing local resilience. The EbA II project, through afforestation of 1,393 ha and sustainable management of 10,000 hectares of forests, achieved an average annual sequestration of 194,053 tons of CO_2. Over five years, these interventions removed approximately 1 million tons of CO_2, with projections indicating more than 10 million

Table 22.3 CO_2 emission reduction/carbon sequestration (tCO_2 e) from year 1 to year 20

CO_2 emission reduction/ sequestration (tCO_2 e)	Year 1	Year 2	Year 3	Year 4	Year 5	Year 6	Year 7	Year 8	Year 9	Year 10	Year 11
Per year	28,875	138,890	208,336	277,781	347,226	389,240	431,268	473,311	515,369	557,442	579,244
Cumulative	28,875	167,765	376,101	653,882	1,001,108	1,390,348	1,821,616	2,294,927	2,810,296	3,367,738	3,946,982

CO_2 emission reduction/sequestration (tCO_2 e)	Year 12	Year 13	Year 14	Year 15	Year 16	Year 17	Year 18	Year 19	Year 20	Annual	Total of one rotation
Per year	602,076	625,988	651,034	677,270	704,754	733,549	763,719	795,332	828,461	711,527	10,329,164
Cumulative	4,549,058	5,175,046	5,826,080	6,503,350	7,208,104	7,941,653	8,705,371	9,500,704	10,329,164	**tCO_2e**	**tCO_2e**

Table 22.4 Annual and total CO_2 sequestration due to project interventions of the EbA II project

Annual carbon sequestration from EbA project	711,527	$MTCO_2e$
Carbon sequestration within the project period (5 years)	**1,001,108**	**$MTCO_2e$**
Total carbon sequestration within 20 years from EbA Project Implementation (One full rotation period of forest crop)	10,329,164	CO_2e
	10.33	$MTCO_2e$

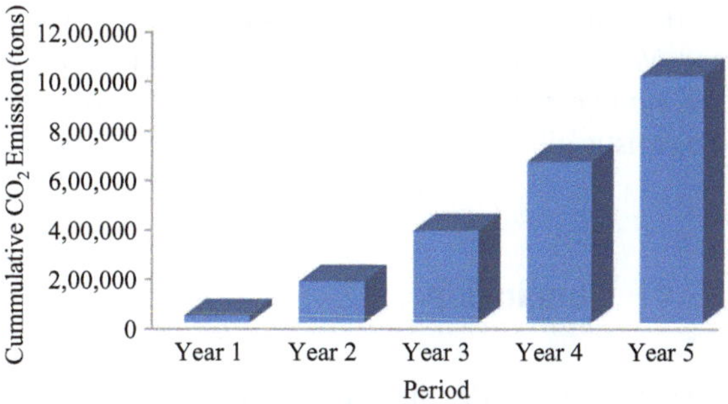

Fig. 22.1 Cumulative CO_2 emission reduction of EbA II project (Tons)

tons over a 20-year forest rotation cycle. These results demonstrate that combining afforestation and Sustainable Forest Management (SFM) can deliver measurable and substantial mitigation benefits within relatively short timeframes.

At the global scale, these findings align with recent assessments that nature-based solutions (NbS) could provide up to 30% of the cost-effective mitigation required to meet the Paris Agreement target of limiting warming to 1.5 °C (Griscom et al., 2017; IPCC, 2021). The sequestration rates recorded here are comparable to those reported from large-scale reforestation and community forestry programs in tropical and subtropical regions (Chausson et al., 2023), demonstrating that EbA in mountainous landscapes can significantly contribute to global carbon neutrality pathways. Such interventions are particularly relevant for Least Developed Countries (LDCs), where financial and technical resources for large-scale engineered solutions remain limited.

Regionally, the results reinforce Nepal's Nationally Determined Contributions (NDCs), which prioritize forest restoration and sustainable management as key mitigation strategies (GoN, 2020). Beyond carbon benefits, EbA measures enhance biodiversity conservation, water regulation, and soil stabilization, providing critical ecosystem services for climate-vulnerable mountain communities (UNEP, 2023). These multifunctional benefits highlight the integrated nature of EbA and its alignment with the UN Decade on Ecosystem Restoration (2021–2030) and Sustainable Development Goals (SDGs), particularly SDG 13 (Climate Action) and SDG 15 (Life on Land).

However, scaling up EbA faces challenges related to financing, governance, and long-term monitoring. Recent debates caution that while NbS are essential, they cannot replace rapid decarbonization in energy, transport, and industrial sectors

(Seddon et al., 2020). Thus, EbA should be viewed as a complementary strategy that simultaneously addresses adaptation and mitigation while delivering socio-economic co-benefits. Ensuring the success of such interventions requires robust community engagement, secure tenure, and sustained policy support (Papa, 2016). Incorporating indigenous knowledge and participatory monitoring systems can further enhance long-term effectiveness (Thompson et al., 2020).

Future research should explore additional dimensions, such as the valuation of socio-economic co-benefits, integration of EbA projects into voluntary carbon markets, and development of standardized frameworks for Monitoring, Reporting, and Verification (MRV). By providing empirical evidence from the Himalayas, this study contributes to the global discourse on scaling nature-based solutions for synergistic climate action and biodiversity conservation.

22.4 Conclusion

This study concludes that although the primary focus of ecosystem-based adaptation projects is climate adaptation for local communities, they also provide a significant contribution to climate change mitigation. Climate change mitigation is one of the significant co-benefits of the EbA II project, contributing to an annual reduction in CO_2 emissions of 711,527 tons and a total reduction of almost 1 million tons within the project's 5-year period. This study demonstrates that a combination of new plantations and sustainable forest management (SFM) practices can serve as a robust strategy for carbon sequestration in Nepal. The EbA II project serves as a powerful example of how targeted ecosystem-based interventions can deliver significant climate mitigation outcomes. The project's success highlights the need to prioritize and expand similar initiatives, fostering partnerships and investments to enhance their impact. Further studies could explore the socio-economic co-benefits of such projects, such as improvements in livelihoods and biodiversity conservation, to develop a more comprehensive understanding of their value. These findings offer valuable insights for policymakers, forest managers, and stakeholders seeking to implement effective climate mitigation measures while promoting sustainable development.

Acknowledgements The authors express their sincere gratitude to Mr. Sharad Babu Pangeni and Ms. Shila Gyawali for their valuable support during project implementation. We also acknowledge Mr. Prakash Lamsal for their expert advice and technical inputs. Special thanks are extended to the field-level staff and the entire EbA II project team for facilitating data collection and providing necessary logistical support, which greatly contributed to the successful completion of this research.

Conflict of Interest
The authors declare that they have no known competing financial interests or personal relationships that could have appeared to influence the work reported in this chapter.

References

Adhikari, S., Bajracharaya, R. M., & Sitaula, B. K. (2009). A review of carbon dynamics and sequestration in wetlands. *Journal of Wetlands Ecology*, 42–46.

Anup, K. C. (2012). *Feasibility analysis of REDD+: Case study of Ghwangkhola Sapaude Babiyabhir Community Forest of Syangja, Nepal* (MSc thesis). Central Department of Environmental Science, Tribhuvan University.

CBD. (2009). *Convention on biological diversity.* https://www.cbd.int/meetings/COP-09

Chausson, A., Turner, B., Seddon, D., Chabaneix, N., Girardin, C. A. J., Kapos, V., & Seddon, N. (2023). Global evidence on the contribution of nature-based solutions to climate mitigation. *Philosophical Transactions of the Royal Society B: Biological Sciences, 378*(1878), 20220108. https://doi.org/10.1098/rstb.2022.0108

DFRS. (2015). State of Nepal's Forests. Forest Resource Assessment (FRA) Nepal, Department of Forest Research and Survey (DFRS). Kathmandu, Nepal

Government of Nepal (GoN). (2020). *Second Nationally Determined Contribution (NDC), 2020.* Kathmandu: Ministry of Forests and Environment, Government of Nepal.

Gregersen, H. M., Ffolliott, P. F., & Brooks, K. N. (2007). *Integrated watershed management: Connecting people to their land and water*. CABI.

Griscom, B. W., Adams, J., Ellis, P. W., Houghton, R. A., Lomax, G., Miteva, D. A., Schlesinger, W. H., Shoch, D., Siikamäki, J. V., Smith, P., Woodbury, P., Zganjar, C., Blackman, A., Campari, J., Conant, R. T., Delgado, C., Elias, P., Gopalakrishna, T., Hamsik, M. R., et al. (2017). Natural climate solutions. *Proceedings of the National Academy of Sciences, 114*(44), 11645–11650. https://doi.org/10.1073/pnas.1710465114

Intergovernmental Panel on Climate Change. (2021). *Climate Change 2021: Mitigation of Climate Change. Contribution of Working Group III to the Sixth Assessment Report of the Intergovernmental Panel on Climate Change.* Cambridge University Press. https://www.ipcc.ch/report/ar6/wg3/

Joshi, R., Singh, H., Chhetri, R., & Yadav, K. (2020). Assessment of carbon sequestration potential in degraded and non-degraded community forests in Terai region of Nepal. *Journal of Forest and Environmental Science, 36*(2), 13–121.

Joshi, R., Singh, H., Chhetri, R., Poudel, S. R., & Rijal, S. (2021). Carbon sequestration potential of community forests: A comparative analysis of soil organic carbon stock in community managed forests of Far-Western Nepal. *Eurasian Journal of Soil Science, 10*(2), 96–104. https://doi.org/10.18393/ejss.825066

Joshi, R., Shrestha, T. K., Mishra, B., Gautam, J., Maharjan, B., Gosai, K. R., Maraseni, T., & Neupane, B. (2023). Assessment of carbon sequestration in private forests across two different physiographic regions of Nepal: Implications for conservation and climate change mitigation. *Scientifica, 2023,* Article ID 6599067, 13 pages. https://doi.org/10.1155/2023/6599067

Karki, S., Joshi, N. R., Udas, E., Adhikari, M. D., Sherpa, S., Kotru, R., Karky, B. S., Chettri, N., & Ning, W. (2016). *Assessment of forest carbon stock and carbon sequestration rates at the ICIMOD Knowledge Park at Godavari.* http://lib.icimod.org/record/32299/files/WP2016_6_Assessment%20of%20forest%20carbon%20stock.pdf

Kumar, N., Singh, H., & Kumar, A. (Eds.). (2021). *Renewable energy and green technology: Principles and practices.* CRC Press.

Kumar, A., Ekka, P., Patra, S., Kumar, G., Kishore, B. S., Kumar, R., & Saikia, P. (2022). Geospatial perspectives of sustainable forest management to enhance ecosystem services and livelihood security. *Advances in Remote Sensing for Forest Monitoring*, 10–42.

Kumari, G., Kotiyal, P. B., Singh, H., Kumar, M., Kumar, N., Malik, A.,… & Singh, S. (2024). Predicting future climate change effects on biotic communities: A species distribution modeling approach. In *Forests and climate change: Biological perspectives on impact, adaptation, and mitigation strategies* (pp. 137–168). Singapore: Springer Nature Singapore.

Marchi, E., Chung, W., Visser, R., Abbas, D., Nordfjell, T., Mederski, P. S., McEwan, A., Brink, M., & Laschi, A. (2018). Sustainable Forest Operations (SFO): A new paradigm in a changing world and climate. *Science of the Total Environment, 634*, 1385–1397.

Onwuka, O. U., & Adu, A. (2024). Subsurface carbon sequestration potential in offshore environments: A geoscientific perspective. *Engineering Science & Technology Journal, 5*(4), 1173–1183.

Papa, R. O. (2016). *Factors influencing public participation in project development in Busia county Kenya* (Doctoral dissertation, University of Nairobi).

Pawson, S. M., Brin, A., Brockerhoff, E. G., Lamb, D., Payn, T. W., Paquette, A., & Parrotta, J. A. (2013). Plantation forests, climate change and biodiversity. *Biodiversity and Conservation, 22*, 1203–1227.

Seddon, N., Chausson, A., Berry, P., Girardin, C. A. J., Smith, A., & Turner, B. (2020). Nature-based solutions in the climate and biodiversity crises. *Nature Climate Change, 10*(8), 659–667. https://doi.org/10.1038/s41558-020-0791-5

Shrestha, S., Karky, B. S., Gurung, A., Bista, R., & Vetaas, O. R. (2013). Assessment of carbon balance in community forests in Dolakha, Nepal. *Small-Scale Forestry, 12*, 507–517. https://doi.org/10.1007/s11842-012-9226-y

Singh, H. (2024). Climate change and forests: Biological perspectives. In *Forests and climate change: Biological perspectives on impact, adaptation, and mitigation strategies* (pp. 1–17). Singapore: Springer Nature Singapore.

Singh, M., Kumar, P., Singh, H., Kumar, A. K. A., Kumar, R., & Kumar, R. (2023). Neolamarckia cadamba: A comprehensive review on its physiological, ecological, phytochemical and pharmacological perspectives. *Ecology Environment and Conservation, 29*, 241–250.

Srivastav, A., & Singh, H. (2024). Agroforestry adoption in the eastern region of Uttar Pradesh, India: A way to mitigate climate change. In *Forests and climate change: Biological perspectives on impact, adaptation, and mitigation strategies* (pp. 445–467). Singapore: Springer Nature Singapore.

Thapa-Magar, K. B., & Shrestha, B. B. (2015). Carbon stock in community managed hill Sal (*Shorea robusta*) forests of central Nepal. *Journal of Sustainable Forestry, 34*(5), 483–501. https://doi.org/10.1080/10549811.2015.1031251

Thompson, K. L., Lantz, T. C., & Ban, N. C. (2020). A review of Indigenous knowledge and participation in environmental monitoring. *Ecology & Society, 25*(2).

United Nations Environment Programme. (2021). *Nepal: Ecosystem-based Adaptation 2019-2022 [Factsheet]*. Nairobi: UNEP. Retrieved from https://wedocs.unep.org/20.500.11822/34879

United Nations Environment Programme. (2023). *State of finance for nature: Tripling investment by 2030*. UNEP. https://www.unep.org/resources/finance-nature-2023

Chapter 23
Social and Economic Impacts of Carbon Neutrality: Impacts and Implications

Darshita Rawat, Ashish Kumar Yadav, and Kumud Dubey

Abstract Carbon neutrality, the equilibrium between anthropogenic greenhouse gas (GHG) emissions and their removal, has become a cornerstone of global efforts to combat climate change and achieve a sustainable future. This chapter examines the social and economic dimensions of carbon neutrality, with a particular emphasis on the critical role of forestry. Integrating sustainable practices and carbon offset strategies serves as a vital tool in addressing both environmental challenges and socio-economic transformations. The chapter begins by examining the social implications of carbon neutrality, focusing on the shift toward sustainable consumption patterns and the adoption of low-carbon technologies in everyday life. It discusses community-level transformations through green job creation, skill development, and active participation in afforestation and renewable energy initiatives. These changes emphasize the potential for fostering equitable and inclusive societal progress. On the economic front, the chapter delves into the emergence of a green economy driven by renewable energy and forestry. It also evaluates the green jobs. The analysis also addresses sectoral shifts from traditional energy sources to renewable energy, the role of forestry in carbon offset mechanisms, and the disparities in advocating for economic policies that support low-income economies. The chapter underscores forestry's multifaceted role in achieving carbon neutrality through carbon sequestration. It is inferred that aligning global and national policies with innovative governance strategies identifies pathways to achieve a sustainable and equitable carbon–neutral future.

Keywords Carbon sinks · Carbon neutrality · Carbon sequestration · Climate change

D. Rawat (✉) · A. K. Yadav
Dev Bhoomi Uttarakhand University, Dehradun, India
e-mail: darshitarawat226@gmail.com

K. Dubey
ICFRE-Eco Rehabilitation Centre, Prayagraj, India

A. K. Yadav
ICFRE-Forest Research Institute, Dehradun, India

H. Singh (ed.), *Forestry for a Carbon-Neutral and Sustainable Future*, Sustainability Solutions, https://doi.org/10.1007/978-3-032-06427-1_23

23.1 Introduction

Over the past century, global warming has intensified significantly, driven by substantial greenhouse gas (GHG) emissions resulting from human activities, both directly and indirectly (Singh & Kumar, 2022). Carbon emissions play a critical role in influencing climate change (Hansen et al., 2013). Global warming is approximately proportional to the cumulative carbon dioxide (CO_2) emissions released into the atmosphere (Rosa & Ribeiro, 2001; Fang et al., 2011), implying that achieving specific temperature targets necessitates corresponding reductions in emissions. Since the 1990s, nations worldwide have intensified efforts to curb the escalating rise in global greenhouse gas (GHG) emissions, recognizing their pivotal role in driving climate change (Le Quéré et al., 2019). A landmark initiative in this regard is the Paris Agreement of 2015, which set ambitious targets to limit the global average temperature increase to well below 2 °C above pre-industrial levels by the end of the twenty-first century, with an aspirational goal of restricting the rise to 1.5 °C (Kuramochi et al., 2018; Kumari et al., 2024). These targets underscore the urgent need for coordinated global action to mitigate the adverse effects of climate change. Global development, heavily reliant on the overexploitation of natural resources since the Industrial Revolution, has driven economic progress but at a substantial environmental cost (Pearson & Foxon, 2012).

The extensive use of fossil fuels, deforestation, and unsustainable land-use changes have significantly contributed to rising greenhouse gas (GHG) concentrations in the atmosphere, leading to unprecedented climate disruptions (Singh et al., 2023). Anthropogenic activities have disrupted natural carbon cycles, exacerbating global warming and threatening ecological balance (Lal, 2004; Prakash & Verma, 2022). Greenhouse gas (GHG) emissions are projected to rise by 50% by 2050, primarily driven by an anticipated 70% increase in energy-related carbon dioxide (CO_2) emissions (Rabaey & Ragauskas, 2014; Tilman et al., 2011). In response to these challenges, achieving carbon neutrality by 2050 has emerged as one of the most pressing global priorities (Chen et al., 2022). This goal necessitates comprehensive reforms in production systems and resource utilization, focusing on minimizing greenhouse gas (GHG) emissions and enhancing carbon dioxide (CO_2) sequestration (Kumar & Singh, 2020).

Transitioning to renewable energy sources, promoting sustainable land-use practices, and adopting innovative carbon capture technologies are critical strategies for mitigating emissions (Fang et al., 2011; Kumar et al., 2021). Zhang et al. (2024a) emphasized the importance of forests and grasslands as crucial ecosystems that play a pivotal role in global carbon cycling and achieving carbon neutrality. These natural landscapes serve as significant carbon sinks, absorbing substantial amounts of atmospheric carbon dioxide (CO_2) through the process of photosynthesis (Hu et al., 2015). Forests, in particular, store carbon in their biomass (including trunks, branches, leaves, and roots) and in the soil, making them vital in mitigating greenhouse gas emissions. Globally, forests are estimated to store approximately 861 gigatons of carbon, accounting for a substantial portion of the Earth's terrestrial carbon stock

(Tadese et al., 2023). Grasslands, though often overlooked, also contribute significantly to carbon sequestration. Their deep-rooted plants store carbon in the soil, often at depths less accessible to disturbance, ensuring long-term carbon storage. This process is further enhanced by the biodiversity of grassland ecosystems, which promotes resilience and stability in carbon capture.

The insights of this chapter revolves around a novel approach that links carbon neutrality with socio-economic evolutions rather than treating it as a purely environmental or technological issue through incorporating concepts such as lifestyle changes, circular economy, green job creation, and forestry-based mitigation strategies, chiefly in the Indian context, the chapter provides a distinctive lens through which localized actions can align with global sustainability goals. This chapter primarily focuses on analyzing the social and economic dimensions of achieving carbon neutrality, by focusing on sustainable development, lifestyle transitions, along with the green employment opportunities, secondly, it also aims to link the global and regional perspectives by evaluating international carbon neutrality commitments alongside India's strategies, emphasizing forestry and green policy initiatives.

23.2 Global Commitments Towards Carbon Neutrality

The concept of carbon neutrality originates from the goal of achieving net-zero greenhouse gas (GHG) emissions, with carbon dioxide (CO_2) being the primary contributor due to human activities (Jia & Lin, 2021; Qin et al., 2021). In this context, the term "carbon" is often used as a proxy for all greenhouse gases (GHGs) when formulating neutrality or net-zero emission targets. As of now, 126 countries have pledged to achieve carbon neutrality by the mid-twenty-first century (Huang & Zhai, 2021). Among these, the United Kingdom became a frontrunner by enacting carbon neutrality into national legislation in 2019. The transition from a carbon neutrality commitment as a vision to tangible actions signals a broader global movement toward sustainable development (Dong et al., 2022). This commitment is particularly significant in the current era, marked by the ongoing impacts of the COVID-19 pandemic and the onset of a new phase of global economic recovery. These circumstances underscore the need to critically evaluate and understand the motivations driving nations to adopt carbon neutrality targets. This analysis is vital for aligning environmental sustainability with economic resilience and recovery strategies in the post-pandemic world.

Governments, organizations, and stakeholders worldwide have made significant commitments to achieving carbon neutrality as a critical step toward addressing the escalating climate crisis. These efforts are guided by international agreements and campaigns aimed at reducing greenhouse gas (GHG) emissions and promoting resilience against the impacts of climate change. The Paris Agreement, adopted in 2015, represents a cornerstone of global climate action. It calls upon all nations to undertake ambitious emission reduction measures and collaboratively adapt to the adverse effects of climate change. A key objective of the agreement is to peak global greenhouse gas (GHG) emissions as soon as possible and subsequently achieve a balance between anthropogenic emissions and carbon sink removals during the latter half of the twenty-first century. This target underscores the urgency of transitioning to a carbon–neutral future (Kantemirova et al., 2023). In 2019, the Climate Ambition Alliance: Net Zero 2050 was launched, uniting countries, cities, corporations, and investors committed to achieving carbon neutrality by mid-century (Daukaev et al., 2020). The conclusion of the 25th Conference of the Parties (COP25) to the United Nations Framework Convention on Climate Change (UNFCCC) in Madrid saw 120 countries join the alliance. Despite this promising figure, these nations accounted for less than 25% of global greenhouse gas (GHG) emissions, highlighting the need for broader participation.

Additionally, only a limited number of countries had legislated binding commitments to carbon neutrality. To further enhance global efforts, the Race to Zero campaign was introduced in 2020 as a complement to the Climate Ambition Alliance. This initiative expanded the scope of commitments, incorporating 23 regions, 454 cities, 1,397 companies, 569 universities, and 74 investors pledging to achieve net-zero emissions by 2050. In November 2020, United Nations Secretary-General António Guterres urged the global community to make 2021 the year of a "big leap" towards carbon neutrality, underscoring the urgency of decisive action (Gakaev, 2022). However, the UNEP GHG Emissions Gap Report 2020 revealed a growing disparity between observed greenhouse gas (GHG) emissions and the reductions required to align with the Paris Agreement's targets. Despite an anticipated decline in emissions during the pandemic, the gap widened, underscoring the inadequacy of current measures. Projections based on the Climate Action Tracker as of November 2020 suggested that implementing all announced national measures would limit global warming to 2.9 °C by the end of the century, an improvement over the 3.6 °C projection from 2015 (Gunya & Gakaev, 2022). However, this falls significantly short of the Paris Agreement's goal of limiting temperature rise to well below 2 °C, ideally 1.5 °C, above pre-industrial levels.

Furthermore, existing 2030 targets under the Paris Agreement are on track to result in a 3.2 °C temperature increase by 2100, emphasizing the need for enhanced ambition and concrete action. The pathway to carbon neutrality demands not only bold commitments but also rigorous implementation and global cooperation. Achieving these targets requires the integration of technological innovation, robust policy frameworks, financial investments, and widespread public engagement. With the urgency of climate change accelerating, these global commitments serve as a foundation for transformative action to secure a sustainable and resilient future.

23.3 Social Impacts of Carbon Neutrality

The role of lifestyle in achieving sustainability was first emphasized in the 1987 Brundtland Report, which highlighted the necessity for affluent populations to adopt lifestyles within the planet's ecological limits. The report explicitly highlighted that sustainable global development necessitates shifts in consumption patterns, particularly in areas such as energy use, to align with ecological constraints (Brundtland, 1987). This recognition marked the beginning of a growing scientific interest in the connection between lifestyles and environmental sustainability. Initial studies on the subject emerged following the oil crises of the 1970s, focusing on the relationship between lifestyles and energy consumption (Dillman et al., 1983; Leonard-Barton, 1981; Maréchal, 1977). Over time, this perspective has evolved, with recent reports by the Intergovernmental Panel on Climate Change (IPCC) placing a stronger emphasis on lifestyle changes as a key component of climate mitigation strategies. Research has since expanded to describe and evaluate "sustainable" lifestyles that align with environmental goals (Druckman & Jackson, 2010; Mont et al., 2014; Neuvonen et al., 2014). These studies aim to identify actionable strategies for achieving such lifestyles (Tukker et al., 2008) and encourage widespread adoption of low-carbon behaviors among individuals and communities (Goodall, 2010). The growing focus on sustainable lifestyles underscores their potential to significantly reduce greenhouse gas emissions, conserve natural resources, and mitigate climate change. However, transitioning to such lifestyles requires coordinated efforts involving policy interventions, education, technological innovation, and cultural shifts, making it a complex yet crucial aspect of achieving global sustainability goals.

23.4 Drivers of Societal Shifts Toward Achieving Carbon Neutrality and Low-Carbon Transformations

Four key drivers of technological innovation—political economy redistribution, new narratives, and transformative learning—act at various leverage points within systems to promote change, drawing on Meadows' framework (2008). These drivers collectively influence and facilitate transformative societal shifts toward achieving carbon neutrality. A detailed overview of these drivers is presented in Table 23.1.

23.5 Transition to Sustainable Consumption Patterns

Transitioning to a pattern that aligns well with a sustainable lifestyle is essential for reducing the environmental impact of modern lifestyles. Implementing a zero-waste philosophy by adhering to the principles of the five R's—Refuse, Reduce, Reuse, Recycle, and Rot—offers a systematic approach to minimizing carbon emissions. Refusing single-use plastics, reducing unnecessary purchases, reusing and maintaining items for prolonged utility, composting organic waste, and recycling in alignment with local regulations collectively contribute to a lower carbon footprint (Blanke & Golombek, 2021). Such practices not only mitigate the waste burden on landfills but also conserve resources and energy, promoting a circular economy.

Table 23.1 Several factors drive the shift toward greener and carbon neutrality

Technological revolutions	Technological innovations have a significant impact on the magnitude of barriers to societal transformations and serve as powerful drivers for reshaping the structure of complex systems. Rapidly altering the dynamics and flows within societal frameworks guide technological advancements, which play a pivotal role in initiating change, determining the trajectory, and influencing the pace of transitions within societal systems (Chan et al., 2020)
Political economy reorganization	The distribution of wealth and financial power serves as a fundamental driver of transformative processes. The political economy governs the utilization and allocation of both material and intangible resources within and across societal groups. It elucidates the interconnections and interactions between the state and the institutions responsible for orchestrating economic transformations, shaping the trajectory of societal and systemic change (Fazey et al., 2018)
New and modern narratives	The development of new narratives is a pivotal driver of transformative societal change, as it fosters the cognitive shifts required to reshape collective mindsets. These narratives have the potential to alter perceptions, mobilize individual actions within the context of a collective purpose, and redefine systems of meaning. Influencing societal values and motivations catalyzes behavioral change and the adoption of innovative approaches that underpin systemic transformation (Linnér & Wibeck, 2020, 2021)
Transformative learning	Transformative learning plays a crucial role in driving change within complex and dynamic societal structures. It facilitates long-term shifts in perspectives, enhances awareness of paradigm transitions, and fosters the evolution of mindsets. Empowering individuals to transcend existing paradigms, transformative learning inspires profound systemic changes and supports the reconfiguration of societal frameworks (Winkler et al., 2022)

Sustainable food consumption is another critical aspect of sustainable lifestyles (Lubowiecki-Vikuk et al., 2021a, 2021b). Mobility, housing, and food are crucial aspects of lifestyle changes that significantly influence carbon neutrality in a growing economy. Shifting towards sustainable mobility solutions, such as adopting electric vehicles, utilizing public transportation, and promoting non-motorized travel, reduces transportation-related emissions (Tasneem Oyshi, 2024).

Energy-efficient housing, which incorporates renewable energy sources and sustainable building materials, reduces the carbon footprint associated with residential energy consumption. Similarly, promoting plant-based diets, minimizing food waste, and supporting local food systems can substantially decrease emissions from agricultural production and transportation (Boukid, 2024). Together, these interconnected lifestyle changes contribute to a holistic approach to achieving carbon neutrality while fostering sustainable economic growth (Table 23.2). Prioritizing seasonal, locally sourced produce reduces the environmental costs associated with transportation and storage. The high resource demand for animal husbandry can be significantly reduced by adopting a predominantly plant-based diet, which in turn reduces water usage, greenhouse gas emissions, and land degradation. Supporting local farming systems further strengthens regional economies while reducing emissions from long-distance food transportation. Water conservation is integral to sustainable consumption. With only a fraction of Earth's freshwater available for human use, efficient usage patterns, such as turning off taps during non-essential activities and adopting shorter shower durations, help preserve this finite resource. Additionally, ensuring proper waste disposal, avoiding the introduction of non-biodegradable items into waterways, and participating in waterway cleanups all bolster efforts to protect freshwater ecosystems and mitigate associated carbon emissions.

Table 23.2 Case study for lifestyle changes to implement carbon neutrality (Koide et al., 2021)

Case study	Category	Changes to lifestyle changes
Mobility	Vehicle sharing	Vehicle pooling/ride sharing for similar routes
	Vehicle efficiency	Use of vehicles that use conventional/ renewable electricity, hybrid vehicles, and Non-plug hybrid vehicles
	Modal shift	The modal shift involves combining transport modes to enhance efficiency and sustainability while minimizing environmental impact
	Demand reduction	Teleworking, online homecoming visits, and less frequent shopping
	Compact city	Buying homes close to the workplace
	Micro leisure	Domestic/local vacation
Housing	Eco house	Life cycle carbon minus house, promotion of zero energy buildings ·
	Renewable energy and appliances	Solar power system with photovoltaic (PV) panels above the house, IH cooking heater, solar water heater, heat pumps, temperature control on AC's, LED lights
	Housing sufficiency	Small and compact houses, nudging clothes, and reusing clothes
Food	Protein/dietary shift	Vegan diet, self-catering, Pescatarian diet, alternative meat, poultry diet
	Food loss	Reducing and minimizing food loss by utilizing the remaining food to feed stray animals
	Seasonal/local food	Emphasizing the use of fresh, local, and seasonal food

23.6 Adoption of Low-Carbon Technologies in Daily Life

Adopting low-carbon technologies is crucial in reducing greenhouse gas emissions and promoting sustainable development. Active transportation options, such as cycling and walking, offer effective low-carbon alternatives to traditional vehicles, which significantly contribute to air pollution and climate change (Brand et al., 2021). Cycling not only minimizes vehicular emissions but also requires less energy-intensive manufacturing compared to cars, while promoting health benefits for users. Transitioning to renewable energy sources, such as solar, wind, and geothermal power, represents a transformative shift away from fossil fuel dependency. Integrating solar panels into residential and commercial settings offers a sustainable alternative for electricity generation, thereby reducing reliance on fossil fuels such as coal and oil. Similarly, utilizing energy-efficient appliances and smart technologies, such as programmable thermostats and LED lighting, helps optimize energy consumption and reduce waste. Water protection measures align with the adoption of low-carbon technology by reducing the carbon intensity of water management processes. Using non-toxic household products and reusable alternatives helps reduce pollution and minimize energy-intensive water treatment needs. Furthermore, promoting public transportation infrastructure and reducing individual car usage exemplifies a scalable strategy for emission reduction in urban contexts. Embedding these technologies into daily routines significantly contributes to achieving a low-carbon lifestyle while mitigating the adverse effects of climate change for individuals and communities.

23.7 Green Jobs and Skill Development

The term "green jobs" has gained increasing prominence in recent years. However, the focus on sectors providing such opportunities and the identification of the specific skills required vary significantly across published studies (Brand et al., 2021) and between different countries (Sulich & Rutkowska, 2020). Green jobs encompass a diverse range of roles spanning agriculture, manufacturing, research and development, administrative organizations, and service sectors, all united by their objective to safeguard and restore environmental quality. Through the adoption of innovative and sustainable strategies, green jobs play a fundamental role in curbing carbon emissions and mitigating the impacts of climate change (Martinez-Fernandez et al., 2010). These roles are intricately linked to economic activities that prioritize the efficient utilization of resources, including energy, materials, and water, thereby minimizing waste and reducing the overall ecological footprint (Hariram et al., 2023). Integrating advanced technologies and practices, such as renewable energy systems, circular economy principles, and eco-friendly infrastructure, green jobs contribute not only to environmental conservation but also to fostering economic growth and creating resilient communities. Green jobs and employment opportunities that contribute to environmental sustainability have emerged as a critical co-benefit of promoting renewable energy in India. According to a study by Zhang et al. (2024b, 2024c), a low-carbon economy can drive significant employment growth while addressing pressing environmental challenges. The renewable energy sector, particularly solar power, offers immense potential for creating labor-intensive jobs compared to wind energy. Solar energy projects, especially in rural and off-grid regions, align well with India's demand for decentralized energy solutions. These projects provide not only employment but also access to clean energy, improving quality of life and reducing dependence on fossil fuels. This sector can engage local communities in sustainable energy production, fostering rural economic growth. Similarly, green transportation initiatives, such as the adoption of electric vehicles and enhancements to public transit, are creating a demand for skilled professionals in manufacturing, maintenance, and infrastructure development (Ahuchogu et al., 2024). India's push for green jobs necessitates robust skill development programs. This includes vocational training in renewable energy technologies, sustainable agriculture practices, and eco-friendly construction methods. Enhancing technical and managerial capabilities will ensure a steady supply of skilled workers to meet the growing demands of a green economy (Table 23.3).

23.8 Economic Impacts of Carbon Neutrality

Moving toward a low-carbon economy offers three major advantages: reducing greenhouse gas emissions, enhancing economic productivity, and fostering inclusive and equitable growth. The "green transition" involves a comprehensive transformation of energy, production, consumption, and resource management systems, with the following key aspects.

23.8.1 Adoption of Renewable Energy Sources

Transitioning to renewable energy entails a significant shift from fossil fuels to sustainable energy sources. These include solar energy, harnessed through photovoltaic systems; wind energy, generated by turbines; biomass, derived from organic materials; geothermal energy, utilizing heat from the Earth's interior; ocean energy, including tidal and wave power; hydropower from dams and rivers; and energy recovery from landfill gas and municipal solid waste. This transition reduces carbon footprints and diversifies energy portfolios, contributing to energy security and environmental sustainability.

Table 23.3 Insight into literature papers related to green jobs

Area of interest	Terms associated with "Green Jobs"	Case study	References
Circular economy Green economy	Eco-initiatives, Green jobs, Profitability advantage	Offering insights into the characteristics of European (SME's) Small and Medium-sized Enterprises. SME's concerning the implementation of processes specific to the circular economy	Darmandieu et al. (2022)
Green energy	Flexible double scheduling Multi-objective optimization Hybrid algorithm Green shop scheduling	Developing an optimized framework for scheduling green work plans to maximize efficiency in working hours and minimize total labor costs while enhancing the overall green index	Liu et al. (2021)
Labor market, Employment policies, Labor market, Online recruitment	Political economy Economic policies European Green Deal Recovery plan New Green Deal Job market Big data analysis	An evaluation of the potential of the European Environment Agreement and the Recovery Plan to create meaningful opportunities for green employment within Spain's tourism sector	Arnedo et al. (2021), Song et al. (2021)
Renewable energy	Renewable energy sources Fair energy transition Discrete choice	Assessment of the proportion of renewable energy sources in total electricity consumption	Martínez-Cruz and Núñez (2021)
Recycling and waste management Environmental protection	Green enterprises (TOPSIS) Technique for order preference by similarity to ideal solution Waste management Waste reduction	Investigating the significance and contributions of green enterprises, green jobs, and green social initiatives in waste management practices	Vesere et al. (2021)

(continued)

Table 23.3 (continued)

Area of interest	Terms associated with "Green Jobs"	Case study	References
Public management Sustainable development	Sustainable and durable development Efficiency Green jobs Green employment	Designing a framework for accurately assessing green job creation	Sulich and Zema (2018)
Green management of human resources	Sustainable human resource management Eco-friendly training programs Management of sustainable workforce resources	Extracting key insights from the literature with a focus on the conceptualization, implementation, and outcomes of sustainable human resource management	Yong et al. (2020)

23.8.2 *Enhancement of Energy Efficiency*

Increasing energy efficiency involves the widespread use of advanced products and services designed to minimize energy consumption. Examples include high-efficiency appliances, lighting systems, and industrial equipment; fuel-efficient and electric vehicles; and building technologies such as advanced insulation, energy-efficient windows, and HVAC (Heating, Ventilation, and Air Conditioning) systems. Smart grid technologies further optimize energy storage, distribution, and consumption, thereby reducing losses and enabling the integration of renewable energy sources.

23.8.3 *Implementation of Pollution and Greenhouse Gas Mitigation Processes*

The transition includes adopting industrial and agricultural processes that reduce or eliminate pollutants and greenhouse gas emissions. This encompasses carbon capture, utilization, and storage (CCUS) technologies, as well as transitioning to clean manufacturing methods and promoting a circular economy focused on recycling and reusing materials to minimize waste and reduce resource extraction.

23.8.4 Conservation of Natural Resources

Sustainable resource management is a cornerstone of the green economy. This includes promoting organic farming practices to reduce chemical use, sustainable forestry for maintaining biodiversity and forest health, effective land and soil management to prevent degradation, and wildlife conservation to protect ecosystems. Advanced water management practices, such as precision irrigation and wastewater recycling, ensure efficient use and preservation of freshwater resources.

23.8.5 Compliance with Environmental Standards and Regulations

Businesses and industries are increasingly aligning their practices with environmental regulations and international standards, such as ISO 14001, as well as emissions reduction targets set under the Paris Agreement. Adopting environmentally compliant products and services reduces legal and reputational risks while fostering sustainable practices.

23.9 Role of Forestry in Achieving Carbon Neutrality

The increasing concentrations of atmospheric carbon dioxide (CO_2) have been identified as a crucial driver of global climate change, intensifying the urgency for mitigation strategies. In alignment with the objectives of the Paris Agreement, numerous signatory nations have committed to achieving "carbon neutrality," a state in which equivalent CO_2 removal mechanisms offset CO_2 emissions. Among the diverse approaches to carbon sequestration, the capture of atmospheric CO_2 by trees stands out as a highly effective and economically viable strategy. Forest ecosystems serve as natural reservoirs for storing terrestrial carbon (Sullivan et al., 2020). Through the natural process of photosynthesis, trees absorb CO_2 and store it in various carbon pools, including live biomass (such as trunks, branches, leaves, and roots), detritus (dead organic matter), wood products (such as timber and paper), and soils (Kothandaraman et al., 2022). These pools serve as long-term carbon reservoirs, mitigating the impacts of anthropogenic emissions (Joshi et al., 2021). Enhancing afforestation, reforestation, and sustainable forest management practices not only supports carbon neutrality but also offers co-benefits, including biodiversity conservation, improved soil health, and increased ecosystem resilience (Joshi et al., 2020). Accordingly, utilizing forest-based carbon sequestration symbolizes a critical component of global climate change mitigation frameworks.

Nature-based solutions (NBS), also referred to as natural climate solutions, are increasingly prioritized in strategies aimed at achieving carbon neutrality through carbon offsetting activities (Raymond et al., 2023). These solutions leverage natural ecosystems to address climate challenges while providing co-benefits for biodiversity, water regulation, and community well-being. A central component of NBS involves enhancing carbon sequestration by creating or augmenting carbon sinks in ecosystems such as forests, grasslands, wetlands, and agricultural landscapes (Roe et al., 2019). Forests, in particular, play a fundamental role in NBS due to their exceptional capacity to absorb and store atmospheric carbon dioxide (CO_2) through photosynthesis (Psistaki et al., 2024).

Sequestering carbon in biomass, soils, and forest detritus not only mitigates greenhouse gas concentrations but also enhances ecosystem resilience and productivity. Similarly, other ecosystems, such as grasslands and wetlands, contribute to carbon storage by accumulating organic matter in soils and peat, underscoring the multifunctionality of NBS. A growing consensus among scientists and policymakers highlights tree growth as one of the most cost-effective and efficient strategies for long-term carbon capture and storage (Bayen et al., 2016). This approach offers scalability and adaptability, making it an attractive option for countries seeking to meet their climate goals and aiming for net zero. Tree planting transcends its traditional role as an ecosystem restoration effort, serving as a critical mechanism for enhancing terrestrial carbon sequestration (Wang et al., 2019). The role of tree planting serves as a vital complement to other carbon neutrality strategies, including the transition to renewable energy sources and the implementation of energy-efficient technologies, by addressing residual emissions that remain unavoidable despite technological advancements. Integrating tree-planting initiatives into comprehensive climate policies enables the global community to leverage the inherent cost-effectiveness and scalability of these efforts (Singh, 2024). This holistic approach ensures that the pursuit of carbon neutrality is not only attainable but also equitable, fostering a sustainable balance between ecological restoration and socio-economic progress.

Offsetting anthropogenic CO_2 emissions in tree planting plays a significant role in advancing global carbon neutrality objectives. In addition to its carbon capture capabilities, tree planting contributes to biodiversity conservation, enhances soil health, and sustains vital ecosystem services, including hydrological regulation and erosion prevention (Dhawi & Aleidan, 2024). When strategically implemented through afforestation and reforestation initiatives, tree planting offers scalable solutions that integrate seamlessly with broader climate mitigation and adaptation frameworks. These efforts are particularly indispensable in regions grappling with deforestation and land degradation, where restoring forest ecosystems can yield profound environmental and socio-economic dividends. As the global community accelerates its response to the climate crisis, tree planting has emerged as a cornerstone of nature-based solutions. Bolstering terrestrial carbon sinks, such initiatives provide a sustainable pathway to mitigate greenhouse gas emissions, enhance ecosystem resilience, and foster planetary health for future generations.

23.10 Role of Plantations in Carbon Sequestration and Ecosystem Management

Plantations are increasingly recognized as one of the most rapid and cost-efficient mechanisms for capturing atmospheric CO_2 and storing it across diverse carbon pools, including aboveground biomass, belowground biomass, detritus, soil, and wood products (Justine et al., 2017; Yen et al., 2020). However, the carbon sequestration potential of any plantation system is governed by several critical factors, including species composition, growth characteristics, sequestration efficiency, age, structural complexity, management practices, site history, and prevailing climatic conditions.

The variability in carbon storage capacity across different plantation types underscores the importance of tailored strategies to optimize carbon capture, such as *Populus deltoides, Eucalyptus tereticornis, Dalbergia sissoo, Mangifera indica, Litchi chinensis, Prunus salicina, Populus deltoids, Bambusa tulda, Dendrocalamus longispathus* (Arora & Chaudhry, 2014; Arora et al. 2014; Devi & Singh, 2021; Kanime et al., 2013). Beyond their direct role in carbon sequestration, plantations contribute to climate mitigation indirectly by alleviating the demand for timber from old-growth forests. This reduces logging pressure on mature forests, which play a critical role in conserving carbon stocks due to their large, long-lived trees (Lutz et al., 2018; Mildrexler et al., 2020). Providing alternative sources of timber and biomass plantations helps preserve the ecological integrity and carbon storage capacity of natural forests, making them an essential component of sustainable forestry management. Short-rotation tree plantations, in particular, exhibit a remarkable ability to accumulate significant quantities of carbon within a relatively short timeframe (Böttcher & Lindner, 2010; Singh et al., 2023). These plantations, comprising fast-growing species such as Casuarina and bamboo, are designed explicitly for high wood production and carbon sequestration efficiency (Chauhan & Chauhan, 2011; Tripathi et al., 2024). They are intensively managed to maximize biomass yield, often outperforming natural forests in terms of productivity. This high biomass yield reduces the pressure on natural forests for timber extraction, contributing to forest conservation and climate mitigation (Prasad et al., 2012).

Moreover, wood products derived from such plantations, including construction materials, furniture, and baskets, serve as long-term carbon storage pools, locking atmospheric CO_2 in durable forms (Behera et al., 2020). Thus, the sustainability of short-rotation plantations is contingent on their end-use. If the harvested wood is used as fuel and subsequently burned, the carbon stored in the biomass is rapidly released back into the atmosphere, negating the sequestration benefits (Arora & Chaudhry, 2014). This highlights the need for strategic utilization of plantation-derived biomass, prioritizing applications that extend carbon storage duration and minimize emissions.

23.11 Global to Regional Transitions in Carbon Neutrality

Countries worldwide have articulated commitments to achieving carbon neutrality as part of their broader climate mitigation strategies, however, the timelines and mechanisms adopted exhibit considerable heterogeneity. These divergences are primarily influenced by each nation's stage of economic development, historical responsibility for greenhouse gas emissions, current emission intensities, and the socio-political feasibility of rapid decarbonization. Developed countries, having already undergone industrialization, possess greater financial and technological capacity to transition swiftly towards net-zero pathways. As a result of these gaps in economic and technological readiness, countries have adopted differentiated timelines for achieving net-zero emissions, respective to their national circumstances and capabilities. Whereas, the European Union aims to reach net-zero emissions by 2050. Similarly, the United States has committed to a 2050 target, supported by massive climate investments under the Inflation Reduction Act (2022) and rapid acceleration of clean energy projects.

On the other hand, China, as the world's largest emitter, has pledged to achieve carbon neutrality by 2060 through initiatives such as the Belt and Road Initiative, significant coal reduction strategies, and the electrification of transportation and industries. In contrast, India has announced a net-zero target for 2070, acknowledging its development needs and concerns about equity. India's strategy is centered around renewable energy expansion (targeting 500 GW by 2030), promotion of green hydrogen, and behavioral transformation through the globally endorsed Lifestyle for Environment (LiFE) other initiative to adapt and mitigate climate change.

This spectrum of target years and transition mechanisms underscores the differentiated responsibilities and capabilities emphasized under the principle of Common But Differentiated Responsibilities and Respective Capabilities (CBDR-RC) as articulated in the UNFCCC. While developed nations leverage advanced technologies, capital markets, and institutional mechanisms to accelerate decarbonization, emerging economies like India must navigate complex trade-offs between development and environmental sustainability. India's approach, though longer-term, reflects a pragmatic balance between emission reduction and inclusive growth. It highlights how regional pathways to carbon neutrality can be aligned with national priorities such as energy security, employment generation, and rural development. Furthermore, India's initiatives, such as decentralized renewable energy, afforestation, and behavior-driven climate action, offer scalable models for other nations in the Global South. The global-to-regional transition narrative thus reinforces that a one-size-fits-all approach to carbon neutrality is neither feasible nor equitable. Instead, changing strategies grounded in local socio-economic realities is essential to achieving globally coordinated yet regionally relevant climate goals.

23.12 Conclusion

Achieving carbon neutrality is a complex, multidimensional goal with far-reaching social, economic, and environmental implications. The success of this endeavor hinges on a collective recognition of its urgency and a willingness to embrace transformative shifts across all sectors of society. Central to this transition is the adoption of sustainable consumption patterns and lifestyle choices that reduce individual and collective carbon footprints. Evidence from case studies reveals that behavioral changes such as reducing energy use, embracing plant-based diets, and minimizing waste, when paired with low-carbon technologies like electric vehicles, energy-efficient appliances, and smart systems, can significantly advance carbon reduction efforts. From an economic perspective, the shift to renewable energy and energy-efficient systems creates both opportunities and challenges. While green energy fosters job creation and skill development, enhancing industrial and domestic energy efficiency also leads to cost savings and optimized resource use. Pollution control and emissions mitigation measures, supported by robust policies, reinforce environmental health and regulatory compliance. However, these initiatives must be guided by equity-focused policies to prevent disproportionate burdens on marginalized communities and ensure inclusive access to low-carbon technologies and infrastructure. Natural ecosystems, particularly forests, play a crucial role in sequestering carbon. Strategies such as afforestation, reforestation, and agroforestry not only mitigate climate change but also support biodiversity and provide economic opportunities, especially in rural areas. Sustainable forest management and plantation models designed for carbon sequestration offer scalable and cost-effective climate solutions. Despite substantial progress, key barriers remain, such as scaling up green technologies, bridging knowledge and technology gaps in developing nations, and overcoming behavioral inertia in high-consumption societies. A cohesive global effort is essential, requiring integrated policy frameworks, long-term investments, and inclusive strategies that align with local needs and capabilities. Hence, it is concluded that carbon neutrality is not a singular goal but a dynamic process that requires harmonizing technological innovation, societal transformation, and environmental stewardship. With decisive leadership, global collaboration, and a steadfast commitment to sustainability and equity, the path toward a carbon–neutral future is not only possible, it is imperative.

References

Ahuchogu, M. C., Sanyaolu, T. O., & Adeleke, A. G. (2024). Workforce development in the transport sector amidst environmental change: A conceptual review. *Global Journal of Research in Science and Technology, 2*(01), 061–077.

Arnedo, E. G., Valero-Matas, J. A., & Sánchez-Bayón, A. (2021). Spanish tourist sector sustainability: Recovery plan, green jobs and wellbeing opportunity. *Sustainability, 13*(20), 11447.

Arora, P., & Chaudhry, S. (2014). Carbon sequestration in tree plantations at Kurukshetra in Northern India. *American International Journal of Research in Formal, Applied & Natural Sciences, 5*(1), 65–70.

Arora, G., Chaturvedi, S., Kaushal, R., Nain, A., Tewari, S., Alam, N. M., & Chaturvedi, O. P. (2014). Growth, biomass, carbon stocks, and sequestration in an age series of Populus deltoides plantations in Tarai region of central Himalaya. *Turkish Journal of Agriculture and Forestry, 38*(4), 550–560.

Bayen, P., Bognounou, F., Lykke, A. M., Ouédraogo, M., & Thiombiano, A. (2016). The use of biomass production and allometric models to estimate carbon sequestration of Jatropha curcas L. plantations in western Burkina Faso. *Environment, Development and Sustainability, 18*, 143–156.

Behera, L., Ray, L. I., Ranjan Nayak, M., & Mehta, A. (2020). Carbon sequestration potential of Eucalyptus spp.: A review. *E-Planet, 18*(1), 79–84.

Blanke, M. M., & Golombek, S. D. (2021). Innovative strategy to reduce single-use plastics in sustainable horticulture by a refund strategy for flowerpots. *Sustainability, 13*(15), 8532.

Böttcher, H., & Lindner, M. (2010). Managing forest plantations for carbon sequestration today and in the future. In *Ecosystem goods and services from plantation forests* (pp. 43–76). Routledge.

Boukid, F. (2024). Holistic benefits of plant-based foods for sustainable agri-food systems. *Current Opinion in Food Science*, 101184.

Brand, C., Dons, E., Anaya-Boig, E., Avila-Palencia, I., Clark, A., de Nazelle, A., Gascon, M., Gaupp-Berghausen, M., Gerike, R., Götschi, T., Iacorossi, F., & Panis, L. I. (2021). The climate change mitigation effects of daily active travel in cities. *Transportation Research Part D: Transport and Environment, 93*, 102764.

Brundtland, G. H. (1987). Our common future world commission on environment and development.

Calvin, K., Dasgupta, D., Krinner, G., Mukherji, A., Thorne, P. W., Trisos, C.,... & Hauser, M. (2023). *IPCC, 2023: Climate change 2023: Synthesis report. Contribution of working groups I, II and III to the sixth assessment report of the intergovernmental panel on climate change* [Core Writing Team, H. Lee and J. Romero (eds.)]. IPCC, Geneva, Switzerland.

Chan, K. M., Boyd, D. R., Gould, R. K., Jetzkowitz, J., Liu, J., Muraca, B., Naidoo, R., Olmsted, P., Satterfield, T., Selomane, O., Singh, G. G., & Brondízio, E. S. (2020). Levers and leverage points for pathways to sustainability. *People and Nature, 2*(3), 693–717.

Chauhan, S. K., & Chauhan, R. (2011). Short rotation forestry species for economic and environmental benefits: country report (India). In *IUFRO Symposium on "Short Rotation Forestry: Synergies for Wood Production and Environmental Amelioration"* (February 10 12, 2011) PAU, Ludhianan (India).

Chen, L., Msigwa, G., Yang, M., Osman, A. I., Fawzy, S., Rooney, D. W., & Yap, P. S. (2022). Strategies to achieve a carbon neutral society: A review. *Environmental Chemistry Letters, 20*(4), 2277–2310.

Darmandieu, A., Garcés-Ayerbe, C., Renucci, A., & Rivera-Torres, P. (2022). How does it pay to be circular in production processes? Eco-innovativeness and green jobs as moderators of a cost-efficiency advantage in European small and medium enterprises. *Business Strategy and the Environment, 31*(3), 1184–1203.

Daukaev, A., Abumuslimov, A., & Gakaev, R. (2020). Geological and geomorphological conditions for the formation of landslide processes in the mountainous part of the Chechen Republic and measures to fight them. In *The second Eurasian RISK-2020 conference and symposium* (pp. 64–65).

Devi, A. S., & Singh, K. S. (2021). Carbon storage and sequestration potential in aboveground biomass of bamboos in North East India. *Scientific Reports, 11*(1), 837.

Dhawi, F., & Aleidan, M. M. (2024). Oasis agriculture revitalization and carbon sequestration for climate-resilient communities. *Frontiers in Agronomy, 6*, 1386671.

Dillman, D. A., Rosa, E. A., & Dillman, J. J. (1983). Lifestyle and home energy conservation in the United States: The poor accept lifestyle cutbacks while the wealthy invest in conservation. *Journal of Economic Psychology, 3*(3–4), 299–315.

Dong, H., Liu, Y., Zhao, Z., Tan, X., & Managi, S. (2022). Carbon neutrality commitment for China: From vision to action. *Sustainability Science, 17*(5), 1741–1755.

Druckman, A., & Jackson, T. (2010). The bare necessities: How much household carbon do we really need? *Ecological Economics, 69*(9), 1794–1804.

Fang, J., Zhu, J., Wang, S., Yue, C., & Shen, H. (2011). Global warming, human-induced carbon emissions, and their uncertainties. *Science China Earth Sciences, 54*, 1458–1468.

Fazey, I., Moug, P., Allen, S., Beckmann, K., Blackwood, D., Bonaventura, M., Burnett, K., Danson, M., Falconer, R., Gagnon, A. S., Harkness, R., & Wolstenholme, R. (2018). Transformation in a changing climate: A research agenda. *Climate and Development, 10*(3), 197–217.

Gakaev, R. (2022). Carbon sequestration in landscapes of the Chechen Republic. *Reliability: Theory & Applications, 17*(SI 3 (66)), 193–196.

Goodall, C. (2010). *How to live a low-carbon life: The individual's guide to tackling climate change*. Routledge.

Gunya, A., & Gakaev, R. (2022). Landscape analysis of exogenic processes distribution in mountain regions of the Chechen republic. *Reliability: Theory & Applications, 17*(SI 3 (66)), 124–128.

Hansen, J., Kharecha, P., Sato, M., Masson-Delmotte, V., Ackerman, F., Beerling, D. J., Hearty, P. J., Hoegh-Guldberg, O., Hsu, S. L., Parmesan, C., Rockstrom, J., & Zachos, J. C. (2013). Assessing "dangerous climate change": Required reduction of carbon emissions to protect young people, future generations and nature. *PLoS One, 8*(12), e81648.

Hariram, N. P., Mekha, K. B., Suganthan, V., & Sudhakar, K. (2023). Sustainalism: An integrated socio-economic-environmental model to address sustainable development and sustainability. *Sustainability, 15*(13), 10682.

Hu, H., Wang, S., Guo, Z., Xu, B., & Fang, J. (2015). The stage-classified matrix models project a significant increase in biomass carbon stocks in China's forests between 2005 and 2050. *Scientific Reports, 5*(1), 11203.

Huang, M. T., & Zhai, P. M. (2021). Achieving Paris Agreement temperature goals requires carbon neutrality by middle century with far-reaching transitions in the whole society. *Advances in Climate Change Research, 12*(2), 281–286.

Jia, Z., & Lin, B. (2021). How to achieve the first step of the carbon-neutrality 2060 target in China: The coal substitution perspective. *Energy, 233*, Article 121179.

Joshi, R., Singh, H., Chhetri, R., & Yadav, K. (2020). Assessment of carbon sequestration potential in degraded and non-degraded community forests in Terai region of Nepal. *Journal of Forest and Environmental Science, 36*(2), 113–121.

Joshi, R., Singh, H., Chhetri, R., Poudel, S. R., & Rijal, S. (2021). Carbon sequestration potential of community forests: A comparative analysis of soil organic carbon stock in community managed forests of Far-Western Nepal. *Eurasian Journal of Soil Science, 10*(2), 96–104.

Justine, M. F., Yang, W., Wu, F., & Khan, M. N. (2017). Dynamics of biomass and carbon sequestration across a chronosequence of masson pine plantations. *Journal of Geophysical Research: Biogeosciences, 122*(3), 578–591.

Kanime, N., Kaushal, R., Tewari, S. K., Raverkar, K. P., Chaturvedi, S., & Chaturvedi, O. P. (2013). Biomass production and carbon sequestration in different tree-based systems of Central Himalayan Tarai region. *Forests, Trees and Livelihoods, 22*(1), 38–50.

Kantemirova, M., Salamova, A. S., & Gishlakaev, S. (2023). Commitment to carbon neutrality. In *SHS web of conferences* (Vol. 172, p. 02021). EDP Sciences.

Koide, R., Kojima, S., Nansai, K., Lettenmeier, M., Asakawa, K., Liu, C., & Murakami, S. (2021). Exploring carbon footprint reduction pathways through urban lifestyle changes: A practical approach applied to Japanese cities. *Environmental Research Letters, 16*(8), Article 084001.

Kothandaraman, S., Dar, J. A., Bhat, N. A., Sundarapandian, S., & Khan, M. L. (2022). Tree plantation: A silver bullet to achieve carbon neutrality? In *Land degradation neutrality: Achieving SDG 15 by forest management* (pp. 205–227). Springer Nature Singapore.

Kumar, M., & Singh, H. (2020). Agroforestry as a nature-based solution for reducing community dependence on forests to safeguard forests in rainfed areas of India. In *Nature-based solutions for resilient ecosystems and societies* (pp. 289–306). Singapore: Springer Singapore.

Kumar, N., Singh, H., & Kumar, A. (Eds.). (2021). *Renewable energy and green technology: Principles and practices*. CRC Press.

Kumari, G., Kotiyal, P. B., Singh, H., Kumar, M., Kumar, N., Malik, A., ... & Singh, S. (2024). Predicting future climate change effects on biotic communities: A species distribution modeling approach. In *Forests and climate change: Biological perspectives on impact, adaptation, and mitigation strategies* (pp. 137–168). Singapore: Springer Nature Singapore.

Kuramochi, T., Höhne, N., Schaeffer, M., Cantzler, J., Hare, B., Deng, Y., Sterl, S., Hagemann, M., Rocha, M., Yanguas-Parra, P. A., Mir, G. U. R., Wong, L., El-Laboudy, T., Wouters, K., Deryng, D., & Blok, K. (2018). Ten key short-term sectoral benchmarks to limit warming to 1.5 C. *Climate Policy, 18*(3), 287–305.

Lal, R. (2004). Agricultural activities and the global carbon cycle. *Nutrient Cycling in Agroecosystems, 70*(2), 103–116.

Le Quéré, C., Korsbakken, J. I., Wilson, C., Tosun, J., Andrew, R., Andres, R. J., Canadell, J. G., Jordan, A., Peters, G. P., & van Vuuren, D. P. (2019). Drivers of declining CO2 emissions in 18 developed economies. *Nature Climate Change, 9*(3), 213–217.

Leonard-Barton, D. (1981). Voluntary simplicity lifestyles and energy conservation. *Journal of Consumer Research, 8*(3), 243–252.

Linnér, B. O., & Wibeck, V. (2020). Conceptualising variations in societal transformations towards sustainability. *Environmental Science & Policy, 106*, 221–227.

Linnér, B. O., & Wibeck, V. (2021). Drivers of sustainability transformations: Leverage points, contexts and conjunctures. *Sustainability Science, 16*(3), 889–900.

Liu, C., Yao, Y., & Zhu, H. (2021). Hybrid salp swarm algorithm for solving the green scheduling problem in a double-flexible job shop. *Applied Sciences, 12*(1), 205.

Lubowiecki-Vikuk, A., Dąbrowska, A., & Machnik, A. (2021a). Responsible consumer and lifestyle: Sustainability insights. *Sustainable Production and Consumption, 25*, 91–101.

Lubowiecki-Vikuk, A., Đerčan, B. M., & de Sousa, B. M. B. (2021b). Sustainable development and leisure services: Changes and trends. In *Handbook of sustainable development and leisure services* (pp. 1–20). Springer International Publishing.

Lutz, J. A., Furniss, T. J., Johnson, D. J., Davies, S. J., Allen, D., Alonso, A., Anderson-Teixeira, K. J., Andrade, A., Baltzer, J., Becker, K. M., Blomdahl, E. M., & Zimmerman, J. K. (2018). Global importance of large-diameter trees. *Global Ecology and Biogeography, 27*(7), 849–864.

Maréchal, P. (1977). Crise de l'énergie et évolution des modes de vie. *Rapport du CRÉDOC*.

Martínez-Cruz, A. L., & Núñez, H. M. (2021). Tension in Mexico's energy transition: Are urban residential consumers in Aguascalientes willing to pay for renewable energy and green jobs? *Energy Policy, 150*, Article 112145.

Martinez-Fernandez, C., Hinojosa, C., & Miranda, G. (2010). Greening jobs and skills: Labour market implications of addressing climate change.

Meadows, D. H. (2008). Thinking in systems: A primer. *Sustainability Institute*.

Mildrexler, D. J., Berner, L. T., Law, B. E., Birdsey, R. A., & Moomaw, W. R. (2020). Large trees dominate carbon storage in forests east of the cascade crest in the United States Pacific Northwest. *Frontiers in Forests and Global Change, 3*, Article 594274.

Mont, O., Neuvonen, A., & Lähteenoja, S. (2014). Sustainable lifestyles 2050: Stakeholder visions, emerging practices and future research. *Journal of Cleaner Production, 63*, 24–32.

Neuvonen, A., Kaskinen, T., Leppänen, J., Lähteenoja, S., Mokka, R., & Ritola, M. (2014). Low-carbon futures and sustainable lifestyles: A backcasting scenario approach. *Futures, 58*, 66–76.

Pearson, P. J., & Foxon, T. J. (2012). A low carbon industrial revolution? Insights and challenges from past technological and economic transformations. *Energy Policy, 50*, 117–127.

Prakash, S., & Verma, A. K. (2022). Anthropogenic activities and Biodiversity threats. *International Journal of Biological Innovations, IJBI, 4*(1), 94–103.

Prasad, J. V. N. S., Srinivas, K., Rao, C. S., Ramesh, C., Venkatravamma, K., & Venkateswarlu, B. (2012). Biomass productivity and carbon stocks of farm forestry and agroforestry systems of leucaena and eucalyptus in Andhra Pradesh, India. *Current Science*, 536–540.

Psistaki, K., Tsantopoulos, G., & Paschalidou, A. K. (2024). An Overview of the role of forests in climate change mitigation. *Sustainability, 16*(14), 6089.

Qin, Z., Deng, X., Griscom, B., Huang, Y., Li, T., Smith, P., Yuan, W., & Zhang, W. (2021). Natural climate solutions for China: The last mile to carbon neutrality.

Rabaey, K., & Ragauskas, A. J. (2014). Editorial overview: Energy biotechnology. *Current Opinion in Biotechnology, 27*, V–VI.

Raymond, C. M., Lechner, A. M., Havu, M., Jalkanen, J., Lampinen, J., Antúnez, O. G., Olafsson, A. S., Gulsrud, N., Kinnunen, A., Backman, L., Kulmala, L., & Järvi, L. (2023). Identifying where nature-based solutions can offer win-wins for carbon mitigation and biodiversity across knowledge systems. *npj Urban Sustainability, 3*(1), 27.

Roe, S., Streck, C., Obersteiner, M., Frank, S., Griscom, B., Drouet, L., Fricko, O., Gusti, M., Harris, N., Hasegawa, T., Hausfather, Z., & Lawrence, D. (2019). Contribution of the land sector to a 1.5 C world. *Nature Climate Change, 9*(11), 817–828.

Rosa, L. P., & Ribeiro, S. K. (2001). The present, past, and future contributions to global warming of CO_2 emissions from fuels. *Climatic Change, 48*(2), 289–307.

Singh, H. (Ed.) (2024). *Forests and climate change: Biological perspectives on impact, adaptation, and mitigation strategies.* Springer (ISBN: 978-981-97-3904-2). https://doi.org/10.1007/978-981-97-3905-9

Singh, H., Kumar, N., Singh, R., & Kumar, M. (2023). Assessing the climate change impact on the habitat suitability of the range-restricted bird species (*Catreus wallichii*) in the Indian Himalayan ecosystem. *Environmental Science and Pollution Research, 30*(57), 121224–121235.

Singh, M., Kumar, P., Singh, H., Kumar, A. K. A., Kumar, R., & Kumar, R. (2023). Neolamarckia cadamba: A comprehensive review on its physiological, ecological, phytochemical and pharmacological perspectives. *Ecology Environment and Conservation, 29*, 241–250.

Singh, H., & Kumar, M. (2022). Climate change and its impact on Indian Himalayan forests: Current status and research needs. In *Climate change: Impacts, responses and sustainability in the Indian Himalaya* (pp. 223–242). Cham: Springer International Publishing.

Song, K., Kim, H., Cha, J., & Lee, T. (2021). Matching and mismatching of green jobs: A big data analysis of job recruiting and searching. *Sustainability, 13*(7), 4074.

Sulich, A., & Zema, T. (2018). Green jobs, a new measure of public management and sustainable development. *European Journal of Environmental Sciences, 8*(1).

Sulich, A., & Rutkowska, M. (2020). Green jobs, definitional issues, and the employment of young people: An analysis of three European Union countries. *Journal of Environmental Management, 262*, Article 110314.

Sullivan, M. J., Lewis, S. L., Affum-Baffoe, K., Castilho, C., Costa, F., Sanchez, A. C., Ewango, C. E., Hubau, W., Marimon, B., Monteagudo-Mendoza, A., Qie, L., & Vargas, P. N. (2020). Long-term thermal sensitivity of Earth's tropical forests. *Science, 368*(6493), 869–874.

Tadese, S., Soromessa, T., Aneseye, A. B., Gebeyehu, G., Noszczyk, T., & Kindu, M. (2023). The impact of land cover change on the carbon stock of moist afromontane forests in the Majang Forest Biosphere Reserve. *Carbon Balance and Management, 18*(1), 24.

Tasneem Oyshi, J. (2024). Transformative changes in urban transportation policies through sustainability and climate change mitigation lens: Case studies from Europe.

Tilman, D., Balzer, C., Hill, J., & Befort, B. L. (2011). Global food demand and the sustainable intensification of agriculture. *Proceedings of the National Academy of Sciences, 108*(50), 20260–20264.

Tripathi, A., Yadav, S., Nishtha, Nkengnamai, M., & Thakur, A. (2024). Bamboo: A fast-growing species to mitigate carbon footprint. In *Forests and climate change: Biological perspectives on impact, adaptation, and mitigation strategies* (pp. 469–487). Springer Nature Singapore.

Tukker, A., Charter, M., Vezzoli, C., Sto, E., & Andersen, M. M. (2008). *Perspectives on radical changes to sustainable consumption and production* (Vol. 1). Greeenleaf Publishing.

Vesere, R., Kalnins, S. N., & Blumberga, D. (2021). Role of green jobs in the reduction of waste and waste management. *Rigas Tehniskas Universitates Zinatniskie Raksti, 25*(1), 1128–1141.

Wang, Y., Liu, L., Yue, F., & Li, D. (2019). Dynamics of carbon and nitrogen storage in two typical plantation ecosystems of different stand ages on the Loess Plateau of China. *PeerJ, 7*, Article e7708.

Winkler, H., Lecocq, F., Lofgren, H., Vilariño, M. V., Kartha, S., & Portugal-Pereira, J. (2022). Examples of shifting development pathways: Lessons on how to enable broader, deeper, and faster climate action. *Climate Action, 1*(1), 1–20.

Yen, T. M., Huang, K. L., Li, L. E., & Wang, C. H. (2020). Assessing carbon sequestration in plantation forests of important conifers based on the system of permanent sample plots across Taiwan. *Journal of Sustainable Forestry, 39*(4), 392–406.

Yong, J. Y., Yusliza, M. Y., & Fawehinmi, O. O. (2020). Green human resource management: A systematic literature review from 2007 to 2019. *Benchmarking: An International Journal, 27*(7), 2005–2027.

Zhang, W., Wang, L., Chen, J., & Zhang, Y. (2024a). Preferential flow in soils: Review of role in soil carbon dynamics, assessment of characteristics, and performance in ecosystems. *Eurasian Soil Science, 57*(5), 814–825.

Zhang, R., Wang, S., & Yuan, C. (2024b). Shock or opportunity? Unveiling the effect of low-carbon transition on employment. *Journal of Environmental Management, 359*, Article 120885.

Zhang, Y., Zhao, Y., Chen, Q., Zhu, Y., Liu, B., Zhang, X., & Yin, X. (2024c). Capacity of forests and grasslands to achieve carbon neutrality in China. *Forests, 15*(6), 1060.

MIX
Papier aus verantwortungsvollen Quellen
Paper from responsible sources
FSC® C105338

If you have any concerns about our products,
you can contact us on
ProductSafety@springernature.com

In case Publisher is established outside the EU,
the EU authorized representative is:
Springer Nature Customer Service Center GmbH
Europaplatz 3, 69115 Heidelberg, Germany

Printed by Libri Plureos GmbH
in Hamburg, Germany